Conformation Theory

ORGANIC CHEMISTRY

A SERIES OF MONOGRAPHS

Edited by

ALFRED T. BLOMQUIST

Department of Chemistry, Cornell University, Ithaca, New York

Volume 1. Wolfgang Kirmse. CARBENE CHEMISTRY. 1964

Volume 2. Brandes H. Smith. BRIDGED AROMATIC COMPOUNDS. 1964

Volume 3. Michael Hanack. CONFORMATION THEORY. 1965

In preparation

Donald J. Cram. FUNDAMENTALS OF CARBANION CHEMISTRY

Kenneth B. Wiberg. OXIDATION IN ORGANIC CHEMISTRY, PART A

CONFORMATION THEORY

Michael Hanack

CHEMISCHES INSTITUT
DER UNIVERSITÄT TÜBINGEN
TÜBINGEN, GERMANY

Translated from the German manuscript
by
Helmut C. Neumann

1965

ACADEMIC PRESS New York and London

ACADEMIC PRESS INC.
111 Fifth Avenue, New York, New York 10003

United Kingdom Edition published by
ACADEMIC PRESS INC. (LONDON) LTD.
Berkeley Square House, London W.1

LIBRARY OF CONGRESS CATALOG CARD NUMBER: 63-21402

PRINTED IN THE UNITED STATES OF AMERICA

Preface

The present book is based on the lectures which the author delivered in the years 1961–1962 at the University of Tübingen. The structure and content of the lecture outline had to be rearranged to give the book a more systematic form. A short introductory chapter on the historic development of stereochemistry was included for those readers who are not well acquainted with the detailed problems of stereochemistry.

The enormous number of publications in recent years in the field covered by this book led in the outset to a certain restriction in the choice of material. Because of the author's personal interest in the field, the chapters on the conformation of monocyclic and polycyclic compounds without hetero atoms have been covered in greater detail than the chapters on the conformation of heterocyclic compounds and acyclic diastereomers. A special section on the conformation and reactivity of cyclohexane compounds was included to emphasize the importance of stereochemistry in study of reaction mechanisms.

All pertinent literature through 1963 has been covered to the best of my ability, but in the selection of literature, complete coverage was not intended. A number of additional references from 1964 were included in proof.

I especially wish to thank my honored teacher, Prof. W. Hückel, (Tübingen) for studying the manuscript and for the great interest he showed in the preparation of this book. In addition, I greatly appreciate the assistance of Prof. V. Prelog (Zürich), Prof. W. Klyne (London), Prof. G. Fodor (Budapest), Prof. R. Riemschneider (Berlin), and Dr. W. Masschelein (Brussels) for checking parts of the manuscript, and for their inspiration and suggestions. I am very grateful to Dr. H. C. Neumann (Rensselaer, New York), who undertook the translation of the German manuscript into English. Mr. K. Görler and Miss G. Wentrup (Tübingen) were most helpful in the proofreading.

MICHAEL HANACK

Tübingen

Contents

Chapter 1—Introduction: Short Historical Survey of the Development and Methods of Stereochemistry

Chapter 2—General Discussion

Chapter 3—Monocyclic Compounds

I. Cyclopentanes

II. Conformation of Cyclohexane and Substituted Cyclohexanes

Chapter 7—Heterocyclic Compounds

Chapter 8—Acyclic Diastereomers

CHAPTER 1

Introduction: Short Historical Survey of the Development and Methods of Stereochemistry[1]

In order to understand the frequently quite complicated train of thought of modern stereochemistry, a short presentation of the fundamentals and definition of the older "classical" stereochemistry is necessary. The results and methods of modern stereochemistry, especially that branch known as conformation theory or conformational analysis, will be covered. Conformational analysis deals with the detailed spatial structure of the molecule.

Whereas stereochemistry originally[2] almost exclusively confined itself to the field of stereoisomerism, i.e., describing and explaining isomeric possibilities due to spatial atomic arrangement, today stereochemistry has become one of the most important branches of theoretical organic chemistry. There is hardly any branch of theoretical organic chemistry in which steric problems are not involved. The following are examples: reactions at an asymmetric carbon atom; the problem of steric hindrance and steric inhibition of resonance. Then there is the explanation of the mechanism of organic reactions. In this latter instance, the important role that steric factors play in the interpretation of reaction mechanism is quite frequently ignored.

---

[1] Compare G. Wittig, "Stereochemie." Akademische Verlagsges. Leipzig, 1930; St. Goldschmidt, "Hand- und Jahrbuch der Chemischen Physik," Vol. IV. Leipzig, 1933; K. Freudenberg, "Handbuch der Stereochemie." Leipzig and Vienna, 1933; R. L. Shriner, R. Adams, and C. S. Marvel, "Organic Chemistry: An Advanced Treatise" (H. Gilman, ed.), Vol. 1, Chapter 4. New York, 1943; G. H. Wheland, "Advanced Organic Chemistry," 2nd ed. Wiley, New York, 1949; E. L. Eliel, "Stereochemistry of Carbon Compounds." McGraw-Hill, New York, 1962.

[2] The expression *stereochemistry* originated with V. Meyer, *Ber.* **21**, 789 (1888); **23**, 568 (1890).

1. Characteristics of Optically Active Compounds

In 1847–1848 Pasteur began his basic investigations on tartaric acid. He was able to isolate a tartaric acid which could rotate the plane of polarized light strongly to the right and another tartaric acid which could rotate it equally strongly to the left. The latter was equivalent in all other physical properties to the former. As a result, a pair of optical antipodes were discovered for the first time. Optical antipodes are substances that have identical composition, molecular weight, and functional groups but differ in one physical property: the ability of each antipode to rotate the plane of polarized light a definite amount and in an opposite direction from that of the other, whether in gaseous, liquid, or solid form.

By mixing both tartaric acids (antipodes) Pasteur obtained the known third tartaric acid (racemic acid) which was optically inactive, i.e., did not rotate the plane of polarized light. This inactive form (racemate) frequently has a different crystalline form, solubility, and melting point from the active forms. Pasteur further established that optical antipodes occur in enantiomorphic forms.[3] His investigation of the sodium ammonium salt of racemic acid revealed two optically active tartaric acids, the crystals of which were mirror images. Pasteur concluded that in the molecular state as well both forms of tartaric acid exist as image and mirror image, i.e., that the molecules are also asymmetric.

2. Principles of the Tetrahedral Theory

The actual founders of stereochemistry were the Dutch chemist van't Hoff[4] and the French chemist LeBel,[5] who formulated exactly for the first time the idea of mirror image isomerism. Although their work appeared almost at the same time, a number of characteristic differences exist in the views of these investigators, as follows.

(*a*) Van't Hoff conceived a model of a tetrahedron starting with the single atom, and formulated the tetrahedral theory. First he established that all compounds which show optical activity in the liquid or dissolved state have at least one carbon atom in the molecule, bound to four different substituents. Such a carbon atom is called "asymmetric" because it exhibits no plane of symmetry.

All the characteristic conditions of symmetry explaining the appearance

[3] This enantiomorphism in crystalline form is not present for all optical antipodes.

[4] J. H. Van't Hoff, "Dix années dans l'histoire d'une théorie." 1887.

[5] J. A. LeBel, *Bull. Soc. Chim. France* [2] **22**, 337 (1874).

or absence of optical activity are derived from the tetrahedral model of the asymmetric carbon atom: carbon is situated in the middle of a regular tetrahedron; the atoms bound to carbon are located in the corners. Based on geometric considerations, the tetrahedral distribution of substituents is the only model with which we can explain the numbers of isomers of a compound $CR_1R_2R_3R_4$ (I).

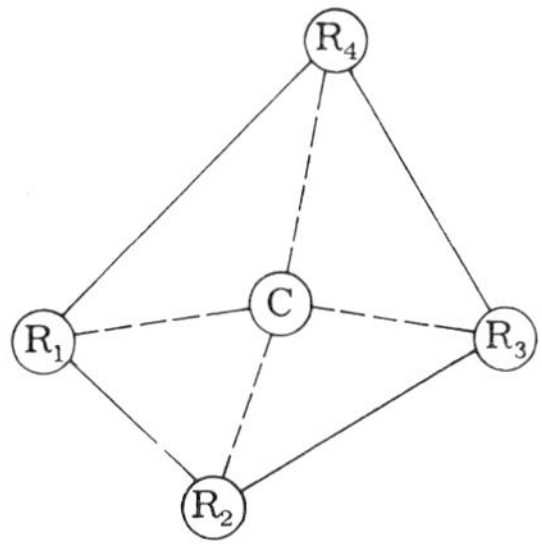

I

Another possibility in strictly geometric terms is for the location of the four substituents around a central atom: the four attached groups would take the four corners of the base plane of a quadratic pyramid, whose peak is formed by the central atom, carbon (II). If the height

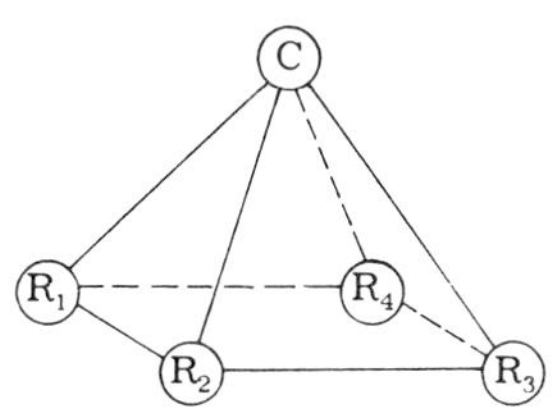

II

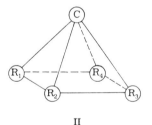

of this pyramid is zero, then all atoms will lie in one plane. For the molecule $CR_1R_2R_3R_4$, the pyramidal or the planar structure requires more than two stereoisometric forms (IIIa-d). With the tetrahedral structure, however, only two forms are possible: these stand in the relation of image and mirror image, and are not superimposable (IVa, IVb).

In actuality, of the many depicted compounds of the type $CR_1R_2R_3R_4$, there exist only two forms. Thus the molecule must have a tetrahedral structure.

(*b*) In contrast to van't Hoff, LeBel did not start with a definite model, but tried to explain the occurrence of mirror image isomers from the symmetry of the entire molecule. In his view, the tetrahedral model was only one possibility out of many that would explain the appearance of a center of asymmetry, i.e., LeBel made no predictions concerning the true position of the atoms in the molecule which could be checked experimentally.

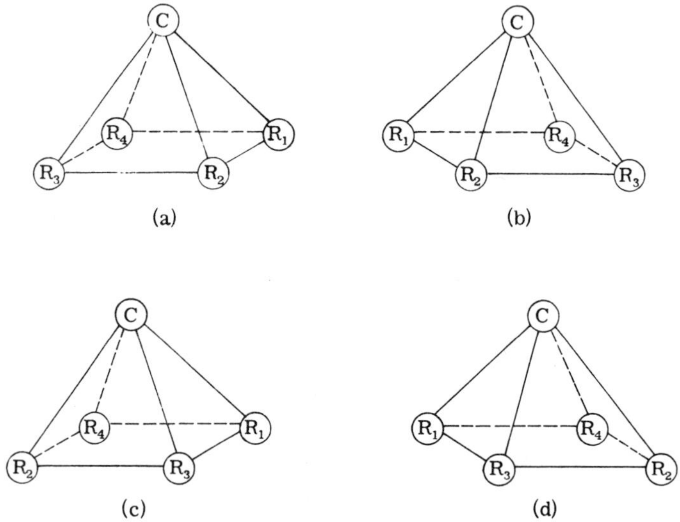

III

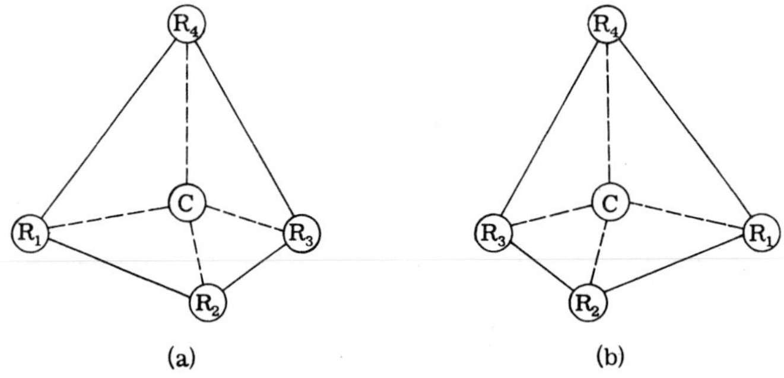

IV

3. Experimental Confirmation of the Tetrahedral Theory

The orientation of four substituents as mirror images was clearly demonstrated by E. Fischer[6] in 1914. By interchanging two substituents,

[6] E. Fischer, *Ber.* **47**, 3181 (1914).

he was able to transform an optically active compound into its antipode of equal but opposite rotation. Dextrorotatory isopropylmalonic acid amide was transformed into the corresponding levorotatory acid in this instance[7]:

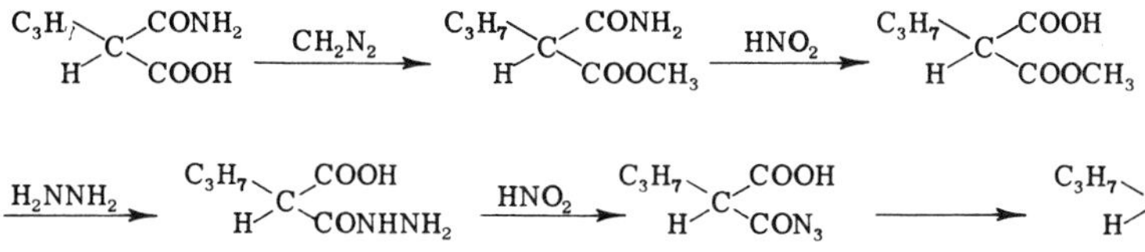

A further proof for the tetrahedral theory, also demonstrated by E. Fischer[8] for the first time, lies in the fact that the optical activity disappears when two identical substituents are placed on the central atom. The molecule is transformed from $CR_1R_2R_3R_4$ to $CR_1R_1R_3R_4$.

The above mentioned pyramidal model of the carbon atom, despite two equal substituents, would not lose its optical activity, because with these models asymmetric mirror image isomerism could still appear (Va, Vb).

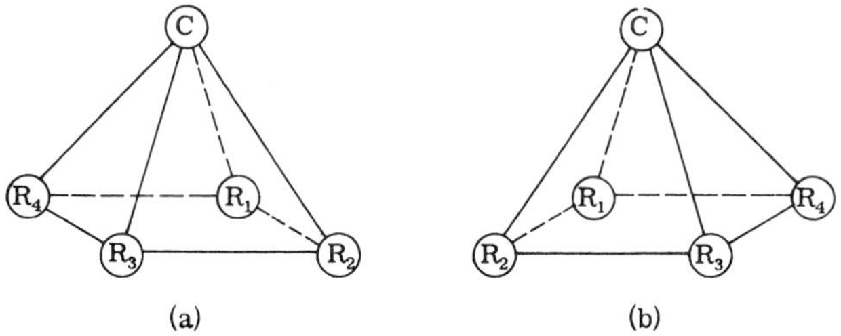

V

In conclusion, it can be said that the tetrahedral theory of van't Hoff has become useful not only for quaternary aliphatic carbon atoms, but also for tertiary olefinic carbon atoms. With *X*-ray and electron diffraction, it has been possible in more recent times to support the chemical proof with physical evidence and to confirm the tetrahedral structure of carbon experimentally.

The tetrahedral configuration of the carbon atom is further explained by the electronic theory. The hybrid bond orbitals of Pauling[9], the four *s* and *p* orbitals of carbon, in which these different orbitals change

[7] Further proof was obtained with acetonylglycerin by E. Baer and H. O. L. Fischer *J. Am. Chem. Soc.* **67**, 944 (1945).

[8] E. Fischer, *Ann. Chem.* **402**, 364 (1914).

[9] L. Pauling, "The Nature of the Chemical Bond," 3rd ed., p. 81 ff. Cornell Univ. Press, Ithaca, New York, 1960.

to similar new electron orbitals, require a complete equality as well as tetrahedral distribution of the four valences of the carbon atom.

Modern stereochemistry goes a step further and does not restrict itself to the quaternary or tertiary carbon atom. It is possible on the basis of the van't Hoff tetrahedral structure to develop ideas concerning the molecular fragments which appear during a reaction, and thereby to make predictions concerning the steric behavior of the molecule during chemical reactions. As a result of their flat structure and formed plane of symmetry, the carbonium ions are always connected with racemization. The carbanions do not lie in a plane because the free electron pair holds the position of a substituent. Nevertheless they do not form mirror image isomers because they are not a stable configuration. The theory requires, also, that the carbon radical must have a planar structure.

4. Number of Stereoisomers with Multiple Centers of Asymmetry

Because a compound with an asymmetric carbon atom exists in only two mirror-image forms, the number of isomers with multiple centers of asymmetry can be easily determined. If a molecule contains a number of different asymmetric centers A, B, C, ... of + and − configuration, respectively, then it can be expected that with the increase in the number of asymmetric centers, the number of isomers will increase exponentially.

With two asymmetric carbon atoms A and B the following four forms can be formed:

$$(1)\ \begin{matrix}+A\\+B\end{matrix}\qquad (2)\ \begin{matrix}-A\\-B\end{matrix}\qquad (3)\ \begin{matrix}+A\\-B\end{matrix}\qquad (4)\ \begin{matrix}-A\\+B\end{matrix} = 4 = 2^2$$

Four optically active compounds are obtained, of which 1 and 2 as well as 3 and 4 are optical antipodes. Each of these pairs of antipodes can be combined into a racemic form. On the other hand, the pairs 1 and 3, 2 and 4, 1 and 4, 2 and 3 are not mirror images. They rotate the plane of polarized light, but not necessarily in the same magnitude or in opposite directions. Since they each contain one carbon atom of the identical configuration and one carbon of opposite configuration, they cannot be superimposed by any symmetry operation. This type of stereoisomer is known as a *diastereoisomer* (diastereomer), differing from optical antipodes not only in optical rotation but also in other physical properties such as melting point, boiling point, refractive index, etc. Characteristic differences have been observed also in chemical properties. The separation of racemates into their optical components depends upon these differences.

In the case of three asymmetric centers, there are $8 = 2^3$ stereoisomeric molecular forms possible, with four asymmetric centers, $16 = 2^4$. In the presence of n different asymmetric centers, 2^n stereoisomeric forms occur.

5. *Meso* Forms

Among the compounds whose molecules have two or more asymmetric carbon atoms, there are also examples which cannot be separated into optical isomers. One or more possibilities of isomerism disappear in these cases and this leads to an exception of the above formulated rule. These nonseparable compounds are called "*meso*-forms" after *meso*-tartaric acid, which Pasteur discovered and tried in vain to separate into optically active components. Even in this case, a simple explanation is possible from the tetrahedral theory: Four projected formulas are possible for tartaric acid.[9a] For these projection formulas a definite direction of rotation is set by designating the arrangement of the four substituents in the molecule $CR_1R_2R_3R_4$ in clockwise fashion (VI, VII)

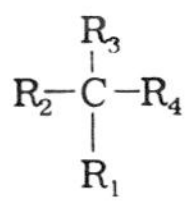

VI

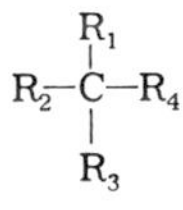

VII

For (VI), the direction of rotation $R_1 \rightarrow R_2 \rightarrow R_3 \rightarrow R_4$ is clockwise; for (VII), the direction $R_1 \rightarrow R_2 \rightarrow R_3 \rightarrow R_4$ is counterclockwise. The four possible projected formulas for tartaric acid would be as shown in formulas (VIII–XI).

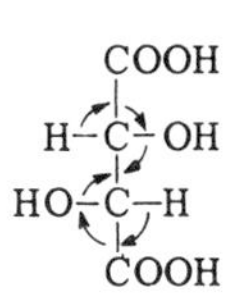

VIII

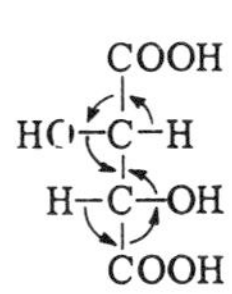

IX

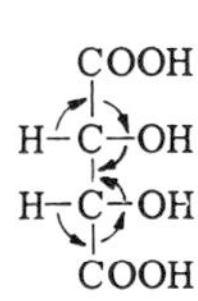

X

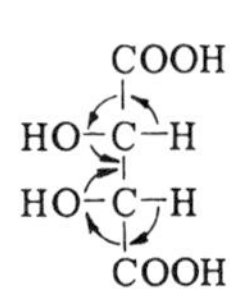

XI

By comparing formulas (VIII–XI) with (VI) and (VII) ($R_1 = H$, $R_2 = COOH$, $R_3 = OH$) it is seen that in (VIII) and (IX) both halves of the molecule have the same rotation, and the structure of (+)-tartaric acid (VIII) as well as the mirror image, (−)-tartaric acid (IX), are

[9a] For a description of configurations, cf. the sequence-rule of R. S. Cahn and C. K. Ingold, *J. Chem. Soc.* p. 612 (1951).

obtained. In formulas (X) and (XI), *meso*-tartaric acid, there is an opposite rotation in the upper and lower halves of the molecule, i.e. the spatial orientation of the substituents on the two asymmetric centers is in relation of image to mirror image. The result is an intramolecular compensation, i.e., the optical rotation of one half of the molecule is nullified by the opposite rotation of the other half, with a net rotation of zero. Because this compensation is intramolecular, the *meso* forms cannot be separated into their antipodes. Formulas X and XI are identical because, differing from VIII and IX, they can be superimposed. Also, in the case of complicated molecules, it is possible to see from the projection formula if nonseparable *meso* forms are present.

6. The Principle of Free Rotation

One of the most important additional hypotheses, directly connected with the tetrahedral theory, is the principle of "free rotation," which is also ascribed to van't Hoff. Visualize two carbon atoms bound by a single bond and the substituents in the corners of the tetrahedra, such as the compound 1,2-dichloroethane. The chlorine atoms can assume an infinite number of positions in relation to each other by turning each tetrahedra. This results in an equally large number of stereoisomers. Isomers of this type, however, have never been observed. One can conclude from this that carbon atoms bound by a single bond have free rotation around this bond, i.e., the possible isomers are readily converted one into the other. The fact that preferred positions are reached during equilibrium (restricted rotation) will not be discussed at this point.

7. Molecular Asymmetry

However, by preventing free rotation, the possibility exists that even molecules without an asymmetric carbon atom exhibit structures which cannot be superimposed on their mirror images. The asymmetric atom is no longer a restriction for the occurrence of an asymmetric molecule. All compounds exhibiting this molecular asymmetry are those whose asymmetry is not the result of the asymmetry of a single atom, but the result of an asymmetry of the whole molecule. The general geometrical condition for this is the lack of a plane or a center of symmetry.

Some examples of compounds with molecular asymmetry follow.

a. Mirror Image Isomers of Cyclic Compounds

Certain cyclic compounds without an asymmetric carbon atom exist in optically active forms. For example, of the eight diastereoisomers of inositol, there is one which is asymmetric and separable into optical antipodes. A plane of symmetry can be drawn through the other seven possible forms. More generally, cycloparaffins with substituents at various locations *cis* to each other are *meso* forms because a plane of symmetry can be drawn through the molecules (XII). The *trans* compounds (XIII), where this is not possible, can be separated into optical antipodes. This fact has often been used in the elucidation of the configuration of *cis-trans* isomers of cyclic compounds. The only

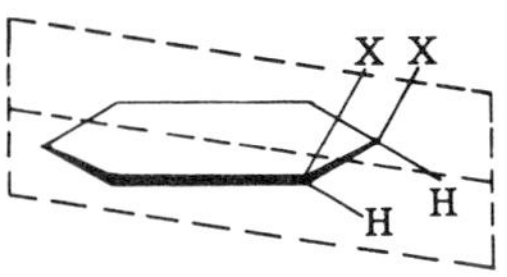

XII

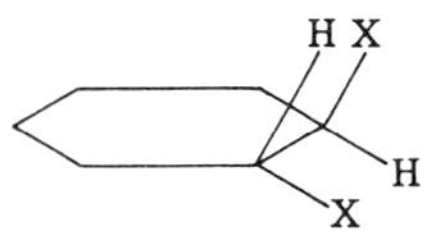

XIII

exceptions are those *trans* compounds whose substituents are diametrically opposed. In this case, diastereo *meso* forms can form.

b. Isomerism of the Spiranes[10]

The planes of the rings in spiranes are perpendicular to each other because of the tetrahedral structure of the quaternary carbon atom. If the spiranes are appropriately substituted, they can be resolved into their optical antipodes,[11] since no further plane of symmetry can be found.

c. Atropisomerism[12]

(i) For Diphenyl Derivatives

Atropisomeric compounds are those whose molecular asymmetry results from the restriction of free rotation.[13] They were first observed in diphenyl derivatives; later in derivatives of dinaphthyl, dianthryl, dipyryls, etc. The best example of this type of compound is dinitrodiphenic acid (2,2′-dinitrobiphenyl-6,6′-dicarboxylic acid)(XIV), which

[10] Compare W. Hückel, "Theoretische Grundlagen der org. Chemie," Vol. I, 8th ed. p. 58. Akad. Verlagsges., Leipzig, 1955.

[11] Compare E. R. Buchman, D. H. Deutsch, and G. I. Fujimoto, *J. Am. Chem. Soc.* **75**, 6228 (1953).

[12] Compare Eugen Müller, "Neuere Anschauungen der organischen Chemie," 2nd ed., p. 81 ff. Springer, Berlin, 1957.

[13] Summaries: R. Adams and H. C. Yuan, *Chem. Rev.* **12**, 261 (1933); R. L. Shriner, R. Adams, and D. S. Marvel *in* "Organic Chemistry" (H. Gilman, ed.), 2nd ed., Vol. I, p. 343 ff. Wiley, New York, 1943; E. L. Eliel, ref. 1.

has been resolved into its optical antipodes. Because of the large size of the substituents, there is an attempt to avoid each other. As a result, the benzene rings assume a tilted or perpendicular position in relation to each other. Since the two benzene rings are no longer in one plane, the provision for molecular asymmetry has been made. This coaxial position is fixed more or less strongly, depending upon the size of the substituents. The substituents can no longer rotate past each other.

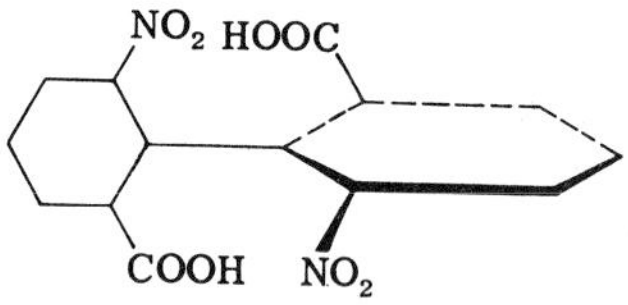

XIV

Dinitrodiphenic acid, coaxial position.

Even at higher temperatures, no measurable racemization takes place in this case. Racemization slowly takes place in compounds in which free rotation is partially restored, as, for instance, in *o*-nitrodiphenic acid (XV). This is verified by the fact that optical activity disappears when *o,o′* groups can react intramolecularly. This is possible only

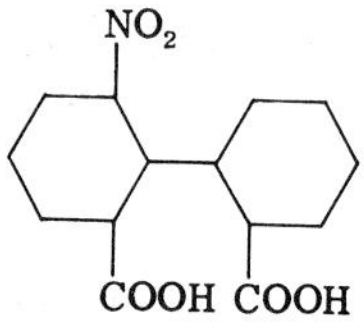

XV

with a planar arrangement of the rings. Meisenheimer[14] was able to show this with optically active 2,2′-diacetylaminodiphenyl-6,6′-dicarboxylic acid (XVI). Saponification of the acetyl groups under mild conditions leads to the inactive dilactam (XVII).

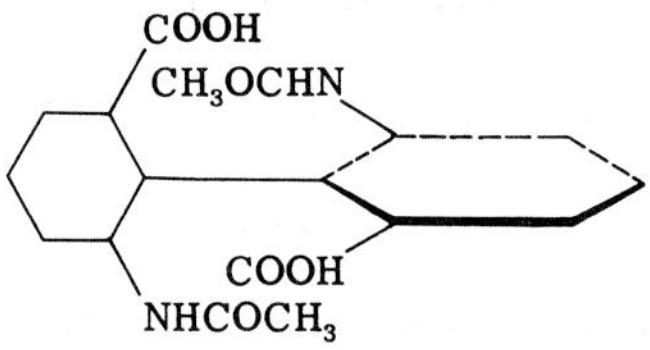

XVI

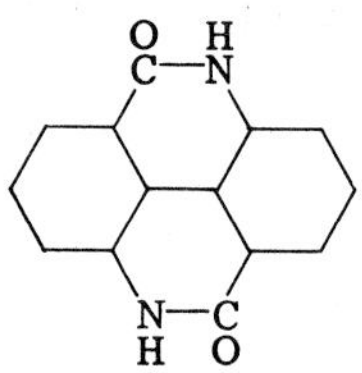

XVII

[14] J. Meisenheimer and M. Höring, *Ber.* **60**, 1425 (1927).

(ii) For "Ansa" Compounds

A further group of atropic isomers can be found among the "ansa" compounds. Molecular asymmetry is brought about by the fact that with suitable substituents on the benzene ring, the latter can no longer rotate through the alicyclic ring. It is assumed that the alicyclic ring is not too large. Lüttringhaus[15] succeeded in resolving the decamethylene ether of 2-bromohydroquinone-5-carboxylic acid (XVIII), for example, into stable optical antipodes.

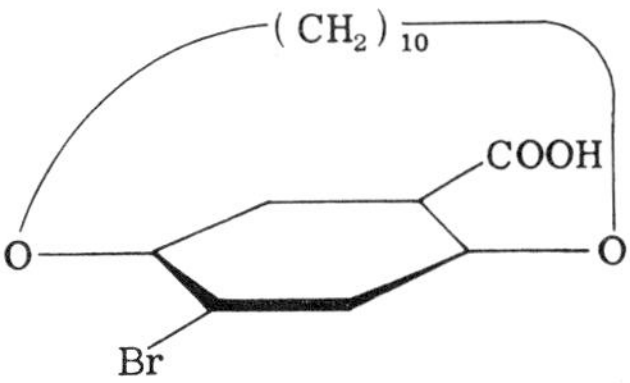

XVIII

8. *Cis-Trans* Isomerism

A further limitation or prevention of free rotation can take place when carbon chains close into rings. Thereby the position of characteristic atomic groups is more or less fixed. The isomerism caused by this is called *cis-trans* isomerism (geometrical isomerism) (XIXa,b). This type of isomerism exists because the substituents are located on the same side (*cis* compounds) or on opposites sides (*trans* compounds) of a certain molecular plane. *Cis-trans* olefins are also classed as geometrical isomers (XXa,b). The position of the substituents is fixed by the double bond, which allows no free rotation:

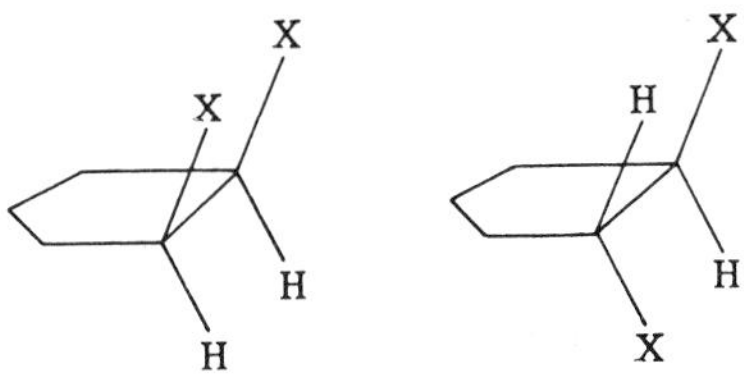

cis (a) *trans* (b)

XIX

cis (a) *trans* (b)

XX

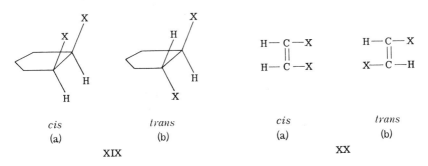

[15] A. Lüttringhaus and H. Gralheer, *Ann. Chem.* **550**, 67 (1941).

9. Strain Theory of Baeyer[16]

The original wording of the tetrahedral theory by van't Hoff and LeBel needs another extension, which Baeyer undertook in 1885 in his "strain theory." In this theory he attempted to show the dependence of ring formation on the number of ring components.

By using the regular tetrahedron as an atomic model, the regular tetrahedral position of the four substituents bound to the central atom can be looked at as the lowest energy state. The angle between two of the valence bonds of the carbon atom is 109°28′ and is designated the "tetrahedral angle." If compounds exist in which other than the regular tetrahedral position is possible, with an angle smaller or greater than the tetrahedral angle between bonds, there results an increase of the energy content of that molecule. Molecules of this type, which are formed with distortion of the tetrahedral angle, exist among cyclic compounds. The expenditure of energy necessary to cause this distortion is called "angle strain." It can be shown with models that this strain decreases as the size increases in cyclic compounds starting with cyclopropane, and reaches a minimum with five-membered and six-menbered rings. On going to larger ring compounds, the tetrahedral angle is spread. The ring strain increases and reaches the same strain with a 17-membered carbon ring that was present with a 3-membered carbon ring. The assumption for this observation according to a model, as Baeyer proposed it, is that all carbon atoms lie in a plane.

An experimental proof of the strain theory set up in this way is possible by measuring the heat of combustion, which is a measure of the energy content of a compound. The energy content is noted by the change in the heat of combustion in relation to the strain. Up to the six-membered ring, the measured values of the heat of combustion agree with the theory of Baeyer. In relation to the normal value of 157.4 kcal. per CH_2 group, the value per CH_2 group starting with a three-membered ring becomes smaller[17] (Table I).

For a six-membered ring and all rings of higher numbers, the deviation from normal value is practically constant, even though according to Baeyer's hypothesis the difference should increase. The cyclic molecules from the six-membered ring on up must, therefore, be practically without strain. The measurement of the heat of combustion presented a basic extension to Baeyer's strain theory. In order to bring this theory into conformity with the experimental results, a change in the original view was made.

[16] A. von Baeyer, *Ber.* **18**, 2277 (1885).

[17] In linear chain structural isomers, the heat of combustion of similarly bonded compounds is practically the same and increases in a homologous series 157.4 kcal per CH_2 group.

TABLE I[a]

Number of carbons in ring	3	4	5	6	7	8
Heat of combustion per CH_2 group (25°; constant volume in kcal/mole)	166.6	163.9	158.7	157.4	158.3	158.6
Difference from tetrahedral angle in planar position	+49°28′	+19°28′	+1°28′	−10°32′	−19°6′	−25°32′
Difference in kilocalories in comparison to normal value of 157.4 kcal/mole	9.2	6.5	1.3	0.0	0.9	1.2

[a] S. Kaarsemaker and J. Coops, *Rec. Trav. Chim.* **71**, 261 (1952); J. W. Knowlton and F. D. Rossini, *J. Research Natl. Bur. Standards* **43**, 113 (1949); R. Spitzer and H. M. Huffman, *J. Am. Chem. Soc.* **69**, 221 (1947); H. van Kamp, Dissertation, Vrije Universiteit, Amsterdam, 1957; J. Coops, H. van Kamp, W. A. Lambregts, B. J. Visser, and H. Dekker, *Rec. Trav. Chim.* **79**, 1226 (1960).

10. Extension of the Strain Theory by Sachse[18], Mohr[19], and Hückel[20]

a. Monocyclic Compounds

The Baeyer strain theory, which assumed a planar structure for cyclic aliphatic compounds, was set up in refutation of the tetrahedron theory.

If we drop the arbitrary hypothesis of Baeyer, and put the requirements of the tetrahedral theory as the prime requisite, the regular tetrahedral position represents the most energetically favorable one. As a result, we arrive at the models of Sachse[18], which have a strain-free construction. Up to a five-membered ring, the centers of the carbon atoms lie approximately in a plane. From a six-membered ring on up, the strain-free construction of the rings is possible only by giving up the planar structure.

[18] H. Sachse, *Ber.* **23**, 1363 (1890); *Z. Physik. Chem.* (*Leipzig*) **10**, 203 (1892).

[19] E. Mohr, *J. Prakt. Chem.* [2] **98**, 315 (1918).

[20] W. Hückel, "Der gegenwärtiger Stand der Spannungstheorie," *in* "Fortschritte der Chemie, Physik und physikalische Chemie" (v. A. Eucken, ed.), Vol. 19, No. 4. Bornträger, Berlin; "Theoretische Grundlagen der organischen Chemie," 8th ed., Vol. I, p. 77 ff. Akad. Verlagsges., Leipzig, 1955.

By this viewpoint, the cyclohexane ring no longer is planar, and can be constructed strain-free in two forms (XXI; XXII).

One structure (XXI) is designated as the "chair" or "rigid" form. Free rotation is prevented in this form and the ring atoms are arranged in two parallel planes, each of which contains three atoms.

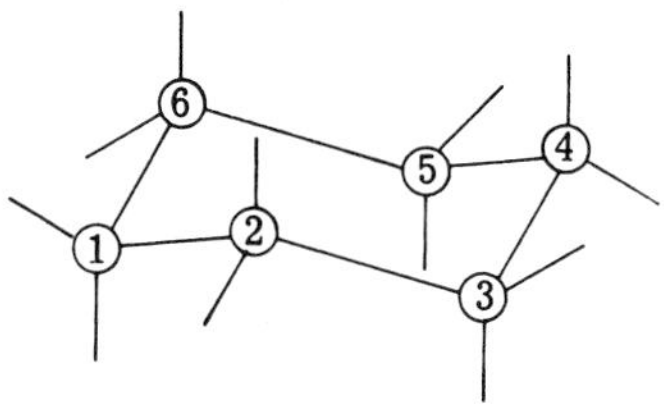

XXI

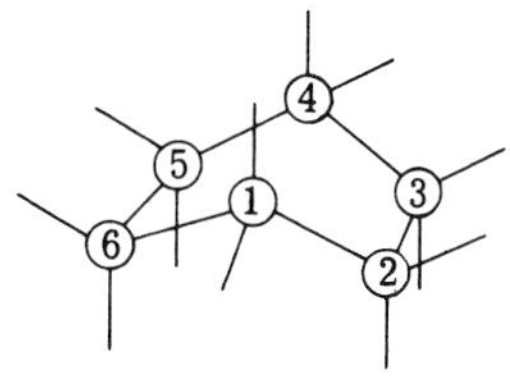

XXII

The second structure (XXI), called the "boat" or "flexible" form, is not rigid. The free rotation is prevented only in part and many possible forms can be made from its model. (For details see Chapter 2, page 44.)

If the principle of free rotation as used for acyclic compounds is applied, then two cyclohexanes as well as two monosubstituted cyclohexanes should exist. This type of isomerism has never been observed in the case of cyclohexane. As a result, the Sachse concept of the spatial structure was rejected for a long time; and, in accord with Baeyer's theory, cyclohexane and all other rings were regarded as planar.

In 1918 Mohr[19] showed for the first time with a model that both forms of the cyclohexane ring were easily interchangeable without requiring a great amount of activation energy. In order to interchange, in XXI, carbon atom 1 flips up and in XXII, carbon atom 4 flips down. In this process only a very small distortion of the tetrahedra angles occurs. The free rotation of the methylene groups around a ring bond as axis must therefore be retained up to a certain point. With this auxiliary hypothesis of Mohr, the number of actually observed isomers of cyclohexane is found to agree completely with the number expected. Similar considerations were made by Mohr[21] for the cycloheptane ring. (See page 158.)

Experimental confirmation was soon obtained for the Sachse–Mohr theory. The difference in behavior of the *cis-trans* isomeric 1,2-diols of 5-, 6-, and 7-membered rings were investigated. Böeseken *et al.*[22]

---

[21] E. Mohr, *J. Prakt. Chem.* [2] **103**, 316 (1922).

[22] J. Böeseken, *Ber.* **46**, 2612 (1913); **55**, 3758 (1922); **56**, 2411 (1923); see also H. G. Derx, *Rec. Trav. Chim.* **41**, 312 (1922); J. Böeseken and P. H. Hermans, *ibid.* **40**, 525 (1921).

were able to show that these diols formed acetals with acetone only if the hydroxyl groups of the 1,2-diols were sufficiently close together:

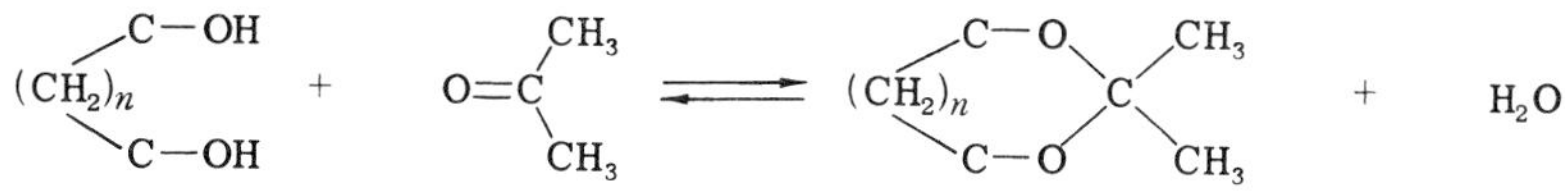

In the planar, rigid, 5-membered compounds, equilibrium in the case of *cis*-1,2-diols favors entirely the formation of the acetone compound. For the *trans*-diols, the OH groups cannot be brought into a plane (as is necessary for the formation of the acetone compound) and no cyclic acetal is formed.

The *cis*-1,2-diol of the 6-membered ring compound still permits the formation of an acetone compound although the equilibrium is less favorable. The *trans*-diol displays no tendency to form an acetone compound. (Compare p. 125.)

For the 7-membered ring as well as for all higher polymethylene compounds, a coplanar position of the *trans* substituents is also possible due to the higher mobility of the ring. No difference in the reactivity between *cis*- and *trans*-cycloheptane-1,2-diol is discernable. In addition, the rise in conductivity of boric acid, which only occurs when the OH groups of 1,2-diols are closer together, as well as the intramolecular formation of anhydrides of cyclopentane and cyclohexane-1,2-dicarboxylic acids, has been used for the proof of the Sachse–Mohr theory.

b. Polycyclic Compounds

Mohr[19] first called attention to the fact that, in the diamond lattice, the ideal tetrahedral carbon arrangement is present. All rings with an even number of carbon atoms from 6 up can be sliced out of this entirely strain free model of a spatial lattice (XXIII).

In addition to the monocyclic compounds, it is possible to have bicyclic and polycyclic systems free of strain. Mohr compiled a series of such strain free models. Of these we may mention *cis*- and *trans*-decalin (XXIV; XXV), [2.2.2]-bicyclooctane (XXVI), [3.2.1]-bicyclooctane (XXVII), [3.3.1]-bicyclononane (XXVIII), and adamantane (XXIX).

The prediction of Mohr that *trans*-decalin could exist was doubted because of the long-standing view that cycloparaffins had a planar structure. If cycloparaffins were planar, then no *trans*-decalin, or a very unstable one, would be possible.

As a result of Mohr's prediction, Hückel[23] succeeded in isolating the two very stable isomers of decalin. The *cis* form can only be converted to the *trans* form under very stringent conditions.

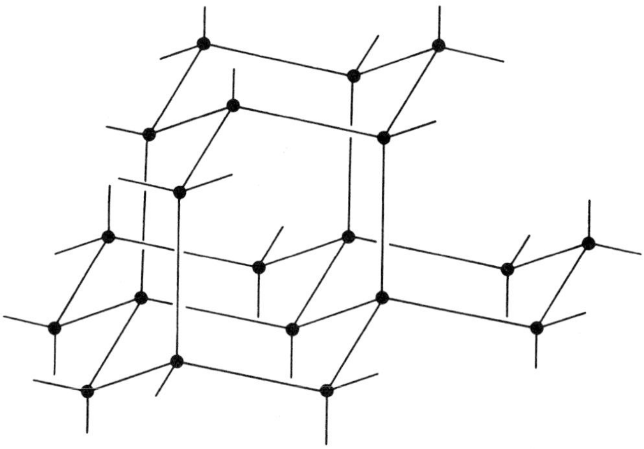

XXIII

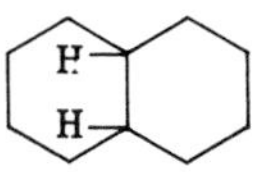

cis-Decalin

XXIV

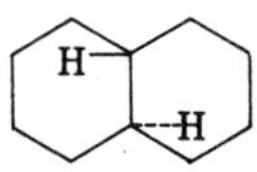

trans-Decalin

XXV

[2. 2. 2]-Bicyclooctane

XXVI

[3. 2. 1]-Bicyclooctane

XXVII

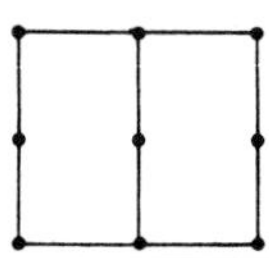

[3. 3. 1]-Bicyclonane

XXVIII

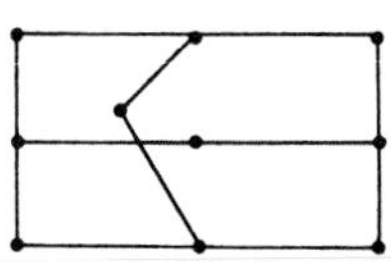

Adamantane

XXIX

The α- and β-substituted decalins (XXX; XXXI) must exist each in four stereoisomeric forms, which are racemic because each substituted decalin contains three asymmetric carbon atoms (designated with an asterisk.)

[23] (a) W. Hückel, *Ann. Chem.* **441**, 1 (1925); (b) **451**, 109 (1926).

Of the α and β compounds, the following are possible: *cis-cis* form,[24] *cis-trans* form, *trans-cis* form, and *trans-trans* form. Hückel isolated the four racemic forms of α- and β-decalol as well as α- and β- decalylamine. He assigned them definitely to the *cis* or *trans* series. Hückel[25]

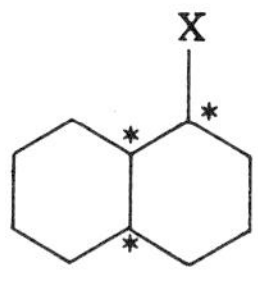

XXX

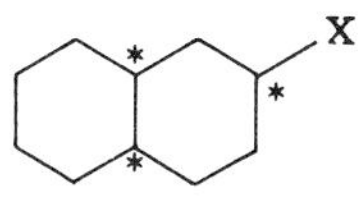

XXXI

was also able to show that the five-membered ring could attach to the six-membered ring in *cis* or *trans* fashion. *Trans*-hydrindane shows a small strain (see p. 174).

c. Bicyclic Systems with Bridgeheads

The improved strain theory of Sachse and Mohr fully confirmed the possibility of forming bicyclic ring compounds with bridgeheads. Camphor (XXXII) is shown as an example:

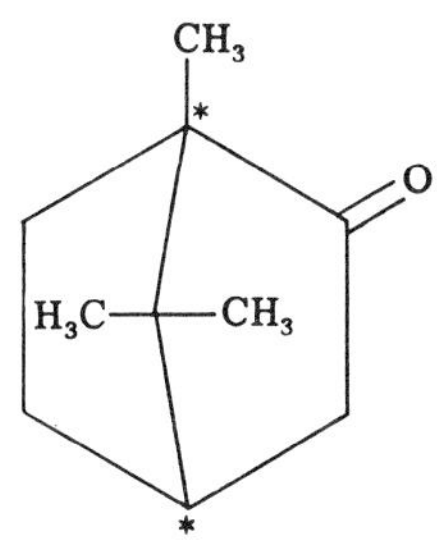

XXXII

Four optically active forms or two racemic forms should exist because two asymmetric carbon atoms are present, but only two optical antipodes are known (one racemic form). The general theory for figuring the number of possible isomers fails in this case. By means of models, this fact can be easily understood[26]: only in the case of the boat form

---

[24] The first "*cis*" relates to the type of ring bond; the second "*cis*" to the relative position of the substituent to the C_9–C_8 valence.

[25] W. Hückel, *Ann. Chem.* **451**, 132 (1926).

[26] J. Bredt, Wüllner anniversary publication, 1905.

it is possible to place the bridge, $>C(CH_3)_2$, between the *para*-carbons of the cyclohexane ring. In this model the two *cis*-oriented valences of the *para*-carbons approach each other to the extent that the formed bridge can be effected without too much strain (XXXIII).

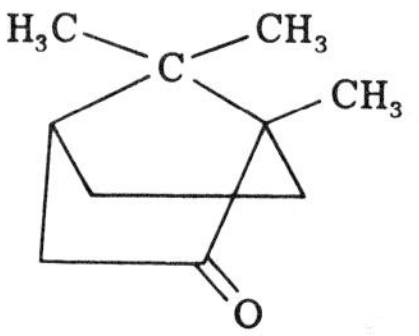

XXXIII

If an attempt were made to place the bridge in *trans* position, this would only be possible with complete breakdown of the tetrahedral configuration, and the strain would be too great. In agreement with theory, no *trans*-camphor exists. The number of possible isomers of compounds whose asymmetric carbon is in a bicyclic system is therefore reduced further.

11. Use of the Strain Theory with Unsaturated Cyclic Compounds

In the van't Hoff model, the double bond is represented by two tetrahedra joined on an edge. This model of various unsaturated monocyclic and bicyclic compounds shows basically very strong strain, as in the case of *trans*-camphor.[27]

Up to now, despite the deficiencies of the van't Hoff model for double bonds, no strongly strained systems of this type are known. Cyclohexene exists only in a nonstrained *cis* form. Cycloheptene is only stable in its *cis* form. The mobile, higher-membered rings, however, exist in *cis* and *trans* forms: cyclooctene, cyclononene etc., are known in *cis* and *trans* forms.

According to the theory, bicyclic compounds with bridge heads are unstable if a double bond is located at the bridge head[28] (Bredt's rule). The formation of such a system does not take place; but an isomeric compound may be formed through a rearrangement, e.g., camphenilol

[27] Compare W. Hückel, "Theoretische Grundlagen der organischen Chemie," Vol. I, p. 105. Akad. Verlagsges., Leipzig, 1955.

[28] J. Bredt and M. Savelsberg, *J. Prakt. Chem.* [2] **97**, 1 (1918); see review article by F. S. Fawcett, *Chem. Rev.* **47**, 219 (1950).

(XXXIV) does not form camphenilene (XXXVI) after splitting out water, but a methyl group migrates (santene shift) and santene (XXXV) is formed.

Camphenilol XXXIV — Santene XXXV — Camphenilene (nonexistent) XXXVI

Bredt's rule has also been applied to heterocyclic ring systems.

An explanation of Bredt's rule has been given by the electron theory[29] as well as by the original strain theory: with a double bond at the bridgehead of a bicyclic system, it can be assumed from steric reasons that a strong deformation of the σ-electron system, usually planar trigonal in a simple double bond, takes place.[30] The resulting change in form and spatial position of the p_z-orbital leads to an uncoupling of the π-electron system, followed by an increase in the energy content. Instability of the double bond results.

For large rings as well as for condensed rings, Bredt's rule is not applicable. A compound such as $\Delta^{1,9}$-octalin (XXXVII), with the double bond at the ring junction, is known. The value for its heat of combustion shows no particular strain.

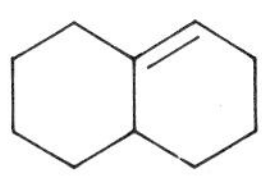

Δ 1, 9-Octalin
(XXXVII)

[29] V. Prelog, P. Barman, and M. Zimmermann, *Helv. Chim. Acta* 32, 1284 (1949).

[30] With all compounds of the form >C=C<, of the four available electron orbitals of carbon, only three, s, p_x, and p_y orbitals, are used. Through hybridization, they become the three trigonal (sp^2) hybrid orbitals. Through overlapping of the (sp^2) orbitals of two carbon atoms, the σ bond is obtained. The fourth orbital of each carbon atom (p_z) pairs up for the π bond, whereby, the p_z-orbitals lie perpendicular above and below the plane of the three σ orbitals. If the p_z-orbitals lie parallel to each other, the maximum overlapping takes place. This is the case when all six atoms of a molecule lie in one plane:

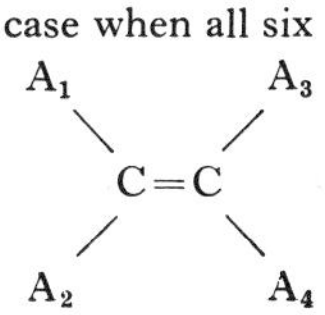

12. Influence on the Ring Strain by Substituents

The relationship between strain and number of members in a cyclic system described up to this point are valid for nonsubstituted rings.

Beesley, Ingold, and Thorpe[31] tried to enlarge the strain theory in this direction, in that they incorporated in their consideration the spatial volume of the substituents in regard to the strain of the ring system.

The regular tetrahedral configuration can occur only with four identical substituents. With different substituents, a change from the tetrahedral arrangement takes place. The changes caused in the tetrahedral angles, due to the volume of the substituents, brings a change in the strain of the entire system. The mathematical comprehension of the influence of the volume of the ligands by Thorpe and Ingold leads only partially to an explanation of observed abnormal ring strain.[32]

The basic concept of the strain theory is not affected by such supplementary ideas. However, the possibility exists that, under certain circumstances, the regular tetrahedral arrangement can be changed sufficiently according to the type of substituents. The compound then may still contain a nearly normal amount of energy.

13. Application of the Strain Theory to Heterocyclic Compounds

Compared to measurements made on carbocyclic systems, application of the strain theory to heterocyclic systems rests on very limited experimental data. Heats of combustion have been measured in only a few cases. From physical measurements of the approximate valence angle of nitrogen and oxygen, the tetrahedral angle is also taken as $\sim 109°$ for purposes of calculation. As for carbocyclic rings, five- and six-membered rings of heterocyclic compounds are especially stable. Rings with more members are free of strain. The NH-group, as well as oxygen, seems to be sterically equivalent to the methylene group ($-CH_2-$). In general, the same strain and isomeric conditions will prevail in heterocyclic compounds as in carbocyclic systems. The conditions are also the same for condensed heterocycles. As with decalin and hydrindane, a series of *cis-trans* isomers are known, of which some

[31] R. M. Beesley, C. Ingold, and J. F. Thorpe, *J. Chem. Soc.* p. 1081 (1915); C. Ingold and J. F. Thorpe, p. 322 (1919). For a review of older literature, see W. Hückel, "Strain Theory," p. 68.

[32] Compare W. L. German and A. I. Vogel, *J. Chem. Soc.* p. 1108 (1937).

examples are given below. Other examples are *cis*- and *trans*-decahydroisoquinoline and *cis*- and *trans*-perhydroindole.

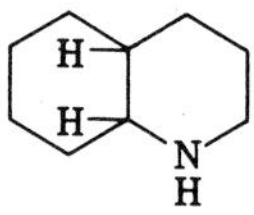

cis-Decahydroquinoline

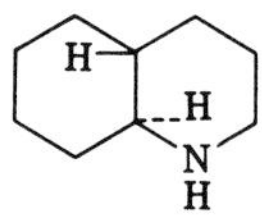

trans-Decahydroquinoline

Of the strain-free heterocyclic bridged cyclic compounds, some examples are hexamethylenetetramine (XXXVIII), azaadamantane (XXXIX), and morphane (XL).

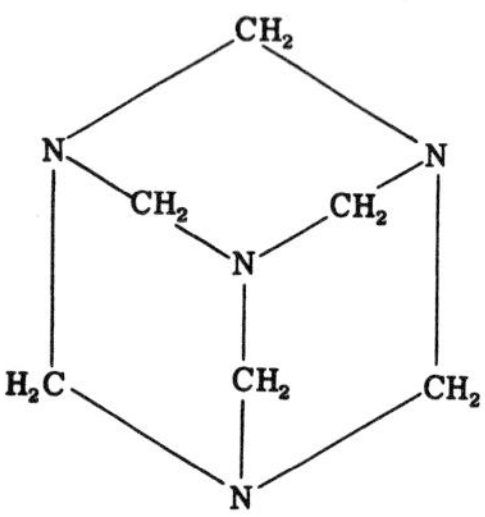

XXXVIII

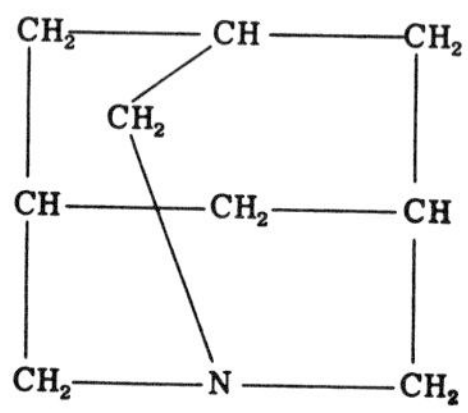

XXXIX

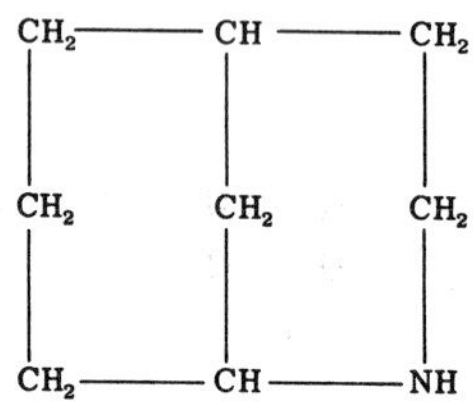

XL

CHAPTER 2

General Discussion

1. The Concept of "Conformation" and "Conformational Analysis"

a. Conformation

With the introduction of the idea of free rotation (cf. Chapter 1) it was determined that, in a compound in which two carbon atoms are connected by a single bond, the substituents situated at the corners of the tetrahedra can assume an infinite number of positions relative to each other.

By turning one tetrahedral model in relation to the other, as for example *meso*-tartaric acid, various forms are possible and only two fit the requirements of a *meso* form. In these two forms the top and bottom half of the molecule are related as image to mirror image, and an inner compensation results. In one form the same substituents are situated on top of each other [plane of symmetry (I)], in the second the substituents are diagonally opposed [center of symmetry (II)].

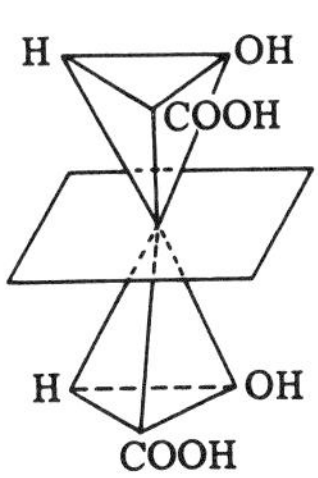

I

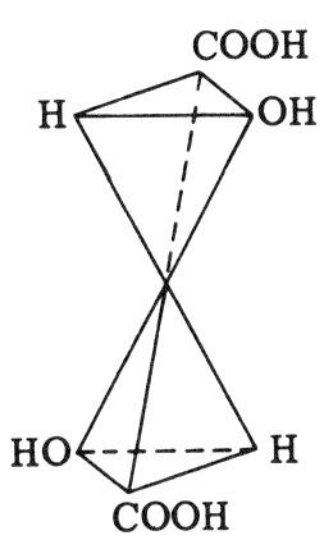

II

The various positions into which the substituents can be brought by rotating the C—C axis are not all of the same energy level. The reason is that not only valence-bound atoms but also further removed atoms

of the same molecule have an effect on each other (see below). Depending on the nature of the substituent, one or more arrangements having the preferred minimum of potential energy are possible. Because of the existence of only one form corresponding to the model, it must be assumed that the energy difference and time of residence of the atoms in the possible positions is so small, that the isolation of such "rotational isomers" is not possible.

The various arrangements which a molecule can assume through free rotation are designated as conformations of the molecule; *meso*-tartaric acid with mirror image symmetry (I) and with a center of symmetry (II) is an example of two different conformations of the same configuration. By interchanging two substituents (e.g., H and OH) on the same carbon atom, a different configuration results. This is *d,l*-tartaric acid of which several conformations are again possible.

The expression "conformation," which is now used exclusively in the English literature, was introduced in 1929 by Haworth[1] into the chemical literature.

In the German literature, besides "conformation," the expression "constellation," introduced by Ebel[2a] in 1933, is used (Freudenberg, Prelog,[2b] Hückel).

Both expressions, "conformation" and "constellation," are identical and are used in the same sense.[3] Aside from these terms, in the physical-chemistry literature especially there appears the designation "rotational isomers" (*Rotationsisomere*), introduced by Wittig[4] in 1930. More exactly, only such conformations are indicated which are relatively stable in the physical sense.[5]

---

[1] W. N. Haworth, "The Constitution of Sugars." p. 90. Arnold, London, 1929.

[2] (a) F. Ebel, *in* "Stereochemie" (K. Freudenberg, ed.), p. 825. Deuticke, Vienna, 1933; (b) V. Prelog, *J. Chem. Soc.* p. 420 (1950).

[3] In the German literature it has not been possible so far to agree on one expression. Both "conformation" and "constellation" are found, and even used together in the same journal. We shall use the expression "conformation" in the following chapters. Besides "conformation" we also use the expression "form"; however in this volume—and this is especially emphasized—the terms "conformation" and "form" are interchangeable. An attempt has been made to differentiate between the expressions "conformation" and "form" [compare, for example, F. V. Brutcher, Jr., T. Roberts, S. J. Barr, and N. Pearson, *J. Am. Chem. Soc.* **81**, 4915 (1959); also H. H. Lau, *Angew. Chem.* **73**, 423 (1961)], whereby "conformation" is only used when the molecule exists in the state of minimum potential (e.g., the chair conformation of cyclohexane) and all other spatial arrangements (e.g., the boat form of cyclohexane) are designated as "forms." To repeat, we shall use both expressions interchangeably, without making this distinction. See also E. A. Braude and C. J. Timmons, *J. Chem. Soc.* p. 3767 (1955) (comment).

[4] G. Wittig, "Stereochemie," p. 317. Akad. Verlagsges., Leipzig, 1930.

[5] Compare C. W. Shoppee, *J. Chem. Soc.* p. 1138 (1946).

Aside from the above description, an exact definition of the concept of conformation is given in various ways:

> The term "conformations" denotes the different arrangements in space of the atoms in a single classical organic structure (configuration), the arrangements being produced by the rotation or twisting (but not breaking) of bonds (Klyne[6]).
>
> By "conformation" is meant any arrangement in space of the [illegible]

[illegible]

(i) Concept of "Conformational Analysis"

Investigations carried out in the last 10 to 15 years in the field of stereochemistry dealing with the elucidation of the more detailed three-dimensional structure mentioned briefly above are included under the concept of "conformational analysis" (German: *Konstellationsanalyse*).[10] Included are simple and complicated aliphatic, alicyclic, and heterocyclic compounds. Barton[11] introduced the concept of "conformational analysis" in his fundamental investigations. The basic idea of conformational analysis is that certain chemical and physical properties of organic compounds are related to preferred conformations.

---

[6] W. Klyne, *Progr. Stereochem.* 1, Chapter 2 (1954).

[7] W. G. Dauben and K. S. Pitzer, *in* "Steric Effects in Organic Chemistry" (M. S. Newman, ed.), Chapter 1. Wiley, New York, 1956.

[8] D. H. R. Barton and R. C. Cookson, *Quart. Rev.* (*London*) **10**, 44 (1956).

[9] See also V. Prelog, *J. Chem. Soc.* p. 420 (1950); E. L. Eliel, *J. Chem. Educ.* **37**, 126 (1960).

[10] See references 6–9. W. Klyne and P. B. D. de la Mare, *Progr. Stereochem.* **2** (1958); **3** (1962). Also compare: (a) S. J. Angyal and J. A. Mills, *Rev. Pure Appl. Chem.* **2**, 185 (1952). (b) H. D. Orloff, *Chem. Rev.* **54**, 347 (1954). (c) K. Bláha, *Chemie* (*Prague*) **9**, 223 (1957). (d) T. N. Nazarov and L. D. Bergelson, *Usp. Khim.* **26**, 3 (1957). (e) W. Hückel and M. Hanack, *Ann. Chem.* **616**, 18 (1958). (f) H. H. Lau, *Angew. Chem.* **73**, 423 (1961). (g) D. H. R. Barton, *Experientia Suppl.* **2**, 121 (1955); Theoretical organic chemistry, Kekulé Symposium London, 1958, p. 127. *Suomen Kemistilehti* **A32**, 27 (1959); *Svensk Kem. Tidskr.* **71**, 256 (1959).

[11] (a) D. H. R. Barton, *J. Chem. Soc.* p. 340 (1948); p. 1027 (1953); (b) *Experientia* **6**, 316 (1950).

(ii) Methods of Conformational Analysis

The methods of determination of the conformation of a molecule can be separated into the following two large groups.

In the first group belong the physical methods such as electron diffraction,[12] x-ray diffraction,[13] determination of dipole moments,[14] ultraviolet and infrared spectra,[15] Raman spectra,[16] microwave spectra,[17] the Kerr effect,[18] nuclear magnetic resonance,[19] and measurements of optical rotatory dispersion.[20]

Also attributed to physical methods are thermodynamic calculations of the energy content of compounds. These are made from the theoretical standpoint, from which conclusions on the conformation can be drawn.[21]

In the second group belong chemical methods. These concern experiments from which predictions of the spatial structure of the molecule can be made, based on certain reactions of model compounds. For this purpose many reactions can be called upon, but only a few have been investigated for their usefulness in conformational analysis. Mention

[12] See for example (a) O. Hassel and B. Ottar, *Acta Chem. Scand.* **1**, 929 (1947); (b) O. Hassel, *Research (London)* **3**, 504 (1950); (c) *Quart. Rev. (London)* **7**, 221 (1953); (d) V. A. Atkinson and O. Hassel, *Acta Chem. Scand.* **13**, 1737 (1959).

[13] Compare (a) O. Hassel, *Research (London)* **3**, 504 (1950); (b) R. W. G. Wyckoff, "Crystal Structures." Vol. 3, Chapter 15. Wiley (Interscience), New York, 1953.

[14] See for example A. Tulinskie, A. DiGiacomo, and C. P. Smyth, *J. Am. Chem. Soc.* **75**, 3552 (1953).

[15] See for example E. A. Braude and E. S. Waight in *Progr. Stereochem.* **1**, Chapter 4 (1954); M. Hanack, *Forsch. Fortschr.* **34**, 259 (1960).

[16] See for example S. Mizushima, "The Structure of Molecules and Internal Rotation." Academic Press, New York, 1954, Chapter 1, p. 162 ff.

[17] See for example D. R. Lide, Jr., *J. Chem. Phys.* **30**, 37 (1959). Compare D. J. Millen in P. B. D. de la Mare and W. Klyne, *Progr. Stereochem.* **3**, Chap. 4 (1962).

[18] See for example C. G. Le Fèvre and R. J. W. Le Fèvre, *J. Chem. Soc.* p. 3549 (1956); p. 3458 (1957).

[19] See for example J. D. Roberts, "Nuclear Magnetic Resonance, Application to Organic Chemistry." McGraw-Hill, New York, 1959; J. A. Pople, W. G. Schneider, and H. J. Bernstein, "High-resolution Nuclear Magnetic Resonance." McGraw-Hill, New York, 1959.

[20] See for example C. Djerassi, "Optical Rotatory Dispersion, Application to Organic Chemistry." McGraw-Hill, New York, 1960; G. G. Lyle and R. E. Lyle, *in* "Determination of Organic Structures by Physical Methods" (J. C. Nachod and W. D. Phillips, eds.), Vol. 2, p. 1 ff. Academic Press, New York, 1962.

[21] See for example (a) K. S. Pitzer, *Chem. Rev.* **27**, 39 (1940); (b) *Science*, **101**, 672 (1945); (c) K. S. Pitzer, and C. W. Beckett, *J. Am. Chem. Soc.* **69**, 977 (1947); (d) C. W. Beckett, K. S. Pitzer, and R. Spitzer, *ibid.* **69**, 2488 (1947); (e) K. S. Pitzer, *Discussions Faraday Soc.* **10**, 66 (1951); (f) S. Mizushima, "The Structure of Molecules and Internal Rotation." Academic Press, New York, 1954; (g) J. B. Hendrickson, *J. Am. Chem. Soc.* **83**, 4537 (1961); **84**, 3355 (1962).

may be made of kinetic conformational analyses, which employ reaction rates to ascertain the conformation: included are solvolysis of carboxylates and tosylates of alicyclic alcohols,[22a,b] rates of oxidation of such alcohols,[22a] as well as their rates of esterification.[23]

These methods do not always produce unequivocal results, because the connection between molecular structure and rate of reaction can be very involved.[10e]

A further method of chemical conformational analysis is based on thermodynamics. By this method, conclusions concerning the preferred conformation are drawn from the state of equilibrium existing between stereoisomeric compounds.[22a,24]

The explanation of finer conformational effects on the reaction rate, the so-called "conformational transmission" effects, must also be included in the realm of conformational analysis.[25] (See Chap. 5, p. 225.)

Conformational analysis was first used successfully on systems of condensed cyclohexane rings, as in steroids and triterpenoids. These compounds, in contrast to compounds in which a number of conformations are possible, are easier to deal with because their rigid structure usually permits only one conformation.

2. Conformations of Simple Hydrocarbons[26]

The various conformations which can be obtained from a certain configuration by rotation of the C—C axis, will not be of equal energy content. The size of the energy difference between two or more conformations determines the adoption of relatively stable arrangements. There will be preferred conformations of a molecule. Each molecule will try to take on the conformation with the lowest energy level. The energy difference between the individual conformational isomers (also called "rotamers" or "conformers," following a suggestion offered by

---

[22] (a) S. Winstein and N. J. Holness, *J. Am. Chem. Soc.* **77**, 5562 (1955); (b) W. Hückel *et al.*, *Ann. Chem.* **624**, 142 (1959); compare also E. A. S. Cavell, N. B. Chapman, and M. D. Johnson, *J. Chem. Soc.* p. 1413 (1960).

[23] See for example E. L. Eliel and C. A. Lukach, *J. Am. Chem. Soc.* **79**, 5986 (1957); compare also E. L. Eliel, *J. Chem. Educ.* **37**, 126 (1960).

[24] Compare (a) E. L. Eliel and C. A. Lukach, *J. Am. Chem. Soc.* **79**, 5986 (1957); (b) E. L. Eliel and R. S. Ro, *ibid.*, 5992, 5995; (c) *Chem. & Ind.* (*London*) p. 251 (1956). (d) Compare also E. L. Eliel, *J. Chem. Educ.* **37**, 126 (1960); E. M. Philbin and T. S. Wheeler, *Proc. Chem. Soc.* p. 167 (1958).

[25] D. H. R. Barton, *Experientia Suppl.* **2**, 121 (1955); D. H. R. Barton and A. J. Head, *J. Chem. Soc.* p. 932 (1956); D. H. R. Barton, A. J. Head, and P. J. May, *ibid.* p. 935 (1957); C. Djerassi, O. Halpern, V. Halpern, and B. Riniker, *J. Am. Chem. Soc.* **80**, 4001 (1958); D. H. R. Barton, F. McCapra, P. J. May, and F. Thudium, *J. Chem. Soc.* p. 1297 (1960).

[26] Compare J. C. McCoubrey and A. R. Ubbelohde, *Quart. Rev.* (*London*) **5**, 364 (1951); ref. 16; D. J. Millen, *in* P. B. D. de la Mare and W. Klyne, *Progr. Stereochem.* **3**, 138 (1962).

A. Dreiding) is very small. Only a few cases are known (compare page 40) where the energy barrier between the conformational isomers is high enough to allow them to be isolated. In any case, definite physical properties can be attributed to the various conformational isomers which can be determined mainly on the basis of their differing spectroscopic behavior.

a. Ethane: "Staggered" and "Eclipsed" Conformation

[illegible]

[illegible] is called the "staggered" conformation (IIIb, d, f; the molecule has been depicted in the Newman projection).

They are formed separately by rotating one methyl group in an angle ϕ of 120° in relation to the other. The angle of rotation is defined more accurately by the so-called dihedral angle ϕ, between two designated hydrogen atoms: Consider two planes, one of which is determined by a designated C—H bond at atom C-1 and the C_1-C_2 bond, and the other by a designated C—H bond at atom C-2 and the C_1-C_2 bond; the dihedral angle is determined by these two planes.

Additionally, three conformations (IIIa, c, e) of equal energy result when the methyl groups are rotated so that hydrogen atoms lie behind each other along the C_1-C_2 axis. These forms are called "eclipsed" conformations. In this position the hydrogen atoms of both methyl groups are closer together than in the *staggered* form. Because of the higher repulsive force for the former, the energy content of the *eclipsed* form is higher.

It has been definitely proven[27] from vibrational spectra of ethane that the *staggered* forms are the most stable, i.e., correspond to a minimum potential energy. For some time it had been assumed that the *eclipsed* form was more stable, e.g., that the minimum potential resulted from the attraction of the hydrogen atoms.

Between the maximum of energy in the *eclipsed* conformation and the minimum in the *staggered* conformation an infinite number of positions are possible. If we rotate one methyl group in ethane 360° around the C_1-C_2 axis relative to the other methyl group, the potential

[27] L. G. Smith, *J. Chem. Phys.* **17**, 139 (1949); J. Romanko, T. Feldman and H. L. Welsh, *Canad. J. Phys.* **33**, 588 (1955); see also R. S. Wagner and B. P. Dailey, *J. Chem. Phys.* **23**, 1355 (1955).

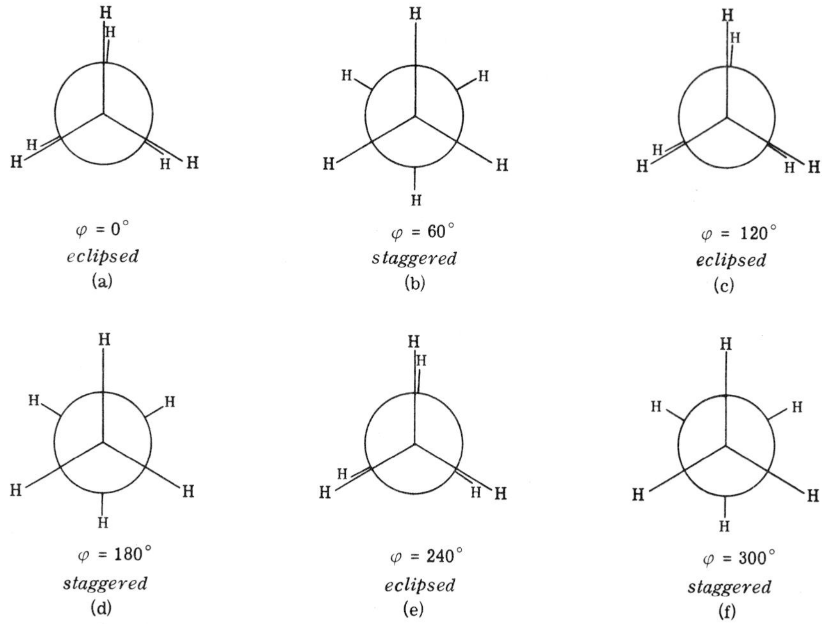

III

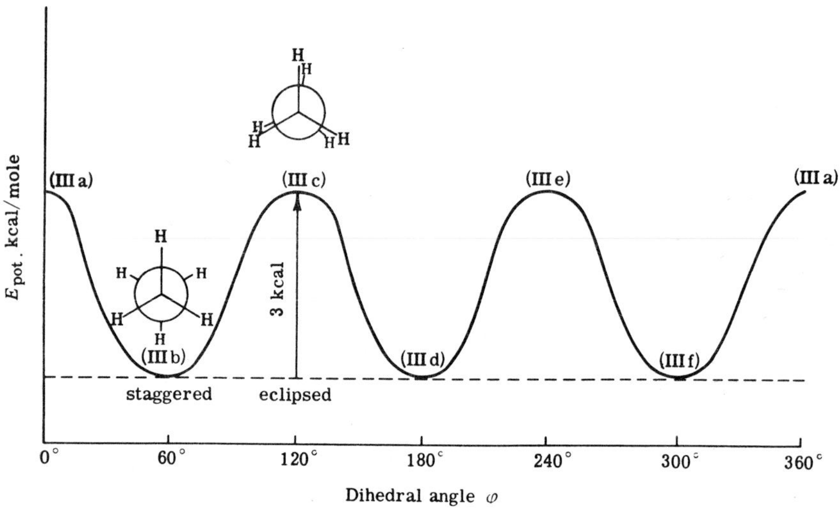

FIG. 1. Potential energy curve for rotation of one methyl group in ethane. Compare with structures (IIIa–f).

energy will change in the form of a sine wave (potential curve) (Fig. 1).[28] An energy barrier, which is repeated three times, stands in the way of the rotating methyl group. To pass this barrier an activation energy must be applied. The height of this energy barrier was calculated byEyring[29] in 1932 as 0.3 kcal./mole. on the basis of simple exchange energies. From spectroscopic data and from exact measurement of the specific heat,[30] the value was soon seen to be too low, and was subsequently set at 2.8 to 3.04 kcal./mole.[31] The specific heat is the combined contributions of the individual degrees of freedom of motion of the molecule. From these measurements it became apparent that the degree of freedom of the inner rotation around the C—C bond cannot be entirely neglected at room temperature.

A value of 3 kcal./mole indicates that the rotation around the C—C bond axis cannot be free, while the value of 0.3 kcal./mole determined by Eyring[29] would still allow unrestricted rotation. If the energy barrier were $E \leqslant kT$ ($RT \sim 0.6$ cal.) at room temperature, then free rotation is possible as a result of thermal energy. If $E \gg kT$, then the rotation around the C—C bond axis is practically frozen except for torsional vibration. The value of ~3 kcal./mole found for ethane indicates that the rotation can be neither entirely free nor entirely hindered. Since the value is in the order of magnitude equal to kT, ethane must represent a "hindered rotator."

b. *n*-Butane: "Staggered," "Gauche," "Eclipsed" Conformations

The conformations of *n*-butane can be evolved in a manner similar to those of ethane (IVa–f).

As in the case of ethane, those positions are preferred in which the substituents (in this case H and CH_3) lie over the gap. Because the substituents are different, there are differences in potential energy of the possible *staggered* forms:

---

[28] For symmetrical groups such as CH_3- or CCl_3-, the potential function V_ϕ can be calculated according to the formula:

$$V_\phi = (V_0/2)(1 - \cos n_0) ,$$

where V_0 is the height of the maxima as related to the energy minima and n_0 is the number of maxima and minima in a rotation of 360°.

[29] H. Eyring, *J. Am. Chem. Soc.* **54**, 3191 (1932).

[30] See for example E. Teller and B. Topley, *J. Chem. Soc.* p. 876 (1935); G. B. Kistiakowsky, J. R. Lacher, and F. Stitt, *J. Chem. Phys.* **6**, 407 (1938); **7**, 289 (1939).

[31] Compare also K. S. Pitzer, *Discussions Faraday Soc.* **10**, 66 (1951); J. D. Kemp and K. S. Pitzer, *J. Am. Chem. Soc.* **59**, 276 (1937). Recent measurement: D. R. Lide, *J. Chem. Phys.* **29**, 1426 (1958), 3.040 kcal./mole from microwave spectra (CH_3CD_3 = 2.880 kcal./mole). For the thermodynamic treatment of the inner rotation see: A. Pacault and P. Bothorel, *Bull. Soc. Chim. France* p. 217 (1956).

(1) The true *staggered* form, in which the two methyl groups are as remote as possible from each other in a *trans zigzag* position (IVd), is also called the *trans* or *anti* conformation.
(2) Two energetically equal metastable conformations (IVb and f) are the ones in which the methyl groups approach each other spatially without lying on top of each other. They are formed from the *staggered* conformation by a rotation of 120° around the C—C bond axis in either direction. These two enantiomorphic forms are called *gauche*, *skew*, or *syn* conformations. In the two *gauche* forms, the methyl groups approach each other more closely than the van der Waals radii allow. They are therefore of higher energy content than the true *staggered* form. The latter has the lowest energy content.

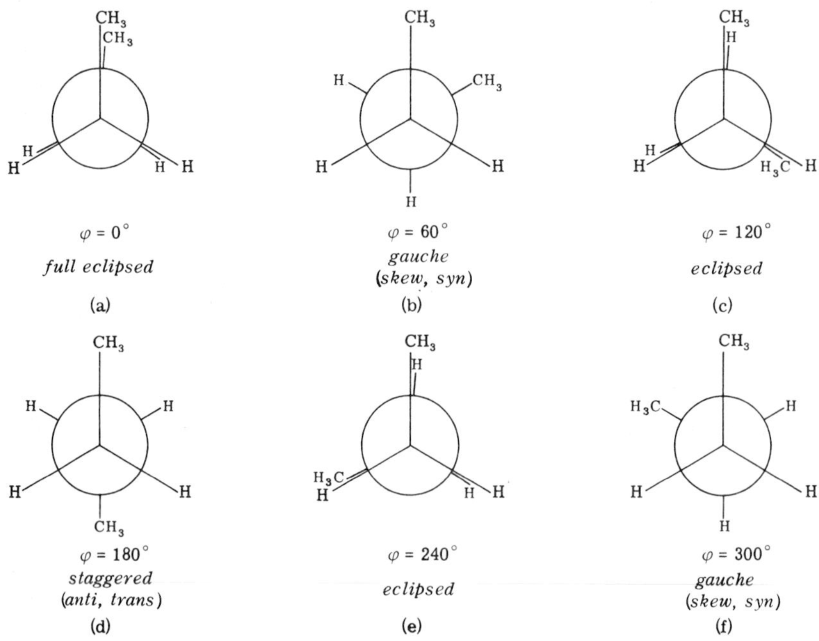

IV

The corresponding situation applies to the three energetically less likely *eclipsed* conformations. If the two methyl groups lie on top of each other, viewed in the direction of the C—C bond axis, then the fully *eclipsed* conformation (IVa) is obtained. In the other two *eclipsed* forms of equal energy content, the methyl group lies over a hydrogen or the reverse (IVc and e). The repulsion is not as strong and the energy content, therefore, is a little lower. In Fig. 2 the energy barrier is shown in relation to the dihedral angle as compared with ethane.

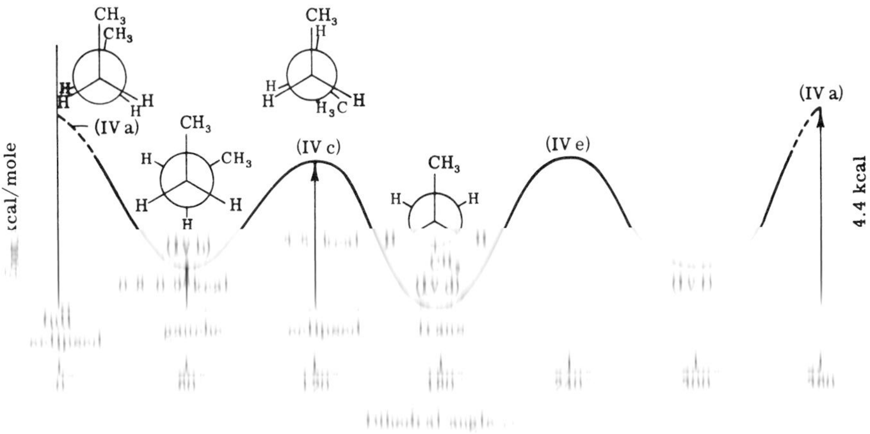

structures (IV a–f).

One can see in Fig. 2 that (b), (d), and (f) represent the energy minima and (a), (c), and (e) the energy maxima. These arrangements are not of the same energy content. Investigations by infrared and Raman spectroscopy have indicated[32] that the *staggered* conformation are also, in this case, of the lowest energy level. A spectrum with few bands was obtained from solid butane at low temperature. This could only be attributed to a *staggered* (*trans*) form with a center of symmetry. At higher temperatures in the liquid or gas phase, the unsymmetrical *gauche* conformations became apparent by an increase in the number of absorption bands. The differences in energy of *fully eclipsed*, *eclipsed*, and *gauche* forms relative to the *trans* form could be calculated from the change in the rates of intensity of suitable lines in the infrared and Raman spectra at various temperatures, and from thermodynamic data.[33] These amount to ~3.5 kcal./mole for the *eclipsed* form and 0.8–0.9 kcal./mole for the *gauche* form (Fig. 2). The difference between the fully *eclipsed* form with the greatest energy content and the *trans* form lies above 4 kcal./mole. A more exact value has not been found as yet.

From these values it can be seen that the *trans* conformation is preferred. From the energy difference of the three conformers, it was calculated that at room temperature about 80% of the *n*-butane molecule is in *trans* conformation and about 20% in the *gauche* conformation.

[32] (a) D. H. Rank, N. Sheppard, and G. J. Szasz, *J. Chem. Phys.* **17**, 83 (1949); (b) N. Sheppard and G. J. Szasz, *ibid.* **17**, 86 (1949); (c) D. W. E. Axford and D. H. Rank, *ibid.* **17**, 430 (1949).

[33] See ref. 21a and e; see also K. Ito, *J. Am. Chem. Soc.* **75**, 2430 (1953); W. B. Person and G. C. Pimentel, *J. Am. Chem. Soc.* **75**, 532 (1953).

c. Baeyer and Pitzer Strain

Baeyer strain (angle strain) was defined as the increase in energy of cyclic compounds which arises from the deformation of the optimum valence angle (with sp^3-hybridized carbon atoms at 109°28′, sp^2-hybridized carbon atoms at 120°). (Compare Chapter 1, page 12.)

A further increase in energy, especially for ring compounds, arises through the possibility of forming a conformation of unfavorable energy content: that increase in energy which a molecule undergoes when it passes by rotation from an energetically favorable *staggered* conformation into an energetically less favorable *gauche* or *eclipsed* conformation is designated as Pitzer strain (or bond opposition strain) of the molecule.[34,35] The origin of Pitzer strain lies in a repulsion of neighboring, nonbonded atoms.

Pitzer strain becomes noticeable especially in the case of cyclic compounds with a low number of members, where the substituents are forced from the *staggered* conformation into the *eclipsed* conformation. In cyclopentane, for example, the appearing Pitzer strain causes a deviation from planar structure[36] (page 73). A deformation of the tetrahedral angle thus also results, and leads to Baeyer ring strain. Despite the increase in energy caused by this, the nonplanar form of cyclopentane corresponds to a lower potential energy than the planar one because of its energetically more favorable conformation (see Chapter 3, I).

d. Higher Normal Hydrocarbons

The higher *normal* hydrocarbons can be broken down into a series of *n*-butane systems for conformational analysis. Each successive C—C bond is then analyzed as was the central bond of the butane molecule.

There are seven definite conformations conceivable for *n*-pentane (*trans* and *gauche*). (The remaining nonexistent *eclipsed* conformations are not mentioned in the following because of their high energy.)

---

[34] Compare M. Kobelt, P. Barman, V. Prelog, and L. Ruzicka, *Helv. Chim. Acta* **32**, 256 (1949); V. Prelog, M. Fausy El-Neweihy, and O. Häfliger, *ibid.* **33**, 1937 (1950). Compare K. S. Pitzer, ref. 21b; H. Kühn, *J. Chem. Phys.* **15**, 843 (1947).

[35] The expression "Pitzer strain" was introduced by W. v. E. Doering: See W. v. E. Doering and M. Farber, *J. Am. Chem. Soc.* **71**, 1514 (1949). Consider also the concept of I-Strain, p. 82.

[36] Compare for example K. S. Pitzer, *Science* **101**, 672 (1945); J. E. Kilpatrick, K. S. Pitzer, and R. Spitzer, *J. Am. Chem. Soc.* **69**, 2483 (1947); K. S. Pitzer and W. E. Donath, *ibid.* **81**, 3213 (1959). See Chapter 3, page 72).

As could be determined[16,32a] from the Raman spectra, liquid *n*-pentane exists as two conformers. One of them is the energetically favored *trans* form (TT-form, see p. 70). As calculated from the *n*-butane system, methyl and ethyl groups are located in a *trans* position (V):

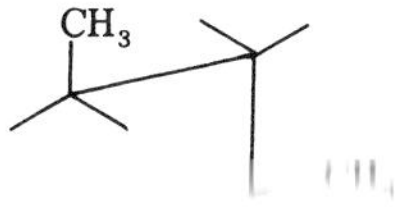

V

The second form present at equilibrium is the next most stable form with a *gauche* (TG) arrangement (VI)[36a].

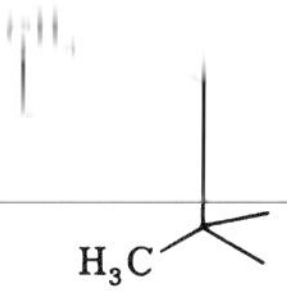

VI

All further possible conformations contain more *gauche* forms and are thus less stable.[37]

The energy difference between the two conformers of *n*-pentane present at equilibrium was determined to be about 0.5 to 0.7 kcal./mole.[32b,38]

(*i*) n-*Hexane*

Whereas *n*-pentane contained two axes of inner rotation, *n*-hexane has three. The number of possible conformers becomes larger. There are twelve *staggered* and *gauche* conformations conceivable. As in the case of *n*-pentane, Mizushima,[16] Sheppard and Szasz[32b] were again able to recognize the *trans* (TTT) form as the predominant one at equilibrium.

[36a] R. A. Bonham, L. S. Bartell, and D. A. Kohl, *J. Am. Chem. Soc.* **81**, 4765 (1959).

[37] As an illustration of the possible conformations, it is best to build models (ball and stick models) which rotate freely around the C—C bond axes. Possible arrangements, not easily recognized with perspective formulas, are readily apparent from such models. Atomic models which take into account the atomic radii (such as Stuart–Briegleb or Fisher–Hirschfelder atomic models) (see also Section 5) are less suitable for such studies.

[38] K. Ito, *J. Am. Chem. Soc.* **75**, 2430 (1953); Compare also K. S. Pitzer and E. Catalano, *ibid.* **78**, 4844 (1956).

Here too, the energy difference between *trans* and *gauche* form lies about 0.5 kcal./mole.[36a,39]

(ii) Paraffins of Longer Carbon Chains

From measurements of the Raman spectra of *n*-paraffins from C_4 to C_{12} in the liquid and solid states, Mizushima *et al.*[16] ascertained the conformations of these hydrocarbons. It was possible to determine that all the investigated hydrocarbons were preferentially present in the *trans* conformation. The conformation of the entire molecule represented a coplanar, zigzag chain. Here the energy difference between *trans* and *gauche* forms amounted to about 0.8 kcal./mole.[39]

e. Steric Partition Function

The number of possible conformations grows with the increasing carbon number of the chain. With the help of the steric partition function introduced by Pitzer,[40] a probability can be set up of the conformation that the molecule will assume. Only those three conformations of each degree of freedom of the inner rotation around a C—C bond are considered which distinguish themselves by an energy minimum (one *trans* and two *gauche* forms). A certain amount of steric energy is attributed to each conformation.

The steric partition function according to Pitzer[40] is:

$$Q^{st} = \sum_i \exp(-E_i/RT) \tag{1}$$

The summation covers all conformations of the molecule; E_i is the steric energy of the ith conformation limited by the intramolecular repulsion forces.

The total number of energetically favorable conformations of a normal paraffin with n carbon atoms amounts to: 3^{n-3}. It should be pointed out that terminal methyl groups cannot produce different conformations during inner rotation.

By using the steric partition function for *n*-butane, the following is obtained:

There are three possible conformations (one *trans* and two *gauche*), whereby the *gauche* conformation contains higher energy relative to the *trans* conformation in the order of $a = 0.8$–0.9 kcal./mole. (Compare Section 2,b.)

---

[39] Compare W. B. Person and G. C. Pimentel, *J. Am. Chem. Soc.* **75**, 532 (1953).

[40] K. S. Pitzer, *J. Chem. Phys.* **8**, 711 (1940), and ref. 21a.

If we set $E_i = 0$ for the *trans* conformation, then the steric partition function for *n*-butane is (1 *trans*, 2 *gauche* conformations):

$$Q^{st} = 1 + 2e^{-a/RT} \tag{2}$$

For branched paraffins, e.g., 2-methylbutane, one obtains[41]:

$$Q^{st} = 2e^{-a/RT} + e^{-2a/RT} \tag{3}$$

[illegible]

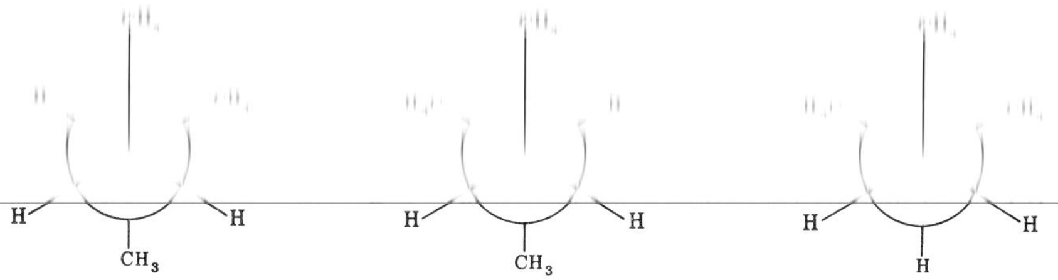

trans-gauche *trans-gauche* *gauche-gauche*

VII

The conformation of lowest energy of a branched paraffin has an energy level different from zero which is designated as E^{st},[41] when related to the conformation of lowest energy of a normal paraffin. For the comparison of the inner movement of branched and normal paraffins, the factor, $\exp(-E^{st}/RT)$, is extracted from the sum of the steric partition function Q^{st}:

$$Q^{st} = \exp(-E^{st}/RT) \sum_i \exp(-E_i'/RT) = \exp(-E^{st}/RT) \cdot Q^{st'} \tag{4}$$

Where E_i' represents the steric energy of the conformations, reduced by the amount of the conformation with lowest energy.

The intramolecular repulsion forces are thereby resolved into two parts: the exponent $(-E^{st}/RT)$ describes the general increase in potential energy of the molecule brought about by the repulsion; the sum $Q^{st'}$ describes the different probability in the occupation of the various conformations.

In order to obtain the energies of the conformations, which are necessary according to Eq. (4) for calculating the steric partition function, a counting method[40] was developed by Becker.[41]

---

[41] F. Becker, *Z. Naturforsch.* **14a**, 547 (1959).

[42] For nomenclature, compare Section 5.

The carbon chain of any molecule is increased by one CH_2 group each time. The starting point for *n*-paraffins is *n*-butane with three conformations, whose steric energies ($a = 0.8$–0.9 kcal./mole) are set equal to the number of neighboring CH_3 groups in the projection formulas of Newman.[42]

If *n*-butane is lenghtened by one CH_2 group, the number of conformations triples. For each starting conformation (one *trans*, two *gauche*), there are obtained one new *trans* and two new *gauche* conformations.

In general the following rules apply[41]:

(1) For a *trans* extension (t) (chain extension in the same direction as the previous extension step) the energy does not increase. It makes no difference if this follows a *trans* or *gauche* extension.
(2) When a *gauche* extension follows (t) (extension of an angle of 120°) then the energy grows about the parameter a, because an additional repulsion of the same type as for the *gauche* conformation of *n*-butane is added on.
(3) When a *gauche* extension follows a *gauche* extension, namely g_r on g_l or g_l on g_r[43] then the energy also increases by the amount a. Since the coils of the chain form in opposite directions (+ 120° or −120° respectively), they do not influence each other.
(4) When g_r follows g_r or g_l follows g_l an overlapping of the effect of the radii of the CH_3 or CH_2 groups appears at the ends of the chain of five C atoms. These types of conformations are therefore doubtful.[40]

Methyl branched paraffins are obtained by the simultaneous replacement of two or three H atoms on the terminal methyl group by CH_3. Here also, the number of possible conformations grows by a factor of three, since methyl groups, as a result of their threefold symmetry, yield no new conformations. The number of possible conformations for any branched paraffin comes to 3^{n-3-v}; where n is the number of C atoms and v the amount of branching. This method developed by Becker[41] also permits the calculation for complicated compounds. The energy of the starting conformation rises in definite increments. It is frequently expressed by the parameter "a".

f. Substituted Aliphatic Hydrocarbons

The substitution of an H atom by halogen in ethane does not change the energy barrier significantly. By spectroscopic methods, values of

[43] For further discussion of g_r and g_l, see Section 5, page 68.

3.33 kcal./mole in ethyl fluoride, 3.5 kcal./mole in ethyl chloride and 3.5 kcal./mole in ethyl bromide were found.[44]

Disubstituted ethanes, for example, 1,2-dichloroethane, whose isomerism of rotation has been investigated carefully by various physical methods, exist as well in a number of conformations. The *trans* form is preferred to the two *gauche* forms (VIII):

VIII

For these compounds as well, the *trans* form only exists in the solid state; in liquid and gaseous state the equilibrium is shifted in the direction of the more energy rich *gauche* form.[45] Gaseous 1,2-dichloroethane exists at 22° about 73 % in the *trans* form and 27 % in the *gauche* form.[46]

The energy difference between the *trans* and *gauche* conformations of the gaseous form of 1,2-dichloroethane was found to lie between 1.03 and 1.32 kcal./mole.[47] For gaseous 1,2-dibromoethane, it lies between 1.40 and 1.77 kcal./mole.[47,48]

In the liquid state, the energy differences are less. The reason lies in a stabilization of the *gauche* form relative to the *trans* form through intermolecular interactions.[49] The unsymmetrical *gauche* forms, in contrast to the *trans* forms, possess a dipole moment. A stabilization results

[44] P. H. Verdier and E. B. Wilson, *J. Chem. Phys.* **29**, 340 (1958); E. Catalano and K. S. Pitzer, *J. Phys. Chem.* **62**, 873 (1958); R. S. Wagner, P. B. Dailey, and N. Solimene, *J. Chem. Phys.* **26**, 1593 (1957); cf. N. Sheppard, *in* "Advances in Spectroscopy," Vol. 1. Interscience, New York, 1959; R. S. Wagner, P. B. Dailey, *J. Chem. Phys.* **26**, 1588 (1957); D. R. Lide, *J. Chem. Phys.* **30**, 37 (1959).

[45] See for example S. Mizushima *et al.*, *J. Chem. Phys.* **17**, 591 (1949); *J. Phys. Chem.* **56**, 324 (1952); *J. Chem. Phys.* **21**, 1411 (1953). See also ref. 16.

[46] J. Ainsworth and J. Karle, *J. Chem. Phys.* **20**, 425 (1952).

[47] Ref. 16, page 38 ff.; H. J. Bernstein, *J. Chem. Phys.* **17**, 258 (1949); W. D. Gwinn and K. S. Pitzer, *J. Chem. Phys.* **16**, 303 (1948); K. Kuratani, T. Miyazawa, and S. Mizushima, *ibid.* **21**, 1411 (1953).

[48] H. J. Bernstein, *J. Chem. Phys.* **18**, 897 (1950); for 1,2-dichloropropane see R. A. Oriani and C. P. Smyth, *J. Chem. Phys.* **17**, 1174 (1949); Y. Morino, I. Miyagawa, and T. Haga, *ibid.* **19**, 791 (1951); for 1,2-difluoroethane: P. Klaboe and J. R. Nielsen, *J. Chem. Phys.* **33**, 1764 (1960).

[49] Ref. 16, pages 43–44, 49, 50, 67.

from electrostatic forces in the liquid phase. For tetrachloroethane, $CHCl_2CHCl_2$[50] and tetrabromoethane,[51,52] the *gauche* form has proven to be the more stable (energy difference 0.75–1.8 kcal./mole). Tetrachloroethane even exists in the *gauche* conformation in the solid state.[50]

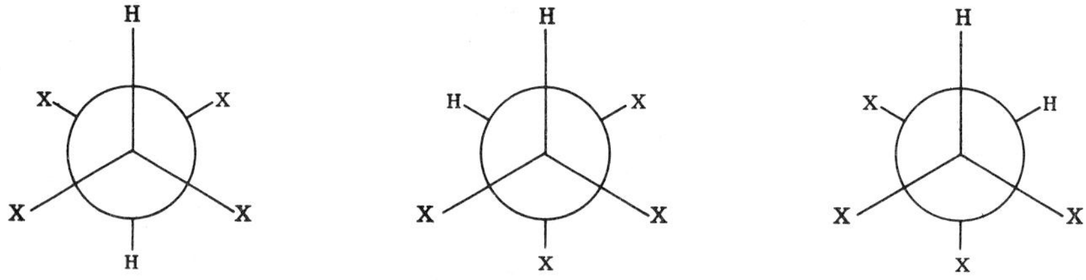

trans (a) *gauche* (b) *gauche* (c)

IX

For 1,2-dichloroethane the energy differences between *trans* and *gauche* forms lie between 0 and 0.05 kcal.mole[53]; for 1,2-dibromoethane between 0.65 and 0.76 kcal./mole.[53,54]

The height of the energy barrier of the *eclipsed* form of 1,2-dichloroethane (H and Cl superimposed in the direction of the C—C bond axis), can be ascertained by investigation of a symmetrical molecule such as $H_3C{-}CCl_3$ to be 2.7–2.9 kcal./mole.[55] The value of this energy barrier, which is of the same order of magnitude as the energy barrier in ethane, leads to the conclusion that here, at least, the repulsion between H and H, as well as H and Cl, is about the same. As a result, the above value

[50] K. Naito, I. Nakagawa, K. Kuratani, I. Ichishima, and S. Mizushima, *J. Chem. Phys.* **23**, 1907 (1955); J. P. Zietlow, F. F. Cleveland, and A. G. Meister, *ibid.* **24**, 142 (1956); compare R. E. Kagarise and D. H. Rank, *Trans. Faraday Soc.* **48**, 395 (1952); H. S. Gutowsky, G. G. Belford, and P. E. McMahon, *J. Chem. Phys.* **36**, 3353 (1962).

[51] R. E. Kagarise, *J. Chem. Phys.* **23**, 207 (1955); **24**, 300 (1956); K. Krebs and J. Lamb, *Proc. Roy. Soc.* **A244**, 558 (1958).

[52] For compounds of the type $X_2CH{-}CHX_2$, we differentiate as well between a *trans* form (IXa), where the substituents X are furthest removed from each other and two *gauche* forms (IXb and c).

[53] See ref. 47; D. H. Rank, R. E. Kagarise, and D. W. E. Axford, *J. Chem. Phys.* **17**, 1354 (1949).

[54] Y. Morino, S. Mizushima, K. Kuratani, and M. Katayama, *J. Chem. Phys.* **18**, 754 (1950); ref. 47.

[55] T. R. Rubin, B. H. Levedahl, and D. M. Yost, *J. Am. Chem. Soc.* **66**, 279 (1944); J. C. Evans and H. J. Bernstein, *Can. J. Chem.* **33**, 1746 (1955); E. Catalano and K. S. Pitzer, *J. Phys. Chem.* **62**, 838 (1958); K. S. Pitzer and J. L. Hollenberg, *J. Am. Chem. Soc.* **75**, 2219 (1953).

of 3.5 kcal./mole given for ethyl chloride also agrees well.[56] It can thus be assumed that the height of the energy barrier for the *eclipsed* form of 1,2-dichloroethane also lies around 3 kcal./mole. This indicates that for halogen substituted hydrocarbons as well, a ready interconversion of the *trans* and *gauche* conformations takes place at normal temperature.[57]

The energy barrier for the fully *eclipsed* conformation (Cl and Cl superimposed in the direction of the C—C axis) is comparatively higher. Measurements on CF_3—CH_2Cl, CF_3—$CHCl_2$, and CF_3—CCl_3[58a] led to values of about 6.0 kcal./mole; whereas 10.8 kcal./mole were found for hexachloroethane by means of electron diffraction.[58,58a]

If the distance between the substituents that interfere with each other increases, then, because of the short range of the repulsion forces, the height of the potential barrier decreases. From electron diffraction data, it can be seen that, in contrast to hexachloroethane, the energy difference for trichloromethyltrichlorosilane, Cl_3C—$SiCl_3$, amounts to only 4.3 kcal./mole. For hexachlorodisilane, Cl_3Si—$SiCl_3$, it amounts to only 1.0 kcal./mole. This indicates an unhindered free rotation.[58]

Aside from the above-mentioned exceptions (tetrachloroethane and tetrabromoethane) to the rule that *trans* conformations are more stable than *gauche* conformations, other compounds can exhibit a reversal of the stability ratio due to special structural circumstances. For ethylene chlorohydrin, ClH_2C—CH_2OH, the *gauche* form is more stable than the *trans* form under all conditions. In the solid state only the *gauche* form exists. In the liquid or gas state an equilibrium exists between gauche and *trans* form. The *gauche* form is 0.95 kcal./mole lower in energy than the *trans* form.[59] The reason for the special stability of the *gauche* form relates back to the formation of an intramolecular hydrogen bond from the OH group to the Cl atom. This is possible only for the *gauche* form (Xa), and not for the *trans* form (Xb)[16]:

In general, all 2-substituted ethanols (XCH_2—CH_2—OH; X = OH, NH_2, F, Cl, Br, —OCH_3, —$NHCH_3$, —$N(CH_3)_2$, —NO_2) are

[56] The value of 3.33 kcal./mole given above for the energy barrier of ethyl fluoride checks well with the measurement on methyl fluoroform, which was found to be 3.0 kcal./mole. (H. S. Gutowsky and H. B. Levine, *J. Chem. Phys.* **18**, 1297 (1950); See also Catalano and Pitzer, ref. 55).

[57] A review of the various methods of determining the energy barrier can be found in M. K. Wilson and V. A. Crawford, *Ann. Rev. Phys. Chem.* **9**, 339 (1958); D. J. Millen, ref. 26.

[58] Y. Morino and E. Hirota, *J. Chem. Phys.* **28**, 185 (1958).

[58a] Compare A. Danti and J. L. Wood, *J. Chem. Phys.* **30**, 582 (1959).

[59] S. Mizushima, T. Shimanouchi, T. Miyazawa, K. Abe, and M. Yasumi, *J. Chem. Phys.* **19**, 1477 (1951).

stabilized in their *gauche* form by intramolecular interaction between the substituent X and the OH group.[60] Whether the stabilization of the ethylene halohydrins actually results from a hydrogen bond or from electrostatic forces, has not been determined.[60]

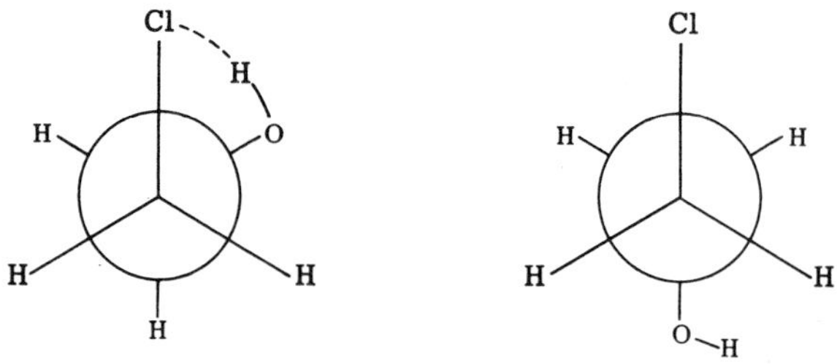

gauche
(a)

trans
(b)

X

For 1,2-dicyanoethane (succinodinitrile), as well, the *gauche* form is the more stable one[61] in the liquid and solid states. As a result of the strongly polar nitrile groups, as in the case of the tetrabromo- and tetrachloroethane (see page 38), the *gauche* form has a large dipole moment and is stabilized[62] by intermolecular interactions.

For all conformers dealt with up to now, the energy barrier has been too small to permit their individual isolation. For this to be possible, the height of the energy barrier must be between 16 and 20 kcal./mole. Examples in which this is the case, and for which rotational isomeric forms have been separated, have been shown among the atropisomers (see Chapter 1, page 9). Energies in the range of 20 to 50 kcal./mole were necessary here to racemize[63–65] the compounds (2-nitro-6-carboxy-6′-ethoxydiphenyl; 2,2′-diamino-6,6′-dimethyl diphenyl).

[60] M. Kuhn, W. Lüttke, and R. Mecke, *Z. Anal. Chem.* **170**, 106 (1959).

[61] G. J. Janz and W. E. Fitzgerald, *J. Chem. Phys.* **23**, 1973 (1955); W. E. Fitzgerald and G. J. Janz, *J. Mol. Spectr.* **1**, 49 (1957).

[62] For investigations on polymethylene halides (for which the *gauche* form is also the more stable one), see J. K. Brown and N. Sheppard, *Proc. Roy. Soc.* **A231**, 555 (1955).

[63] See for example F. W. Cagle, Jr. and H. Eyring, *J. Am. Chem. Soc.* **73**, 5628 (1951).

[64] A. C. D. Newman and H. M. Powell, *J. Chem. Soc.* p. 3747 (1952). They calculated the necessary activation energy for the racemization of tri-*o*-thymotid at 16 kcal./mole. The optical activity is caused by blocking the rotation around a single bond in a twelve membered ring. (Thymotinic acid = 2-hydroxy-6-methyl-3-isopropyl benzoic acid); L. V. Dvorken, R. B. Smyth, and K. Mislow, *J. Am. Chem. Soc.* **80**, 486 (1958).

[65] Compare also G. Wittig and H. Zimmermann, *Chem. Ber.* **86**, 629 (1953); D. M. Hall and E. E. Turner, *J. Chem. Soc.* p. 1242 (1955); K. Mislow, *Angew. Chem.* **70**, 683 (1958); E. L. Eliel, "Stereochemistry of Carbon Compounds," pp. 156 ff. McGraw-Hill, New York, 1962.

Prevention of free rotation around a single bond of course is not limited to the C—C bond. For other single bonds as well, for example, the single C—N bond of aliphatic amines or the C—O bond of alcohols, this also holds.[66]

With the help of microwave spectroscopy, which makes an especially accurate determination of the potential barriers possible, the investigation of the hindrance of the free rotation for such bonds[66] has made great progress.[66]

4. Reason for the Appearance of Potential Barriers during the Inner Rotation of the Molecule

The nature of repulsive forces which exist between the substituents of a molecule capable of rotation is not yet fully known.[67] These repulsive forces lead to the appearance of potential barriers and result in hindrance to rotation. All theoretical calculations of such small energy differences lead to difficulties because of the lack of exactness in the methods of approximations. For a more qualitative explanation, it is sufficient to assume that the reasons for the appearance of potential barriers during the rotation originate from electrostatic attractions and repulsions of the participating electrons and nuclei in the bonds concerned.

On the basis of various interpretations, a whole series of experiments were undertaken in order to explain the appearance of the potential barriers in particular. Some of these older theories do not agree very well with values now known for the height of the potential barriers. A few of these explanatory experiments will now be dealt with briefly.

The first theoretical attempt at a quantum mechanical treatment of ethane was undertaken by Eyring,[29] who tried to calculate the interactions of the hydrogen atoms of the methyl groups. The results, which gave a much too low potential barrier (0.3 kcal./mole) (see page 29) seemed at the time to be in agreement with experimental results. Later quantum mechanical calculations, which attempted to verify the presently accepted potential energy barrier of 3.0 kcal./mole, did not lead to satisfactory results. An entire series of assumptions, such as lack of cylindrical symmetry of the C—C bond, was introduced. This could

[66] Ref. 1b, p. 69 ff.; N. W. Luft, *Z. Elektrochem.* **59**, 46 (1955); J. D. Swalen, *J. Chem. Phys.* **13**, 1739 (1955); R. W. Kilb, *ibid.* **23**, 1736 (1957); D. R. Lide, *ibid.* **26**, 343 (1957); D. R. Lide and D. E. Mann, *ibid.* **28**, 572 (1958); **29**, 914 (1957).

[67] A review of some of the theories concerning the origin of the potential barriers is found in E. B. Wilson, Jr., *in* "Advances in Chemical Physics" (I. Prigogine, ed.). Wiley (Interscience), New York, 1959; see also: E. B. Wilson, Jr., *Proc. Natl. Acad. Sci. U.S.* **43**, 816 (1957); D. J. Millen, l.c. ref. 26.

have led to the dependence of the energy of the C—C bond on the angle of rotation.[68]

Because the quantum mechanical calculations always resulted in a lower value[69] for the potential barrier, explanatory investigations of the hindrance of rotation on the basis of the dipole moment of the bonds and steric repulsion forces were undertaken. It was assumed that through a dipole-dipole interaction, as for instance in the case of halogens or other polar groups, an influence on the potential barrier was present.[16] Such a dipole-dipole interaction is undoubtedly present, but a satisfactory explanation for the energy difference between rotational isomers cannot be obtained therefrom. One should recall that the height of the potential barrier does not depend much on the polarity of the bond, as was described[69] in Section 2,f. An electrostatic interaction of the quadrupole moments of the bonds has also been discussed.[70]

Steric repulsive forces between the atoms involved have been, as mentioned above, used in an attempted explanation. These can be estimated from the repulsive forces between separated atoms, in which one takes as a basis the data which follow from the kinetic gas theory. Difficulties of the same magnitude appear, especially for hydrogen.

For a much more approximate calculation of these steric repulsive forces, the van der Waals atomic radii can be used. Here as well, because of the oversimplification, quite inaccurate values are obtained.[71a] Even at present, there is no uniform theory concerning the origin of the potential barrier. Two recent attempts at an explanation follow briefly:

As already explained previously an attempt was made to relate the potential barrier, at least in part, to the properties of the σ-bonds, whose cylindrical electron distribution is disturbed. The potential barrier then originates from exchange interactions between these unsymmetrical electron distributions. Pauling[71b] developed a theory on the basis of this view, in which the potential barrier comes about from exchange interactions of the *sp* hybrid bond orbitals which possess a small amount of *d* and *f* character. We shall not go into the details here; but owing to the additional acquisition of f and d character during hybridization, the *staggered* form is so much preferred over the *eclipsed* form through exchange interactions, that it suffices for an explanation of the observed energy differences between the two forms.

---

[68] E. Gorin, J. Walter, and H. Eyring, *J. Am. Chem. Soc.* **61**, 1876 (1939).

[69] Compare also E. A. Mason and M. M. Kreevoy, *J. Am. Chem. Soc.* **77**, 5808 (1955); D. E. Mann and E. K. Plyler, *J. Chem. Phys.* **21**, 1116 (1953).

[70] E. N. Lasettre and L. B. Dean, Jr., *J. Chem. Phys.* **17**, 317 (1949).

[71a] L. Pauling, "The Nature of the Chemical Bond," 3rd ed., p. 81 ff. Cornell Univ. Press, Ithaca, New York, 1960.

[71b] L. Pauling, *Proc. Natl. Acad. Sci. U. S.* **44**, 211 (1958).

Eyring[72] and co-workers developed an essentially different theory to explain the potential barrier in ethane and other rotamers. For this, they set up three main principles of the chemical bond: "1. An electron seeks lowered potential energy by moving close to a nucleus; 2. An electron seeks lowered kinetic energy by lengthening its path by moving over several nuclei; 3. An electron seeks lowered kinetic energy by avoiding orbital bending."[72]

The third point, also designated the principle of minimum bending of orbitals, is available as an explanation of the potential barrier. Ethane can be regarded as a system of three four-nuclear paths, in which the electrons are somewhat delocalized because of hyperconjugation.

In the staggered form of ethane, the four nuclei lie in one plane. Thereby the smoothest path, or the path of lowest kinetic energy, can be adopted because of hydrogen atoms in the *trans* position.

In the eclipsed form only a semicircular path is possible for the electrons when the four nuclei are in a plane. This path is not as favorable as that in the *staggered* form. Since the *staggered* form possesses three such smoother paths, it is preferred over the *eclipsed* form. "The straighter or smoother paths in the *staggered* conformation of the ethane-like molecules account for the greater stability of this form, and illustrate the idea that electrons hate to go around corners."[72]

In conclusion the statement should be repeated that at present there is no final theory which is able to explain all experimentally found differences in energy among conformers. No theory can predict the height of the potential barriers of conformers.

4. The Conformations of Cyclic Hydrocarbons

The conformations of cyclic hydrocarbons will be covered in the following for cyclohexane, the mediums-sized rings, and decalin. The smaller rings such as cyclopentane, as well as condensed ring systems such as hydrindane, perhydrophenanthrene, the steroids, etc. will be covered in detail in Chapters 3, 4, and 5.

The rules of conformational theory developed in Section 2 for aliphatic hydrocarbons, whereby certain conformers are preferred to others, can be applied to cyclic systems. It must be borne in mind though, that the preferred conformation cannot always be attained, especially by smaller rings. Because of their rigidity, the CH_2 groups are forced into an unfavorable arrangement (see the discussion of cyclobutane and

[72] H. Eyring, G. H. Stewart, and R. P. Smith, *Proc. Natl. Acad. Sci. U.S.* **44**, 259 (1958); G. H. Stewart and H. Eyring, *J. Chem. Educ.* **35**, 550 (1958).

cyclopentane Chapter 3, page 72). The Pitzer-strain that must result (see page 32) leads to an increase in the energy content, in addition to that brought about by Baeyer-strain (see also Section 4,b).

a. Cyclohexane

For cyclohexane, two nonplanar forms can be constructed, as Sachse and Mohr had already recognized. They are free of Baeyer strain (see Chapter 1, page 14):

(1) the chair form (also called the rigid form) (XI).

(2) the boat form, (also called the flexible form) (XII). (See also Chapter 6).

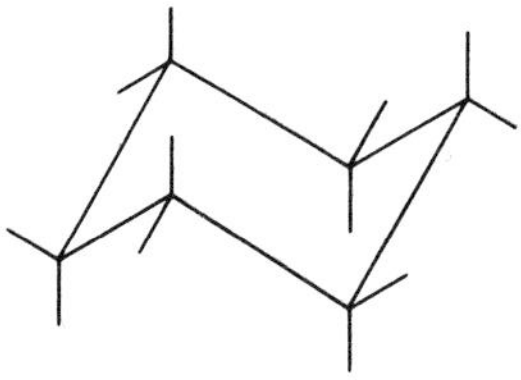

XI

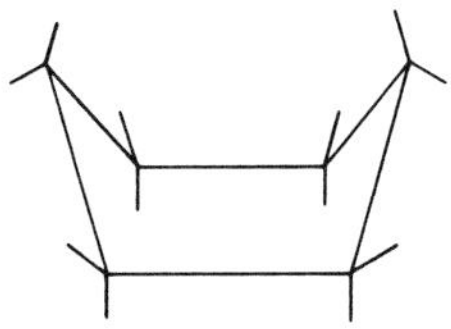

XII

The chair form is converted to the boat form by the flipping up of one carbon atom and vice versa. In the process of flipping from chair to boat form, the valence angles are slightly deformed, and the cyclohexane molecule goes through a maximum of potential energy. The height of this energy barrier was determined at 10–11 kcal./mole with the help of nuclear magnetic resonance spectroscopy.[72a] Figure 3 shows the height of this energy barrier in comparison to the potential energy of the chair and boat form of cyclohexane.

The Baeyer strain gives no information concerning the relative stabilities of the chair and boat conformation. Upon inspection of models of both forms, however, it becomes clear at once that the chair and boat conformation differ in their Pitzer strain. A difference in energy must exist.

If we apply to cyclohexane the rules developed for conformations, the following results: the chair form (XI) has no Pitzer strain, and is thus more stable than the boat form. In the chair form all the hydrogen atoms of the adjacent CH_2 groups occupy the energetically preferred *staggered* conformation relative to each other.

[72a] F. R. Jensen, D. S. Noyce, C. H. Sederholm and A. J. Berlin, *J. Am. Chem. Soc.* **84**, 386 (1962); R. K. Harris and N. Sheppard, *Proc. Chem. Soc.* p. 418 (1961); F. A. L. Anet, M. Ahmad, and L. D. Hall, *Proc. Chem. Soc.* p. 145 (1964); see also F. A. Bovey, F. P. Hood, III, E. W. Anderson, and R. L. Kornegay, *ibid.* p. 146 (1964); L. W. Reeves and K. O. Strømme, *Can. J. Chem.* **38**, 1241 (1960): for cyclohexylbromide 10.85 kcal./mole; G. van Dyke Tiers, *Proc. Chem. Soc.* p. 389: (1960): for perfluorocyclohexane 9.9 kcal./mole; J. D. Roberts, *Angew. Chem.* **75**, 20 (1963): for 1,1-difluorocyclohexane 12 kcal./mole. See also W. B. Moniz and J. A. Dixon, *J. Am. Chem. Soc.* **85**, 1771 (1961).

The boat form, on the other hand, has Pitzer strain and is less stable because two pairs of adjacent CH_2 groups occupy the energetically less favorable *eclipsed* conformation (XII). A further increase in energy occurs from an adverse interaction of the hydrogen atoms in the 1,4-positions, which approach each other closely in the boat form.

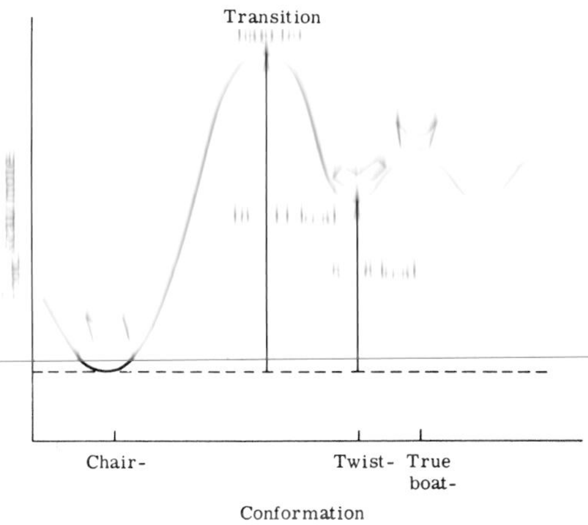

FIG. 3. Potential energy curve for transformation cyclohexane from the chair to the boat form.

Various attempts were undertaken to calculate the energy difference between the chair and boat form. A few of these are given below.[73,74]

(*i*) In the boat form, two C—C bonds occur in the *eclipsed* conformation. If we use the value found for the energy barrier, of ethane, ~3 kcal./mole (see Section 2,a), then the energy difference between chair and boat forms should be twice as great, or about 6 kcal/mole.[21b,c]

(*ii*) An effort was made by Barton[11] to calculate energy differences between both conformations semi-empirically from the interaction energies of all atoms influencing each other sterically. This method goes back to that of Dostrovsky, Hughes and Ingold.[75] However, values which fluctuate between 1.3 and 6.8 kcal./mole are obtained because of

[73] Besides the methods mentioned here, see also: W. D. Kumler, *J. Am. Chem. Soc.* **67**, 1901 (1945); C. W. Shoppee, *J. Chem. Soc.* p. 1138 (1946); C. W. Shoppee, *Ann. Rep. Prog. Chem.* (*Chem. Soc. London*) **43**, 200 (1946). Compare also: A .I. Kitaygorodsky, *Tetrahedron* **9**, 183 (1960).

[74] It is very difficult to make an exact calculation. In order to calculate the maxima and minima of the potential barriers, it would be necessary to know the interaction potential of each atom for every conformation. For the calculation of energies of the conformations of cyclohexane with the aid of computers, cf. J. B. Hendrickson, *J. Am. Chem. Soc.* **83**, 4537 (1961).

[75] I. Dostrovsky, E. D. Hughes, and C. K. Ingold, *J. Chem. Soc.* p. 173 (1946); compare also F. H. Westheimer and J. E. Mayer, *J. Chem. Phys.* **14**, 733 (1946).

the uncertainty with which the van der Waals radii of the hydrogen atoms were known.

(*iii*) A detailed investigation of the geometry and energy of the equilibrium between the chair and boat form of cyclohexane was carried out by Hazebroek and Oosterhoff.[76] On the basis of a value of 2.75 kcal. mole[77] for the energy difference between the *staggered* and *eclipsed* form of ethane, the energy difference between the chair and boat form is calculated as 5.5 kcal./mole.

Considering the model of the boat form of the cyclohexane ring, it becomes clear that the boat form represents only one of the flexible conformations. It represents the most unfavorable form from an energy standpoint. Besides this form, innumerable other flexible forms are possible, which were designated "stretched" forms by Hazebroek and Oosterhoff.[76,78] In the model, the eclipsing of the CH_2 groups and the interaction of the 1,4 hydrogen atoms has been somewhat reduced. This produces a minimum of energy relative to that of the true boat form (Fig. 3).[78a] These "stretched," "skew," "twist," or "half-rotated in-between" forms (XIII) were already anticipated by Sachse[79] and

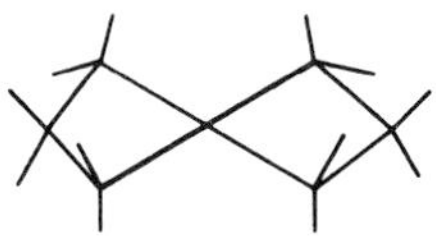

XIII

Mohr[79a] in their basic work. They were discussed by Wightman,[80] Hückel,[80a] Brodetzky,[81] and Cohen Henriquez.[82]

Hazebroek and Oosterhoff[77] were able to show by their calculations

[76] P. Hazebroek and L. J. Oosterhoff, *Discussions Faraday Soc.* **10**, 87 (1951).

[77] See R. Spitzer and H. M. Huffmann, *J. Am. Chem. Soc.* **69**, 211 (1947).

[78] G. Lamaty, "Propriétés des divers cycles hexaniques," Thèse, Montpellier, 1959; designates the flexible in-between forms as croisées-forms; see also J. Jullien and G. Lamaty, *J. Chim. Phys.* **55**, 2 (1958).

[78a] Cf. also S. J. Angyal and R. M. Hoskinson, *J. Chem. Soc.* p. 2991 (1962); J. B. Hendrickson, *J. Org. Chem.* **29**, 991 (1964).

[79] H. Sachse, *Ber.* **23**, 1363 (1890); *Z. Physik. Chem.* (*Leipzig*) **10**, 203 (1892).

[79a] E. Mohr, *J. Prakt. Chem.* [2] **98**, 315 (1918).

[80] W. A. Wightman, *J. Chem. Soc.* p. 1421 (1925).

[80a] W. Hückel, "Der gegenwärtiger Stand der Spannungstheorie"; also *in* "Fortschritte der Chemie, Physik und physikalische Chemie" (v. A. Eucken, ed.), Vol. 19, No. 4. Bornträger, Berlin; "Theoretische Grundlagen der org. Chemie," 8th ed., Vol. 1, p. 77 ff. Akad. Verlagsges., Leipzig, 1955.

[81] G. Brodetzky, *Proc. Leeds Phil. Lit. Soc.* **1**, 370 (1929).

[82] P. Cohen Henriquez, *Proc. Acad. Sci. Amsterdam* **37**, 532 (1934); see also J. A. Blekkingh, *Rec. Trav. Chim.* **68**, 345 (1949).

that, at normal temperatures, the chair form is always preferred, in spite of the fact that twist forms exist which are lower in energy (4.75 kcal./mole) than the boat form (5.5 kcal./mole). At higher temperatures, however, they obtain a greater contribution of twist conformations, whereby the higher entropy of the twist form cannot be neglected.

Other attempts have been made to calculate the energy difference between the chair form and the twist form. According to a method developed by [illegible] the difference between chair and twist form was calculated at [illegible] kcal./mole.[83]

With the help of model studies and exact measurements of the dihedral angles (see page [illegible]) Allinger[85] arrived at a value [illegible] kcal./mole. With the assistance of a computer, Hendrickson[85a] calculated the energy difference between the chair and the boat form at [illegible] kcal./mole. However, the energy difference between the chair and the twist conformation came to only [illegible] kcal./mole, making the twist form of cyclohexane [illegible] kcal./mole more stable than the boat form (Fig. [illegible]).

The calculations of such energy differences depends mainly on approximations. But as will be shown in Chapter 6 for substituted cyclohexane systems, twist forms can occur.[86] These intermediate forms, of course, hinder every conformational analysis and have not been considered further in most publications before 1957.

If we take as a basis the value 5 kcal./mole, then the ratio C (twist)/C (chair) at 25° would be less than 0.001, i.e., the portion of flexible intermediate forms can be neglected.[84]

(*iv*) A further method, originating with Turner,[87] resolves the cyclohexane molecule into six *n*-butane systems (XIV) and thereby permits the calculation of the energy of the chair and boat form. In the chair form, the six *n*-butane systems are arranged in the *gauche* conformation. Related to the arbitrarily assumed zero point of energy of the *trans* con-

[83] K. E. Howlett, *J. Chem. Soc.* p. 4353 (1957); see also ref. 79 and E. A. Mason and M. H. Krevoy, ref. 69; H. O. Pritchard and F. H. Sumner, *J. Chem. Soc.* p. 1041 (1955).

[85] N. L. Allinger, *J. Am. Chem. Soc.* **81**, 5727 (1959); compare also G. A. Bottomley and P. R. Jefferies, *Australian J. Chem.* **14**, 657 (1961).

[85a] J. B. Hendrickson, *J. Am. Chem. Soc.* **83**, 4537 (1961).

[86] Compare (a) C. Sandris and G. Ourisson, *Bull. Soc. Chim. France* p. 1524 (1958); (b) W. Hückel *et al. Ann. Chem.* **624**, 142 (1959); (c) W. Hückel and Y. Riad, *Ann. Chem.* **637**, 33 (1960); (d) Compare also J. Levisalles, *Bull. Soc. Chim. France* p. 551 (1960); (e) R. E. Reeves and F. A. Blouin, *J. Am. Chem. Soc.* **79**, 2261 (1957); (f) R. E. Reeves, *Ann. Rev. Biochem.* **27**, 15 (1958). See also ref. 84 and 85; (g) J. A. Mills, *Advan. Carbohydrate Chem.* **10**, 1 (1955); (h) S. J. Angyal and J. A. Mills, *Rev. Pure Appl. Chem.* **2**, 185 (1952); (i) J. F. Biellmann, R. Hanna, G. Ourisson, C. Sandris and B. Waegell, *Colloq. Inter. Chim. Probl. Stereochim., Montpellier*, 1959, in *Bull. Soc. Chim. France* p. 1429 (1960).

[84] Compare R. D. Stolow, *J. Am. Chem. Soc.* **81**, 5806 (1959).

[87] R. B. Turner, *J. Am. Chem. Soc.* **74**, 2118 (1952).

formation, the chair form possesses an energy of $6 \times 0.8 = 4.8$ kcal./mole (compare Section 2a). In the boat form there are four *n*-butane systems arranged *gauche* and two *eclipsed* conformations. The exact value for the energy barrier of an *eclipsed* arrangement in the *n*-butane system is not known (see page 31), so that various values have been used

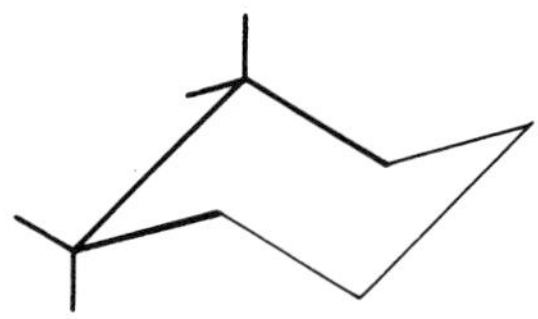

XIV

for the calculation.[88] If the value is set between 4.4 and 6.1 kcal./mole[89] (see page 31), then the results for the energy of the true boat form are: $(4 \times 0.8) + 2 \times (4.4 \text{ to } 6.1) = 12.0$ to 15.4 kcal./mole. Then the energy difference between chair and boat form is 7.2–10.6 kcal./mole.

Allinger[89a] determined the energy difference between chair and twist form experimentally in a quite original piece of work.

As will be explained in more detail in Chapter 3, certain substituents can be found for the cyclohexane ring which are so large spatially that they can exist only in one position, namely *equatorial*. Such a substituent is the *tert*-butyl group. Whereas both substituents for the *cis* isomer of 1,3-di-tert-butyl cyclohexane are in an equatorial arrangement, one *tert*-butyl group should go into *axial* position in the corresponding *trans*-1,3-isomer. This is detailed on page 106. An *axial* position of the *tert*-butyl group is not possible for spatial reasons, and would result in a molecule of very high energy. The molecule would therefore tend to seek a lower energy level by transition into the twist form. Thereby both *tert*-butyl groups can go into *equatorial* position. By measuring the equilibrium between the *cis* and *trans* isomers (on palladium-charcoal catalyst), one therefore measures the equilibrium between chair and twist form. The energy difference (enthalpy difference) was determined at 5.9 ± 0.6 kcal./mole from the temperature dependence of the equilibrium constant. Further experimental methods for the determination of the energy difference betweeen the chair and the twist form of

[88] See ref. 87. The value originally used by Turner, 3.6 kcal./mole, is certainly too low. A critical discussion of all energy values employed can be found in ref. 78.

[89] This value, which seems to represent the true value more closely, is used by W. G. Dauben and K. S. Pitzer in ref. 7; compare also ref. 38.

[89a] N. L. Allinger and L. A. Freiberg, *J. Am. Chem. Soc.* **82**, 2393 (1960).

cyclohexane were used by Johnson and others.[89b] They ascertained from measurements of the heats of combustions of a pair of stereoisomeric lactones, in which the chair as well as the twist form was fixed, and from calorimetric and heat of vaporization measurement on *trans-syn-trans-* and *trans-anti-trans*-perhydroanthracene (see page 216) a value of 4.8 to 5.5 kcal./mole, which agreed very well with the value found by Allinger.[89a]

In summary, on the basis of calculations and experimental results, it may be stated that the energy difference between the chair form and the twist form of cyclohexane may be assumed to be between 5 and 6 kcal./mole (Fig. 3). Because the twist form is more flexible in comparison to the chair form, the higher entropy value of 5 e.u.[89a] must be considered. The difference in free energy (ΔF) is easily determined at 3.5 to 4.5 kcal./mole (at 25°C) from this information.

The result arising from all calculations, that the chair form of cyclohexane is more stable than the boat form is today substantiated by physical methods such as infrared[90] and Raman spectroscopy,[91] X-ray and electron diffraction,[92] as well as by thermodynamic measurements.[93]

b. Medium Sized Rings[94]

(See also Chapter 3, p. 164.)

Among the higher ring homologs of cyclohexane, the so-called medium sized rings distinguish themselves by special properties in their conformations. They will be considered briefly at this time.

[89b] W. S. Johnson, J. L. Margrave, V. J. Bauer, M. A. Frisch, W. N. Hubbard, and L. H. Dreger, *J. Am. Chem. Soc.* **82**, 1255 (1960); **83**, 606 (1961); J. L. Margrave, M. A. Frisch, R. G. Bautista, R. L. Clarke, and W. S. Johnson, *ibid.* **85**, 546 (1963).

[90] R. S. Rasmussen, *J. Chem. Phys.* **11**, 249 (1943); D. A. Ramsay and G. B. M. Sutherland, *Proc. Roy. Soc.* **A190**, 245 (1947).

[91] (a) K. W. Kohlrausch and H. Wittek, *Z. Physik. Chem. (Leipzig)* **48**, 177 (1941); (b) H. Gerding, E. Smit, and R. Westrik, *Rec. Trav. Chim.* **61**, 561 (1942); compare also (c) S. Mizushima, Y. Morino, and R. Fujishiro, *J. Chem. Soc. Japan* **62**, 587 (1941); ref. 21.

[92] (a) O. Hassel and H. Viervoll, *Arch. Math. Naturv.* **47**, No. 13 (1944); (b) *Acta Chem. Scand.* **1**, 149 (1947); M. Davis and O. Hassel, *ibid.* **17**, 1181 (1963).

[93] J. G. Aston, S. C. Schumann, H. L. Fink, and P. M. Doty, *J. Am. Chem. Soc.* **63**, 2029 (1941); also refs. 21 and 36.

[94] Compare (a) J. D. Dunitz and V. Prelog, *Angew. Chem.* **72**, 896 (1960); (b) V. Prelog, *Colloq. Intern. Chim. Org. Probl. Stéréochim. Montpellier*, 1959, in *Bull. Soc. Chim. France*, p. 1433 (1960); (d) R. A. Raphael, *Proc. Chem. Soc.* p. 97 (1962); (e) J. Sicher, *in* P. B. D. de la Mare and W. Klyne, *Progr. Stereochem.* **3**, Chapter 6 (1962). V. Prelog, *in* "Pure and Applied Chemistry," Vol. 6, p. 545. Butterworths, London, 1963.

The eight- to eleven-membered rings are called the medium rings[95] and differ characteristically from common and large rings[95] in their physical and chemical properties.[96]

Aside from the specific physical and chemical properties (e.g. the medium-size rings show a maximum density with the 10-membered ring),[97] it has been known for some time that the yield in the various ring closure reactions, is at a minimum for medium-sized rings.

For the determining factor in their difficult formation, Stoll[97] considered "compression" of the van der Waals radii as the cause. This occurs when the methylene groups are twisted out of the plane of the ring owing to the ring closure. The effective volumes of the H atoms then collide inside the ring; this leads to an increased strain and greater difficulty of ring closure.

Prelog[96b] first recognized that Pitzer-strain is an essential factor in the stability of medium-sized rings. The ring strain of medium-sized rings is currently regarded as being dependent on various contributions.[94] These are the deformation of the valence angle, or Bayer-strain, the bond-opposition forces or Pitzer-strain, and the transannular interactions.

We can state in general, that the physical and chemical properties in the area of the medium rings show decided maxima or minima. They are no longer monotonous functions of the number of carbons in the ring.

Such a dependence on the ring size is shown, for example, by the experimentally obtained heat of combustion. Values are known accurately up to cycloheptadecane.[98] From the heat of combustion, the strain of the system is revealed. It can be seen from this that 9- and 10-membered rings have a relatively large strain energy (Fig. 4). Six-membered

---

[95] The classification goes back to H. C. Brown: cf. H. C. Brown, R. S. Fletcher, and R. B. Johannesen, *J. Am. Chem. Soc.* **73**, 212 (1951), footnote 21, who originally classified the 12-membered ring with the medium rings. But compare: E. L. Eliel, "Stereochemistry of Carbon Compounds," p. 189. McGraw-Hill, New York, 1962.

Small rings: 3 and 4 membered.
Common rings: 5 to 7 membered.
Medium rings: 8 to 11 membered.
Large rings: 12 and higher membered.

[96] (a) Compare H. C. Brown *et al.* in ref. 95, explanation of the reactivity due to I-strain effect; (b) V. Prelog, *J. Chem. Soc.* p. 420 (1950); (c) compare also M. Kobelt *et al.* Ref. 34; (d) K. Ziegler, *in* "Houben–Weyl Methoden der organischen Chemie," 4th ed. (E. Müller, ed.), Vol. 4, Part 2, p. 815. Thieme, Stuttgart, 1955; (e) R. Huisgen, *Angew. Chem.* **69**, 341 (1957); (f) N. L. Allinger and S. Greenberg, *J. Am. Chem. Soc.* **81**, 5733 (1959).

[97] M. Stoll and G. Stoll-Comte, *Helv. Chim. Acta* **13**, 1185 (1930); L. Ruzicka, M. Kobelt, O. Häfliger, and V. Prelog, *ibid.* **32**, 544 (1949).

rings or rings containing more than 12 members have practically no strain.

In order to explain these facts, the spatial structure of the medium-sized ring compounds was studied more thoroughly. Due to the lack of experimental data, it was necessary to depend first upon hypothetical

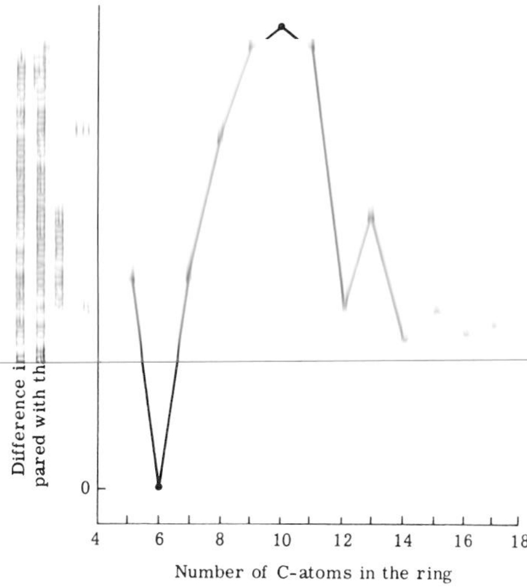

FIG. 4. Strain in cyclic hydrocarbons.

working models. It must be recognized that the spatial structure of medium-sized rings is much more difficult to derive from theoretical considerations than that of the six-membered rings. Models not obtained by direct experimental methods are therefore viewed with reservations. Such difficulties arise during the calculation of the energetically favorable conformations of these compounds due to the following: deviation from the normal valence angle (Baeyer strain) and the transannular interactions must be considered in addition to Pitzer strain in the energy considerations. The Pitzer strain has been the only consideration up to now in conformational studies. As we have seen, an estimation of the Pitzer strain is possible only in a very rough way. This also holds true for Baeyer strain and especially for transannular interactions.

Only recently has a clearer concept concerning the spatial structure of medium-sized rings been developed, of which some details follow.

[98] (a) Refs. 36 and 78; (b) S. Kaarsemaker, and J. Coops, *Rec. Trav. Chim.* **71**, 261 (1952); (c) H. van Kamp, Thesis, Vrije Universiteit, Amsterdam, 1957; (d) J. Coops, H. van Kamp, W. A. Lambregts, B. J. Visser, and H. Dekker, *Rec. Trav. Chim.* **79**, 1226 (1960).

Cyclooctane[99]: As observed from a model, the preferred conformation of cyclooctane is not as easily found as for cyclohexane, which is the case in general for medium-sized rings. Because of the necessity of having to accept unfavorable conformations in the construction of the molecule, as well as the possibility of transannular hydrogen repulsion, the many possible conformations of medium-sized rings must be subjected to a careful conformational analysis. If possible, the energies of the various forms have to be determined by the methods described with cyclohexane and the experimentally obtained values are then to be compared with them.

As a result of exact model studies and measurements of the dihedral angles,[100] from which, with the help of known potential functions, every CH_2 conformation can be estimated, only two conformations can be considered for cyclooctane, according to Allinger.[85]

The two conformations are the center-symmetrical so-called crown form (XVa), in which the carbon atoms lie alternately in two parallel planes, and the boat form (XVb).

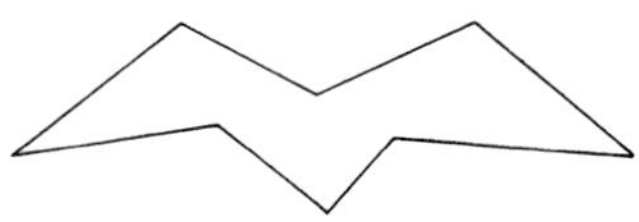

Crown conformation

(a)

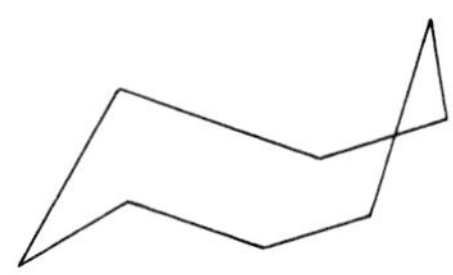

Boat conformation

(b)

XV

All other forms able to be constructed with a model distinguish themselves by a higher energy content and are therefore less probable.[99b]

The dihedral angles for the crown form are alike and equal to 96°. The energy content was estimated at 14.4 kcal./mole[85] (related to cyclohexane). Estimating the energies presents difficulties in this case, just as was described for cyclohexane. The exchange potentials of the hydrogen atoms are only known with questionable accuracy.[101] By deformation of the crown form in various ways, other lower energy

---

[99] (a) Ref. 85. (b) R. Pauncz and D. Ginsburg, *Tetrahedron* **9**, 40 (1960); (c) H. E. Bellis and E. J. Slowinsky, Jr., *Spectrochim. Acta* **10**, 216 (1957); *ibid.* **13**, 1103 (1960). (d) N. L. Allinger and S. Hu, *J. Am. Chem. Soc.* **83**, 1664 (1961); (e) N. L. Allinger and S. Greenberg, *ibid.* **84**, 2394 (1962); (f) N. L. Allinger, S. P. Jindal, and M. A. Da Rooge, *J. Org. Chem.* **27**, 4290 (1962); (g) N. L. Allinger and S. Greenberg, *J. Am. Chem. Soc.* **81**, 5733 (1959); (h) compare also: R. Kolinski, H. Piotrowska, and T. Urbanski, *J. Chem. Soc.* p. 2319 (1958); N. J. Leonard, T. W. Milligan, and T. L. Brown, *J. Am. Chem. Soc.* **82**, 4075 (1960).

[100] See p. 27. Compare also L. Pauling, *Proc. Natl. Acad. Sci. U.S.* **35**, 495 (1949).

[101] R. Pauncz and D. Ginsburg[99b] calculated a higher value for the energy content of the crown form on the basis of certain distances between the H atoms.

forms can be found, such as one with a pair of the dihedral angles spread to 130°, and one pair diminished to 60°. The energy content then comes to only 11.1 kcal./mole.[85]

For the boat form (XVb) the energy values amount to about the same magnitude if the transannular hydrogen atom repulsion forces are neglected. Nevertheless, the existence of this form is hardly probable; because, as is seen from the model, the hydrogen atoms on the bottom side repel each other. If we twist the model so that these are separated, then the hydrogen atoms on the top side get too close to each other. From this transannular hydrogen atom repulsion, the energy level of the boat form is probably too high. We cannot count on its existence. Similar considerations can be made for the other forms of cyclooctane mentioned above.

In summary it may be stated that a slightly deformed crown form (XVa) most likely is the most stable conformation of the cyclooctane ring. Investigations of the infrared spectra[102a–d] and n.m.r.-spectra[102e] of cyclooctane lead to the conclusion that one conformation which probably is a deformed crown form is greatly preferred. The crown form has been determined for azacyclooctane[103] by X-ray investigations.

A decrease in the energy content of the conformations of cyclooctane can be produced by a spreading of the tetrahedral angles of the carbon atoms. This spreading of the C—C—C valence angle has been proven experimentally by means of X-ray diffraction with derivatives of cyclononane, cyclodecane and with cyclododecane itself.[104] The Baeyer strain, no doubt, is increased by changing the tetrahedral angles. At the same time, in the case of cyclooctane however, probably a decrease of the Pitzer strain takes place.[105]

According to Allinger,[85] spreading of the tetrahedral angle by 4° would probably increase the energy (per carbon) by 0.28 kcal./mole. By decreasing the Pitzer strain, the energy per carbon atom drops 0.43 kcal./mole, and for the ring as a whole, therefore, 1.2 kcal./mole.

---

[102] (a) G. Chiurdoglu, Th. Doehaerd, and B. Tursch, *Chem. & Ind. (London)* p. 1453 (1959); (b) E. Billeter and H. H. Günthard, *Helv. Chim. Acta* **41**, 338, 686 (1958); (c) E. Billeter *et al., ibid.* **40**, 2046 (1957); (d) G. Chiurdoglu, Th. Doehard, and B. Tursch, *Colloq. Intern. Chim. Org. Probl. Stereochim., Montpellier*, 1959, in *Bull. Soc. Chim. France*, p. 1322 (1960); (e) F. A. L. Anet and J. S. Hartman, *J. Am. Chem. Soc.* **85**, 1204 (1963); (f) For thermodynamic properties of cyclooctane see: H. L. Finke, D. W. Scott, M. E. Gross, J. F. Messerly and G. Waddington, *J. Am. Chem. Soc.* **78**, 5469 (1956).

[103] H. C. Mez, cited in J. D. Dunitz and V. Prelog, *Angew. Chem.* **72**, 896 (1960).

[104] J. D. Dunitz and H. M. M. Shearer, *Proc. Chem. Soc.* p. 348 (1958); *Helv. Chim. Acta* **43**, 18 (1960); R. F. Bryan and J. D. Dunitz, *ibid.* **43**, 3 (1960). See also refs. 109 and 110.

[105] Compare C. A. Evans, *Trans. Faraday Soc.* **42**, 719 (1946); N. L. Allinger, ref. 96.

The energy content of the crown form of cyclooctane would then be set at 10.5 kcal./mole, which agrees with the value of 9.9 kcal./mole found from the heat of combustion.[98b,c,d]

Cyclononane: The number of possible conformers of cyclononane increases greatly. It is very difficult to treat higher rings in the same way as cyclooctane because the estimation of the energy values is fraught with continually greater uncertainties as a result of increase in flexibility of the ring system.

For cyclononane, the conformation has been represented as shown[106] (XVI).

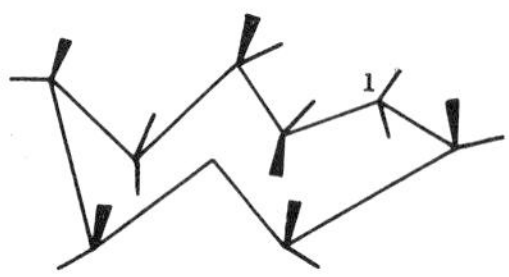

XVI

According to this there are eight hydrogen atoms (*quasi-axial*), (compare Chapter 3, p. 160) four above and four below the ring, arranged in such a way that they point toward the center of the ring. On these carbon atoms there are eight more *quasi-equatorial* hydrogen atoms which are directed outwards. One carbon atom [marked by 1 in XVI] has both hydrogen atoms point outwards and thus take up a position midway between *quasi-axial* and *quasi-equatorial.*

An X-ray analysis of the cyclononane structure[107] in cyclononylamine hydrobromide gave a slightly different arrangement of the hydrogen atoms. The conformation of cyclononane is asymetric and characterized by a preponderance of nearly *gauche* conformations. The dihedral angles lie between 30 and 90°; only one dihedral angle of 105° was measured. The result is that only six hydrogen atoms, three above and three below the ring, are directed into the ring (XVII).[108]

An estimate of the Pitzer strain of cyclononane from the determined dihedral angles gave values between 9.5 and 13.3 kcal./mole relative to cyclohexane. The experimental values determined from heats of combustion came to 12.9[98b] and 12.25[98c,d] kcal./mole (see Fig. 4). Corrections which result from experimentally obtained deviations of the tetrahedral angle between 111° and 124° were not considered in these

[106] A. T. Blomquist and P. R. Taussig, *J. Am. Chem. Soc.* **79**, 3505 (1957). Compare also: A. T. Blomquist and J. C. Meinwald, *J. Org. Chem.* **23**, 6 (1958).

[107] R. F. Bryan and J. D. Dunitz, *Helv. Chim. Acta* **43**, 3 (1960).

[108] Compare also: V. Prelog, H. H. Kägi, and E. H. White, *Helv. Chim. Acta* **45**, 1658 (1962).

energy estimations. The Baeyer strain contributes to the total strain of the ring of cyclononane because the deviation from the tetrahedral angle is large.

Experimental facts for the conformation of cyclodecane, as for cyclononane, have been obtained recently from X-ray analysis of

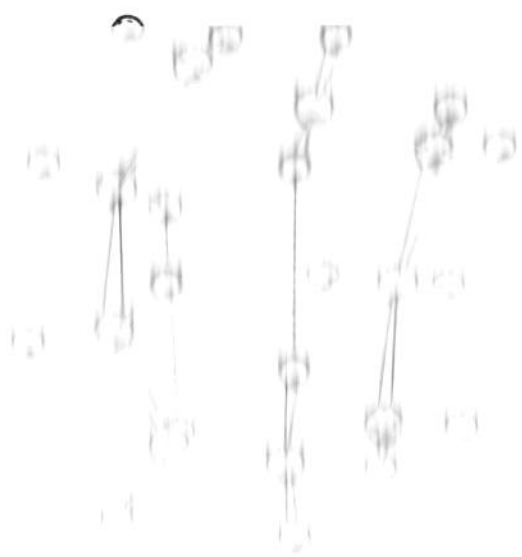

(XVII)

trans-1,6-diaminocyclodecane dihydrochloride by Dunitz[109,110] and collaborators. The results point to the fact that a substantial deviation from the basic crown form, up to now the common working hypothesis, must be present.[85,102b] It has been found from the ascertained dihedral angles that cyclodecane has practically no Pitzer strain. Certain tetrahedral angles are spread considerably, however, which seems to indicate a decided Baeyer strain. Structure (XVIII) shows a schematic representation of the conformation of cyclodecane. This conformation is characterized by great stability (see also p. 171).

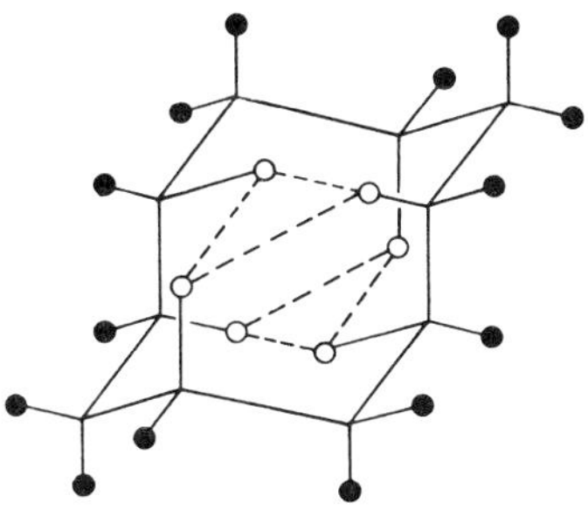

XVIII

[109] E. Huber-Buser and J. D. Dunitz, *Helv. Chim. Acta* **43**, 760 (1960).

[110] E. Huber-Buser, J. D. Dunitz, and K. Venkatesan, *Proc. Chem. Soc.* p. 463 (1961); E. Huber-Buser and J. D. Dunitz, *Helv. Chim. Acta* **44**, 2027 (1961); J. D. Dunitz and K. Venkatesan, *ibid.* **44**, 2033 (1961); W. Nowacki and H. M. Mladeck, *Chimia* **15**, (1961); compare also: V. Prelog, W. Küng and T. Tomljenovic, *Helv. Chim. Acta* **45**, 1352 (1962).

c. Decalins

We can visualize the conformations of decalin in the same way as was done for cyclohexane. In this case also, by applying the rules of conformation, an estimate of the energies of the possible conformational isomers can be undertaken.

Decalin, a strain-free bicyclic ring system, exists in two isomers. The two cyclohexane rings are attached *cis* and *trans* to each other:

cis-Decalin *trans*-Decalin

The geometric isomers of decalin predicted by Mohr[111] from the nonplanar models (the planar models of Baeyer would permit no isomers of decalin—see Chapter 1, page 15) were prepared by Hückel[112], and the first case of a *trans* linkage of two carbocyclic rings was found. The proof of *cis-trans* isomerism was furnished by Hückel,[80a] and the number of isomeric decalin derivatives was established. He undertook the determination of the configuration of the isomeric derivatives, as well as the first measurements of the heats of combustion for a comparison of the energy content of *cis*- and *trans*-decalin. *Cis*-decalin appeared to have a higher energy content. Mohr[111] in his basic work on the stereoisomeric decalins, described various strain-free forms,[113] which were first discussed by Windaus and Hückel[114] and later by Wightman from models.[115]

Hassel[116] again subjected the decalins to an investigation of models. *Trans*-decalin (XX) possesses in its rigid two-chair form, as already indicated by Hückel, the same as cyclohexane in its chair form, a lower Pitzer strain than a *cis*-decalin, if this can be visualized as in two-boat form. By using the same line of thought for the two-boat form of

[111] E. Mohr, *J. Prakt. Chem.* [2] **98**, 315 (1918).

[112] W. Hückel, *Nachr. kgl. Ges. Wiss.* p. 43 (1923); also footnote 23 in Chap. 1.

[113] It may be mentioned, that the original formulation of Mohr clearly required a number of strain-free forms, for the *cis*- as well as the *trans*-decalin. Mohr explicitly emphasized (elsewhere note 1, page 322 in ref. 111) that besides the two-boat form of *cis*-decalin and the two-chair form of *trans*-decalin shown in diagrams (elsewhere note 1, page 321 in ref. 111) "there are still other forms of this molecule conceivable" (as quoted). He repeated this claim in a further publication concerning the *cis-trans* isomerism of decalin [E. Mohr, *Ber.* **55B**, 230 (1922)].

[114] A. Windaus and W. Hückel, *Nachr. kgl. Ges. Wiss.* p. 1 (1921).

[115] See ref. 80. Wightman constructed four *cis* and three *trans* forms.

[116] O. Hassel, *Tidsskr. Kjemi Bergvesen Met.* 3, 91 (1943).

cis-decalin as in the case of the boat form of cyclohexane, the bonds at C_2-C_3, C_6-C_7 and C_9-C_{10} result in the eclipsed conformation, leading to an increase of the Pitzer strain. In addition, an interaction results from the spatial proximity of the hydrogen atoms on C_1 and C_4 as well as on C_5 and C_8 (XIX).

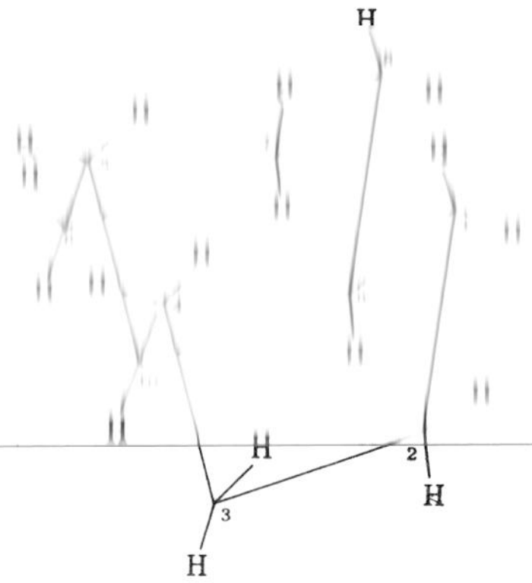

XIX

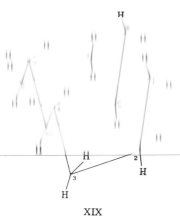

From this it follows that for *cis*-decalin as well the energetically preferred conformation must be the two-chair form (XXI). Hassel[117] had come to this conclusion in 1943 and proved it in 1946 by electron diffraction,[117] i.e., *cis*-decalin exists in the energetically preferred two-chair conformation (XXIa, XXIb). This result was confirmed by LeFèvre[18] by measurement of the molecular polarization.

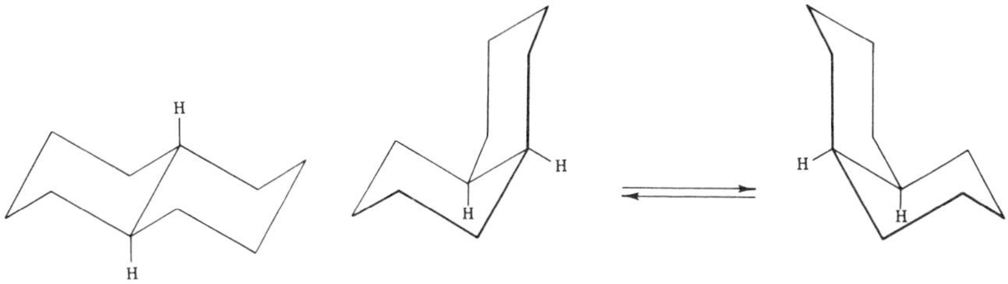

XX XXIa XXIb

As inspection of a model clearly indicates, *trans*-decalin (XX) is rigid; it cannot be converted into its mirror image by flipping of the chair form. However, this is readily possible in the case of *cis*-decalin: (XXIa) is converted to its mirror image (XXIb) by flipping of the chair.

[117] O. Bastiansen and O. Hassel, *Nature* **157**, 765 (1946) and ref. 116.

Cis-decalin, therefore, is a nonresolvable *d,l*-pair, as is later given in detail on page 106 for *cis*-1,2-dimethylcyclohexane.

With the help of nuclear magnetic resonance, the difference in flexibility of *cis*- and *trans*-decalin may be determined directly.[118] Even at low temperatures,[119] the spectrum of *cis*-decalin shows only one sharp signal due to the rapid conversion of one two-chair form into the other, whereby the hydrogens of the CH_2 groups become equivalent in NMR. In contrast to this, rigid *trans*-decalin shows a broad signal, due to the spin-spin coupling of the non-equivalent *axial* and *equatorial* hydrogens. NMR can also provide information concerning the energy barrier in the conversion of one two-chair form of *cis*-decalin into the other. It appears to lie below 6 kcal./mole and, therefore, is lower than that of cyclohexane (page 45).[119] The two-chair form is also possible for every substituted *cis*-decalin (see Chapter 4, page 180f.) and is easily converted into the other two-chair conformation. To predict other than chair forms we have to be careful, at least with the basic views; it has not yet been determined if a conversion into the mobile boat form results at higher temperatures.[120] The anomalies observed at higher temperatures for *cis*-decalin in studies of the temperature dependence of the specific heat and viscosity,[121] which might have indicated a transformation into the mobile boat form, could not be confirmed by recent measurements.[122,123]

By applying the method described briefly under cyclohexane (see page 45), the energy difference between *cis*-and *trans*-decalin in the two-chair-conformation and the difference between the two-chair-conformation and the two-boat-conformation of *cis*-decalin were calculated by Barton.[11] *Trans*-decalin proved to be more stable than *cis*-decalin. The energy difference was calculated to be between 0.52 and 8.23 kcal./mole. Between the two forms of *cis*-decalin, an energy difference from 2.87

[118] J. Musher and R. E. Richards, *Proc. Chem. Soc.* p. 230 (1958); compare also J. Homer and L. F. Thomas, *ibid.* p. 139 (1961).

[119] W. B. Moniz and J. A. Dixon, *J. Am. Chem. Soc.* **83**, 1671 (1961).

[120] Compare D. H. R. Barton, *J. Chem. Soc.* p. 340 (1948).

[121] W. F. Seyer, *J. Am. Chem. Soc.* **75**, 616 (1953); W. F. Seyer and C. H. Davenport, *ibid.* **63**, 2425 (1941); W. F. Seyer and J. D. Leslie, *ibid.* **64**, 1912 (1942).

[122] J. P. Mc Cullough, H. L. Finke, J. F. Messerly, S. S. Todd, T. C. Kincheloe, and G. Waddington, *J. Phys. Chem.* **61**, 1105 (1957); C. Boelhouwer and H. I. Waterman, *Rec. Trav. Chim.* **77**, 411 (1958); cf. also P. Staudhammer and W. F. Seyer, *J. Am. Chem. Soc.* **80**, 6491 (1958); J. Levisalles and J.C.N. Ma, *Bull. Soc. Chim. France* p. 1597 (1962).

[123] That *cis*-decalin also exists in the two-chair conformation, and that this fact has been shown experimentally since 1946 is especially stressed here. There appears off and on, partly in publications, partly in text books, the opinion that *cis*-decalin exists in the two-boat form.

to 7.28 kcal./mole was estimated. The two-chair form, therefore, is the more stable.[83]

Turner[87] carried over his method described for cyclohexane (see page 47) to decalin. Since he visualized decalin as built up of *n*-butane systems, he could calculate the energy differences between *trans* (XX) and *cis* (XXI) in the two-chair form, as well as the two-boat form (XIX) of *cis*-decalin. As can be easily seen from a model, (XX) contains six [illegible] methods described below.) The energy difference between (XXI), and (XIX) is more difficult to estimate. In addition to four *trans*, eight *gauche* and four *eclipsed* *n*-butane interactions in the two-boat form (XIX) of *cis*-decalin, there is an amount of energy, which cannot be estimated, that arises by a 60°-rotation from the stable *trans* conformation. By neglecting this amount, Turner calculated the energy difference between (XXI) and (XIX) to be at least 8.8 kcal./mole.

The value of 2.4 kcal./mole for the energy difference between *cis*- and *trans*-decalin was obtained by Johnson[126] by use of the same parameter in a still more simplified way[7]: *trans*-decalin (XXII) as the more stable form, was arbitrarily given the energy value of zero; *cis*-decalin (XXIII) now differs from *trans*-decalin by three *gauche* interactions.[127] The energy difference then is $3 \times 0.8 = 2.4$ kcal./mole.

From the temperature dependence of the equilibrium *cis* $\rightleftarrows$ *trans*-decalin, Allinger[128] calculated 2.72 kcal./mole for the energy (enthalpy)

---

[124] 2.07 kcal./mole is obtained after correcting for the different heats of evaporation of the two isomeric decalins.

[125] The heat of isomerization was obtained from the heats of combustion of pure *cis*- and *trans*-decalin: G. F. Davies and E. C. Gilbert, *J. Am. Chem. Soc.* **63**, 1585 (1941). They found for *cis*-decalin 1499.7 kcal./mole; for *trans*-decalin 1497.6 kcal./mole. Almost the same values (*cis*: 1499.9 kcal./mole; *trans*: 1497.1 kcal./mole; difference 2.8 kcal./mole) were found by W. Hückel. Compare also G. S. Parks and J. A. Hatton, *J. Am. Chem. Soc.* **71**, 2773 (1949).

[125a] F. D. Rossini, cited in T. Miyazawa and K. S. Pitzer, *J. Am. Chem. Soc.* **80**, 60 (1958).

[126] W. S. Johnson, *J. Am. Chem. Soc.* **75**, 1498 (1953).

[127] For a better understanding it is recommended that a model be constructed. The *gauche* interactions shown in XXIII for *cis*-decalin then become clear at once.

[128] N. L. Allinger and J. L. Coke, *J. Am. Chem. Soc.* **81**, 4081 (1959).

difference (liquid phase). This agrees well[129] with the more recent measurements of Rossini from heats of combustion, 3.1 kcal./mole (vapor phase).[129a]

We disregarded entropy differences between *cis*- and *trans*-decalin in our considerations. Qualitative predictions of entropy differences

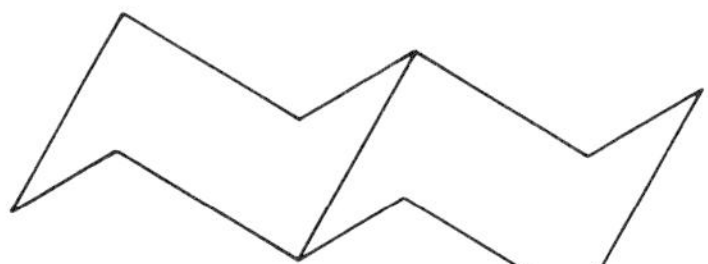

XXII

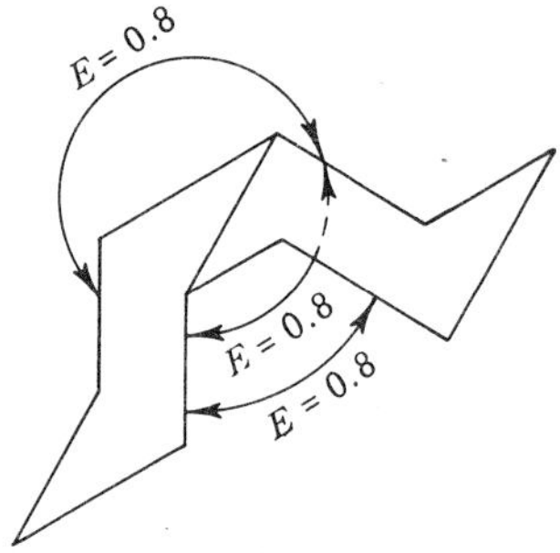

XXIII

are based mostly on such views as the rigidity or the symmetry of the molecule. The reasons for the varying stability of differently substituted *cis-trans* isomeric decalins were generally related back to the higher entropy that is a higher mobility of the *cis* fused system.[130]

The symmetry number of both *cis*- and *trans*-decalin is 2, because both have a twofold axis of symmetry. Because the *cis*-isomer is a *d,l*-pair (see page 58) (*trans*-decalin possesses a center of symmetry and is, therefore, optically inactive), it has a higher entropy than *trans*-decalin by an amount of $R \ln 2 = 1.38$ e.u.[128] The experimentally found value for the entropy difference is lower (0.55 e.u.[128] liquid phase), indicating a higher degree of order for *cis*-decalin than for *trans*-decalin in the liquid phase. The value of the entropy difference indicates that *cis*- and *trans*-decalin do not differ in their rigidity; this difference cannot be attributed to a greater flexibility of *cis*-decalin.[131]

The conclusion, that *trans*-decalin, due to a more favorable conformation, possesses the greater stability,[132] follows from all of these reported

---

[129] Calculation of the energy difference by the above described method gives Allinger[128] 2.7 kcal./mole. He uses 0.9 kcal./mole for the *gauche* interaction.

[129a] D. M. Speros and F. D. Rossini, *J. Phys. Chem.* **64**, 1723 (1960).

[130] N. L. Allinger, *J. Org. Chem.* **21**, 915 (1956).

[131] See ref. 128, cf. N. L. Allinger, M. Nakazaki, and V. Zalkow, *J. Am. Chem. Soc.* **81**, 4074 (1959).

[132] The conditions of the energy differences are no longer as clear for the 9-substituted decalins (compare Chapter 4, page 199).

values of the energy differences, which compared with the calculated values, agree very well.

Incorporation of the knowledge gained from decalin for condensed cyclohexane rings in general leads to the rule that the more stable conformation of a condensed cyclohexane ring system is the one which possesses the largest number of chair conformations[11] (see Chapter 5).

5. Illustrations and Nomenclature for Conformational Analysis

a. Illustrations

Stereochemical problems, especially those of conformational analysis, become most clear by using three-dimensional models. A large number of otherwise difficult to understand facts can be easily demonstrated by construction of such ball and stick models. True scale atomic models due to their lack of clear visibility, are less suitable for such consideration. With a little skill such ball and stick models can be easily constructed; but several of these models are described and can be purchased.[133] Uniformly elastic neoprene can be used for the balls, so that the strain set up during construction is distributed to all valence angles (*Petersen models*).[134]

For blackboard and paper, which permit only two-dimensional representations, various limitations had to be introduced. Two methods of representation can be chosen.

(1) The so-called "perspective" formulas, in which two adjoining carbon atoms are viewed obliquely from the side (XXIV).

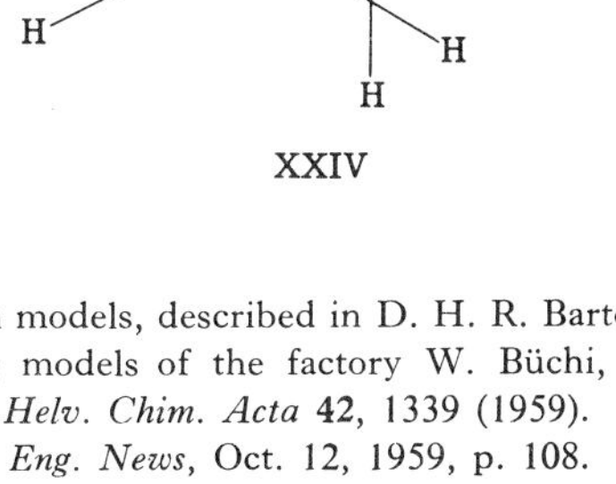

XXIV

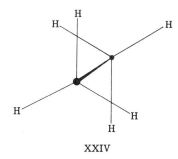

[133] For example: Barton models, described in D. H. R. Barton, *Chem. & Ind. (London)* p. 1136 (1956); Dreiding models of the factory W. Büchi, Flawil, SG, Switzerland, compare A. S. Dreiding, *Helv. Chim. Acta* **42**, 1339 (1959).

[134] Described in *Chem. Eng. News*, Oct. 12, 1959, p. 108.

(2) The so-called "projection" formulas, in which one carbon atom is viewed in the direction of the C—C bond, so that the second carbon atom lies directly behind the first one. (XXV):

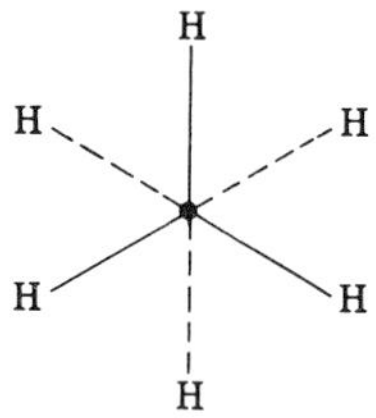

XXV

Both types of reproduction have advantages and disadvantages. The perspective formulas allow good recognition of all bonds. They are often not too obvious and are difficult to draw.[135] A further problem lies in the fact that these formulations also demand a certain amount of ability in spatial observation, which is difficult for some readers to follow, especially when it involves complicated cyclic compounds.

The projection formulas permit a good understanding of the relations of the geometrical angles, and they are easy to draw. The disadvantage lies in the fact that some bonds between the carbon atoms are not visible.[136]

The projection formulas which have enjoyed wide usage, especially for the representation of conformational relations of open-chain compounds, were introduced by Hermans[137] in their original form (see

[135] Some rules for drawing perspective formulae are found in G. Snatzke "News from Chemistry and Technology," *Angew. Chem.* **8**, 110 (1960). For the determination of interatomic distances with the aid of the vector method see E. J. Corey and R. A. Sneen, *J. Am. Chem. Soc.* **77**, 2505 (1955); compare also R. Pauncz and D. Ginsburg, *Tetrahedron* **9**, 40 (1960).

[136] The well known Fischer projections are not suitable for conformational problems. They do help in the representation of various configurations, but they always show the molecules in the *eclipsed* conformation (XXVI).

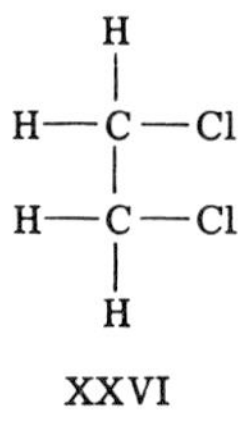

XXVI

For the representation of configurations compare R. S. Cahn, C. K. Ingold, and V. Prelog, *Experientia* **12**, 81 (1956).

[137] P. H. Hermans, *Z. Physik. Chem.* (*Leipzig*) **113**, 337 (1924), especially p. 351.

XXV). They are still used in this form at times. In addition, there is a somewhat modified method of representation being used increasely, which originated with Newman[138,139] and which has replaced the old Hermans projection. The Newman type of projection formulas, which are also suitable for representing cyclic compounds, are clearer and give a better idea of the spatial arrangements than the older Hermans formulas.

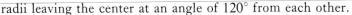

The Newman projection formula for open chain compounds appears as shown (XXIX). The carbon atom is again viewed in the direction of the C—C bond axis. The second carbon atom lies directly behind the first one [XXIX; the Hermans projection (XXVIII) is drawn next to it for comparison].

The carbon atom nearest to the viewer is designated by 1; the one behind it by 2. Carbon atom 2 is shown symbolically by a circle, on which are the three valences. Carbon atom 1 is represented by three radii leaving the center at an angle of 120° from each other.

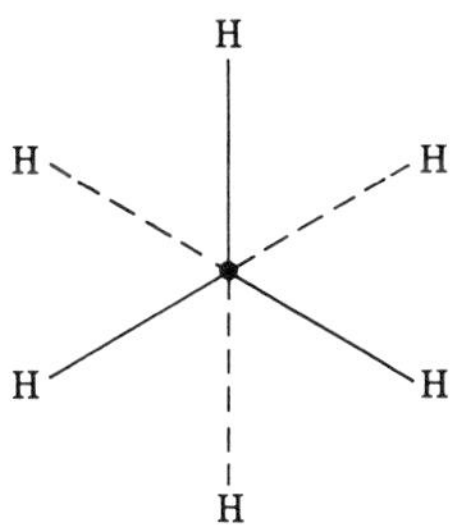

XXVIII

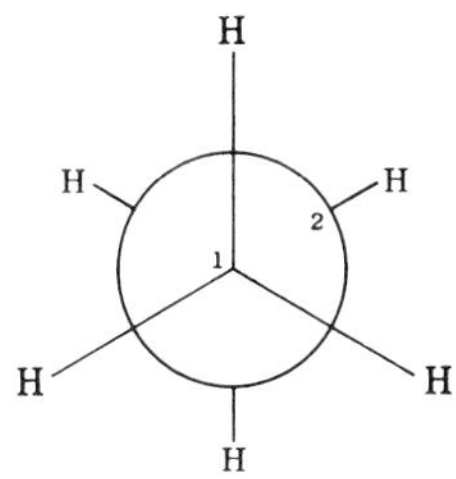

XXIX

[138] M. S. Newman, *Record Chem. Progr.* (*Kresge-Hooker Sci. Lib.*) **13**, 111 (1952); compare also M. S. Newman; "Steric Effects in Organic Chemistry," p. 4. Wiley, New York, 1956.

[139] At times we can find still another method of representation: W. Hückel, *Deut. Apotheker-Zt.* **95**, 302 (1955). For the *eclipsed* form, the substituents appear at the corners of two superimposed triangles; for the *staggered* form in the corners of a "Star of David" (XXVIIa, b).

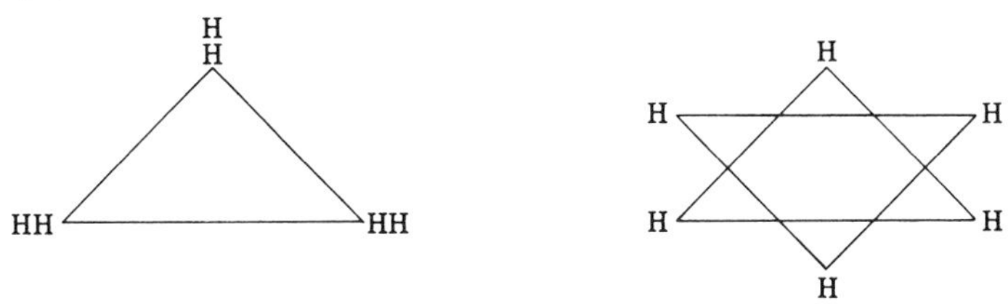

(a) (b)

XXVII

This type of notation permits an easy representation of various forms occurring in open-chain compounds (such as *erythro* and *threo* forms; compare Chapter 8, page 332). The addition of halogen (X_2) to an olefinic double bond is represented as shown (XXX).

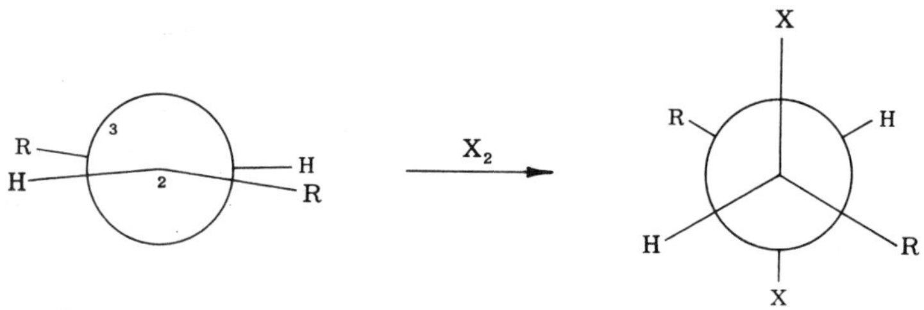

XXX

When R = CH_3 (*trans*-2-butene), carbon atom 2 again lies in front of 3 (circle). The bonds, which carry H and R, are bent toward each other slightly to make them more visible. The resulting dibromide can be represented as a definite form (*meso* form, compare Chapter 8, page 331).

As already mentioned above, the Newman projection can also be used for cyclic compounds, although it does not replace the perspective formula in this case.

As an example cyclohexane is shown with perspective formula shown for comparison (XXXI and XXXII).

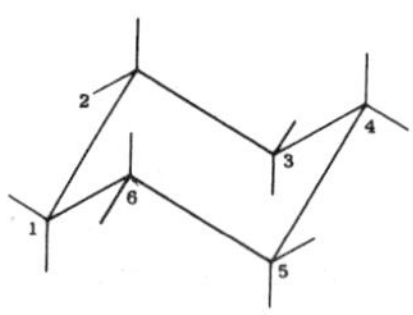

XXXI

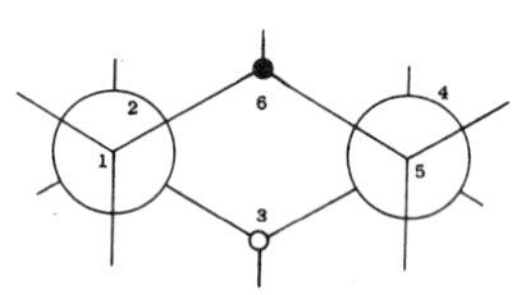

XXXII

Carbon atom 1 is again characterized by the three radii, carbon atom 2 by the circle with three valences.[140] The other carbon atoms in (XXXII) are designated as 3, 4, 5, and 6. Four bonds cannot be represented: one on carbon 6 (equatorial, see Chapter 3, page 87), which points directly at the observer; the other on carbon 3 (equatorial) which points

[140] For better visualization, a spatial model can help in this case. If a cyclohexane ring is built in the chair form and viewed in such a way that carbon atom 1 and 2 lie one behind the other, the depicted projection (XXXII) can be seen. In the case of unsubstituted cyclohexane, it makes no difference, of course, which carbon atom is designated as 1.

away from the observer. The other bonds are between carbon atoms 1 and 2, as well as 5 and 4. To illustrate the arrangement on carbon atoms 3 and 6, various symbols can be introduced:  for an invisible bond which points away from the observer;  for a nonvisible bond which points toward the observer.

Naturally this type of projection can also be applied to condensed ring systems such as *cis*- and *trans*- decalin (XXXIV and XXXVI).

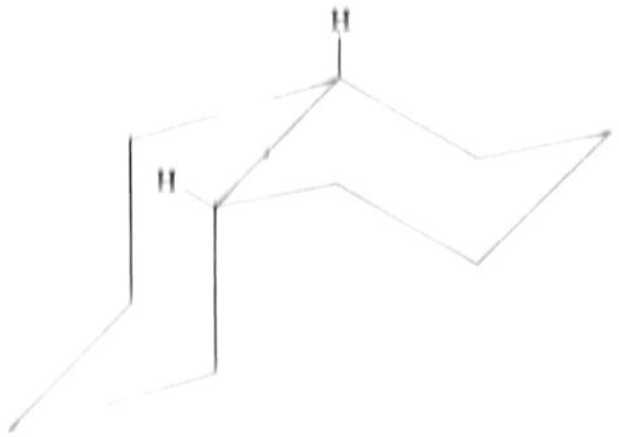

(XXXIII)

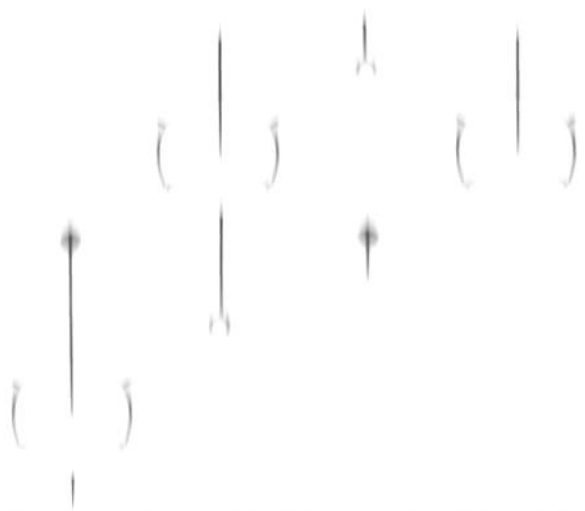

(XXXIV)

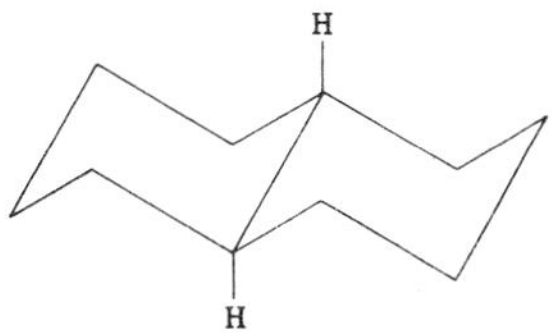

(XXXV)

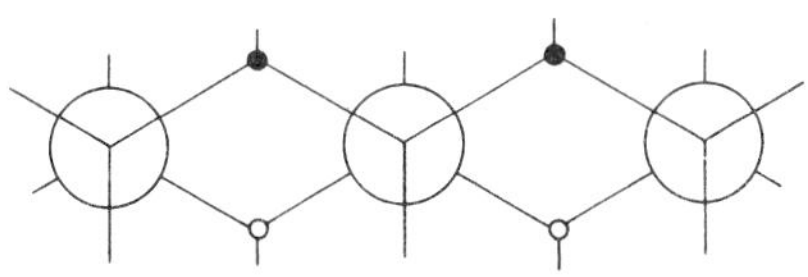

(XXXVI)

For multiple condensed ring systems, such as the steroid structure, representation by projection formulas becomes quite confusing. For smaller rings also, such as the five-membered ring, or for medium-sized rings, the perspective representation is preferred. Thus both types of representation, the perspective and projection formulas, are used according to suitability.

The problem of representation of various conformers is hereby not exhausted. Theoretically there are an infinite number of forms for each compound. It is well to recall the flexible forms of cyclohexane which can only be described clearly when it is possible to represent them diagramatically in a simple form. For the six-membered ring, six describable forms exist according to this viewpoint[86f] (XXXVII a to f.) At least four carbon atoms always lie in a plane.

For the twist form XXXVII f) a somewhat different formulation may be used (XXXVIII).

All remaining spatial arrangements can be represented when needed by the designation of the dihedral angle φ between valences on neighboring carbon atoms.

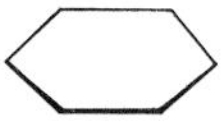

Plane form

(a)

Envelope form

(b)

Chair form

(c)

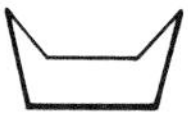

Boat form

(d)

Half chair form

(e)

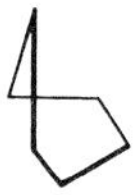

Twist-(skew, stretched)-form

(f)

XXXVII

Twist form

XXXVIII

b. Nomenclature

During the discussion of the conformation of ethane and butane, the distinguishable conformations which arise through rotation around the C—C bond axis of the molecule were designated with various names. The nomenclature of these conformations is not uniform. (One should here again stress that an infinite number of conformations exist between these distinghishable conformations.) Numerous names and symbols have been introduced or suggested for these forms. Those names used in the English and German literature have been summarized in Table II and III.

Independent of this vacillating nomenclature, the possible conformations can be characterized uniformly by indicating the degrees or radians of the angle which two chosen substituents form with each other during the rotation around the C—C bond axis (in XXXIX, R and R').[141,142] The fully *eclipsed* form thereby obtains the symbol φ°. If we now turn the substituents on carbon atom 2 (in the Newman

[141] A. P. Terentiev and V. M. Potapov, *Tetrahedron* **1**, 125 (1957).

[142] For a further type of description of aliphatic conformational isomers see R. Riemschneider, *Z. Naturforsch.* **116**, 38 (1956).

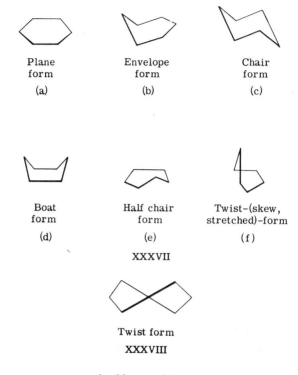

projection) clockwise 60° around the central bond, then we obtain the other forms, which receive the symbols φ^1 to φ^5 (XXXIX).

The symbols φ^0 to φ^5 thereby replace the nonuniform nomenclature and can also be applied to cyclic compounds. This is done by starting with a definite carbon atom, then allocating each carbon atom of the ring its corresponding symbol (Table II).

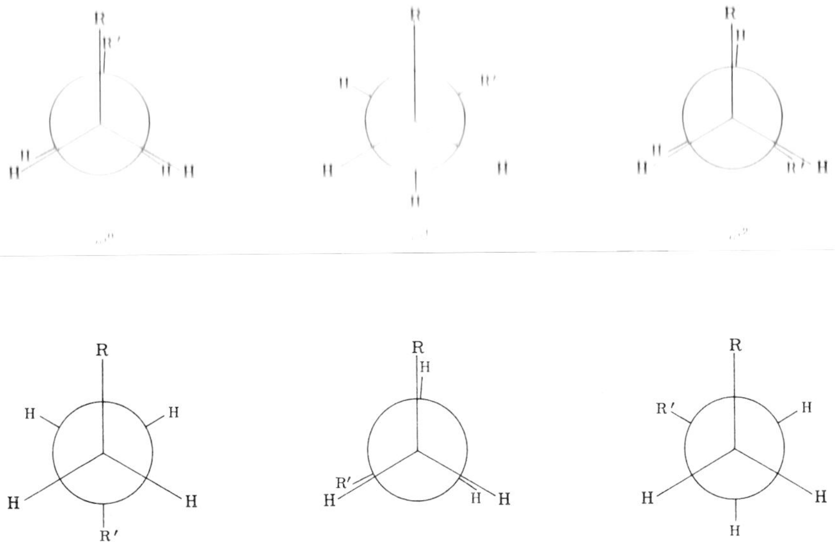

φ^3 φ^4 φ^5

XXXIX

TABLE II

English names[a]	German names	Symbol
fully eclipsed	planar-syn,[b] ekliptisch, verdeckt, Atom-Atom	φ^0
gauche, skew	schief-syn,[b] syn, windschief	φ^1
partially eclipsed	schief-anti,[b] teilweise verdeckt	φ^2
fully staggered, trans, anti	planar-anti,[b] gestaffelt, auf Lücke, anti, trans, Atom-Lücke	φ^3
partially eclipsed	schief-anti,[b] teilweise verdeckt	φ^4
gauche, skew	schief-syn,[b] windschief	φ^5

[a] The French nomenclature in general follows the English. For *eclipsed* the expression "vis-à-vis" is also used.

[b] According to V. Prelog.

In addition, the *gauche* forms can be designated by the notation g_l (*gauche* "left") and g_r (*gauche* "right") if the symbols are not used. These correspond to φ^5 and φ^1 in XXXIX. For branched-chain paraffins such as 2-methyl butane, the notation *trans-gauche* (t_{gl} and t_{gr}) as well as *gauche-gauche* ($g_l g_r$) are used (XL).

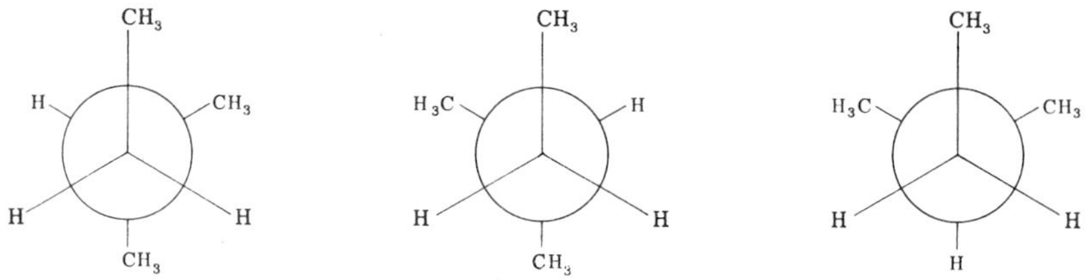

XL

trans-gauche (t_{gr}) *trans-gauche* (t_{gl}) *gauche-gauche* ($g_l g_r$)

Nevertheless, for a still more accurate description of the conformations, the nomenclature in use up to now has been defined more clearly by Klyne and Prelog.[143] The dihedral angle φ is made positive when it is measured clockwise from one substituent in the front to a substituent at the rear (cf. XXXIX); φ is negative when it is measured counter clockwise. The partial conformation of two atoms connected by a single bond is defined by the dihedral angle between the most preferred substituents. The choice of the most preferred substituents follows the sequence rule of Cahn, Ingold, and Prelog.[144] For further definition the following supplementary rules are necessary. If all substituents on one of the two substituted atoms are the same, then the smallest possible dihedral angle is chosen for definition of the conformation. If two substituents on a tetrahedral atom are alike, then the third substituent, different from these, determines the conformational designation, independent of the sequence rule. These conventions are applicable to tetrahedral, trigonal, and digonal atoms.

If the exact dihedral angles are not known, the following nomenclature is proposed by Prelog and Klyne[143] in order to describe the dihedral angle approximately. The terms "periplanar" and "clinal" are

[143] V. Prelog, private communication; W. Klyne and V. Prelog, *Experientia* **16**, 521 (1960).

[144] R. S. Cahn, C. K. Ingold, and V. Prelog, *Experientia* **12**, 81 (1956). According to the sequence rule, the substituents bound to an asymmetric carbon atom are arranged according to decreasing atomic numbers. If the relative preference of two groups cannot be decided this way, the preference is decided by the comparison of the atomic numbers of the next atoms of this group. If this still does not lead to the goal, then the next atoms after those are compared.

introduced. "Periplanar" describes the approximately planar position, "clinal" describes the inclined position. The expression "syn" is used when dihedral angle φ is $< 90°$, "anti" when φ is $> 90°$. The plus and minus sign ($\pm$) are used in the above given sense. The resulting terms are shown in Table III.

TABLE III

Dihedral angle		Abbreviation
$0° \pm 30°$	$\pm$ syn-periplanar	($\pm$ sp)
$+60° \pm 30°$	+ syn-clinal	(+sc)
$+120° \pm 30°$	+ anti-clinal	(+ac)
$180° \pm 30°$	$\pm$ anti-periplanar	($\pm$ap)
$-120° \pm 30°$	− anti-clinal	(−ac)
$-60° \pm 30°$	− syn-clinal	(−sc)

For long chains or multiple branched hydrocarbons, the question of nomenclature is more complicated. The recommended procedure in such a case is the notation introduced by Shimanouchi and Mizushima,[16,145] according to which the conformation of complicated saturated compounds can be represented in the following two ways.

(1) A carbon atom is examined and the direction of its four bonds is then designated as a, b, c, and d (XLI):

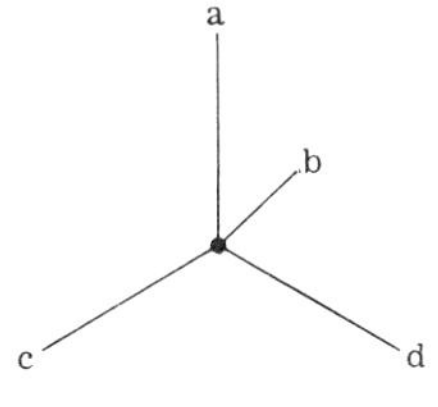

XLI

The direction of any other C—C bond present in the same molecule must be parallel to one of those (a, b, c, or d) of the original carbon atom. In this way, by giving the direction of all C—C bonds, the carbon skeleton can be unequivocally characterized.

[145] J. Shimanouchi and S. Mizushima, *Sci. Papers Inst. Phys. Chem. Res.* (*Tokyo*) **40**, 467 (1943).

Thus, for example, the *trans* form of *n*-butane is represented by a, b, a; the one *gauche* form by a, b, c and the second *gauche* form by a, b, d (XLII).

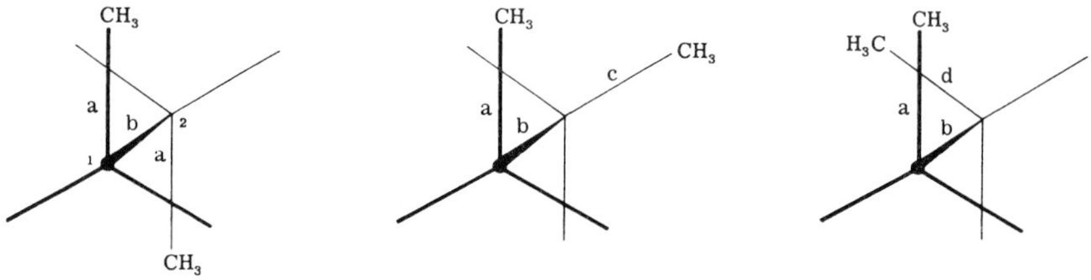

XLII

The side chains are placed in parentheses, such as isobutane,

$$CH_3-CH(CH_3)-CH_3 = a(c)b.$$

For longer chain compounds the notation corresponds, for example for *n*-hexane: a b a b a; a b a b c; a b a b d; etc.

(2) The second possibility for the notation of conformers uses the symbols T (*trans*), as well as G and G′ (*gauche*) for the corresponding forms of *n*-butane. In order to differentiate between the two conformations G and G′ we start with the *trans* form. G is the conformation which arises from the rotation of carbon atom 2 counterclockwise through 120°; G′ the conformation through rotation of 120° clockwise. (Compare XLII: a b a = T; a b c = G; a b d = G′.)

The fully *staggered* form, for example of *n*-pentane or of 1,3-dichloropropane, would receive the symbol TT (a b a b in the first notation). The other forms are designated with TG (corresponding to a b a c); TG′ (corresponding to a b a d); GG (corresponding to a b c d); G′G′ (corresponding to a b d c); GG′ (corresponding to a b c a); G′G (corresponding to a b d a).[146]

The advantage of the second notation lies in the fact that certain properties of symmetry can be recognized more easily, i.e., mirror image isomerism. (For example, TG′ and TG or G′G′ and GG are mirror image isomers.)

For the notation of *n*-hexane started above, the corresponding symbols read: TTT (corresponding to a b a b a); TTG (corresponding to a b a b c); TTG′ (corresponding to a b a b d); etc.

[146] The relations here also are most easily clarified by the use of models. After rotation of the second carbon atom (see XLII) to obtain the G or G′ form, it is best to rotate the entire molecule until bond "a" on carbon atom 1 assumes the position as shown in XLI. The positions of the other C—C bonds can then be noted correctly. We see, for example, in the case of *n*-pentane that the GG′ and G′G conformation must be unstable, because the terminal methyl groups approach each other very closely.

For isoparaffins, the first notation is again recommended (side chains in parentheses). Otherwise, for an unequivocal representation, additional symbols must be introduced.

6. Conformational Analysis and Its Application

The views brought forth up to now concerning the possible conformations of certain compounds, which led to the conclusion that every molecule exists in definite preferred conformations, proceeded mainly from physical suppositions. The reasons for the occurrence of preferred conformations was characterized by the presence of repulsion forces of nonbonded atoms. This is also the basis for their energy level. Other effects such as a deformation of the tetrahedral angle, formation of hydrogen bonds, or electrostatic forces due to bond dipoles were described as contributing to the presence of preferred conformations.

For characterization of an organic compound, the conformation therefore is joined with the constitution and the configuration as a third important criterion.

In the view of the organic chemist, an important question now arises: which chemical and physical properties of a molecule can be connected with the geometry and the population of the various conformations in which the molecule can exist?

The effect of preferred conformations on certain physical properties, such as the position of absorption bonds in the infrared (and ultraviolet) spectra, has already been mentioned. A detailed description of the connection between conformation and physical properties will be presented in the following chapters.

The effect of preferred conformations on the chemical properties of a molecule, besides other effects, have become a reality through the intensive research of the last ten years in this field. It has long outgrown its significance as just a special branch of stereochemistry.

The significance of conformational effects in reaction mechanisms is especially emphasized as, for example, during substitution or cleavage reactions (coplanarity of participating centers) or the relative stability of epimers during equilibrium reactions.

In the following chapters we shall attempt to show in more detail the result of conformational effects on the properties of the different compound classes. A complete description of all results available up to now would be far beyond the scope of this book. We shall restrict ourselves to characteristic and clear examples from the viewpoint of conformation.

CHAPTER 3

Monocyclic Compounds

Chapter 2 dealt with the general rules of conformation. In this chapter monocyclic compounds without hetero atoms (for the latter, see Chap. 7, page 301) will be discussed in detail.

Though cyclopropane can exist only in a planar form, there is a deviation from the planar form present in cyclobutane as determined from spectroscopic data and entropy measurements.[1] More recent electron diffraction measurements corroborate the evidence, although quantitative data concerning the size of the deviation was not available.[2] As will be described later for cyclopentane, the reasons for the deviation from the planar arrangement may be found not only in the *eclipsed* relationship of the adjacent CH_2 groups, but also in 1,3-interactions of the carbon atoms.[2a]

I. CYCLOPENTANES

1. Conformation of the Cyclopentanes

a. Cyclopentane

The problem of the conformation of the cyclopentane ring has been discussed thorougly in the recent literature,[3] although there is older information[4] available.

[1] J. D. Dunitz and V. Schomaker, *J. Chem. Phys.* **20**, 1703 (1952); G. W. Rathjens, Jr. and W. D. Gwinn, *J. Am. Chem. Soc.* **75**, 5629 (1953); G. W. Rathjens, N. K. Freeman, W. D. Gwinn, and K. S. Pitzer, *ibid.* **75**, 5634 (1953).

[2] A. Almenningen, O. Bastiansen, and P. N. Skancke, *Acta Chem. Scand.* **15**, 711 (1961).

[2a] Compare also: J. M. Conia, J. L. Ripoll, L. A. Tushaus, C. L. Neumann, and N. L. Allinger, *J. Am. Chem. Soc.* **84**, 4982 (1962); G. C. Fonken and S. Shiengthong, *J. Org. Chem.* **28**, 3435 (1963); F. Lautenschlaeger and G. F. Wright, *Can. J. Chem.* **41**, 863 (1963); J. B. Lambert and J. D. Roberts, *J. Am. Chem. Soc.* **85**, 3710 (1963); A. Wilson and D. Goldhamer, *J. Chem. Educ.* **40**, 504 (1963).

[3] See for example ref. 7 and 18 in Chap. 2; (a)B. Curnutte and W. H. Shaffer, *J. Mol. Spectr.* **1**, 239 (1957); (b) K. S. Pitzer and W. E. Donath, *J. Am. Chem. Soc.* **81**, 3213 (1959); (c) F. V. Brutcher, Jr., T. Roberts, S. J. Barr, and N. Pearson, *ibid.* **81**, 4915

As explained earlier (see Chap. 1, Table I, page 13), the deviation of the C—C bond angle from the tetrahedral angle in cyclopentane is so small (1°28′) that it can have no noticeable effect on the energy content. Yet if one compares the heat of combustion per CH_2 group of cyclopentane (158.7 kcal./mole) with that of cyclohexane (157.4 kcal./mole) or the normal value per CH_2-group in open-chain aliphatic compounds (157.5 kcal./mole), a definite increase of the heat of combustion of cyclopentane is observed. From the heats of combustion data a higher energy content is estimated for cyclopentane in comparison to open-chain compounds which amounts to 6–7 kcal./mole.

The higher energy level of cyclopentane cannot be related to Baeyer ring strain, but seems to result from a sterically unfavorable arrangement of adjacent methylene groups. As can be seen from a model, the hydrogen atoms of the adjacent methylene groups are nearly completely *eclipsed*. Considerable Pitzer strain results (I). (Compare Chap. 2, page 32)

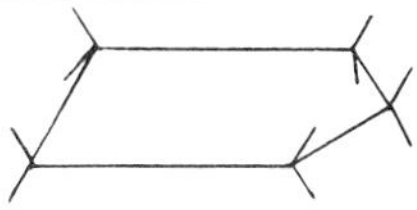

I

The increased value of 6–7 kcal./mole obtained from the heat of combustion is not as high as would be expected from five *eclipsed* methylene groups, if the basic value of ~3 kcal./mole found for ethane is used. By means of simple addition, 14-15 kcal./mole are obtained. A planar arrangement of the cyclopentane ring cannot be very stable because of the *eclipsed* arrangements of the methylene groups. This has been shown during the discussion of Pitzer-strain (see page 32). Because of the repulsive forces, the neighboring hydrogen atoms will seek to avoid each other, which must lead to a puckered arrangement of the ring structure. This happens even though the deviation from the normal valence angles has been increased.[5]

(1959); also F. V. Brutcher, Jr. and W. Bauer, Jr. *Science* **132**, 1489 (1960); *J. Am. Chem. Soc.* **84**, 2233, 2236 (1962); (d) J. P. McCullough, *J. Chem. Phys.* **29**, 966 (1958); (e) J. P. McCullough, R. E. Pennington, J. C. Smith, I. A. Hossenlopp, and Guy Waddington, *J. Am. Chem. Soc.* **81**, 5880 (1959); compare also (f) C. W. Shoppee, R. H. Jenkins, and G. H. R. Summers, *J. Chem. Soc.* p. 3048 (1958); (g) J. Fishman and C. Djerassi, *Experientia* **16**, 138 (1960).

[4] See for example J. G. Aston, ref. 93 in Chap. 2; K. S. Pitzer, ref. 36 in Chap. 2; O. Hassel and H. Viervoll, *Tidsskr. Kjemi Bergvesen Met.* **6**, 31 (1946); F. A. Miller and R. G. Inskeep, *J. Chem. Phys.* **18**, 1519 (1950); J. N. Haresnape, *Chem. & Ind. (London)* p. 1091 (1953).

[5] Compare S. A. Barker and R. Stephens, *J. Chem. Soc.* p. 4550 (1954). Barker and Stephens consider the puckering of the five-membered ring so slight, that it can be disregarded in most cases.

According to Pitzer the cyclopentane ring is puckered to such an extent,[6,7] that one carbon atom sticks out about 0.2Å from the plane formed by the other four carbon atoms (II).

More recent calculations have shown[3b,c] that the energy minimum of the cyclopentane molecule is attained when one carbon atom twists out of the plane by 0.5 Å.

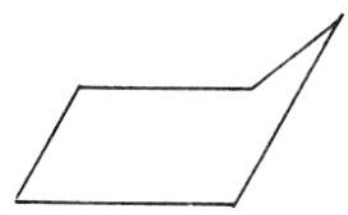

II

The carbon atom protruding from the plane is not fixed in that position,[8] rotation occurs in such a way that all of the carbon atoms attain this preferred position in successive order (pseudorotation). In other words, the nonplanar cyclopentane cannot be represented by a single structure. "First, an ordinary vibration in which the amount of puckering oscillates about a most stable value and second, a pseudo one-dimensional rotation in which the phase of the puckering rotates around the ring."[9] Entropy measurements of gaseous cyclopentane can be interpreted[3b,10] in favor of such an assumption. Initially a deviation from the planar arrangement of the carbon skeleton of cyclopentane could not be determined with certainty by means of electron diffraction,[11] but more recent measurements have confirmed the puckered structure.[12]

Bastiansen and Hassel[13] proposed a structure in which one carbon atom lies above and one below the plane of the other three carbon atoms. Other investigators[14] mention only a very slight deviation from the planar structure, without describing the form any further.

---

[6] K. S. Pitzer, *Science* **101**, 672 (1945); J. E. Kilpatrick, K. S. Pitzer, and R. Spitzer, *J. Am. Chem. Soc.* **69**, 2483 (1947).

[7] Compare also R. Spitzer and H. M. Huffmann, *J. Am. Chem. Soc.* **69**, 211 (1947).

[8] Cf. W. G. Dauben and K. S. Pitzer, *in* "Steric Effects in Organic Chemistry" (M. S. Newman, ed.), Chap. 1. Wiley, New York, 1956.

[9] See Pitzer and co-workers.[6]

[10] Compare J. G. Aston, ref. 93 in Chap. 2; J. G. Aston, H. L. Fink, and S. C. Schumann, *J. Am. Chem. Soc.* **55**, 341 (1943).

[11] O. Hassel and H. Viervoll, *Acta Chem. Scand.* **1**, 149 (1947).

[12] A. Almenningen, O. Bastiansen, and P. N. Skancke, *Acta Chem. Scand.* **15**, 711 (1961). From these investigations a normal C—C bond length of 1.539 ± 0.003 Å and a C—H bond length of 1.095 ± 0.010 Å were found. For an analysis of the IR and Raman spectra cf. F. A. Miller and R. G. Inskeep, *J. Chem. Phys.* **18**, 1510 (1950).

[13] O. Bastiansen, O. Hassel, and L. K. Lund, *Acta Chem. Scand.* **3**, 297 (1949). Electron diffraction measurements of decafluorocyclopentane.

[14] H. Tschamler and H. Voetter, *Monatsh. Chem.* **83**, 302, 835, 1228 (1952).

The pseudorotation of the cyclopentane ring was at first rejected by a number of investigators.[15] For example, Miller and Inskeep[4] were able to determine that the cyclopentane molecule does not have the symmetry of a regular, planar pentagon. From analysis of the infrared and Raman spectra these workers could not confirm the pseudorotation assumed by Pitzer. LeFèvre[15] also determined a puckered structure by means of the Kerr effect, but was not able establish a pseudorotation.[16] Nevertheless, the original view of Pitzer concerning the pseudorotation of the cyclopentane ring seems to have been confirmed. It could be ascertained from very exact thermodynamic measurements of cyclopentane and various substituted cyclopentanes,[17] and comparison with available spectroscopic data that one of the degrees of freedom of the puckered cyclopentane ring was associated with a free "pseudorotation."

A detailed discussion of the possible forms of cyclopentane and substituted cyclopentanes has been undertaken by Brutcher.[18] Two forms can be differentiated (compare also p. 221): (1) the so-called *envelope* form (C_s), in which one carbon atom projects out of the plane of the other four (III); (2) the so-called *half-chair* form (C_2), in which three neighboring carbon atoms lie in one plane, while the other two are twisted in such a way, that one lies above and the other below the plane [these carbons are equidistant from this plane (IV)].

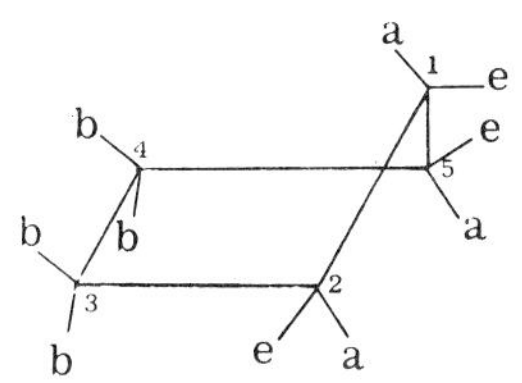

III

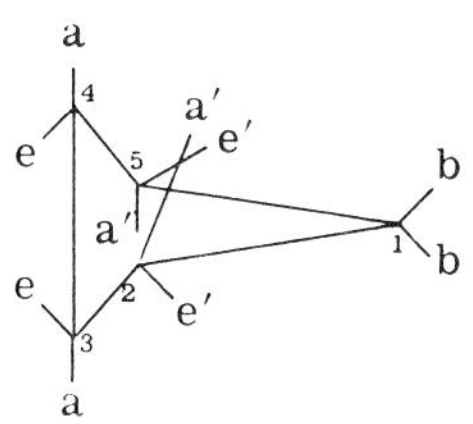

IV

On the bais of these two forms, three types of bonds can be recognized (III and IV). They are classical *axial* and *equatorial* bonds (*a* and *e*) as found in cyclohexane (see page 87), the *quasiaxial* and *quasiequatorial* bonds (*a'* and *e'*) and the so-called bisectional bonds (*b*), which take up

[15] See F. A. Miller[4]; C. G. LeFèvre ref. 18 in Chap. 2; *Chem. & Ind.* (*London*) p. 54 (1956); B. Curnutte and W. J. Shaffer[3a].

[16] Cf. however, M. Löw, *Acta Chim. Acad. Sci. Hung.* **25**, 425 (1960).

[17] J. P. McCullough[3d,3e]; J. P. McCullough, D. R. Douslin, W. N. Hubbard, S. S. Todd, J. F. Messerly, I. A. Hossenlopp, F. R. Frow, J. P. Dawson, and Guy Waddington, *J. Am. Chem. Soc.* **81**, 5884 (1959); Compare also: R. N. Selby and J. G. Aston, *J. Am. Chem. Soc.* **80**, 5070 (1958).

[18] F. V. Brutcher, Jr. *et al.*[3c]; compare also K. S. Pitzer[3b]; F. V. Brutcher, Jr., Th. Roberts, S. J. Barr, and N. Pearson, *J. Am. Chem. Soc.* **78**, 1507 (1956).

a position between the *axial* and *equatorial* bonds. The latter form an angle of 54°44′ with the plane.

Both conformations, the envelope and half-chair, do not vary greatly in their energy levels[3b] and can be attained[19] by pseudorotation of any cyclopentane molecule. This was also determined by Hendrickson[20] by machine computation of the energies.

b. Derivatives of Cyclopentane

Brutcher[3c] and Pitzer[3b] arrived at a somewhat modified view for substituted cyclopentane rings with the aid of infrared spectroscopic investigations, nuclear magnetic resonance and dipole moment measurements, as well as on the basis of theoretical considerations. Whereas the nonsubstituted cyclopentane ring has essentially free "pseudorotation," one of the two possible conformations, either the *envelope* or the *half-chair* form is stabilized[3d,3e,17] by replacement of a hydrogen atom by a bulkier substituent. The energy barrier for pseudorotation then is not zero.

As mentioned previously in detail, the Pitzer strain of the entire system is lowered because one carbon atom twists out of the plane of the five-membered ring. The envelope conformation is thus formed. The distance that carbon atom 1 (in III) twists out of the plane depends upon the nature and size of the substituent attached to the equatorial position. Every substituent which increases the energy barrier between two neighboring bonds will force the five-membered ring into the envelope form according to Pitzer.[3b] Thus the nonbonded interactions of the methylene groups on C-2 and C-5 (if the substituent is located on C-1) are at a minimum. The conformation then corresponds very nearly to a *staggered* ethane form on both sides. Methylcyclopentane, for example, can, therefore, exist preferably in conformation (V), because the interactions between the methyl group and the neighboring H atoms are lowest in this conformation. Conformation (V) is stabilized by 0.9 kcal./mole compared to the half-chair form,[3b] in good agreement with the value of 0.75 kcal./mole obtained from entropy measurements[21] of methylcyclopentane. The half-chair form would be less suitable in this case. If the substituent is placed on C-3 or C-4 of the *half-chair*

[19] Compare however C. G. LeFèvre and R. J. LeFèvre, *J. Chem. Soc.* p. 3549 (1956), who have demonstrated a preference of the C_2 form of cyclopentane. The fixation of one position does not agree with the experimentally derived and calculated entropy, which only makes sense with the help of the pseudorotation. Compare ref. 3d, 3e, and 17.

[20] J. D. Hendrickson, *J. Am. Chem. Soc.* **83**, 4537 (1961); cf. also A. Kitaygorodsky. *Tetrahedron* **9**, 183 (1960).

[21] J. E. Kilpatrick, H. G. Werner, C. W. Beckett, K. S. Pitzer, and F. D. Rossini, *J. Res. Natl. Bur. Stand.* **39**, 523 (1947).

form (IV) only one side would correspond to a *staggered* ethane while the other side would remain mostly in the *eclipsed* arrangement.

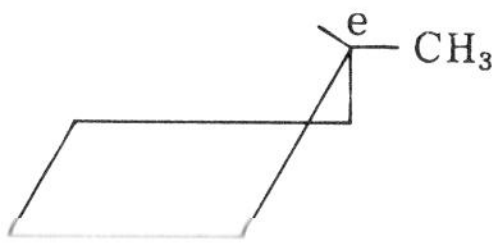

V

The envelope form may be illustrated with an additional example[22,3b]: *cis*-1,3-dimethylcyclopentane[23] is, as determined from heats of formation,[24] 0.59 kcal./mole lower in energy than the *trans* isomer. This difference becomes understandable if one considers the envelope conformation common to both isomers. In *cis*-1,3-dimethylcyclopentane both methyl groups at C-1 and C-3 assume the *equatorial* position (VI), while in the case of the *trans*-isomer one methyl group must assume the unfavorable *axial* position (VII). It will be shown later (see page 189f.) that the energy difference between *axial* and *equatorial* positions of cyclohexane is greater because the *equatorial* position of cyclohexane is preferred to the *equatorial* position of cyclopentane.

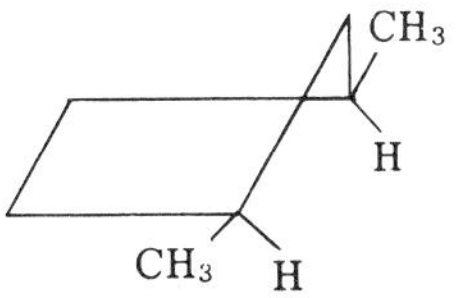

VI

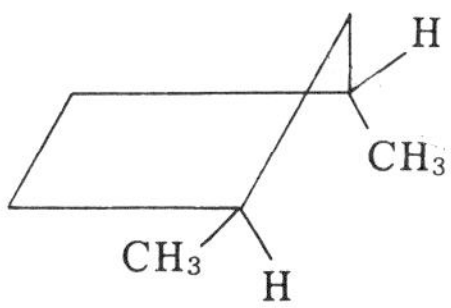

VII

To date, structural investigations on substituted cyclopentanes have been carried out on only a few examples. It remains to be seen whether the model representations described here, which indicate preference for the *envelope* form, are confirmed.

LeFèvre[25] was of the opinion, based on measurements of the molar Kerr constants of halogenated cyclopentanes, that the preferred form was the one in which there were not four carbon atoms in one plane, e.g. the envelope-form. On the other hand, infrared studies of cyclo-

---

[22] J. N. Haresnape, *Chem. & Ind. (London)* p. 1091 (1953).

[23] For the configurational assignment see S. F. Birch and R. A. Dean, *J. Chem. Soc.* p. 2477 (1953). (Cf. also page 86).

[24] M. B. Epstein, G. M. Barrow, K. S. Pitzer, and F. D. Rossini, *J. Res. Natl. Bur. Stand.* **43**, 245 (1949).

[25] C. G. LeFèvre and R. J. W. LeFèvre, *J. Chem. Soc.* p. 3549 (1956); 3458 (1957).

pentanol[26] disclosed the presence of two C—(OH) stretching vibrations in the region of 1065 and 996 cm^{-1}, which were assigned to the *quasi-equatorial* and *bisectional* positions of the OH group (cf. page 75). This leads to the conclusion that cyclopentanol exists in two conformations. However, according to these investigations,[26a] 2-alkyl-substituted cyclopentanols appear to prefer the envelope form.[26e]

c. Cyclopentanone

The *half-chair* form of cyclopentane (IV) compares with the *half-chair* form of cyclohexene (see page 147). The greatest puckering is reached when C-3 and C-4 (IV) assume a staggered conformation. There is as noted above, one carbon atom above the plane formed by three carbon atoms and the remaining carbon atom the same distance below this plane.[27]

The half-chair form of the five membered ring is preferred by such molecules whose energy barrier between neighboring bonds is lowered by substitution. The oxygen of the carbonyl group would be an example of such a substituent. X-ray investigations of ethylene carbonate are reported[28] in which, however, the same potential barrier is assumed for the oxygen as for a methylene group.[28] Cyclopentanone was calculated[29] as 1.9 kcal/mole lower in energy than cyclopentane by Pitzer[3b] with consideration of the greater angle strain caused by the carbonyl group.

Experimental data for cyclopentanone and substituted cyclopentanones are more numerous than for cyclopentane and its derivatives.[30] The results of infrared and ultraviolet spectroscopy, as well as measurements of the dipole moment and rotatory dispersion of substituted cyclopentanones are best interpreted by assumption of a *half-chair* conformation for cyclopentanone.[31] Measurements of molar Kerr constants and

---

[26] (a) W. Hückel and J. Kurz, *Ann. Chem.* **645**, 194 (1961); (b) G. Chiurdoglu and W. Masschelein, *Bull. Soc. Chim. Belges* **71**, 59 (1962); (c) W. Hückel and R. Bross, *Ann. Chem.* **664**, 1 (1963); (d) W. Hückel, *Bull. Soc. Chim. France* p. 8 (1963); (e) compare also: K. Kozima and W. Suetaka, *J. Chem. Phys.* **35**, 1516 (1961).

[27] Brutcher[3c] calculated that for a *staggered* conformation of C-3 and C-4, the distance of the carbon atoms from the plane is 0.4 Å.

[28] C. J. Brown, *Acta Cryst.* **7**, 92 (1954).

[29] For the heat of combustion of cyclopentanone see P. Sellers and S. Sunner, *Acta Chem. Scand.* **16**, 46 (1962).

[30] From the microwave-spectrum G. Erlandson, *J. Chem. Phys.* **22**, 563 (1954), determined that cyclopentanone is not planar.

[31] See for example (a) F. V. Brutcher, Jr.[3c]; (b) F. V. Brutcher, Jr., T. Roberts, S. J. Barr, and N. Pearson, *J. Am. Chem. Soc.* **78**, 1507 (1956); also (c) J. Fishman[3g]; (d) C. W. Shoppee[3f]; compare also (e) E. J. Corey, *J. Am. Chem. Soc.* **75**, 2301 (1953); (f) R. Mecke, R. Mecke, and A. Lüttringhaus, *Chem. Ber.* **90**, 975 (1957); (g) W. D. Kumler and A. C. Huitric, *J. Am. Chem. Soc.* **78**, 3369 (1956); (h) C. Djerassi, J. Osiecki,

molecular anisotropism indicate a *half-chair* form of cyclopentanone[32,33] as well.

d. Cyclopentene

There are less detailed investigations available concerning the conformation of cyclopentene.

The presence of one double bond in the cyclopentane ring leads to a considerable strain, as shown by the heats of hydrogenation of cyclopentene and cyclohexene. The heat of hydrogenation of cyclopentene is 1.68 kcal./mol lower than that of cyclohexene,[34] but it must be recognized that cyclopentane is at a higher energy level, viz. in comparison with cyclohexene, cyclopentene is less favorable by ~5 kcal./mol. Although the problem of the stability of a double bond in a five-membered ring has been approached from many aspects,[35] the literature permits no general conclusions concerning the energy relations in cyclopentene.

Pitzer *et al.*[36] proposed a planar arrangement for the carbon atoms

R. Rinicker, and B. Rinicker, *ibid.* **80**, 1216 (1958); (i) C. Castinel, G. Chiurdoglu, M. L. Josien, J. Lascombe, and E. Valanduyt, *Bull. Soc. Chim. France* p. 807 (1958); (j) Th. Bürer and H. H. Günthard, *Helv. Chim. Acta* **40**, 2054 (1957); (k) R. Zbinden and H. K. Hall, Jr., *J. Am. Chem. Soc.* **82**, 1215 (1960); (l) G. Allen, P. S. Ellington, and G. D. Meakins, *J. Chem. Soc.* p. 1909 (1960); (m) M. I. Batuev, A. A. Akhrem, A. Kamernitzkii, and A. D. Matveeva, *Isvest. Akad. Nauk SSSR*, p. 1138; cf. *Chem. Abstr.* **1961**, 27129; (n) K. M. Wellman, E. Bunnenberg, and C. Djerassi, *J. Am. Chem. Soc.* **85**, 1870 (1963); cf. also O. H. Wheeler and E. G. de Rodriguez, *J. Org. Chem.* **29**, 718 (1954).

[32] C. G. LeFèvre and R. J. W. LeFèvre, *J. Chem. Soc.* p. 3549 (1956); p. 3458 (1957).

[33] (a) C. G. LeFèvre, R. J. W. LeFèvre, and B. B. Rao, *J. Chem. Soc.* p. 2340 (1959); (b) cf. also M. Löw, *Tetrahedron Letters* 1, 3 (1960).

[34] (a) J. B. Conn, G. B. Kistiakowsky, and E. A. Smith, *J. Am. Chem. Soc.* **61**, 1868 (1939); see also (b) R. B. Turner and W. R. Meador, *ibid.* **79**, 4133 (1957).

[35] See especially H. C. Brown, J. H. Brewster, and H. Shechter, *J. Am. Chem. Soc.* **76**, 467 (1954). Brown set up the following rule: reactions will proceed in such a manner as to favor the formation or retention of an *exo*-double bond in the 5-membered ring and to avoid the formation or retention of an *exo*-double bond in the 6-membered ring systems. Because this rule led to misunderstandings, H. C. Brown, *J. Org. Chem.* **22**, 439 (1957), restated it once more: double bonds which are *exo* to a 5-membered ring are less reactive and more stable than related double bonds *exo* to a 6-membered ring. J. B. Bream, D. C. Eaton, and H. B. Henbest, *J. Chem. Soc.* p. 1974 (1957), (further references there); R. B. Turner and R. H. Garner, *J. Am. Chem. Soc.* **79**, 253 (1957) for example came to the opposite conclusion, that the endocyclic double bond of a five-membered ring is more stable. See also ref. 22b in Chap. 2; B. R. Fleck, *J. Org. Chem.* **22**, 439 (1957); G. Chiurdoglu and Y. Rypens, *Bull. Soc. Chim. Belges* **67**, 185 (1958); W. J. Bailey and W. F. Hale, *J. Am. Chem. Soc.* **81**, 651 (1959); A. C. Cope, D. Ambros, E. Ciganek, C. F. Howell, and Z. Jacura, *ibid.* **81**, 3153 (1959); E. Gil-Av and J. Shabtai, *Chem. & Ind. (London)* p. 1630 (1959).

[36] (a) C. W. Beckett, N. K. Freeman, and R. S. Pitzer, *J. Am. Chem. Soc.* **70**, 4227 (1948) for further references; (b) compare also: F. V. Brutcher, Jr., and E. Longenecker-James, *Diss. Abstr.* p. 1398 (1963).

of cyclopentene on the basis of thermodynamic investigations and an analysis of the available infrared and Raman spectra. On the other hand, they were able to show from calculations of the strain energy of various possible molecular models, that a deviation from the planar structure of 0.3 Å by the carbon atom lying transannular to the double bond is possible without any noticeable change in energy.

Similar investigations were also carried out[36b,37] for mono- and dimethylated cyclopentenes, but the results do not permit an unequivocal statement concerning the detailed structure of the carbon skeleton.[38] With the aid of NMR spectroscopy[38a] and microwave spectras[38b] it also has been shown, that cyclopentene has a nonplanar conformation.

2. Conformation and Reactivity of the Five-Membered Ring

The steric relationships of the cyclopentane ring set forth in Section 1 were obtained mostly from physical measurements and give impetus for confirming the established conformations by chemical means. One would hope to find a relationship between the reactivity and the conformation of five-membered ring compounds.

No unequivocal correlation between reactivity and conformation has been found in comparison with those derived for the six-membered rings (Chap. 5, p. 234). The available experimental results are relatively few. Whereas one group of investigators obtained experimental data,[34b,35,39,40] supporting a structure only slightly divergent from a planar structure, others have come to the conclusion that similarities exist between the reactivities of five-membered and the six-membered rings. These results are related to a similar geometry at the reaction center of both rings. That is, the five-membered ring is regarded as being so puckered that it results in the formation of various individual conformations.[41,42]

---

[37] H. J. Hrostowski and G. C. Pimentel, *J. Am. Chem. Soc.* **75**, 539 (1953).

[38] Compare F. V. Brutcher, Jr. and N. Pearson, *Chem. & Ind.* (*London*) p. 1295 (1957).

[38a] I. J. Lawrenson and F. A. Rushworth, *Proc. Phys. Soc.* (*London*) **72**, 791 (1958).

[38b] G. W. Rathjens, Jr., *J. Chem. Phys.* **36**, 2401 (1962).

[39] W. Hückel and M. Hanack, *Ann. Chem.* **616**, 18 (1958).

[40] W. Hückel, R. Bross, O. Fechtig, H. Feltkamp, S. Geiger, M. Hanack, M. Heinzel, A. Hubele, J. Kurz, M. Maier, D. Maucher, R. Neidlein, and R. B. Rashingkar, *Ann. Chem.* **624**, 142 (1959); compare also E. A. S. Cavel, N. B. Chapman, and M. D. Johnson, *J. Chem. Soc.* p. 1413 (1960).

[41] See for example F. V. Brutcher, Jr., T. Roberts, S. J. Barr, and N. Pearson, *J. Am. Chem. Soc.* **78**, 1507 (1956).

[42] J. Weinstock, S. N. Lewis, and F. G. Bordwell, *J. Am. Chem. Soc.* **78**, 6172 (1956).

a. Conformational Analysis by Means of Elimination and Substitution Reactions

The work of the investigators mentioned last has concerned itself mainly with reactions in which at least two ring carbon atoms and the other atoms participating in the reaction lie in one plane in the transition state. Examples include the opening and closing of epoxide rings,[41] as well as certain elimination reactions of the E2 type[42] (cf. page 236).

An example of an elimination reaction which permits conclusions[42] as to the conformation of the cyclopentane ring is the reaction of *trans*-1,2-dibromocyclopentane with iodide ion to form the olefin. It was possible to demonstrate that *trans*-1,2-dibromocyclopentane reacts more rapidly than *trans*-1,2-dibromocyclohexane. This was done by a determination of the relative rates of bromine elimination of a homologous series (4-, 5-, 6-, 7-, and 8-membered ring) of *trans*-1,2-dibromides.

The iodide-catalyzed elimination of bromine, as an E2 reaction, goes through a transition state in which the four atoms, Br—C—C—Br (see VIII), exist in one plane (coplanar transition state)[43,44] (see also page 236).

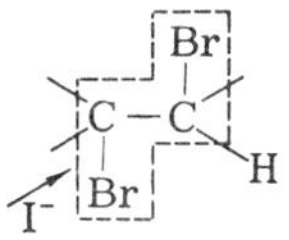

VIII

The two halogen atoms can assume the necessary conformation readily in an open-chain compound such as 1,2-dibromoethane. A relatively unhindered rotation around the C—C bond is possible in this case. The bromine atoms are in a coplanar arrangement in *trans*-

[43] W. G. Young, D. Pressman, and C. D. Coryell, *J. Am. Chem. Soc.* **61**, 1640 (1939); S. Winstein, D. Pressman, and W. G. Young, *ibid.* **61**, 1645 (1939); D. H. R. Barton and E. Miller, *ibid.* **72**, 1066 (1950); G. H. Alt and D. H. R. Barton, *J. Chem. Soc.* p. 4284 (1954); S. J. Cristol, J. Q. Weber, and M. C. Brindell, *J. Am. Chem. Soc.* **78**, 598 (1956).

[44] The claim of a coplanar transition state [see F. O. Rice and E. Teller, *J. Chem. Phys.* **6**, 489 (1938)] implies the following concerning the geometry of a molecule during elimination and addition reactions: Before breaking a molecule into fragments or before the union of two molecules into a new one, the atoms must be located in the approximate position that they will take up in the reaction products. The four atoms involved in the formation of a double bond must lie in a plane or nearly in a plane on the basis of quantum mechanical considerations for a low activation energy. (See S. Glasstone, K. J. Laidler, and H. Eyring, "The Theory of Rate Processes," p. 90. McGraw-Hill, New York, 1941. M. L. Dhar, E. D. Hughes, C. K. Ingold, A. M. M. Mandour, G. A. Maw, and L. I. Woolf, *J. Chem. Soc.* p. 2093 (1948).

1,2-dibromocyclohexane (both bromine atoms *trans-diaxial*, see page 105) and are therefore especially preferred in an elimination reaction. However, if the substituents are located in the *cis*-1,2-position on the cyclohexane ring (*axial-equatorial* positions, see page 105), then a coplanar transition state is not possible and the rate of elimination is slower.

When applied to the problem of the five-membered ring, these considerations signify the following: the very rapid elimination of *trans*-1,2-dibromocyclopentane points to the fact that a *diaxial-trans* arrangement of the bromine atoms can be approached in the transition state of the five-membered ring without a great expenditure of energy. However, this is only possible if, as described above, the five-membered ring does not occur in a planar structure but is in the puckered envelope conformation.[45]

[45] H. C. Brown referred to I-strain (internal strain) to explain the higher absolute reaction rate of the five-membered ring in substitution and elimination reactions in comparison with other ring systems, especially the six-membered ring. This represents a change of internal strain, which results from the change in coordination number of a ring carbon during a chemical reaction. The change of the coordination number of a ring atom may or may not be favored depending on the number of carbon atoms in the ring. If the coordination number is decreased in a reaction, the reaction is favored. If it is increased, then the reaction proceeds with more difficulty. The I-strain is reduced in the first case, by changing the coordination number from 4 to 3 (or sp^3 to sp^2):

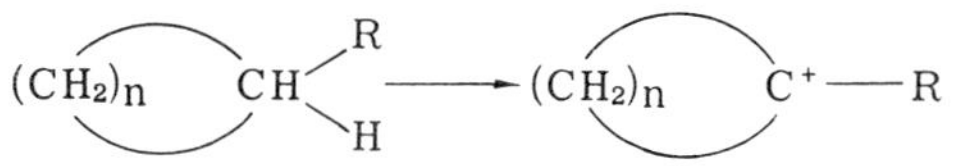

The I-strain is increased in the second case by the increase of the coordination number from 3 to 4 (sp^2 to sp^3) and the reaction proceeds more slowly:

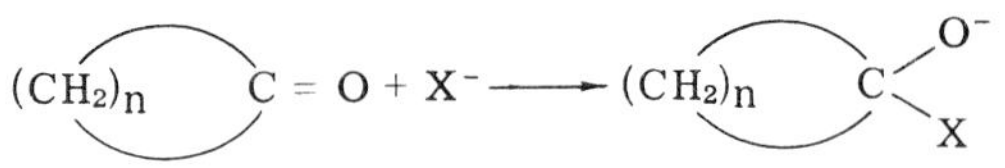

It can be seen that originally only the bond opposition forces were included under I-strain. The rule is therefore only valid for the five- and seven-membered rings, in which considerable Pitzer strain is present. For the six-membered ring the situation is reversed because all the atoms exist in the ideal *staggered* arrangement. A change of the coordination number from 4 to 3 by the loss of a ring substituent results in the loss of the ideal *staggered* arrangement and leads to an increase in I-strain. The reaction would proceed more slowly.

The I-strain in small rings (three- and four-membered rings) is caused mainly by divergence from the normal tetrahedral bond angles.

The concept of I-strain has been modified in recent times so that all strains have been included which originate from: (1) bond opposition forces; (2) distortion of the bond angles; (3) compression of the van der Waals radii. See ref. 95 in Chap. 2; H. C. Brown and M. Borkowski, *ibid.* **74**, 1894 (1952); also ref. 35; H. C. Brown and G. Ham, *J. Am. Chem. Soc.* **78**, 2735 (1956); H. C. Brown, *J. Chem. Soc.* p. 1248 (1956) for further references (see also p. 167). Difference in opinion of I-strain theory, see C. H. Collins and G. S. Hammond, *J. Org. Chem.* **25**, 698 (1960).

A chemical conformational analysis is dependent on certain assumptions which shall be referred to quite often in other sections of this book. The above-mentioned case of *trans*-1,2-dibromocyclopentane depended on the assumption that a coplanar transition state must be traversed. Other investigations of five-membered ring compounds have shown that the coplanar transition state is not always involved in the rate-determining step of elimination reactions. For example, in the elimination reactions of 2-*p*-toluenesulfonyl toluenesulfonates of five- and six-membered ring compounds no great difference in the reaction rate of *cis* and *trans* isomers[46] is observed under certain conditions.

In the case of *trans*-1,2-dibromocyclopentane, the calculated activation energy has a lower level than for open-chain dibromides, and appears to indicate a reaction with a coplanar transition state.[47]

The measured dipole moments of cyclopentane dibromide indicate that the carbon skeleton does not deviate too much from the planar arrangement, i.e. the two bromine atoms are not able to come into an *axial-trans* position. The measured dipole moment of *trans*-1,2-dibromocyclopentane ($\mu = 1.6$)[48] indicates a quasi-*axial* position of the bromine.[49]

Aside from the detailed investigations described for the elimination reaction of dibromides, other results can be evaluated in favor of the conformations of the five-membered ring described above in Section 1. In these studies as well, reactions are employed which go through a coplanar transition state; for instance the opening and closing of the epoxide ring[41,49a] or the solvolysis of *cis*- and *trans*-2-chlorocyclopentyl sulfide.[50] Attempts have been made to draw conclusions as to the conformation of the carbon skeleton or the conformation of the transition state.

---

[46] J. Weinstock, R. G. Pearson, and F. G. Bordwell, *J. Am. Chem. Soc.* **78**, 3468, 3473 (1956); *ibid.* **76**, 4749 (1954); F. G. Bordwell and R. J. Kern, *ibid.* **77**, 1141 (1955), F. G. Bordwell and M. L. Peterson, *ibid.* **77**, 1145 (1955); J. Weinstock and F. G. Bordwell,; *ibid.* **77**, 6706 (1955); see also ref. 22b in Chap. 2.

[47] The activation energy of *trans*-9,10-dibromodecalin, which amounts to 20.9 kcal., should not go unmentioned in this connection. In spite of the definite coplanar *trans* arrangement of the bromine atoms (see page 157) this value is clearly higher than the values cited by Bordwell[42] for the dibromocyclanes. Any conclusions from this have not yet been established with certainty: W. Hückel and H. Waiblinger, *Ann. Chem.* **666**, 17 (1963).

[48] W. D. Kumler and A. C. Huitric, *J. Am. Chem. Soc.* **78**, 3369 (1956); W. D. Kumler, A. C. Huitric, and H. K. Hall, *ibid.* **78**, 4345 (1956).

[49] Measurements of the dipole moment of *trans*-1,2-dibromocyclohexane have given, depending on the solvent, values between 2.16 and 1.76 D. If the bromide atoms are *diaxial*, as in *trans*-9,10-dibromodecalin, then a value of zero is obtained. [Measured value: $\mu = 0.2$ to 0.4 D. (J. A. van der Linden, Dissertation, Leiden, 1958).]

[49a] F. V. Brutcher, Jr. and T. Roberts, *Abstract of Papers, 127th Meeting, Am. Chem. Soc., Cincinnati* 1955, p. 39N.

[50] H. L. Goering and K. L. Howe, *J. Am. Chem. Soc.* **79**, 6542 (1957).

The results of the conformational analysis of the five-membered ring by chemical means have been explained without the aid of the conformations deduced in Section 1 for a major part of the available investigations. There is no longer any objection to the claim that the cyclopentane ring is not planar. The question remains: are the deviations from the planar arrangement shown in Section 1 actually a necessary factor in the reactivity of the cyclopentane ring? Further investigations are obviously needed.

As mentioned above, data are available[51] which cannot be interpreted in terms of a planar cyclopentane ring. The deviation of the ring from the planar structure however was not considered so great that it leads to characteristic differences in the reaction rate of substituents as in the case of *axial* and *equatorial* substituents of the cyclohexane ring[52] (see page 259f.).

W. Hückel *et al.*[40,53] studied the solvolysis of the toluenesulfonates of various alkylated cyclopentanols and determined reaction rates, activation energies, and the resulting reaction products. The results were compared with those of the corresponding alkylated cyclohexanols.[54]

Briefly summarized, the following results were obtained: the solvolyses of cyclopentyl tosylates is significantly faster than analogous cyclohexyl tosylates in all solvents studied.[40,53,55] The ethanolysis of cyclopentyl tosylate itself affords nearly exclusively ether resulting from displacement, whereas cyclohexyl tosylate gives mainly elimination (e.g., cyclohexene). A 2-alkyl substituent *cis* to the toluenesulfonate group increases the reaction rate over the unsubstituted compound by a factor of 3 to 5. A *trans*-2-substituent decreases the rate by one-half to one-quarter.

The rate ratio *cis*/*trans* for the corresponding cyclohexane compounds amounts to about 100, i.e., the *cis*-compound being solvolysed much more rapidly. The *axial* toluenesulfonate group of the *cis* compound

---

[51] The experiments of H. C. Brown *et al.*[45] concerning the reactivity of the cyclopentane ring permit only a limited application for a detailed conformational analysis. Although Brown notes that the cyclopentane ring is considered nonplanar, special assumptions concerning the conformation of the cyclopentane ring were not necessary for the interpretation of his experimental results.

[52] See also A. C. Cope, C. L. Bumgardner, and E. E. Schweizer, *J. Am. Chem. Soc.* **79**, 4729 (1957).

[53] W. Hückel and E. Mögle, *Ann. Chem.* **649**, 13 (1961); W. Hückel and R. Bross, ref. 26c; W. Hückel and M. Hanack, *ibid.* **616**, 18 (1958).

[54] Measurements of the saponification rates of carboxylic acid esters of various alkylated cyclopentanols: G. Vavon, *in* "Traité de chimie organique" (V. Grignard, ed.), Vol. II, p. 934. Masson, Paris, 1936.

[55] See also S. Winstein, B. K. Morse, E. Grunwald, H. W. Jones, J. Corse, D. Trifan, and H. Marshall, *J. Am. Chem. Soc.* **74**, 1127 (1952).

is solvolysed more rapidly for reasons explained on page 260. In the case of the 2-alkyl cyclopentyl toluenesulfonates, the ratio between the *cis* and *trans* compound lies between 7 and 10,[40,53] i.e., the *cis* compound is solvolysed more rapidly. This may be explained by assuming a puckered conformation for the cyclopentane ring with an *quasi-axial* position of the toluenesulfonate group in the *cis*-1,2 compound. *Trans*-2-alkyl cyclopentyl toluenesulfonates react about 30 to 40 times faster than the analogous cyclohexane compounds; but the activation energy is about 3 kcal. lower.

The substitution and elimination reactions occurring concurrently during the solvolysis of the toluenesulfonates (for the mechanism see page 260) is shifted in favor of the substitution reaction for the *trans*-2-alkylcyclopentane compounds. Yet considerable amounts (between 30 and 70%) of unsaturated hydrocarbon are still formed. The explanation could be found in the fact that, as a result of the *quasi-axial* position of the substituents on the cyclopentane ring in comparison to the *equatorial* position on the cyclohexane ring, backside attack of the solvent is favored in an S_N2 reaction with Walden inversion.

The *cis*-2-alkyltoluenesulfonates have about the same activation energies, but react 3–4 times more rapidly than the cyclohexyl compounds. The elimination reaction is favored more strongly than for the cyclohexane ring.

To draw conclusions concerning the conformation of the molecule from the above facts offers similar uncertainties as in the case of dibromocyclopentane. From this it can be said that a conformational analysis by chemical means must always be regarded with the necessary care due to the many inherent assumptions and uncertainties. This is especially true when experimental data is limited, as with cyclopentane compounds.

In the reaction of cyclopentylamines with nitrous acid[56] no parallelism with cyclohexylamines has yet resulted. Nor can any conclusions regarding conformational analysis be drawn from this reaction as in the cyclohexyl series.

b. Conformational Analysis through Other Reactions

Experimental data for other reactions is too limited to be used for conformational analysis with unequivocal value. This is especially true for the equilibrium measurements of stereoisomeric alcohols, a reaction which has been used with success to determine relative thermodynamic stabilities of cyclohexanols (see page 97). In the few examples studied,

[56] W. Hückel and R. Kupka, *Chem. Ber.* **89**, 1694 (1956); W. Hückel and G. Ude, *ibid.* **94**, 1026 (1961).

the equilibrium favors the *trans* compound. However, it is quite dependent on the nature of the substituents which are located next to the hydroxyl group.[57]

Reduction of 2-alkyl-ketones with sodium in alcohol leads predominantly to the *trans*-alcohol.[57] Reduction of ketones with lithium aluminum hydride yields also the thermodynamically more stable *trans*-alcohols in greater amounts[57,58] (see page 269f.).

3. Correlation between Physical Properties and Configuration

The frequently used Auwers-Skita[59] rule for the determination of the configuration of alicyclic compounds has been valid for the 1,2-alkyl cyclopentanes investigated to date[60] (1,2-dimethylcyclopentane; 1,2-diethylcyclopentane). According to the rule for alicyclic *cis-trans* isomers, the *cis* compound has the higher boiling point, the higher refractive index and density but the lower molar refraction. The Auwers-Skita rule does not hold for 1,3-dimethylcyclopentane. The compound originally designated as *cis*-1,3-dimethylcyclopentane on the basis of its higher boiling point, refractive index, and density, could be separated into optical antipodes.[61] This is only possible for a *trans* compound (see page 9). Hence, the Auwers-Skita rule must be reversed in this case.

The rule is also not applicable for cyclopentanols alkylated at position 2. These properties can be related to differences in intermolecular association of the individual alcohols.

II. CONFORMATION OF CYCLOHEXANE AND SUBSTITUTED CYCLOHEXANES

Conformational analysis has been applied most intensively to cyclohexane and its substitution products, as well as to compounds derived

---

[57] (a) G. Vavon and M. Barbier, *Bull. Soc. Chim. France* [4] **49**, 572 (1931); (b) J. B. Umland and B. W. Williams, *J. Org. Chem.* **21**, 1302 (1956); (c) W. Hückel *et al.*, *Ann. Chem.* **616**, 46 (1958).

[58] J. B. Umland and M. I. Jefraim, *J. Am. Chem. Soc.* **78**, 2788 (1956).

[59] (a) K. v. Auwers, *Ann. Chem.* **420**, 91 (1920); (b) A. Skita, *Ber.* **53**, 1792 (1920). Compare the modified conformational rule (page 143).

[60] See G. Chiurdoglu, *Bull. Soc. Chim. Belges* **47**, 363 (1938), *ibid.* **42**, 347 (1933); *ibid.* **53**, 45, 55 (1944); cf. also B. A. Kasanski, A. L. Lieberman, N. I. Tjunkina, and I. M. Kutonetzowa, *Dokl. Akad. Nauk SSSR* **122**, 1025 (1958).

[61] S. J. Birch and R. A. Dean, *J. Chem. Soc.* p. 2477 (1953).

from it, such as decalin or the steroids. Many of the important facts, especially the relationship between reactivity and conformation of a compound, have been obtained by studies of such compounds (see Chap. 5, p. 234). The cyclohexane ring, therefore, occupies a central position in the conformational considerations of alicyclic compounds. Because of the extraordinary number of publications in the field of the stereochemistry of the cyclohexane ring we are forced right at the beginning to certain limitations so as not to exceed the scope and purpose of this book. Therefore, only such investigations which are concerned with the actual conformational analysis of the cyclohexane ring shall be described; that is, such investigations in which the relationship between conformation and reactivity can be recognized with some certainty. As a result of a series of outstanding review articles concerning the conformation of the six-membered ring,[62] the major emphasis will be placed on the newer literature.

1. Cyclohexane

The energy considerations of the cyclohexane ring system described in Chap. 2 (page 44) led to the conclusion that of the two possible conformations (the intermediate forms are neglected at present), the chair form is energetically more preferred than the boat form. This is due to its maximum of *staggered* conformations which the six methylene groups assume in relation to each other. In the following considerations we shall therefore deal with the chair form of the cyclohexane ring, exclusively, as the fundamental form. There will be exceptions under special circumstances, as has already been indicated (page 46). They will be dealt with in greater detail in Chap. 6.

In the chair form of cyclohexane, we differentiate between two types of C—H bonds, i.e., two types of bonds from the ring carbon atom to the substituent. They differ in their geometric relation to the trigonal axis, or the imaginary plane of the cyclohexane ring:

(1) Six *axial* bonds (*a*). They lie parallel to the trigonal axis, or perpendicular to the plane.

(2) Six *equatorial* bonds (*e*). They form an angle of 109°28′ with the trigonal axis, or lie roughly in the plane. Actually they do not lie exactly in the plane, but alternately above and below it (IX and Fig. 1).

---

[62] Cf. references, 6, 7, 10b, 11 in Chap. 2, and also Vol. 2 of Klyne's and de la Mare's *Progr. in Stereochem.*; E. L. Eliel, *J. Chem. Educ.* **37**, 126 (1960); H. H. Lau, *Angew. Chem.* **73**, 423 (1961); compare also E. L. Eliel, "Stereochemistry of Carbon Compounds." McGraw-Hill, New York, 1962.

The definitive difference between these two types of bonds goes back[63] to Hassel[64] and Pitzer.[65]

Hassel[64] designated the bonds situated perpendicular to the plane with "ϵ" (from ϵστηχως = erect), the ones lying approximately in the plane with "κ" (from κϵίμϵνος = lying). Pitzer[66] introduced the terms

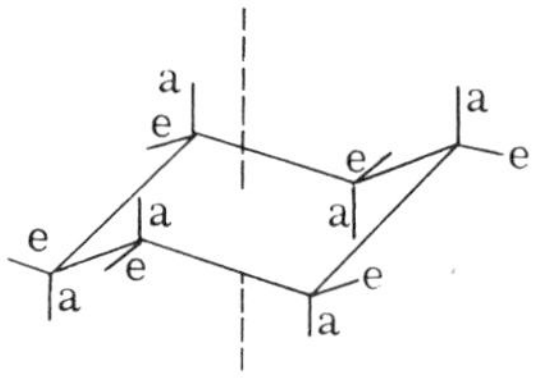

IX

FIG. 1.

polar (p) (pointing to the pole) and *equatorial* (e). The term *polar* (p), which led to ambiguity because a substituent can also be electro-"polar," is no longer used. As a result of a proposal by Barton *et al.*[67] it was

[63] Cf. also K. W. F. Kohlrausch, A. W. Reitz, and W. Stockmair, *Z. Physik. Chem. (Leipzig)* **B32**, 229 (1936).

[64] O. Hassel, *Tidsskr. Kjemi, Bergvesen Met.* **3**, 32 (1943).

[65] C. W. Beckett, K. S. Pitzer, and R. Spitzer, *J. Am. Chem. Soc.* **69**, 2488 (1947).

[66] See for example K. S. Pitzer, *Chem. Rev.* **27**, 39 (1940).

[67] D. H. R. Barton, O. Hassel, K. S. Pitzer, and V. Prelog, *Science* **119**, 49 (1953); *Nature* **172**, 1096 (1953).

replaced[68] by the term *axial* (a), which has been in use since then. The abbreviations "a" for *axial* and "e" for *equatorial* were changed[69] in the British literature to "ax" and "eq" in the year 1959. We shall, however, continue to use the abbreviations "a" and "e" in the following.

2. Monosubstituted Cyclohexanes

a. General

A monosubstituted cyclohexane can exist in two different forms. The substituent can assume the *equatorial* (X) or the *axial* (XI) position.

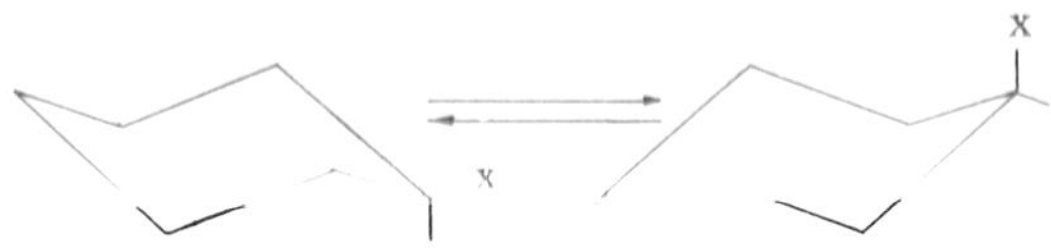

X XI

As shown by (X) and (XI), each substituent can be brought from the e-position to the a-position by conversion of one chair form into the other. In this way the flexible form of the ring is traversed.

As has been mentioned frequently, the energy difference[70] between conformers is too small to permit their individual isolation. Only one form can be obtained in each case. The position in which the substituent is found, i.e., whether *axial* or *equatorial*, or if the compound exists as an equilibrium mixture of both, is a function which depends upon the substituent. Not only steric effects of the substituent, but also electrostatic factors play important roles. Generally the rule holds that at equilibrium that conformation is preferred in which the substituent assumes the *equatorial* position. Further, the temperature as well as the solvent must be considered for the position of the equilibrium.

The thermodynamically more stable position of an *equatorial* substituent on the cyclohexane ring was demonstrated relatively early by

[68] Occasionally one finds *azimuthal* instead of the term *axial*. Cf. for example Eugen Müller, "Neuere Anschauungen der organischen Chemie," 2nd ed., p. 36. Springer, Berlin, 1957.

[69] *Proc. Chem. Soc.* p. 4 (1959).

[70] By "energy difference" we understand the difference in potential energy in the following. It is assumed, in this regard, that the difference in free energy (ΔF) is equal to the potential energy, i.e., ΔH is set equal to the potential energy difference. In addition, it must be assumed that $\Delta S = 0$, which should apply at least for the conformers of methyl cyclohexane. Cf. also E. L. Eliel, "Stereochemistry of Carbon Compounds," p. 120. McGraw-Hill, New York, 1962.

Hassel with electron diffraction studies of cyclohexylchloride,[71] (cf. also page 100). From these investigations it follows that the *axial* position of a substituent is less favored, because the steric interaction of an *axial* substituent with the *axial* hydrogens in positions 3 and 5 is greater than the disturbance of an *equatorial* substituent toward the *equatorial* hydrogen atoms on neighboring carbon atoms (XII). In the case of cyclohexyl chloride, the distance between the *axial* chlorine and the *axial* hydrogens in positions 3 and 5 amounts to 2.57 Å. This amount is smaller than the sum of the van der Waals radii of both atoms. The distance between the *equatorial* chlorine and the *equatorial* hydrogens on the neighboring carbon atom corresponds to the distance in the similar *staggered* form of chlorinated aliphatic hydrocarbons (XII).

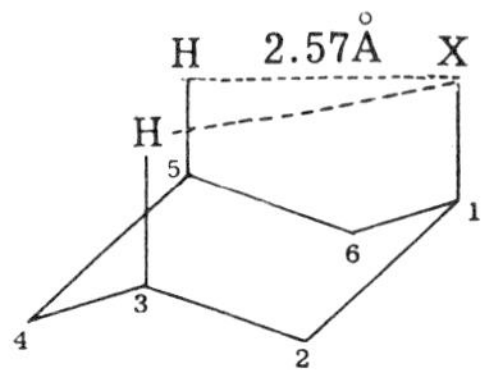

XII

The fact that a substituent such as the methyl group in methyl cyclohexane is less stable in the *axial* position as compared to the *equatorial* position may be clarified also in the following way: In the *axial* conformer (XIII) there are interactions present (shown by the arrows) which correspond to the interactions of the *gauche* form of butane (see page 29). However, in the *equatorial* conformer (XIV) these interactions are not present. The *equatorial* position is, therefore, preferred by the substituent, which has been confirmed for gaseous methylcyclohexane by

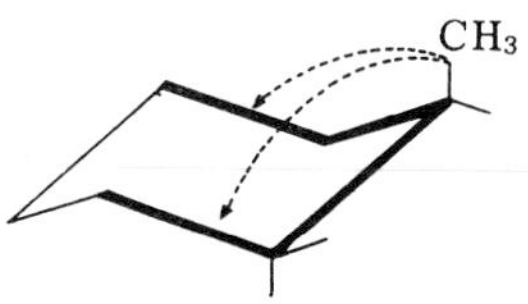

XIII

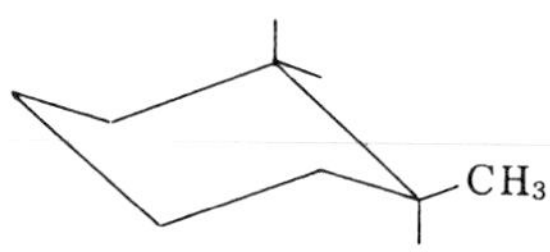

XIV

[71] (a) O. Hassel and H. Viervoll, *Acta Chem. Scand.* **1**, 149 (1947); (b) *Tidsskr. Kjemi Bergvesen* **3**, 35 (1943). (c) For newer, more accurate measurements of cyclohexylchloride with the aid of electron diffraction, see V. A. Atkinson, *Acta Chem. Scand.* **15**, 599 (1961), cf. page 100. (d) Cf. the atomic distance calculations of R. Cornubert, *Bull. Soc. Chim. France* p. 996 (1956).

[72] O. Hassel, *Research (London)* **3**, 504 (1950).

[73] M. Larnaudie, *J. Phys. Radium* **15**, 650 (1954).

electron diffraction[11,72] and by infrared investigations.[73] According to both methods, methylcyclohexane exists almost entirely in the *equatorial* form. In the case in which the distance between the nonbonded atoms is smaller than the sum of the van der Waals radii, strain arises in the molecule that makes itself noticeable by a rise in the energy of the entire molecule. The result can be a certain deformation of the valence angle, as was shown by electron diffraction of various substituted cyclohexanes.[71c,74,75] This effect is known as the "buttressing" effect (compare page 283).

As we shall see further on, the energy difference between the *axial* and *equatorial* positions is relatively small in spite of a preference for the *equatorial* position. It is normally considerably smaller than the activation energy of most reactions. The result is that the preference of a single conformation of a substituent is no longer guaranteed in chemical reactions. The substituent can react entirely in the less stable conformation during the reaction.[76] The same can take place at higher temperatures.

The results of conformational analysis obtained by physical methods, whereby the structure of the molecule is not disturbed, can not always be compared with the results of chemical conformational analysis, at least not in the case of cyclohexane and its derivatives. It is emphasized at this point that for this reason the chemical conformational analysis of monocyclic cyclohexane derivatives is fraught with great uncertainties. For condensed ring systems (decalin, steroids, cf. page 216), which are more rigid, the transformation into a less stable conformation requires a larger expenditure of energy. If, for geometrical reasons the change is impossible, then such uncertainties are fewer.

If we replace the methyl group by a bulkier substituent, then the energy necessary to bring this group into the less stable *axial* position becomes greater. It can be seen from a scale model that it is still possible to have a methyl group in an *axial* position, but it would seem to be less favorable to have an isopropyl group in this position[77] [but compare the free energy difference between an *equatorial* and *axial* methyl and isopropyl group (p. 93)]. If we use a still bulkier substituent, e.g., the *tert*-butyl group, then an *axial* orientation of this substituent is no longer possible because of crowding. The cyclohexane molecule then

---

[74] O. Hassel and B. Ottar, *Acta Chem. Scand.* 1, 929 (1947).

[75] Cf. for example O. Hassel and E. Wang Lund, *Acta Cryst.* 5, 309 (1949); V. A. Atkinson and O. Hassel, *Acta Chem. Scand.* 13, 1737 (1959).

[76] Compare D. H. R. Barton, *Experientia* 6, 316 (1950).

[77] This is only conceivable when the isopropyl group is oriented in such a way that both of its methyl groups are turned away from the *axial* hydrogens in positions 3 and 5, and free rotation is practically eliminated.

exists in a uniform conformation. This thought, originating with Winstein and Holness,[78] had an extremely important significance for chemical conformational analysis. Thereby the possibility of investigation of a cyclohexane molecule that could exist in only one conformation was arrived at for the first time (see page 96).[79]

b. Conformational Equilibria of Monosubstituted Cyclohexanes and the Free Energy Difference between *Equatorial* and *Axial* Substituents

The free energy differences ΔF between *equatorial* and *axial* oriented substituents have been determined experimentally for some substituents in various ways. For others, only rough estimates are available (compare Table I, p. 103).

(*i*) *Alkyl Groups*

For the methyl group, a value in use for a long time, amounts to -1.8 kcal./mole which Pitzer *et al.* determined on the basis of the internal rotation barrier in acyclic hydrocarbons (see p. 27).

As can be seen from XIII, two *gauche* butane interactions are present in the *axial* conformer of methyl cyclohexane. As we have seen on page 31 the energy difference between the *gauche* and *trans* form of butane amounts to 0.8 to 0.9 kcal./mole, so that the energy difference between the *equatorial* and *axial* form of methyl cyclohexane amounts to twice as much or -1.6 to -1.8 kcal./mole. The value of 1.8 kcal./mole is in good agreement with the value obtained from the heat of combustions of the dimethyl cyclohexanes, $\Delta H = 1.94$ kcal./mole,[80] and has been approximately confirmed by the determination of the equilibrium of epimerization of alkylcyclohexanols[80a] (-1.5 to -1.8 kcal./mole).

Allinger and co-workers were able to show[80b] that a semi-quantitative method for the determination of the energies of a particular group in the *axial* position relative to the *equatorial* position may be carried out by calculating the energies resulting from the van der Waals interactions[80c] The interaction energies of every atom pair for which the

---

[78] S. Winstein and N. J. Holness, *J. Am. Chem. Soc.* **77**, 5562 (1955).

[79] Cf. N. L. Allinger and L. A. Freiberg, *J. Am. Chem. Soc.* **82**, 2393 (1960).

[80] E. J. Prosen, H. W. Johnson, and F. D. Rossini, *J. Res. Natl. Bur. Stand.* **39**, 173 (1947).

[80a] E. L. Eliel and M. N. Rerick, *J. Am. Chem. Soc.* **82**, 1367 (1960); cf. also ref. 23 in Chap. 2.

[80b] N. L. Allinger and L. A. Freiberg, *J. Am. Chem. Soc.* **84**, 2201 (1962); N. L. Allinger and W. Szkrybalo, *J. Org. Chem.* **27**, 722 (1962); N. L. Allinger, J. Allinger, L. A. Freiberg, R. Czaja and N. A. LeBel, *J. Am. Chem. Soc.* **82**, 5876 (1960).

[80c] T. L. Hill, *J. Chem. Phys.* **16**, 399 (1948).

interatomic distances in both conformations is different are determined and summed up. However, exact measurements are not possible because neither the van der Waals functions nor the space intervals are known with sufficient accuracy.

With the aid of nuclear magnetic resonance spectroscopy a value of -1.74 kcal./mole was found.[81] A somewhat lower value was found for the isomerization of 1,3,5-trimethylcyclohexane, amounting to $\Delta F_{270} = -1.41$ kcal./mole.[82a] The same value was obtained by equilibrating the isomeric *cis*- and *trans*-1,2-, 1,3-, and 1,4-dimethylcyclohexanes in the presence of hydrogen over a nickel catalyst: $\Delta F_{250} = -1.44$ to -1.35 kcal./mole.[82b] Other investigators also use the lower value of $\Delta F = -1.4$ kcal./mole.[82c] On the basis of -1.8 kcal./mole as the difference in free energy between *equatorial* and *axial* conformers, $\sim$95% methylcyclohexane exists in the *equatorial* form and $\sim$5% in the *axial* form. (For the calculation see page 94f.).

From the equilibration between *cis*- and *trans*-1,3-diethyl- and 1,4- diethylcyclohexane at various temperatures with a palladium catalyst present, the difference in free energy between an *equatorial* and an *axial* ethyl group was determined to be -1.86 kcal./mole[83] ($\Delta H = 1.71$ kcal./mole; $\Delta S = -0.51$ e.u.). Compared to earlier estimations of -2.1 kcal./mole[78], ΔF for the ethyl group differs only slightly from ΔF for the methyl group, if we use as a basis the above-quoted value of -1.8 kcal./mole.

Up to now, higher values for the energy difference between the *equatorial* and *axial* positions of the isopropyl group have been assumed. These varied between -2.5 to -3.5 kcal./mole because greater bulk was attributed to the isopropyl group in comparison to the methyl or ethyl groups.[84] The equilibration of the *cis-trans* isomers of 1,3-diisopropyl cyclohexane over a palladium catalyst at various temperatures resulted in the determination of the free energy of isomerization of *trans*- to *cis*-1,3-diisopropylcyclohexane.[85] Considering the symmetry properties of the molecule (cf. page 107), it was determined that the difference in free energy amounted to -2.10 kcal./mole. In comparison to

[81] A. H. Lewin and S. Winstein, *J. Am. Chem. Soc.* **84**, 2464 (1962).

[82] (a) C. J. Egan and W. C. Buss, *J. Phys. Chem.* **63**, 1887 (1959); (b) C. Boelhouwer, G. A. M. Diepen, J. van Elk, P. Th. van Raaij, and H. I. Waterman, *Brennstoff-Chem.* **39**, 299 (1958); (c) W. Masschelein, "Etude conformationalle de quelques alcoylcyclohexanols secondaires." Thèse, Université libre de Bruxelles, 1960–1961.

[83] N. L. Allinger and S. Hu, *J. Am. Chem. Soc.* **84**, 370 (1962).

[84] Cf. ref. 78; D. S. Noyce and L. J. Dolby, *J. Org. Chem.* **26**, 3619 (1961); H. van Bekkum, P. E. Verkade, and B. M. Wepster, *Koninkl. Ned. Akad. Wetenschap. Proc., Ser. B* **64**, 161 (1961); cf. also R. D. Stolow, *J. Am. Chem. Soc.* **86**, 2170 (1964).

[85] N. L. Allinger, L. A. Freiberg, and S. Hu, *J. Am. Chem. Soc.* **84**, 2836 (1962); N. L. Allinger and S. Hu, *J. Org. Chem.* **27**, 3417 (1962).

the value for ethyl group, this value is only slightly higher. This would indicate that, contrary to earlier assumptions, the isopropyl group actually differs only slightly in its spatial requirements compared to the methyl and ethyl group.

This result was verified by means of nuclear magnetic resonance spectroscopy in which the ethyl group ($\Delta F = -1.74$ kcal./mole) is practically the same as the methyl group, and the isopropyl group possessed only a slightly higher value ($\Delta F = -2.28$ kcal./mole).[81]

(ii) Hydroxyl Groups

More investigations are available concerning the ΔF values of the OH group. The derived values fluctuate between -0.3 and -0.96 kcal./mole depending upon the method employed.

Before going into greater detail concerning the ΔF values of the OH group, we will present some examples of how, in general, a determination of the free energy difference of a substituent can be undertaken, using the OH group as a model.

The conformational equilibrium constant K is given by: $K = N_e/N_a$, where N_e and N_a are the mole fractions of the substituted cyclohexane in the *equatorial* and *axial* conformations, respectively. From the equilibrium constant K, the free energy difference can easily be determined as $\Delta F = -RT \ln K$.

The real problem is the determination of the equilibrium constant. From the investigations of Winstein and Holness[78], and Eliel and coworkers[86,87] based on cyclohexane derivatives which have no conformational equilibrium (i.e., exist in a fixed conformation), the ratio N_e/N_a can be determined. They found a relation between the reaction rate constant, k, for any reaction of a substituted cyclohexane molecule and the individual rate constants k_a and k_e. The observed reaction rate of a compound which exists in two conformers made up of two parts; that part which reacts from the e-position and that part which reacts from the a-position. The ratio is determined by the amounts of the e- and a-form at equilibrium. The rate of reaction for an axial and equatorial substituent is measured by the constants k_e and k_a, i.e.,

$$k = N_e \cdot k_e + N_a \cdot k_a \tag{1}$$

Independent of Winstein, Eliel[86,87] obtained the equivalent relation

$$k = \frac{(k_e \cdot K + k_a)}{(K + 1)} \tag{2}$$

[86] See for example E. L. Eliel and C. A. Lukach, *J. Am. Chem. Soc.* **79**, 5986 (1957).
[87] E. L. Eliel, *J. Chem. Educ.* **37**, 126 (1960).

where K is the conformational equilibrium constant.[88] One obtains from this

$$K = \frac{(k_a - k)}{(k - k_e)} \tag{3}$$

i.e., the equilibrium constant K is expressed by the rate constants k_a, k_e and k.

Derivation of Eq. (2) was explained by Eliel[87] in the following manner: k_e is the specific reaction rate of the *equatorial* conformer, $[E]$ is the concentration of the conformer in which the reacting group assumes the *equatorial* position; k_a is the specific reaction rate and $[A]$ the concentration of the axial conformer. Then the over-all rate of a reaction is:

$$\text{Rate} = k_e[E][P] + k_a[A][P] \tag{4}$$

$[P]$ is the product of all concentrations of other participants in the reaction.

If $[C] = [E] + [A]$ is equal to the stoichometric concentration of a substituted cyclohexane and k is the observed specific rate, then

$$\text{Rate} = k[C][P] \tag{5}$$

When Eq. (4) is equated with Eq. (5), we obtain

$$k[C] = k_e[E] + k_a[A] \tag{6}$$

and since $[C] = [E] + [A]$, we obtain by dividing by $[A]$

$$\frac{k([E] + [A])}{[A]} = k_e \frac{[E]}{[A]} + k_a \tag{7}$$

In addition, $[E]/[A] = K$ (conformational equilibrium constant); if this is substituted into Eq. (7), we have:

$$k(K + 1) = k_e K + k_a$$

or from this Eq. (2):

$$k = \frac{k_e K + k_a}{K + 1} \tag{2}$$

[88] That Eq. (1) and (2) are equivalent becomes apparent when we transform them: since $K = N_e/N_a$, then $K + 1 = (N_e + N_a)/N_a$ and since $N_e + N_a = 1$, it follows that $K + 1 = 1/N_a$. If in Eq. (2) $K + 1$ is replaced by $1/N_a$ and K by N_e/N_a, then Eq. (1) results.

Equation (1) was used by Winstein and Holness[89] for the first time for the determination of the ΔF value of the OH group, namely from the oxidation rates with chromic acid in acetic acid.

The required values of k, k_a, and k_e for Eq. (1), respectively, (2), can be determined as follows: k is the reaction rate constant for the oxidation of cyclohexanol (at a given temperature); for the determination of k_a and k_e, *cis*- and *trans*-4-*tert*-butylcyclohexanol are used. As already described on page 91, the *tert*-butyl group assumes the *equatorial* position exclusively. A conversion to the *axial* position is not possible for steric reasons. This indicates that the conformation of the *cis* and *trans* isomeric 4-*tert*-butylcyclohexanols is set, i.e., the OH group is in a fixed position. As shall be explained in more detail on page 104, the OH group assumes the *axial* position in *cis*-4-*tert*-butylcyclohexanol (XV) and the *equatorial* position in the *trans* compound (XVI).

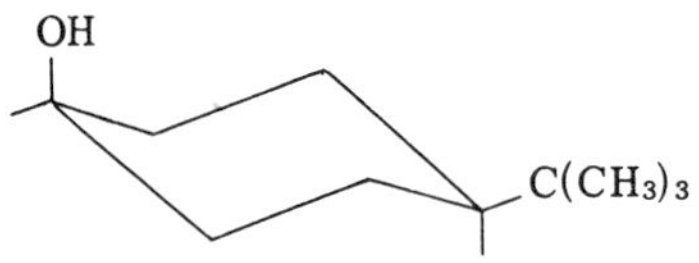

XV

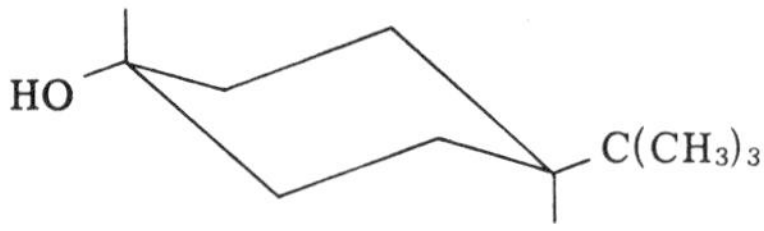

XVI

Now k_a and k_e are the rate constants for the oxidation of these two conformationally fixed alcohols.[90] An average value of —0.8 kcal./mole for ΔF of the OH group was calculated from the measured reaction rate constant of the oxidation of cyclohexanol (k), *cis*-4-*tert*-butylcyclohexanol (k_a) and *trans*-4-*tert*-butyl cyclohexanol (k_e) at various temperatures.

The assumption for such a comparison between the conformationally homogeneous 4-*tert*-butylcyclohexanols on the one hand and cyclohexanol on the other is, naturally, that the *tert*-butyl group in position 4 to the OH group does not exert any uncontrolled polar or steric effects on the reaction rate of the OH-group in any type of chemical reaction.[90]

The fact that the measured reaction rate constants of *cis*-3-, *trans*-4-, *trans*-3-, and *cis*-4-*tert*-butyl- as well as *trans*-3- and *cis*-4-methyl-substituted cyclohexane compounds are practically identical, argues against a long-range polar effect of the alkyl groups across the ring. In other words, there is no polar effect with respect to alkyl substitution[89] (cf. page 114). In addition, to clarify this question the pK_a values of

[89] S. Winstein and N. J. Holness, *J. Am. Chem. Soc.* **77**, 5574 (1955).

[90] Cf. ref. 10e in Chap. 2. Conformational analyses on the basis of conformationally fixed *tert*-butylcyclohexane derivatives will be dealt with in more detail on page **114**.

cyclohexane carboxylic acid were compared[91] with the pK_a values of *cis*-3- and *trans*-4-methylcyclohexane carboxylic acids, and it was found that the pK_a values, i.e., acid strengths, were practically the same.

Furthermore, Eliel[86] was able to show that the rates of acetylation of cyclohexanol and 4,4-dimethyl cyclohexanol are practically identical. This precludes a direct steric effect of substituents at C-4 on the reaction rate of a substituent at C-1. On the other hand, Cornubert,[92] on the basis of an examination of molecular models, referred to the fact that an *equatorial tert*-butyl group at C-4 (XVI) could disturb the *axial* hydrogens at C-3 and C-5. Experimental criteria on the size of such a disturbance, which would lead to a distortion of the ring, do not exist. These problems will be discussed in greater detail page 114.

Aside from the determination of the rates of oxidation of the OH-group there are obviously other reactions that can be used, such as the rate of esterification of the OH-group (see page 115). This can be compared with corresponding rates for the epimeric *tert*-butylcyclohexanols in order to determine the conformational equilibrium constant. The acetylation rate with acetic anhydride in pyridine at 25° was investigated.[86] The specific reaction rates are: cyclohexanol, $k = 8.37 \cdot 10^{-5}$ liter mole^{-1}sec^{-1}; *cis*-4-*tert*-butylcyclohexanol (XV), $k_a = 2.89 \cdot 10^{-5}$ liter mole^{-1}sec^{-1}; *trans*-4-*tert*-butylcyclohexanol (XVI), $k_e = 10.65 \cdot 10^{-5}$ liter mole^{-1}sec^{-1}. Using Eq. (3), K amounts to 2.40, corresponding to a free energy difference of -0.52 kcal./mole between *equatorial* and *axial* OH groups in cyclohexanol.

A more direct approach to the determination of the conformational equilibrium constant consists in the equilibration of *cis*- and *trans*-4-*tert*-butylcyclohexanol.[93] Although both are stable configurational isomers, they can be isomerized, for instance, by aluminum isopropylate and an equilibrium is established between the *cis* and *trans* compound.[94] With the assumption described above that *cis*- and *trans*-4-*tert*-butylcyclohexanol are conformationally homogeneous, the equilibrium between the configurational isomers corresponds to the conformational equilibrium between *equatorial* and *axial* OH group. After equilibration, 79% of the more stable *trans*-(XVI) and 21% *cis*-4-*tert* butylcyclohexanol (XV) were found.[93] The equilibrium constant $K = 3.76$ and a free energy difference of -0.96 kcal./mole for the OH group is

[91] (a) J. F. J. Dippy, S. R. C. Hughes, and J. W. Laxton, *J. Chem. Soc.* p. 4102 (1954); (b) R. D. Stolow, *J. Am. Chem. Soc.* **81**, 5806 (1959).

[92] R. Cornubert, *Bull. Soc. Chim. France* (1956), especially p. 1000, and see the references and formulas XIVa and e on page 1001.

[93] E. L. Eliel and R. S. Ro, *J. Am. Chem. Soc.* **79**, 5992 (1957).

[94] Cf. W. G. Dauben, G. J. Fonken, and D. S. Noyce, *J. Am. Chem. Soc.* **78**. 2579 (1956).

easily calculated from this. (These data refer to the solvent isopropanol at 89°, with the assumption that *tert*-butylcyclohexanol exists as the free alcohol due to the presence of an excess of solvent, and not as an alcoholate.) Chiurdoglu and Masschelein, however, were able to show that the position of the equilibrium depends on the conditions under which it is established.[95] Whereas the catalytic equilibration with Raney nickel in the presence of hydrogen leads to a true thermodynamic equilibrium,[95,96] this is not true of the equilibration with metal alcoholates, even when a large excess of solvent is used. Part of the alcohol is always present in the form of the alcoholate or the alcoholate anion. The catalytic equilibration of *cis*- and *trans*-4-*tert*-butylcyclohexanol with Raney nickel and hydrogen actually gave free energy differences which were smaller than the value indicated above. The value found was $\Delta F = -0.47$ to -0.51 kcal./mole. The equilibration of *trans-cis*-2-decalol (*equatorial* OH group, see page 194) with Raney nickel and hydrogen at 100° also gave $\Delta F = -0.56$ kcal./mole.[97]

Of the physical methods available for the determination of the conformational equilibrium constant, infrared spectroscopy is especially mentioned. This method again is based on a comparison with the conformationally homogeneous epimeric 4-*tert*-butylcyclohexanols.[82c,98] *Cis*- and *trans*-4-*tert*-butylcyclohexanol show different absorption bands of the C—(OH) stretching vibrations with the *cis* compound (*axial* OH) at 955 cm.$^{-1}$ and the *trans*- compound (*equatorial* OH) at 1062 cm.$^{-1}$. For cyclohexanol itself the C—(OH) stretching vibration band lies at 1069 cm.$^{-1}$, which would correspond to that portion of cyclohexanol with an *equatorial* OH group. That portion with an *axial* OH group corresponding to the band at 955 cm.$^{-1}$ could not be assigned in the analysis because of the superposition with skeletal frequencies of the cyclohexane ring. Assuming that the absorption intensities of the C—(OH) stretching vibrations in cyclohexanol and in *trans*-4-*tert*-butylcyclohexanol are the same, a comparison of the relative intensities of cyclohexanol at 1069 cm.$^{-1}$ and *trans*-4-*tert* butylcyclohexanol at 1062 cm.$^{-1}$, gave the amount of *equatorial* cyclohe-

---

[95] G. Chiurdoglu and W. Masschelein, *Bull. Soc. Chim. Belges* **70**, 767 (1961); **70**, 782 (1961); G. Chiurdoglu, H. Gonze, and W. Masschelein, *ibid.* **71**, 484 (1962); W. Masschelein, *Experientia* **19**, 520 (1963).

[96] Cf. also R. J. Wicker, *J. Chem. Soc.* p. 2165 (1956); E. G. Peppiatt and R. J. Wicker, *ibid.* p. 3122 (1955).

[97] W. Hückel and D. Rücker, *Ann. Chem.* **666**, 30 (1963). The ΔF value given was calculated from the ratio for *trans-cis*/*trans-trans*-2-decalol.

[98] (a) R. A. Pickering and C. C. Price, *J. Am. Chem. Soc.* **80**, 4931 (1958); (b) G. Chiurdoglu, L. Kleiner, W. Masschelein, and J. Reisse, *Bull. Soc. Chim. Belges* **69**, 143 (1960); G. Chiurdoglu and W. Masschelein, *ibid.* **69**, 154 (1960); **70**, 29 (1961); **70**, 231, 307 (1961); (c) cf. M. Hanack, *Forsch. Fortschr.* **34**, 259 (1960).

xanol to as 62-66%. From this, the difference in free energy is calculated as −0.3 to −0.4 kcal./mole.[98d]

Nuclear magnetic resonance spectroscopy has also been utilized for the determination of ΔF values. Because *axial* and *equatorial* protons exhibit different chemical shifts,[99] these shifts for *axial* (δ_e) (corresponding to the *equatorial* OH group) and *equatorial* protons (δ_a) (corresponding to the *axial* OH group)[100] may be determined with the help of the conformationally homogeneous *trans*- and *cis*-4-*tert*-butylcyclohexanols. It is assumed that the *tert*-butyl group in position 4 has no influence on the chemical shift of the proton in position 1. As a result, from Eq. (3) we obtain[101]:

$$K = \frac{N_e}{N_a} = \frac{\delta_a - \delta}{\delta - \delta_e}$$

Applied to cyclohexanol, the value $\Delta F = -0.78$ kcal./mole was found.[100]

A determination of the ΔF value based on coupling constants of 3,3,4,4,5,5-hexadeuterocyclohexanol yielded higher values: between −1 and −1.25 kcal./mole.[102]

In conclusion the following values for the free energy differences of the OH group are summarized[103] once more (see Table I, page 103): −0.3 to −0.4 kcal./mole, from a comparison of the infrared absorption intensities of the C—(OH) stretching vibrations of cyclohexanol and *trans*-4-*tert*-butylcyclohexanol[98]; 0.78 kcal./mole from the chemical shift of cyclohexanol in comparison with the epimeric 4-*tert*-butylcyclohexanols[100]; −0.5 to − 0.55 kcal./mole from the rate of acetylation of cyclohexanol, as well as 4,4-dimethylcyclohexanol, with acetic anhydride in pyridine[86] (cf. page 97) in comparison with the epimeric 4-*tert*-butylcyclohexanols; −0.9 kcal./mole from the rate of borate-complex formation for cyclitols[104]; −0.5 to −0.96 kcal./mole from the determination of the equilibration of *cis*- and *trans*-4-*tert*-butylcyclohexanol.[93,95]

As a result of this slight discrepancy in ΔF values, the reports concerning the position of the conformational equilibrium of cyclo-

[98d] Compare also: W. Masschelein, *J. Mol. Spectroscopy* **10**, 161 (1963).

[99] (a) R. U. Lemieux R. K. Kullnig, H. J. Bernstein, and W. G. Schneider, *J. Am. Chem. Soc.* **79**, 1005 (1957); *ibid.* **80**, 6098 (1958); (b) E. L. Eliel, *Chem. & Ind. (London)* p. 568 (1959).

[100] A. H. Lewin and S. Winstein, *J. Am. Chem. Soc.* **84**, 2464 (1962).

[101] E. L. Eliel and M. H. Gianni, *Tetrahedron Letters* p. 97 (1962).

[102] F. A. L. Anet, *J. Am. Chem. Soc.* **86**, 1053 (1962); compare also E. L. Eliel, M. H. Gianni, Th. Williams, and J. B. Stothers, *Tetrahedron Letters* p. 741 (1962).

[103] Cf. also N. B. Chapman, R. E. Parker, and P. J. A. Smith, *J. Chem. Soc.* p. 3634 (1960); N. B. Chapman, J. Shorter, and K. J. Toyne, *ibid.* p. 2543 (1961).

[104] S. J. Angyal and D. J. McHugh, *Chem. & Ind. (London)* p. 1147 (1956).

hexanol fluctuate as well, since the ΔF values have been calculated from these. It is certain that cyclohexanol exists predominantly in the conformation with an *equatorial* OH group. The values given fluctuate between 62 and 90% for the *equatorial* conformer.[82c,89,98a,105]

A dependence of the conformational equilibrium on the solvent has not been observed for cyclohexanol. As has been determined by infrared investigations, the amount of equatorial conformer in ten different solvents lies between 63 and 65%.[106]

(iii) Halogens

For the cyclohexyl halides as well, there are a series of investigations which deal with the position of the conformational equilibrium and the free energy difference between *equatorial* and *axial* halogens.

Again, a number of methods, partly physical and partly chemical, have been used to determine the position of the conformational equilibrium.

The investigations of Hassel *et al.*[71a,107] (page 90) on halogen substituted cyclohexanes have already been mentioned. They were carried out with the help of X-ray and electron diffraction, as well as measurements of dipole moments, whereby cyclohexyl chloride in the vapor state was established to be predominantly in the *equatorial* form. The electron diffraction method is not sensitive enough in most cases to determine small amounts of a conformation in equilibrium. In addition, it cannot be used for investigations in the liquid phase. However, Atkinson could show by means of electron diffraction,[71c] using a novel technique, that 45% of cyclohexylchloride exists in the *axial* conformation in the gas phase ($\Delta F = -0.26$ kcal./mole), which is in good agreement with the results mentioned later.[107a]

The following has been determined with the help of infrared and Raman spectroscopy after coordination of *equatorial* and *axial* C—Cl and C—Br stretching vibration. In the gaseous state[108] for cyclohexyl chloride both frequencies, 742 and 688 cm.$^{-1}$, corresponding to the *equatorial* and *axial* positions, respectively, appear. For cyclohexyl bromide in the gaseous state, the *axial* position seems to be predominant; for cyclohexyl iodide the *equatorial*.

[105] Cf. also R. Cornubert, *Bull. Soc. Chim. France* p. 996 (1956).

[106] G. Chiurdoglu and W. Masschelein, *Bull. Soc. Chim. Belges* **69**, 154 (1960).

[107] O. Hassel, *Research* (*London*) **3**, 504 (1950).

[107a] For cyclohexyl fluoride see: P. Anderson, *Acta Chem. Scand.* **16**, 2337 (1962) ($\Delta F = -0.17$, 57% *equatorial* fluorine).

[108] M. Larnaudie, *J. Phys. Radium* **15**, 650 (1954); cf. also E. Canals, M. Mousseron, R. Granger, and J. Gestaud, *Bull. Soc. Chim. France* **4**, 2048 (1937).

In the liquid state, for cyclohexyl chloride and cyclohexyl bromide,[108,110] both *equatorial* and *axial* forms are present. In these two cases the conformation with *equatorial* position of the halogen predominates. Cyclohexyl fluoride exists in only one conformation[110] in the liquid state.

An equilibrium between *axial* and *equatorial* conformations of cyclohexyl chloride was still observed[110] at slightly below the melting point of the cubic modification, at which, in the case of cyclohexane derivatives of simple structure, a conformational change of the molecule is still possible.[111] Cyclohexyl bromide forms no cubic modification. Cyclohexyl fluoride exists in only one conformation in the cubic modification. At lower temperatures the chloro-, bromo-, and fluorocyclohexane exist in only one conformation, the *equatorial* form.

The measurement of the infrared spectra of cyclohexyl chloride and bromide in various solvents of higher or lower dielectric constant showed no change of the conformational equilibrium in these more qualitative investigations.[110]

Whereas the results described to this point were mostly qualitative in nature, Kozima and Sakashita[112] carried out a quantitative investigation of the conformational equilibrium of cyclohexyl chloride. They calculated the energy difference between the two conformers of cyclohexyl chloride in the gaseous state from infrared investigations at -0.3 to -0.4 kcal./mole indicating that about 40% exists in the conformation with axial chlorine. Chiurdoglu[113] confirmed this value for cyclohexyl chloride and bromide in the liquid state. This value was obtained in all solvents studied, showing that the conformational equilibrium is nearly independent of the solvent.

Other investigations with the help of infrared spectroscopy have lead to slightly higher ΔF values (-0.6[113a] and -0.8 to -1.0[114] kcal./mole) for cyclohexyl bromide. The fraction of *axial* conformer existing in all solvents studied amounts to 20%[115,116] for the bromide.

---

[109] M. Larnaudie, *Compt. Rend. Acad. Sci.* **235**, 454 (1952); **236**, 909 (1953). Deuterocyclohexane exists as a mixture with the deuterium in *axial* and *equatorial* position.

[110] P. Klaeboe, J. J. Lothe, and K. Lunde, *Acta Chem. Scand.* **10**, 1465 (1956).

[111] Cf. O. Hassel and A. M. Sommerfeldt, *Z. Physik Chem.* **B40**, 391 (1938); Chr. Finback, *Arch. Math. Naturvidenskab.* **42**, No. 1 (1938).

[112] K. Kozima and K. Sakashita, *J. Chem. Soc. Japan* **31**, 796 (1958).

[113] G. Chiurdoglu, L. Kleiner, W. Masschelein, and J. Reisse, *Bull. Soc. Chim. Belges* **69**, 143 (1960).

[113a] F. R. Jensen and L. H. Gale, *J. Org. Chem.* **25**, 2075 (1960).

[114] (a) C. G. LeFèvre, R. J. W. LeFèvre, R. Roper, and R. K. Pierens, *Proc. Chem. Soc.* p. 117 (1960). ΔF was derived from the K values mentioned there. Cf. also: (b) P. Klaeboe, J. J. Lothe, and K. Lunde, *Acta Chem. Scand.* **11**, 1677 (1957).

[115] Cf. E. L. Eliel, *Chem. & Ind. (London)* p. 568 (1959).

[116] The value of 20% reported by LeFèvre[114] for the *axial* conformer of cyclohexyl

Infrared spectroscopy is not the only physical method which has been used to undertake a conformational analysis of the cyclohexyl halides. In the literature are found measurements of the molar Kerr constants,[116] as well as investigations conducted with the help of nuclear magnetic resonance spectroscopy.[117] Twenty per cent *axial* cyclohexyl chloride and bromide were found with the equilibrium practically independent of the temperature. The difference in free energy was estimated to be -0.3 to -0.5 kcal./mole.

On the basis of measurements of the rate of substitution and elimination reactions of cyclohexyl bromide (determination of k) and *cis*- and *trans*-4-*tert*-butylcyclohexyl bromide (determination of k_a and k_e) with thiophenolate, ΔF was calculated to be -0.73 kcal./mole.[118,118a] With the aid of Eq. (3), the conformational equilibrium constant K was found to be 3.4.

It is quite striking to note the fact that the value of ΔF is surprisingly low between *equatorial* and *axial* bromine[118] as compared with a methyl group, which is about the same size.[119] Eliel[120] has presented various reasons for this low ΔF value: Because the C—Br bond (1.95 Å) is longer in comparison to the C—C bond of the methyl group (1.54 Å), the distance between an *axial* bromine and an *axial* hydrogen on C-3 or C-5 is a little larger than the distance between an *axial* methyl group and an *axial* hydrogen. Thus, even though Br and CH_3 are of the same size, an *axial* bromine is exposed to slightly less interaction. In addition, due to the greater polarizability of bromine as compared to carbon, London attractive forces arise between the *axial* bromine and the *axial* hydrogen. These are greater than the attractive forces between the *axial* hydrogen and the *axial* CH_3, and they partially compensate for the van der Waals repulsion. We must also consider the fact that the van der Waals interaction of the methyl group is the same in all directions, which is not the case for bromine. [Cf. *cis*-1,4-chlorobromohexane (p. 123), which occurs predominantly in the conformation with *equatorial* chlorine and *axial* bromine.]

---

bromide appears high in comparison with the measurements of the molar Kerr constants of cyclohexyl chloride and bromide. A preponderant *equatorial* conformation was derived from these investigations for both compounds (LeFèvre[15]; cf., on this question, ref. 114).

[117] See ref. 115; A. J. Berliner and F. R. Jensen, *Chem. & Ind. (London)* p. 998 (1960); L. W. Reeves and K. O. Strømme, *Can. J. Chem.* **38**, 1241 (1960).

[118] E. L. Eliel and R. G. Haber, *J. Am. Chem. Soc.* **81**, 1249 (1959).

[118a] Cf. E. J. Corey, *J. Am. Chem. Soc.* **75**, 2301 (1953). Corey estimated the ΔF value of cyclohexyl bromide at 0.7 kcal./mole by comparison with the *gauche* and *trans* form of 1,2-dibromoethane.

[119] L. Pauling, "The Nature of Chemical Bond," p. 81 ff. Cornell Univ. Press, Ithaca, New York, 1960.

[120] E. L. Eliel, *J. Chem. Educ.* **37**, 126 (1960).

In conclusion, a compilation of the ΔF values for various substituents on the cyclohexane ring is given in Table I:

TABLE I

DIFFERENCE IN FREE ENERGY BETWEEN EQUATORIAL AND AXIAL SUBSTITUENTS ON THE CYCLOHEXANE RING

Substituent	$-\Delta F$ (kcal./mole)	References
Methyl	1.97	*a*
	1.90	*b*
	1.8	*c, d, e, f*
	1.5	*g*
	1.35–1.44	*h, i, j*
Ethyl	1.74	*f*
	1.86	*k*
n-Propyl	~2.1	*l*
n-Butyl	~2.1	*l*
Isopropyl	2.10	*m, bb*
	2.28	*f*
tert-Butyl	>5.4	*l*
F	0.25	*n*
	0	*o*
	0.17	*nn*
Cl	0.26	*p*
	0.3–0.4	*q, r*
	0.5	*n*
Br	0.3	*o*
	0.45	*n*
	0.61	*a*
	0.73	*n*
	0.8	*s*
I	0.4	*s, n*
OH	0.4	*j, t, u*
	0.5	*v*
	0.78	*f*
	0.9	*w*
	0.96	*x*
	1.0–1.25	*mm*
SH	0.9	*dd, hh*
NH_2	1.8	*ii*
COOH	1.7	*y, z*
$COOCH_3$	1.0	*aa*
$COOC_2H_5$	1.0–1.2	*u, cc, dd*
OAc	0.36	*u*
Phenyl	2.6-3.1	*g, jj*
OCH_3	0.74	*b*
OC_2H_5	0.98	*b*
OTs	0.6-0.7	*ee*

TABLE I [*cont.*]

Substituent	—ΔF (kcal./mole)	References
CN	0.15–0.25	*kk*
$p-NO_2-C_6H_4-CO$	0.98	*ff*
$-OOC-C_6H_4COO^{--}$	1.2	*l*
HgBr	~0	*gg, ll*

[a] See ref. 121a below. [b] See ref. 121b and c below. [c] See ref. 7 in Chap. 2. [d] See ref. 21b in Chap. 2. [e] See ref. 21d in Chap. 2. [f] A. H. Lewin and S. Winstein, See ref. 100. [g] See ref. 80a. [h] See ref. 82a. [i] See ref. 82b. [j] See ref. 82c. [k] See ref. 83. [l] See ref. 89. [m] See ref. 85. [n] See ref. 117. [o] See ref. 110. [p] See ref. 71c. [q] See ref. 112. [r] See ref. 113. [s] See ref. 114. [t] See ref. 98. [u] Compare ref. 120. [v] See ref. 86. [w] See ref. 104. [x] See ref. 93. [y] See ref. 91b. [z] M. Tichy, J. Jonáš, and J. Sicher, *Collection Czech. Chem. Commun.* 24, 3434 (1959). [aa] N. L. Allinger and R. J. Corby, Jr., *J. Org. Chem.* 26, 933 (1961). [bb] See ref. 121g below. [cc] See ref. 121d below. [dd] See ref. 101. [ee] E. L. Eliel and R. S. Ro, *J. Am. Chem. Soc.* 79, 5995 (1957). [ff] See ref. 121e below. [gg] See ref. 121f below. [hh] See ref. 121h below. [ii] See ref. 121i and j below. [jj] See ref. 121k below. [kk] See ref. 121l and m below. [ll] See ref. 121n below. [mm] See ref. 102. [nn] See ref. 107a.

3. Disubstituted Cyclohexanes

a. General

The situation becomes a little more complicated[72] with the introduction of two substituents in the cyclohexane ring. The introduction of a second substituent makes possible two diastereoisomeric forms, a *cis* and a *trans* compound. In addition, the second substituent can be introduced in the 2, 3, or 4 position in relation to the first substituent.

The relatively simple formulation of the classical planar cyclohexane structure indicated for the *cis* orientation of the substituents, a "true"

[121] (a) F. R. Jensen and L. H. Gale, *J. Org. Chem.* **25**, 2075 (1960); (b) L. J. Dolby and D. S. Noyce, *163rd Meeting, Amer. Chem. Soc., Atlantic City*, 1959, *Abstr. of Papers*, p. 73 p; (c) D. S. Noyce and L. J. Dolby, *J. Org. Chem.* **26**, 3619 (1961); (d) E. L. Eliel, H. Haubenstock, and R. V. Acharya, *J. Am. Chem. Soc.* **83**, 2351 (1961); (e) G. F. Hennion and F. X. O'Shea, *ibid.* **80**, 614 (1958); (f) F. R. Jensen and L. H. Gale, *ibid.* **81**, 6337 (1959); cf. also F. R. Jensen and L. H. Gale, *ibid.* **82**, 145 (1960). (g) N. Mori and F. Suda, *Bull. Chem. Soc. Japan* **36**, 227 (1963); (h) E. L. Eliel and B. P. Thill, *Chem. & Ind. (London)* p. 88 (1963); (i) J. Sicher, J. Jonáš and M. Tichý, *Tetrahedron Letters*, p. 825 (1963); (j) E. L. Eliel, E. W. Della, and T. H. Williams, *ibid.* p. 831 (1963); (k) E. W. Garbisch, Jr., and D. B. Patterson, *J. Am. Chem. Soc.* **85**, 3228 (1963), compare also N. L. Allinger, J. Allinger, M. A. Da Rooge, and S. Greenberg, *J. Org. Chem.* **27**, 4603 (1962); (l) N. L. Allinger and W. Szkrybalo, *ibid.* **27**, 4601 (1962); (m) B. Rickborn and F. R. Jensen, *ibid.* **27**, 4608 (1963); (n) compare also F. R. Jensen, R. J. Oulette, G. Knutson, and D. A. Babbe, *Tetrahedron Letters*, p. 339 (1963).

cis arrangement, with both substituents on one side of the ring. In the case of the *trans* compound in a "true" *trans* arrangement, the substituents were understood to be on opposite sides of the ring. [See (XVII) to (XXI)]. A "true" *cis* position for 1,2 and 1,4 derivatives is no longer possible[122] as a result of the spatial structure of the cyclohexane ring. Let us first consider the 1,2-substituted cyclohexanes (XVII):

For *cis*-1,2 compounds, one substituent is located in the *axial* position, the other is *equatorial*. The *cis* valences form an angle of 60° in relation to each other. As mentioned above, one is not concerned with a "true" *cis* position but with a so-called *meso-cis* position.[123] The *axial* substituent is brought into the *equatorial* position and *vice versa* (XVII) during the conversion of one chair form into the other (without substituents, these of course are identical).

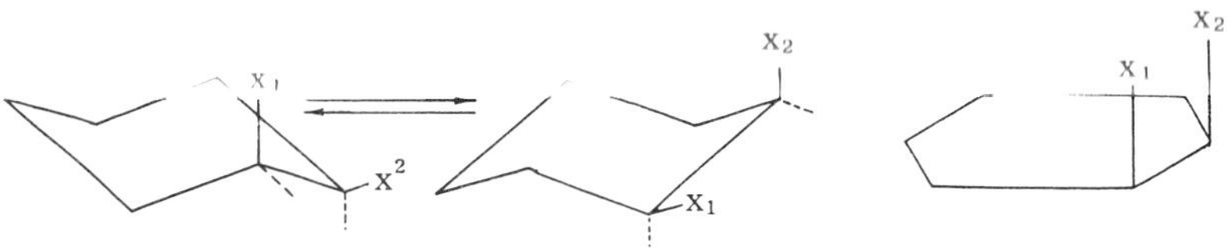

XVII

The *trans*-1,2-*compounds* can theoretically exist in the following conformations. Both substituents assume the *equatorial* position (e, e) or both substituents assume the *axial* position (a, a). In the former case (e, e), the dihedral angle again amounts to 60° (see page 27). This is the so-called *meso-trans* position. In the latter case (a, a) the angle comes to 180°, and the substituents assume the "true" *trans* position. The substituents can again assume both described positions by conversion of one chair form into the other (XVIII).

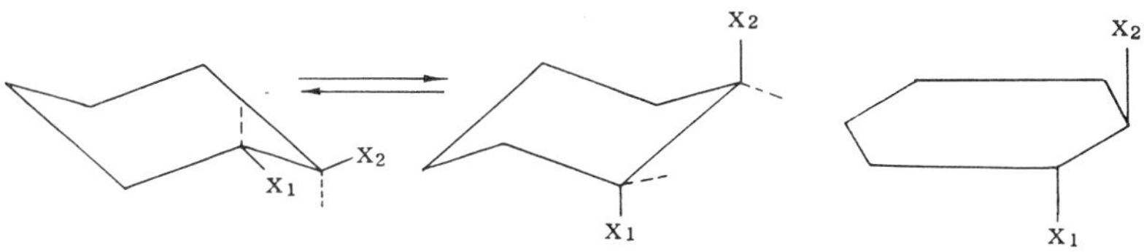

XVIII

The conditions are reversed for the 1,3-substituted cyclohexanes: In the case of the *cis*-1,3 compounds, the two substituents assume the *equatorial* (e, e) or the *axial* (a, a) positions. The projected valence

---

[122] A true *cis* arrangement, that is, a dihedral angle of 0°, would only be possible for the boat form (see page 44).

[123] The expression *meso-cis* or *meso-trans* originates with J. Bredt, *Ann. Chem.* **395**, 26 (1913).

angle amounts to 60° in the first case (*meso-cis* position) and 0° in the second case. In the latter, the substituents assume the "true" *cis* position (XIX).

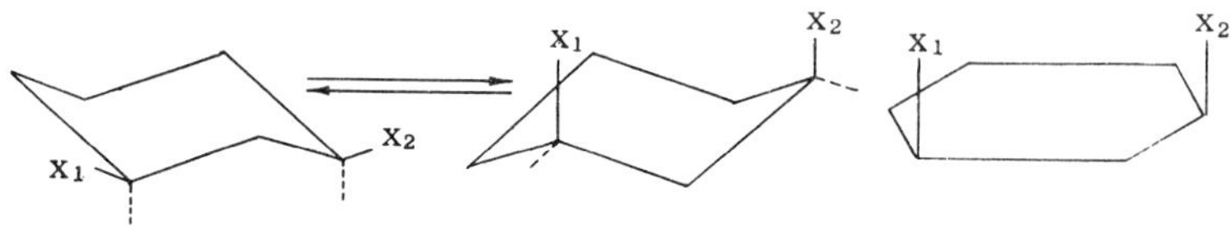

XIX

In the case of *trans*-1,3 compounds, one substituent lies in the *axial* position, and the other in the *equatorial* position, or the reverse. The projected valence angle equals 120° (XX).

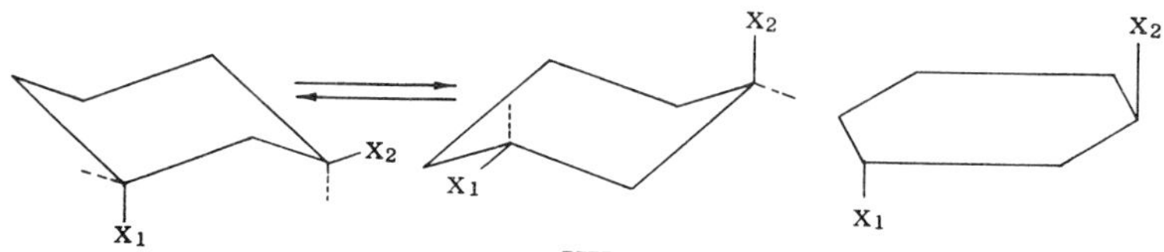

XX

The same conditions exist for 1,4-substituted cyclohexanes as for the 1,2-substituted compounds: *Cis*-1,4: ae→ea; *trans*-1,4: ee→aa (XXI).

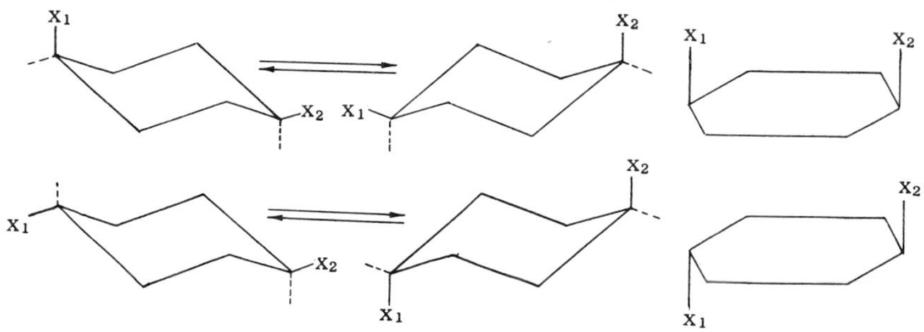

XXI

The classical, planar formulae differ theoretically from the interconvertible, spatial formulae of the cyclohexane ring from still another point of view. Let us consider once more the *cis*-1,2 compound (XVII, $X_1 = X_2$). The *cis*-1,2 compound possesses a plane of symmetry according to the planar model, i.e. it cannot be resolved into optical antipodes. Under the premise that both substituents in the *cis*-1,2 compound (e, a form) are alike, one chair conformation is the mirror image of the other (XXII), and because both conformations are equally stable, one half exists in the one form, the other half in the other form.

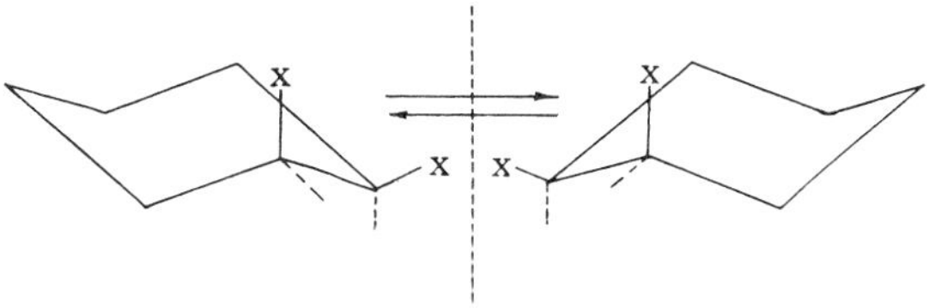

XXII

If the energy barrier between the two chair conformations were large enough, a separation of the two forms into optical antipodes should be possible. The problem in this case is identical with the general problem of separating conformers. As mentioned frequently, the energy barrier between the two forms is too small to allow a separation.

The *trans*-1,2 compound (XVIII, $X_1 = X_2$) also consists of optically active molecules. The e,e form as well as the a,a form exist in mirror image isomers, which can only be converted into each other by exchange of substituents (XXIII).

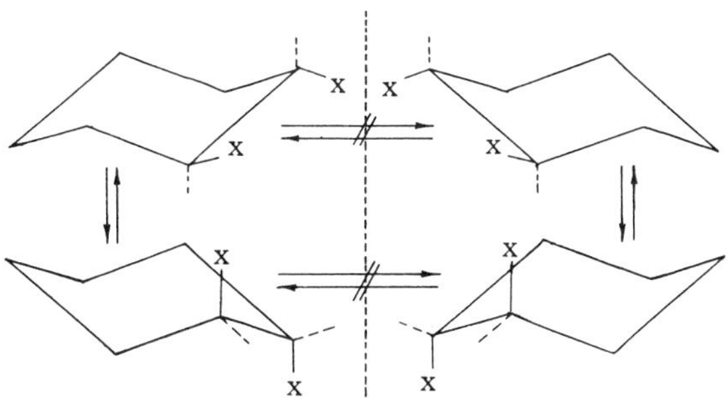

XXIII

By conversion of one chair form into the other, a single e,e form gives rise to a single a,a form and *vice versa*, and not, as in the case of the *cis*-1,2 compound, the mirror image. A resolution into optical antipodes is therefore possible.

The *cis*-1,3 compound (XIX, $X_1 = X_2$) is again optically inactive, but for a different reason than the 1,2 compound. The a,a—as well as the e,e—conformation has a plane of symmetry.

An identical form (e, a), superimposable with the first (a, e), arises in the case of the *trans*-1,3 (XX, $X_1 = X_2$) compound during the conversion of one chair form into the other. As in the case of the *trans*-1,2 compound, a resolution into optical antipodes is possible.

The *cis*-1,4 and *trans*-1,4 compounds (XXI, $X_1 = X_2$) are both optically inactive and have a plane of symmetry in both possible conformations. Identical, superimposable forms arise by the exchange of substituents.

Summarizing, therefore, in the case of two identical substituents, the *meso* form is obtained with *cis*-1,2; *cis*-1,3; *cis*-1,4 and *trans*-1,4 compounds. The *trans*-1,2 and *trans*-1,3 compounds are *d,l* forms.

The *cis*-1,2 compound can be resolved if the two substituents are different, giving rise to optical antipodes, as can readily be seen with a model.

From the point of view of a nonplanar cyclohexane ring, the results easily agree with those of the classical theory as well, without the need for additional assumptions.

b. Conformational Equilibria for Disubstituted Cyclohexane Derivatives

The conformational equilibria described for monosubstituted cyclohexane derivatives, as well as the difference in free energy between *equatorial* and *axial* substituents, can be applied to disubstituted cyclohexane derivatives only in part. This is brought out by the following consideration: The same differences in energy can only arise for a substituent which fulfills the same conditions as it does in a monosubstituted cyclohexane during the conversion from an *equatorial* into an *axial* position. This is further explained by using the dimethylcyclohexanes as an example.[8]

(*i*) *Dimethylcyclohexanes*

For *trans*-1,4-dimethylcyclohexane practically the same steric conditions arise as for methylcyclohexane. Both *equatorially* situated methyl groups are brought into the *axial* position by converting one chair conformation into the other, in such a way that one methyl group lies above the ring, and the other one below the ring. The value of 1.8 kcal./mole, valid for methylcyclohexane, can be doubled in this case. *Diaxial trans*-1,4-dimethylcyclohexane has a higher energy level than the *diequatorial* conformer by 3,6 kcal./mole[65] (XXIV), i.e., the molecule exists exclusively in the *diequatorial* form at room temperature.

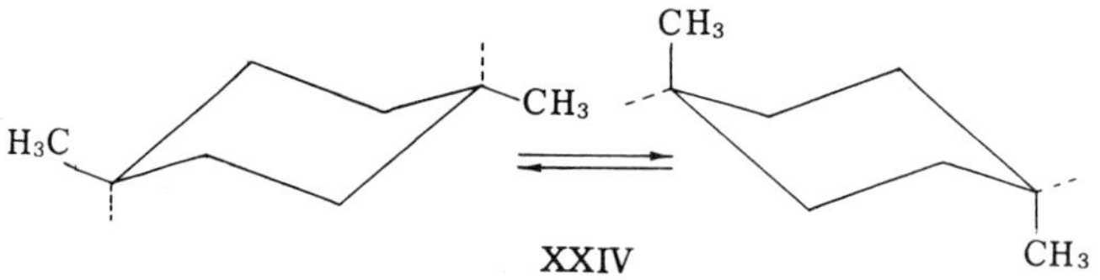

XXIV

Cis-1,2-, *cis*-1,4-, and *trans*-1,3-dimethylcyclohexane are identical in both of the chair conformations. They will therefore not be considered further here.

For *trans*-1,2-dimethylcyclohexane, with the conversion of the e,e conformation into the a,a conformation, the two methyl groups increase the energy level of the molecule by 2 × (− 1.8) kcal./mole, corresponding to four *gauche*-butane interactions. Nevertheless, the energy difference in this case does not amount to − 3.6 kcal./mole, but only to − 2.7 kcal./mole. The two methyl groups are not isolated in the e,e form, but give rise to one *gauche* butane interaction of 0.9 kcal./mole. The energy difference between the two conformers amounts to − (3.6 − 0.9) = − 2.7 kcal./mole (XXV). Therefore, the molecule exists almost exclusively in the *diequatorial* form at room temperature.

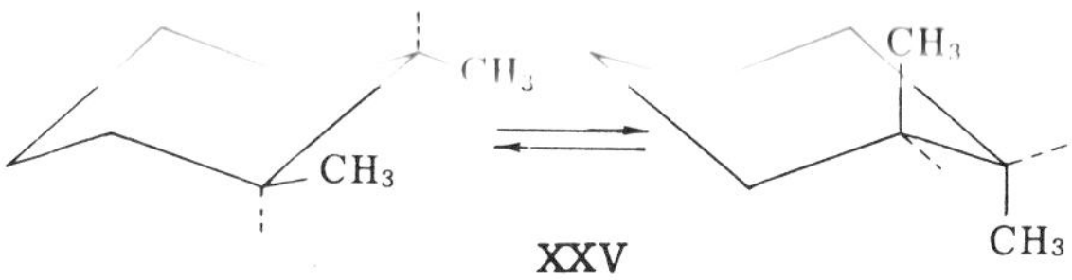

XXV

The energy difference between the two conformations (e,e→a,a) is the largest for *cis*-1,3-dimethylcyclohexane, for which the methyl groups are in the *diaxial* form on the same side of the ring. (XXVI). The energy difference between the a,a form and the e,e form was estimated at 5.4 kcal./mole.[65] From the equilibration of *cis*- and *trans*-1,1,3,5-tetramethylcyclohexane, the energy of the 1,3-*diaxial* dimethyl interaction could be determined to be 3.7 kcal./mole.[124] From this value, the energy difference between the a,a form and e,e form was determined to be 5.5 kcal./mole, in excellent agreement with the calculated value [5.5 kcal./mole = energy of the methyl-methyl interaction (3.7 kcal./mole) + two methyl—hydrogen interactions (2 × 0.9 kcal./mole).

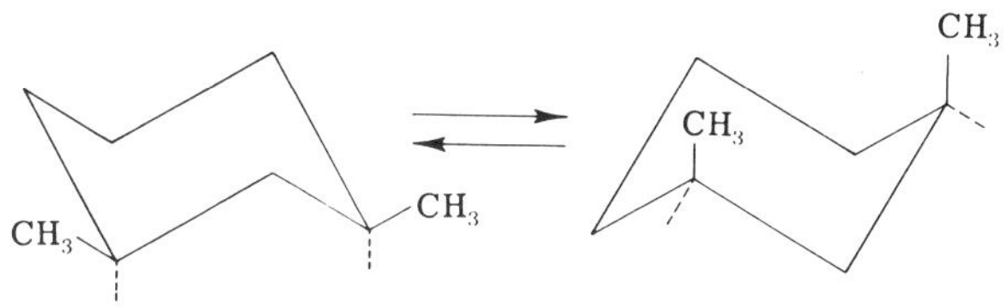

XXVI

The conformation with the *equatorial* position of the substituents is favored for the disubstituted cyclohexane derivatives, just as in the

[124] N. L. Allinger and M. A. Miller, *J. Am. Chem. Soc.* **83**, 2145 (1961).

case of the monosubstituted cyclohexanes. This is demonstrated by the energy differences.

This indicates further that the *cis* and *trans* configurations of a compound vary in their stability, because an energetically more stable conformation can be assigned to each configuration. Or, stated more specifically, of the various dimethylcyclohexanes, the *trans*-1,2 and *trans*-1,4 compounds (e,e conformation) are more stable than the epimeric *cis*-1,2- and *cis*-1,4-dimethylcyclohexanes, because the latter possess a conformation in which one substituent is in an *axial* position. For the 1,3-dimethylcyclohexanes, the *cis* compound is in the more stable conformation (e,e), and this is preferred over the *trans* compound. Here we have a case in which the classical theory cannot be reconciled with modern stereochemistry. Originally it was assumed that all *trans* compounds were more stable, a sensible concept from the viewpoint of a planar model.

Experimental criteria for the stability of epimeric disubstituted cyclohexane derivatives can be obtained by equilibration experiments. Equilibration can be obtained in the case of isomeric dimethylcyclohexane with sulfuric acid or aluminum chloride.[125] Prerequisite for the determination of the equilibrium is a knowledge of the configuration of the dimethylcyclohexanes.

A proof of configuration has not been carried out for the 1,2-dimethylcyclohexanes. The physical data (refractive index, molar refraction, and density, see page 142), however, leave no doubt as to the chosen arrangement. The isomeric 1,3-dimethylcyclohexanes were assigned configurations after it became possible to prepare an optically active 1,3-dimethylcyclohexane, which must have possessed the *trans* configuration.[126] A proof of configuration was obtained for the 1,4-dimethylcyclohexanes by way of the 1,4-dimethylcyclohexenes.[127]

Investigation of the equilibrium point of the isomeric dimethylcyclohexanes,[125] has always led, in agreement with the previous statements,[65,128] to the thermodynamically more stable isomers in preponderant amounts ($> 90\%$). A comparison of the relative stabilities of the isomeric 1,2-, 1,3-, and 1,4-dimethylcyclohexanes (Chiurdoglu *et*

---

[125] A. K. Roebuck and B. L. Evering, *J. Am. Chem. Soc.* **75**, 1631 (1953); G. Chiurdoglu, J. Versluys-Evrard, and J. Decot, *Bull. Soc. Chim. Belges* **66**, 192 (1957). For isomerization with $AlCl_3$ see G. Chiurdoglu, P. J. C. Fierens, and C. Henkart, *ibid.* **59**, 156 (1950).

[126] M. Mousseron and R. Granger, *Compt. Rend. Acad. Sci.* **207**, 366 (1938); *Bull. Soc. Chim. France* **5**, 1618 (1938).

[127] W. Hückel and H. Feltkamp, *Chem. Ber.* **92**, 2851 (1959); compare also A. van Bekhum, A. A. B. Kleis, A. A. Massier, P. E. Verkade, and B. M. Wepster, *Rec. Trav. Chim.* **80**, 588 (1961).

[128] See F. D. Rossini, "Selected Values of Properties of Hydrocarbons." U. S. Govt. Printing Office, Washington, D.C., 1947. K.S.Pitzer[66]—(Entropies, Heats of Isomerization).

al.)[125] gave the following order of decreasing stability: *cis*–1,3 > *trans*-1,4 ≫ *trans*-1,2.[129]

The experimentally determined values for the enthalpy differences and free energy differences between the configurationally isomeric dimethyl cyclohexanes are shown in Table II[129a].

TABLE II

ENTHALPY, FREE ENERGY, AND ENTROPY DIFFERENCES BETWEEN *cis*- AND *trans*-DIMETHYLCYCLOHEXANES

Dimethyl-cyclohexanes	$\Delta H_{exp.}$[a] (kcal./mole)	ΔF[a] (kcal./mole)	$\Delta S_{exp.}$[b] (cal/degree mole)
1.2	1.87	1.61	0.74
1.3	−1.96	−1.55	−1.04
1.4	1.90	1.49	1.58

[a] E. J. Prosen, W. H. Johnson, and F. D. Rossini, ref. 80; J. E. Kilpatrick, H. G. Werner, C. W. Beckett, K. S. Pitzer, and F. D. Rossini, *J. Res. Natl. Bur. Stand.* **39**, 523 (1947); calculated from the values given for the single isomeric dimethyl cyclohexanes.

[b] Calculated from the values given by C. W. Beckett, K. S. Pitzer, and R. Spitzer.[65]

With the most stable conformation of each isomer as a basis, then *cis*-1,2-dimethylcyclohexane (e,a) differs from the *trans*-1,2 compound (e,e) by two *gauche*-butane interactions (*cis*-1,2 possesses three *gauche*-butane interactions, *trans*-1,2 possesses one such interaction). That is to say, the two configurational isomers differ from each other by 3 × 0.9 − 0.9 = 1.8 kcal./mole in their interaction energy. Correspondingly, *trans*-1,3-dimethylcyclohexane (e,a) (two *gauche*-butane interactions) is 1.8 kcal./mole of higher energy than the *cis*-1,3 compound (e,e) (no *gauche*-butane interactions). The same is valid for the isomeric 1,4 compounds, in which the *cis* isomer (a,e) (two *gauche*-butane interactions) differs from the *trans* isomer (e,e) (no *gauche*-butane interactions) by 1.8 kcal./mole. The energy differences estimated in this way are, as shown in Table 2, in good agreement with the experimentally determined enthalpy differences. No longer to be disregarded are the entropy differences[65] between the configurational isomers. They cause the free energy differences to become smaller than the enthalpy differences ($\Delta F = \Delta H - T\Delta S$). The entropy differences are mainly composed of the entropies of mixing, which must be considered for each *d,l* form

[129] For investigations of the dimethylcyclohexanes by nuclear magnetic resonance spectroscopy see N. F. Chamberlain, *Anal. Chem.* **31**, 56 (1959).

[129a] For 1,3- and 1,4-diethylcyclohexane see N. L. Allinger and S. Hu, *J. Am. Chem. Soc.* **84**, 370 (1962).

and for nonequal conformers[65,129a] and the amount which is contributed by the symmetry of the molecule.[65]

In connection with the equilibration, we especially refer to the already mentioned (page 48) investigation by Allinger and Freiberg[130] on the epimeric 1,3-di-*tert*-butylcyclohexanes. Both *tert*-butyl groups are in the *equatorial* conformation for the *cis* configuration. One *tert*-butyl group must take up the *axial* position in the *trans* form, which, as mentioned frequently, would lead to a molecule of extraordinarily high energy (compare page 91). For this reason the *trans*-1,3-compound would not exist in the chair form. In order to lower its energy level, it would assume the twist conformation of cyclohexane, in which the *tert*-butyl groups can both assume *equatorial* positions. The measurement of the equilibrium constant of this hydrocarbon did not lead to any conclusions, concerning the energy difference for a- and e- *tert*-butyl substituents, but rather gave the energy difference between the chair and twist form of cyclohexane (see page 46f.).

In addition for the dialkylcyclohexanes, as for the monosubstituted cyclohexanes, further investigations are available which deal with the position of the conformational equilibria for various substituents.[131] A few selected cases will be discussed.

(ii) Alkylcyclohexanols

The conformations of the alkylcyclohexanols follow[131a] from the general relation between configuration and conformation through the following rule: Of the two substituents, alkyl and hydroxyl, the alkyl group gets preference for the *equatorial* position if there is a choice between an *equatorial* and an *axial* position. The bulkier the alkyl group is, the more this is true. The hydroxyl group is forced into the *axial* position.

A number of alkylcyclohexanols were investigated by Cole[132], and co-workers, Chiurdoglu and Masschelein,[133] and Hückel[134,135,135a] and co-workers with the aid of infrared spectroscopy.

[129a] See also N. L. Allinger and J. L. Coke, *J. Am. Chem. Soc.* **81**, 4080 (1959); N. L. Allinger and S. Hu, *ibid.* **84**, 371 (1962).

[130] N. L. Allinger and L. A. Freiberg, *J. Am. Chem. Soc.* **82**, 2393 (1960).

[131] For a series of older investigations concerning this problem see H. D. Orloff, *Chem. Rev.* **54**, 353 (1954).

[131a] Alkylcyclohexylamines: see H. Booth and N. C. Franklin, *Chem. & Ind. (London)* p. 954 (1963); compare also ref. 121i and 121j.

[132] A. R. H. Cole, P. R. Jefferies, and G. T. A. Müller, *J. Chem. Soc.* p. 1222 (1959).

[133] (a) G. Chiurdoglu, A. Cardon, and W. Masschelein, *Bull. Soc. Chim. Belges* **68**, 388 (1959); (b), G. Chiurdoglu and W. Masschelein, *ibid.* **68**, 484 (1959); **69**, 154 (1960); **70**, 29 (1961); **70**, 231 (1961); **70**, 307 (1961).

[134] W. Hückel and Y. Riad, *Ann. Chem.* **637**, 33 (1960).

[135] W. Hückel and J. Kurz, *Ann. Chem.* **645**, 194 (1961).

[135a] W. Hückel, *Bull. Soc. Chim. France*, p. 1 (1963).

Cole[132] evaluated the differences between the *equatorial* and *axial* O—H stretching vibration bands. The difference between the two is small (*axial* OH group: 3627–3632 cm.$^{-1}$; *equatorial* OH group: 3622–3623 cm.$^{-1}$); nevertheless, they present indications concerning the position of the hydroxyl group. *Cis*-2-, *trans*-3-, and *cis*-4-methylcyclohexanol possess *axial* hydroxyl groups on the basis of their absorption between 3627 and 3632 cm.$^{-1}$, whereas their configurational isomers absorb between 3622 and 3623 cm.$^{-1}$ and possess *equatorial* hydroxyl groups. There is no indication of the presence of less stable conformers from these investigations since the difference in frequency is too small for this[135b].

Chiurdoglu and Masschelein[133] investigated secondary as well as tertiary methyl cyclohexanols in the region of the C—(OH) stretching vibration. They observed two bands and calculated that in the case of *cis*-2-, *trans*-3-, and *cis*-4-methylcyclohexanol, 80–84% is in the conformation with an *axial* OH group.

Trans-2-[135c], *cis*-3-, and *trans*-4-methylcyclohexanol exist exclusively in the conformation with an *equatorial* OH group according to these investigations. By obtaining measurements in various solvents, the C—(OH) stretching vibration could be assigned. Also, from the extent of association of the alcohols, certain conclusions as to the position of the conformational equilibrium can be drawn.

Hückel[134,135] investigated only secondary alkyl cyclohexanols, also in the region of the C—(OH) stretching vibration (940–1070 cm.$^{-1}$). The results of these investigations were that *cis*-2-, *trans*-3-, and *cis*-4-methylcyclohexanol are not conformationally homogeneous. The assignment of the C—(OH) stretching vibration was obtained with the help of hydrogen chloride. As a result of the production of an oxonium ion, the absorption band is shifted to lower wave numbers.

In addition Hückel[135] also investigated the 3-ethyl, 2-isopropyl, 3-isopropyl, and 2-cyclohexyl cyclohexanols. Each of these cyclohexanols shows a close analogy in the region of the C—(OH) stretching vibration with the conformationally homogeneous epimers of the *tert*-butylcyclohexanol (compare page 98). Thus, the conclusion was reached that a broad conformational homogeneity in case of the last-mentioned exists in these alkylated cyclohexanols.

Nuclear magnetic resonance has also been applied for the conformational determination of alkyl cyclohexanols.[136]

[135b] Compare also: H. S. Aaron and C. P. Rader, *J. Am. Chem. Soc.* **85**, 3046 (1963).

[135c] Compare also: F. Šipoš, J. Krupička, M. Tichý, and J. Sicher, *Collection Czech. Chem. Commun.* **27**, 2079 (1962).

[136] R. U. Lemieux, R. K. Kullnig, H. J. Bernstein, and W. G. Schneider, *J. Am. Chem. Soc.* **80**, 6098 (1958); S. Brownstein and R. Miller, *J. Org. Chem.* **24**, 1886 (1959);

With the possibility of comparing the reaction rates, as described on page 94, of conformationally nonhomogeneous compounds with the conformationally fixed epimeric 4-*tert*-butylcyclohexanols, further investigations with alkylated cyclohexanols have been undertaken.[86,89,93,137]

The method used for this, although already described on page 94, will be taken up again in more detail at this point :

The conformational homogeneity of the *tert*-butylcyclohexanols depends according to Winstein[89] on the size of the *tert*-butyl group, whereby a second substituent is fixed in its position.

Cornubert[138] aimed his investigations in the same direction, but did not use a substituent as large as the *tert*-butyl group as a standard. He took the 3- and 4-substituted methylcyclohexanols as a basis for his calculations, and it seems questionable that the hypothesis of conformational homogeneity is met by these compounds. He claimed that the *tert*-butyl group is not a suitable substituent because it has a disturbing influence on the *axial* hydrogens at C-3 and C-5 (or C-2 and C-6) even when in an *equatorial* position, a situation which is not encountered with the methyl group.[39]

Apart from the choice of a standard compound, we can state the following:

If the rate of a certain reaction is dependent only upon the *equatorial* or *axial* conformation of a substituent, the rates of the following compounds must be the same on the basis of their comparable conformations: 1,2-*cis* = 1,3-*trans* = 1,4-*cis*, and 1,2-*trans* = 1,3-*cis* = 1,4-*trans*. The ratio of the rates of the first series to that of the second series, assuming conformational homogeneity, would then give the ratio of the reactivities of the *axial* to the *equatorial* position (compare also Chapter 5, p. 259).

If one disregards the 1,2 compounds at present since they represent a special case, then this hypothesis agrees rather well, with for example, the *cis-trans* isomeric *tert*-butylcyclohexanols, in the following reactions[89]: Oxidation with chromic acid, $k_a:k_e = 2.5:1$; saponification of the acid phthalates in water, $k_a:k_e = 1:9$; solvolysis of the toluenesulfonates in glacial acetic acid, formic acid, and ethanol $k_a:k_e = 3.2$–$3.4:1$.

Cornubert[138] found, for the saponification of the acid phthalates of the corresponding methylcyclohexanols in 25% ethanol,[139,140] values

A. H. Lewin and S. Winstein, *J. Am. Chem. Soc.* **84**, 2464 (1962). Cf. also J. A. Pople, W. G. Schneider, and H. J. Bernstein, "High-Resolution Nuclear Magnetic Resonance" McGraw-Hill, New York, 1959; also A. C. Huitric, W. G. Clarke, Jr., K. Leigh, and D. C. Staiff, *J. Org. Chem.* **27**, 715 (1962).

[137] Compare also E. L. Eliel, E. M. Philbin and T. S. Wheeler, *Proc. Chem. Soc.* p. 167 (1958).

[138] R. Cornubert, *Bull. Soc. Chim. France* p. 996 (1956).

of 1:2.45 for the 3-methyl compound and 1:2.5 for the 4-methyl compound.[139,140]

From these data we can now calculate the amount of each conformation in an equilibrium mixture where both *axial* and *equatorial* forms exist, as was described on page 94f. This, of course, depends on the premise that the differences in the reaction rate depend only on the differences in conformation.

By determining the esterification rates of the epimeric 3- and 4-methylcyclohexanols and the 4,4- and 3,3-dimethylcyclohexanols with acetic anhydride in pyridine (determination of k, see p. 97), in comparison with the esterification rates of the epimeric 4-*tert*-butylcyclohexanols at 25°, the conformational equilibrium constants (K) and free energy differences (ΔF) for these compounds can be calculated[86] (Table III).

TABLE III

Reaction of Alkylated Cyclohexanols with Acetic Anhydride in Pyridine[86]

	$k \times 10^5$ liters $mole^{-1}sec^{-1}$ 25°	K	ΔF (kcal./mole)
Trans-4-*tert*-butylcyclohexanol	10.65	—	—
Cis-4-*tert*-butylcyclohexanol	2.89	0	—
Trans-4-methylcyclohexanol	9.66	6.8	—1.14
Cis-4-methylcyclohexanol	3.76	0.126	1.22
Trans-3-methylcyclohexanol	3.94	0.156	1.11
Cis-3-methylcyclohexanol	10.71	—	—
4,4-Dimethylcyclohexanol	8.43	2.50	—0.55
Cyclohexanol	8.37	2.40	—0.52

It should also be possible to calculate the energy differences listed in Table III from the individual ΔF values of the separate substituents (Table I, page 103), assuming that they are additive.[141] However, no interactions should arise between two substituents. If we use the value of —0.5 kcal./mole determined for the OH group (see p. 97f. and Table III) from the esterification rate of cyclohexanol with acetic anhydride in pyridine and an average value of —1.5 kcal./mole determined for the

[139] A direct comparison of the measurements by Winstein and Cornubert is not possible because of the difference in solvents.

[140] As will be detailed below, page 259, the following rule holds for the reactions mentioned above: *axial* carboxylates are saponified more slowly than *equatorial* ones. *Axial* sulfonates are solvolyzed more rapidly than *equatorial* ones.

[141] Compare E. L. Eliel, ref. 87.

CH_3 group, then we would obtain, for example, for *trans*-4-methylcyclohexanol:

$$\Delta F = \Delta F_{OH} + \Delta F_{CH_3} = -0.5 - 1.5 = -2.0 \text{ kcal./mole}$$

This compares with an experimental value of −1.1 kcal./mole (Table III). For *cis*-4- and *trans*-3-methylcyclohexanol the calculated value amounts to 1.0 kcal./mole,[87] and the agreement with the experimental value is better.

It can be seen from the conformational equilibrium constants worked out by Eliel[86] and listed here, that *trans*-4-methylcyclohexanol exists mostly (~88%) in the *diequatorial* conformation. *Cis*-3-methylcyclohexanol exists exclusively in that conformation. The *cis*-4- and *trans*-3 isomers accordingly take up the conformation (80-90%) in which the OH group is *axial. Cis*-3-methylcyclohexanol exists exclusively in the *diequatorial* conformation. This is understandable, because a *diaxial* position of hydroxyl and methyl groups would be highly unfavorable due to their "true" *cis* position on the same side of the ring. The energy difference between the two conformers of *cis*-3-methylcyclohexanol can be estimated at about 3.8 kcal./mole, whereby an experimental value of 2.4 kcal./mole[142] is taken as the basis for the 1,3-*diaxial* methyl–hydroxy interaction.

When the alkyl group becomes larger, as in the case of cyclohexylcyclohexanol, the conformational equilibrium shifts almost entirely in one direction. For the appropriate isomers it shifts to a *diequatorial* conformation. Otherwise, it shifts in such a way that the alkyl group goes into the *equatorial* position.

The equilibria of the 2-alkylated cyclohexanols were estimated by Winstein[89] for the acid phthalates of the *cis* configuration, of which the following exist with *axial* ester groups: the *cis*-2-methylcyclohexanol, ~70%, and the 2-ethyl, *n*-propyl, and *n*-butyl, ~80%, under the assumption that the *trans*-2-methylcyclohexanol is conformationally homogeneous.

A further possibility of drawing conclusions for the conformation of alkylated cyclohexanols consists in equilibration experiments, as already mentioned on page 97, with a suitable reagent such as aluminum isopropylate,[143] or better yet, equilibration with Raneynickel and hydrogen.[144]

---

[142] E. L. Eliel and H. Haubenstock, *J. Org. Chem.* **26**, 3504 (1961).

[143] E. L. Eliel and R. S. Ro[86]; W. Hückel *et al.*[57]; E. L. Eliel and M. N. Rerick, *J. Am. Chem. Soc.* **82**, 1367 (1960); Cf. also B. Alexandre, R. Cornubert, S. Fagnoni, and W. Kondrachoff, *Bull. Soc. Chim. France* p. 661 (1960).

[144] G. Chiurdoglu and W. Masschelein, *Bull. Soc. Chim. Belges* **70**, 782 (1961).

The epimeric 2- and 4-methylcyclohexanols gave a preponderance of the *trans* configuration (~65%) on equilibration with Raney-nickel and hydrogen. The epimeric-3-methylcyclohexanols gave a preponderance of *cis* configuration (~68%) at equilibrium.[145,146] The results obtained in this way can be understood if we base them on the conformational equilibria outlined above for each isomer. The equilibrium is of the same magnitude for larger alkyl groups such as *tert*-butyl, (see p. 97) or cyclohexyl.

(*iii*) *Criticism of the Quantitative Calculation of Conformational Equilibria from Reaction Rates*

The methods described above in Sections 2,b and 3,b for drawing conclusions concerning conformational equilibria and the difference in free energy of conformers by chemical means from the reaction rates were the basis of a critical review by Hückel and Hanack,[39,40,135a,146a] In these, the choice of the various standard substances by Winstein on one hand and Cornubert (see page 114) on the other hand is critically examined.

The calculation of the energy difference between *axial* and *equatorial* substituents on the cyclohexane ring is given careful attention. Winstein[89] approached the matter as follows: The energy relations of two substituents on the cyclohexane ring are obtained by adding the individual A values of the substituents in question in *equatorial* and *axial* positions. In the example chosen by Winstein,[89] one of the values would apply to the reactive substituent (in this case the esterified acid phthalate group), the other to the second substituent on the ring. The individual values (called A values by Winstein) can now be calculated as follows:

An energy difference of 1.4 kcal./mole is calculated from the reaction rate constants of the conformationally homogeneous *cis–trans* isomeric *tert*-butylcyclohexyl acid phthalates ($k_a:k_e = \sim 1:10$). In order to obtain individual A values, a substituent is used for which the energy difference between *axial* and *equatorial* conformations is known, namely, a methyl group (1.8 kcal./mole—see page 92).

The methyl group is especially suitable, because the saponification rate of the acid phthalate of *trans*-2-methylcyclohexanol (in water at

---

[145] G. Chiurdoglu and W. Masschelein, *Bull. Soc. Chim. Belges* **70**, 767 (1961).

[146] 3-Methylcyclohexanol, like 1,3-dimethylcyclohexane, only recently has been assigned its correct configuration: A. K. Macbeth and J. A. Mills, *J. Chem. Soc.* p. 705 (1945); the configurations originally assumed have to be reversed: S. Siegel, *J. Am. Chem. Soc.* **75**, 1317 (1953); H. L. Goering and C. Serres, Jr., *ibid.* **74**, 5908 (1952); D. S. Noyce and D. B. Denney, *ibid.* **74**, 5912 (1952); see also W. Hückel and J. Kurz, *Chem. Ber.* **91**, 1290 (1958).

[146a] Compare also: H. Kwart and T. Takeshita, *J. Am. Chem. Soc.* **86**, 1161 (1964).

39°) is the same, within the limits of experimental error,[147] as the saponification rate of the acid phthalate of *trans*-4-*tert*-butylcyclohexanol. This condition indicates that the acid phthalate of *trans*-2-methylcyclohexanol is also practically conformationally homogeneous, i.e. has a pure e,e conformation. The corresponding *cis* isomer, however, cannot have the acid phthalate group exclusively in the *axial* position. It should then react ten times slower, yet it reacts two times slower.[140] From this a value of 27% *equatorial* conformer and an energy difference of 0.6 kcal./mole can be calculated. The individual A value for the acid phthalate group is now calculated as the difference of the two energy values, 1.8 kcal./mole (for the a position of the methyl group) and 0.6 kcal./mole, and is found to be 1.2 kcal./mole.

In a corresponding manner the individual values of several alkyl groups were estimated after obtaining the value for the acid phthalate group, in so far as the saponification rates of the acid phthalates are known.[89]

The doubts raised by Hückel and Hanack[39] are directed against the interpretation of the A values, especially against the breakdown into an additive scheme. The measured results used by Winstein[89] do not allow separation of the energy (ΔH) and entropy (ΔS) part of the rates. The calculated A values accordingly are energy differences, because they are computed with a disregard of the entropy contribution. We know from numerous measurements[148] that only in relatively rare cases are differences in the energies of activation alone decisive in observations of rate differences of chemical reactions. Entropy differences are very important.

Energy values of different origin are combined in the resolution into individual A group values. Such a combination is possible because the formula used for the calculation basically is nothing more than an abbreviated Nernst approximation formula ($A \approx \Delta F = F_a - F_e = RT \ln K$). The position of equilibrium between the conformers at various temperatures can also be calculated from this, as already shown in Section 2,b. For example, at 39° about 13% of the methyl groups in methylcyclohexane are *axial*. We obtain 27% *cis*-2-methylcyclohexyl phthalate with an *axial* methyl group, at 39°, from kinetic measurements. In other words, we have more *axial* methyl than in the case of methylcyclohexane itself. By introduction of the phthalate group, as shown above, the energy difference between the two conformers is reduced from 1.8 to 0.6 kcal./mole. The difference of 1.2 kcal./mole is introduced as the individual A group value and made the basis

[147] G. Vavon, *in* "Traité de chimie organique" (V. Grignard, ed.), Vol. II, p. 934. Masson, Paris, 1936.

[148] Hückel, cf. ref. 10 in Chap. 1, Vol. II, page 629.

of further estimations.[89] The value 1.2 kcal./mole would be a constant energy difference between all acid phthalates and alkylated cyclohexanes not carrying an ester group. The constancy of this energy difference indicates that energy differences of stipulated specific interactions between alkyl and phthalate group are not present or must always be the same, independent of the nature and position of the alkyl groups. However, to assume always the same interaction between alkyl and phthalate groups, especially for 1,2 compounds contradicts all other known facts. The following experimental results are mentioned in this respect.[39]

The saponification rate of cyclohexyl phthalate is 3.5 times larger than that of *trans*-2-methylcyclohexyl acid phthalate at 50°. An exclusively *equatorial* position of the phthalate group was assumed for the latter[89] because of the agreement of its saponification rate with that of *trans*-4-*tert*-butylcyclohexyl phthalate. However, all e,a equilibria must lead to smaller reaction rates. The greater rate for cyclohexyl phthalate cannot be explained by such an equilibrium. We can see from this one example that we must account for a specific mutual interaction of the substituents, which, may even produce a deformation of the chair form of the cyclohexane ring.[141] Investigations by Cornubert[149] also indicate a mutual interaction of substituents. In contrast to the results reported by Eliel[86] (see page 114), who concluded the absence of a steric effect of a 4-methyl group on the OH group from the identical acetylation rates of cyclohexanol and 4,4-dimethylcyclohexanol, Cornubert found no correspondence in the saponification rates of the acid phthalates of cyclohexanol and 4,4-dimethylcyclohexanol. (Cyclohexanol, $k = 7.32$ liter $mole^{-1}hr^{-1}$; 4,4-dimethylcyclohexanol, $k = 6.65$ liters $mole^{-1}hr^{-1}$).

Further, the kinetics of the alkaline saponification of the acid phthalates of the epimeric 3- and 4- isopropylcyclohexanols was investigated. These were compared with the saponification rates of *cis*- and *trans*-3-methylcyclohexyl acid phthalates (in 20% alcohol at 70°) (see Table IV).

The similar e,e compounds, *trans*-4-isopropylcyclohexyl and *cis*-3-isopropylcyclohexyl acid phthalate, do not react alike —note the saponification rate constants in Table IV. The former reacts a little more rapidly. The corresponding e,a compounds (*cis*-4-and *trans*-3-) react equally rapidly. We might be tempted, within the same series, to relate the somewhat higher saponification rate of the *trans*-4-isopropylcyclohexyl phthalate, in comparison to the *cis*-3-compound, to a variation in the

[149] R. Cornubert, Y. Fagnoni, and G. Ivanowski, *Compt. Rend. Acad. Sci.* **248**, 2926 (1959); D. Capon, R. Cornubert, S. Fagnoni, and G. Ivanowsky, *Bull. Soc. Chim. France* p. 240 (1961).

conformational equilibrium of both compounds. However, the reasons for this are not quite discernible. Such an explanation can no longer be used in the comparison with the 3-methylcyclohexyl compounds on the basis of the observed saponification rates, because a conformational equilibrium is not probable between the 1,3 compounds (cf.

TABLE IV

SAPONIFICATION RATES OF ACID PHTHALATES

	k (liters mole^{-1}hr^{-1}) 70°
Trans-4-isopropylcyclohexanol	7.6
Cis-4-isopropylcyclohexanol	2.36
Trans-3-isopropylcyclohexanol	2.24
Cis-3-isopropylcyclohexanol	6.3
Trans-3-methylcyclohexanol	3.31
Cis-3-methylcyclohexanol	8.4

p. 113). According to Table III, *cis*-3-isopropylcyclohexyl acid phthalate reacts more slowly than *cis*-3-methylcyclohexyl acid phthalate. These results suggest mutual interaction of the substituents on the cyclohexane ring; they also indicate that for conformational analysis,[150] measurements of saponification rates of acid phthalates have only qualitative significance.

The same reasoning can be applied to other calculations of the quantitative relationship of conformers from various chemical reactions. The ratio of rate constants k_e:k_a is not the same for various reactions; it may even reverse. As already noted,[140] $k_e > k_a$ for the hydrolysis of carboxylic esters, and $k_a > k_e$ for the solvolysis of toluenesulfonates on the basis of certain theoretical concepts (see p. 259).

If we wish to determine the relative ratio e:a quantitatively for a certain compound not only from the properties of the compound itself but also from the properties of its derivatives, then complete agreement cannot be expected. The interaction with the rest of the molecule can cause a different ratio of conformers for the derivatives than for the original substance. Various chemical methods of conformational analysis will not produce the same quantitative relation. We need only compare

[150] On the other hand, compare the measurements of the saponification rates of the epimeric 4-*tert*-butylcyclohexyl acetates with those of the epimeric 2-, 3-, and 4-methylcyclohexyl acetates, which have been used for the determination of conformational equilibria. N. B. Chapman, R. E. Parker, and P. J. A. Smith, *J. Chem. Soc.* p. 3634 (1960).

the ratio e:a, which was obtained from the chromic acid oxidation (of cyclohexanol, 81% e), solvolysis of the toluenesulfonate (90% e), and the saponification of the carboxylic ester[89].

(iv) Dihalocyclohexanes[151]

The introduction of two halogen atoms into the cyclohexane ring can lead to a fundamental shift of the conformational equilibria, at least for the 1,2 compounds. In this way, the rule concerning the preference of *equatorial* substituents can be broken.

Let us consider a *trans*-1,2-dihalocyclohexane. Normally the *diequatorial* position of the substituents would be assumed. Because of their polar character, which leads to an electrostatic repulsion, the substituents attempt to avoid each other and go into the *diaxial* position. This does not occur exclusively, but the equilibrium between the conformers shifts in favor of the *diaxial* form.[152]

The dihalocyclohexanes have been well investigated for the location of their conformational equilibria with the help of various physical methods. These include electron diffraction, measurement of dipole moments, infrared and Raman spectroscopy, and nuclear magnetic resonance.

In the case of *trans*-1,2-dichloro- as well as *trans*-1,2-dibromocyclohexane, it has been determined from electron diffraction investigations that these compounds exist[153] at about 60% in the *diequatorial* form and about 40% in the *diaxial* form in the gas phase.

Both forms are present together in solution, as has been determined for the *trans*-1,2-dihalocyclohexanes from dipole moment measurements.[154] The *diaxial* form has a moment of ~0 D and the *diequatorial* form a moment of ~3.1 D.[155] The *trans*-compound shows a strong dependence on the solvent, whereas the dipole moments of the *cis*-compound are independent of the solvent.

---

[151] Alkylcyclohexyl bromides: compare E. L. Eliel and R. G. Haber, *J. Am. Chem. Soc.* **81**, 1249 (1959); E. L. Eliel, *Chem. & Ind.* (*London*) p. 568 (1959).

[152] See ref. E. J. Corey.[31e]

[153] O. Bastiansen and O. Hassel, *Tidsskr. Kjemi, Bergvesen Met.* **8**, 96 (1946); O. Bastiansen, O. Hassel and A. Munthe-Kaas, *Acta Chem. Scand.* **8**, 872 (1954).

[154] (a) A. Tulinskie, A. diGiacomo, and C. P. Smyth, *J. Am. Chem. Soc.* **75**, 3553 (1953); (b) W. Kweestroo, F. A. Meijer, and E. Havinga, *Rec. Trav. Chim.* **73**, 717 (1954); (c) K. Kozima, K. Sakashita, and S. Maeda, *J. Am. Chem. Soc.* **76**, 1965 (1954); (d) P. Bender, D. L. Flowers, and H. L. Goering, *ibid.* **77**, 3463 (1955).

[155] A. Tulinskie *et al.* [154a] According to this the dipole moment for the pure *diequatorial* form of a *trans*-1,2-dihalocyclohexane lies close to that of a *cis*-1,2-dihalocyclohexane, determined at 3.1 D. The dipole moment of the pure *diaxial* form, because of the inductive effect, is not exactly zero, but estimated at 0.4 D.

Values found for the *trans*-1,2 compounds in benzene are 2.6 D (dichloro-), and 2.1 D (dibromo-); and in carbon tetrachloride, 2.2 D (dichloro-) and 1.7 D (dibromo). The values are 3.1 D (dichloro) and 3.1 D (dibromo) for the *cis* compound, independent of solvent. It follows from the given approximate values that the dipole moment of the *trans*-dibromo compound is smaller than that of the *trans*-dichloro compound. It can be concluded from this that the equilibrium for the *trans*-dibromo compound is shifted more in favor of the less polar, *diaxial* conformation.[156]

The results of measurements of the dipole moments of *trans*-1,2-dihalocyclohexanes have been verified by Raman[154c] and infrared spectroscopy.[157] Both conformations exist simultaneously in the liquid state as well as in solution. The equilibrium between them is quite dependent on the dielectric properties of the solvent. There exists in the liquid phase in the case of *trans*-1,2-dichlorocyclohexane about 50% of the *diequatorial* form; for *trans*-1,2-dibromocyclohexane about 40% of this form exists at room temperature. The free energy difference between the two forms of both compounds was found to be very low, about 0.8 kcal./mole.[157b,158] Nuclear magnetic resonance spectroscopic investigations of *trans*-1,2-dibromocyclohexane indicated ~30% of the *diequatorial* form in nonpolar solvents, whereas ~65% of *trans*-1,2-dichlorocyclohexane exits in the *diequatorial* form.[158a]

The *trans*-1,2-dibromocyclohexane, compared with *trans*-1,2-dichlorocyclohexane, has been found to exist to a larger extent in the *diaxial* conformation. The reason for this must lie in the larger radius of the bromine atom in comparison to the chlorine atom. (Twice the van der Waals radius of bromine equals 3.90 Å in the e,e form, while it equals only 3.60 Å for chlorine.[159]) Thereby the halogen–halogen electrostatic repulsion becomes more important in the *diequatorial* conformation of *trans*-1,2-dibromocyclohexane and the *diaxial* conformation is more favored in relation to the dichloro compound.[157b]

Infrared[157b] and Raman[154c] investigations conclusively confirm that in the solid state, *trans*-1,2-dichlorocyclohexane exists in the *diequatorial* conformation and *trans*-1,2-dibromocyclohexane in the *diaxial* conformation. For *cis*-1,3-dihalocyclohexanes, the *diequatorial* conformation

[156] Cf. the investigations of the Raman spectra by Kozima *et al.* in ref. 154c.

[157] (a) M. Larnaudie, *Compt. Rend. Acad. Sci.* **235**, 154 (1952); *ibid.* **236**, 919 (1953); (b) P. Klaeboe, J. J. Lothe, and K. Lunde, *Acta Chem. Scand.* **11**, 1677 (1957).

[158] Cf. Kozima *et al.* [154c], who had determined the difference at a still lower figure of 0.5 kcal./mole.

[158a] L. W. Reeves and K. O. Strømme, *Trans. Faraday Soc.* **57**, 390 (1961); cf. also R. U. Lemieux and J. W. Lown, *Can. J. Chem.* **42**, 893 (1964).

[159] L. Pauling, "The Nature of the Chemical Bond." Cornell Univ. Press,, Ithaca New York, 1960.

is preferred over the *diaxial* conformation because of the "true" *cis* position of the substituents.[160]

Trans-1,4-dichlorocyclohexane and *trans*-1,4-dibromocyclohexane exist in about equal amounts of *diequatorial* and *diaxial* conformations in the gaseous state, based on electron diffraction investigations.[161] The free energy difference between the two forms is very small ($\sim$0.17 kcal./mole). In the liquid state as well, both conformations are present, as shown by Raman investigations.[162] The equilibrium is strongly dependent on the solvent. In the solid state only the *diequatorial* form exists.[162]

Cis-1,4-chlorobromocyclohexane exists almost entirely in the conformation with *equatorial* chlorine and *axial* bromine, according to electron diffraction investigations in the gaseous state.[163] The same is true for the liquid and solid states, as determined from the infrared investigations.[163]

(v) *Cyclohexane Diols*[164]

With the cyclohexanediols effects arise, which can cause deviations from the normal conformations as was the case for the halogenated cyclohexanes.

Cis- and *trans*-1,2-cyclohexanediols[165] investigated with electron diffraction show that, for the latter, the *diequatorial* as well as the *diaxial* form is present.

An important contribution to the elucidation of the conformations of the diols was obtained by investigations of infrared spectra. The diols, as compared with other substituted cyclohexane derivatives can form intramolecular hydrogen bonds between the hydroxyl groups. Thereby, the above mentioned deviations from the normal conformations can

---

160 Cf. the older investigations of (a) O. Hassel, *Trans. Faraday Soc.* **30**, 874 (1934); (b) O. Hassel and I. G. Gudmundsen, *Z. Physik. Chem.* (*Leipzig*) **40B**, 326 (1938); see also (c) B. Franzus and B. E. Hudson, Jr., *J. Org. Chem.* **28**, 2238 (1963); (d) H. M. van Dort and Th. J. Sekuur, *Tetrahedron Letters*, p. 1301 (1963).

161 V. A. Atkinson and O. Hassel, *Acta Chem. Scand.* **13**, 1737 (1959).

162 K. Kozima and J. Yoshino, *J. Am. Chem. Soc.* **75**, 166 (1953).

163 V. A. Atkinson and K. Lunde, *Acta Chem. Scand.* **14**, 2139 (1960); compare however ref. 160d.

164 Concerning conformational investigations of 2-aminocyclohexanols, see for example G. Drehfahl and G. Heublein, *Chem. Ber.* **94**, 915 (1961); G. Drehfahl, H. Zimmermann, and K. Gehrke *ibid.* **93**, 509 (1960); J. Sicher, M. Horák, and M. Svoboda, *Collection Czech. Chem. Commun.* **24**, 950 (1959); J. Sicher, J. Jonáš, M. Svoboda, and O. Knessl, *ibid.* **23**, 2141 (1958); M. Tichý, J. Šipoš, and J. Sicher, *ibid.* **27**, 2907 (1962); G. Fodor, E. Fodor-Varga, and A. Furka, *Croat. Chem. Acta* **29**, 303 (1957).

165 B. Ottar, *Acta Chem. Scand.* **1**, 521 (1947); cf. also S. Furberg and O. Hassel, *ibid.* **4**, 597 (1950).

arise.[166] The existence of such intramolecular hydrogen bonds, whose detection with infrared spectroscopy affords no difficulties, permits conclusions as to the geometry of the molecule. It is assumed that the strength of the hydrogen bonds

$$-\mathrm{OH}\cdots\underset{\mathrm{H}}{\mathrm{O}}-$$

increases with decreasing distance. A lowering of the frequency ν (O–H) then takes place. The following relation was set up by L.P. Kuhn[167] between the distance $d_{\mathrm{O\cdots H}}$ and the shift in frequency between free and bonded OH groups $[\Delta\nu = \nu(\mathrm{OH})_{\mathrm{free}} - \nu(\mathrm{OH}\cdots\underset{\mathrm{H}}{\mathrm{O}})_{\mathrm{bonded}}]$:

$$\Delta\nu = \frac{250}{d_{\mathrm{O\ldots H}}} - 74 \qquad (d_{\mathrm{O\ldots H}} \text{ in Å})$$

This equation was used to determine the conformations of the diols. It had been found quite useful, even though it rests on a relatively narrow experimental basis.[168]

Only such diols which can bring their OH groups into sufficiently close proximity to each other show a shift in frequency, i.e. form intramolecular bonds. Thus *cis*-1,2-cyclopentanediol shows a $\Delta\nu = 61$ cm^{-1}, while for the *trans* epimer, $\Delta\nu = 0$. *Cis*- and *trans*-1,2-cyclohexandiol show $\Delta\nu = 38$ and 33 cm^{-1}, for *cis*-1,3-cyclohexanediol, $\Delta\nu = 75$ cm^{-1}, and for *trans*-1,3-diol as well as the *cis*- and *trans*-1,4-diols $\Delta\nu = 0$ (all measurements in carbon tetrachloride). Kuhn[167] derived the conformations from these results. *Trans*-1,2-cyclohexanediol accordingly exists in the *diequatorial* form. A *diaxial* conformation would permit no hydrogen bonding, as shown by the calculation of the distances on the basis of bond distances. The *diaxial* conformation accordingly belongs to the *cis*-1,3-diol. It can be said, in general, that that conformation is favored which brings the two hydroxyl groups as close together as possible, whereby the interaction energy between the OH groups is extensively compensated for by the energy resulting from the hydrogen bonds.

The slight differences brought out above in the $\Delta\nu$ value for *cis*- and *trans*-1,2-cyclohexanediol, despite the identical distances of the OH groups, are explained as follows: The dihedral angle of the OH groups

[166] Cf. W. J. Svirbely and J. J. Lander, *J. Am. Chem. Soc.* **72**, 3756 (1950), also contains measurements of dipole moments.

[167] (a) L. P. Kuhn, *J. Am. Chem. Soc.* **76**, 2492 (1952); (b) *ibid.* **76**, 4323 (1954); (c) *ibid.* **80**, 5950 (1958); (d) Cf. also T. Bürer, E. Maeder, and H. H. Günthard, *Helv. Chim. Acta* **40**, 1823 (1957), with further references; compare also M. Tichý, *Chem. Listy* **54**, 506 (1960); H. Buc, *Ann. Chim.* [13] **8**, 409 (1963).

[168] Cf. S. Julia, D. Varech, T. Bürer, and H. H. Günthard, *Helv. Chim. Acta* **43**, 1623 (1960).

amounts to $\varphi = 60°$ in both cases. The resulting formation of a hydrogen bond leads to a rotation around the C—C bond, whereby φ is changed leading to a deformation of the chair form of the ring.[141] As can be seen from a model, less energy will be required to bring the *cis* OH groups together than the *trans*-oriented ones. The *cis* hydroxy groups will therefore be able to come closer together ($\varphi \sim 50°$), i.e., the more easily the groups can approach each other for the formation of a hydrogen bond, the greater is the hydrogen bond energy and the higher is the value of $\Delta\nu$(OH).

The conformations derived from infrared investigations have been verified by nuclear magnetic resonance investigations.[169]

The Kuhn equation was tested[168] on a series of tricyclic diols and found not to fit all cases. A correction term must be applied, by introducing new parameters such as the angle:

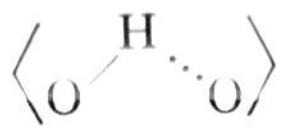

The classical investigations of Böeseken[170] are referred to in this connection (cf. also page 15) concerning the steric assumptions for forming the boric acid–diol complexes. According to these, an approximate coplanarity of the neighboring hydroxyl groups is necessary for the formation of boric acid–diol complexes. This is present neither in the case of the *cis* nor the *trans*-1,2-cyclohexanediol. Neither one is able to form this complex in significant amounts. It has been shown by Reeves[171] that complex formation between diols and copper ammonium ion (Schweitzer's reagent) can only take place when the dihedral angle between the two hydroxyl groups lies between 0° and 60°. If it exceeds 60°, then complex formation is no longer possible (compare p. 328). While neither the *cis* nor the *trans*-1,2-diol can form a complex with boric acid, in agreement with the earlier investigations, such complex formation takes place with copper ammonium ion.[172] The *cis* compound exhibits a somewhat larger conductivity increment than the *trans* compound. This is in agreement with the results of the infrared

---

[169] S. Brownstein and R. Miller, *J. Org. Chem.* **24**, 1886 (1959). R. U. Lemieux and J. W. Lown, *Tetrahedron Letters* p. 1229 (1963); compare also H. Finegold and H. Kwart, *J. Org. Chem.* **27**, 2361 (1962); J. Pitha, J. Sicher, F. Šipoš, M. Tichý, and S. Vašicková, *Proc. Chem. Soc.* p. 301 (1963).

[170] See ref. 22 in Chap. 1; also *Bull. Soc. Chim. France* [4] **53**, 1332 (1933); J. Böeseken and J. v. Giffen, *Rec. Trav. Chim.* **39**, 183 (1920).

[171] (a) R. E. Reeves and J. P. Jung, Jr., *J. Am. Chem. Soc.* **71**, 209 (1949); (b) R. E. Reeves *ibid.* **71**, 212, 215, 2116, (1949); (c) *ibid.* **73**, 957 (1951); (d) R. E. Reeves, *Advan. Carbohydrate Chem.* **6**, 109 (1951).

[172] H. Kwart and G. C. Gatos, *J. Am. Chem. Soc.* **80**, 881 (1958); cf. also E. J. McDonald, *J. Org. Chem.* **25**, 111 (1960).

investigations described above. The *diequatorial* conformation, therefore, belongs to the *trans*-1,2-cyclohexanediol.[173]

These investigations are not accurate enough to discriminate between energy differences of the various conformations. If two hydroxyl groups on the same side of the ring (*cis*-1,3-diol) are in the *diaxial* position then this conformation is estimated to be about 2.8 kcal./mole higher in energy than the *diequatorial* conformation.[174]

(vi) Cyclohexane Carboxylic Acids

(a) Monocarboxylic Acids

The conformations of monosubstituted cyclohexane carboxylic acids were first discussed by Kilpatrick and Morse[174a] for the case of the hydroxycyclohexane carboxylic acids. A distinct difference in dissociation constants exists between the conformers of epimeric cyclohexane carboxylic acid with *axial* and *equatorial* carboxyl groups.[175] With this established fact, predictions can be made concerning the position of the conformational equilibrium of such compounds.[176] Stolow[176a,177] and Tichy, Jonáš and Sicher[176b] investigated the dissociation constants of the conformationally homogeneous epimeric 4-*tert*-butylcyclohexane carboxylic acids as compared with cyclohexane carboxylic acid, and the following order of acidities was determined: *trans*-4-*tert*-butyl > cyclohexyl ≫ *cis*-4-*tert*-butylcyclohexane carboxylic acid. The lower acidity of the *axial* carboxyl group in the *cis*-4-*tert*-butyl compound can be explained by the difference between the solvation of the carboxylate anion in the *equatorial* and *axial* positions.[176] The solvation of the *axial*

[173] For intermediate complex formation in the periodate oxidation of cyclohexane-1,2-diols cf. G. J. Buist, C. A. Bunton, and J. H. Miles, *J. Chem. Soc.* p. 743 (1959). For the glycol cleavage rate of cyclohexanediols with lead tetraacetate see R. Criegee, E. Höger, G. Huber, P. Kruck, F. Marktscheffel, and H. Schellenberg, *Ann. Chem.* **599**, 81 (1956); compare also: F. Fischer and R. Schiene, *J. prakt. Chem.* [4] **22**, 39 (1963).

[174] S. J. Angyal and D. J. McHugh, *Chem. & Ind.* (*London*) p. 1147 (1956).

[174a] M. Kilpatrick and J. G. Morse, *J. Am. Chem. Soc.* **75**, 1846 (1953).

[175] (a) J. F. J. Dippy, S. R. C. Hughes, and J. W. Laxton, *J. Chem. Soc.* p. 4102 (1954); (b) P. J. Sommer, V. P. Arya, and W. Simon, *Tetrahedron Letters* p. 18 (1960).

[176] (a) R. D. Stolow, *J. Am. Chem. Soc.* **81**, 5806 (1959); (b) M. Tichý, J. Jonáš, and J. Sicher, *Collection Czech. Chem. Commun.* **24**, 3434 (1959); (c) H. van Bekhum, P. E. Verkade, and B. M. Wepster, *Proc. Koninkl. Ned. Akad. Wetenschap.* **B64**, 161 (1961); cf. H. H. Lau and H. Hart, *J. Am. Chem. Soc.* **81**, 4897 (1959).

[177] For correlation of dissociation constants and structure of organic compounds see H. C. Brown, D. H. McDaniel, and O. Häfliger in "Determination of Organic Structures by Physical Methods." (E. A. Braude and F. C. Nachod, eds.), pp. 567–655. Academic Press, New York, 1955.

situated anion is hindered[178] by the axial hydrogens in positions 3 and 5.

The measured dissociation constants of the epimeric 4-*tert*-butyl-cyclohexane carboxylic acids were used to determine the conformational equilibrium constant for cyclohexane carboxylic acid.[176a,b] The experimentally determined dissociation constant K of an acid is composed of the individual dissociation constants K_e and K_a for the *equatorial* or *axial* positions of the carboxyl group multiplied by the mole fractions N_a or N_e (see page 94). From this, as described previously for the relationship between rate constant and conformational equilibrium, the conformational equilibrium constants of cyclohexane carboxylic acids can be calculated. The measurements reveal for cyclohexane carboxylic acid that this compound prefers the conformation with an *equatorial* carboxyl group. The free energy difference between the *axial* and *equatorial* conformation was found to be 1.5–1.6 kcal./mole.[176a,b,c] In this special case, the conformational equilibrium or the ΔF value cannot be determined very accurately because the dissociation constant of cyclohexane carboxylic acid lies very close to that of *trans*-4-*tert*-butyl cyclohexane carboxylic acid. The free energy difference of the carboxylate anion was determined to be 2.1–2.4 kcal./mole[176a,b,c] and thus is higher than that of the carboxyl group because of solvation effects.

On the basis of the dissociation constants of the epimeric 4-*tert*-butylcyclohexane carboxylic acids, the conformational equilibria of other 4-alkylated cyclohexane carboxylic acids have been determined.[176a,c,179]

(*b*) *Dicarboxylic Acids*[180]

From measurements of the rates of esterification of the epimeric 1,2-, 1,3-, and 1,4-cyclohexane dicarboxylic acids with methanol, conclusions have been drawn concerning their conformations.[181] The various acids exist predominantly in the following conformations:

---

[178] Cf. also G. S. Hammond and D. H. Hogle, *J. Am. Chem. Soc.* **77**, 338 (1955).

[179] Concerning the conformation of cyclohexane carboxylic esters cf. E. A. S. Cavell, N. B. Chapman, and M. D. Johnson, *J. Chem. Soc.* p. 1413 (1960).

[180] For determination of the configurations see: (a) A. Werner and H. E. Conrad, *Ber.* **32**, 3046 (1899); (b) J. Böeseken and A. Peek, *Rec. Trav. Chim.* **44**, 841 (1925); (c) W. H. Mills and G. H. Keats, *J. Chem. Soc.* p. 1373 (1935).

[181] H. A. Smith and F. B. Byrne, *J. Am. Chem. Soc.* **72**, 4406 (1950). For the alkaline and acidic hydrolysis of cyclohexane dicarboxylic acid esters see: H. A. Smith and T. Fort, Jr., *J. Am. Chem. Soc.* **78**, 4000 (1956); H. A. Smith, K. G. Scrogham, and B. L. Stump, *J. Org. Chem.* **26**, 1408 (1961).

Cis-1,2: e,a; *trans*-1,2:a,a; *cis*-1,3:e,e; *trans*-1,3:e,a; *cis*-1,4:e,a; *trans*-1,4:e,e (see below). The assignments of the conformations of these dicarboxylic acids are based on the assumption that *equatorial* carboxyl groups are esterified more easily than those in an *axial* position. *Cis*-1,3- and *trans*-1,4-cyclohexane dicarboxylic acids are esterified at almost the same rate; the same is true for the *trans*-1,3 and *cis*-1,4-compounds.

Dissociation constants can also be used for a conformational analysis of the dicarboxylic acids.[182] The *cis*- and *trans*-1,2-cyclohexane dicarboxylic acids show a great variation in differences between the pK values of the first and second dissociation constants (ΔpK value), whereas the variation of the ΔpK values for the 1,3-diacids is smaller. The ΔpK value determined for *cis*-1,2-cyclohexane dicarboxylic acid was 2.42 and for *trans*-dicarboxylic acid, 1.75 (in water, 19°).[183] The larger ΔpK value for the *cis*-1,2 compound in comparison to the *trans*-1,2 isomer was rationalized on the basis that the carboxyl groups in the former are closer together (a,e form) than in the latter. *Trans*-1,2 compound could assume at least partly the *diaxial* conformation placing the carboxyl groups further from each other, and causing the ΔpK value to become smaller. The reasons for this will be explained briefly in a simplified manner: Carboxyl groups situated close together influence each other strongly as dipoles. Upon dissociation of the first carboxyl group of a dicarboxylic acid, the second carboxyl group is subjected to a field effect by the appearance of the charge. The field effect of the negative charge must include the ratio of the first and second dissociation constant which manifests itself in such a way that the removal of the proton of the second carboxyl group is made more difficult. In spatially close carboxyl groups, as a result, the second dissociation constant is considerably smaller than the first. The difference in dissociation constants, or ΔpK values, is larger when the carboxyl groups are near to each other, and smaller when they are farther apart.

A reliable method for determining the conformation of *trans*-1,2-cyclohexane dicarboxylic acid consists again in choosing for comparison a 1,2-cyclohexane dicarboxylic acid in which the carboxyl groups are fixed in the *diequatorial* position. Their first and second dissociation constants are compared with those of the *trans*-1,2-cyclohexane dicarboxylic acid. *Trans*-4-*tert*-butyl-*trans*-1,2-cyclohexane dicarboxylic acid (XXVII),

---

[182] Cf. ref. 8 in Chap. 2; D. H. R. Barton and G. A. Schmeidler, *J. Chem. Soc.* p. 1197 (1948). Compare also G. S. Hammond *in* "Steric Effects in Organic Chemistry." (M. S. Newman, ed.), p. 433 ff. Wiley, New York, 1956.

[183] R. Kuhn and A. Wassermann, *Helv. Chim. Acta* 11, 50 (1928).

whose conformation is fixed by the *tert*-butyl group with a *diequatorial* position of the carboxyl groups, was used for this comparison.[184]

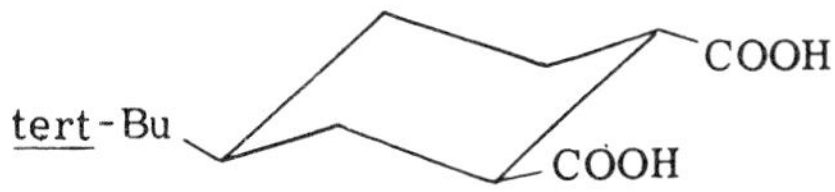

XXVII

A ΔpK value of 1.79 was found for (XXVII), which is practically identical with the ΔpK value of 1.76 found for *trans*-1,2-cyclohexane dicarboxylic acid (water solution, 25°). This is also true for the individual pK values of the first and second dissociation constants of both compounds. It follows, contrary to other views[181] and older measurements of the esterification rate (see previous description) that *trans*-1,2-cyclohexane dicarboxylic acid in aqueous solution exists almost exclusively in the *diequatorial* conformation.

4. Trisubstituted Cyclohexanes

a. General

The introduction of three substituents into the cyclohexane ring increases the number of possible configurational isomers.

Relative to the first substituent (X_1), the two others (X_2 and X_3) can be introduced into positions 2,3-, 2,4-, or 3,5- (XXVIII, XXIX, XXX).[185]

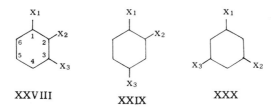

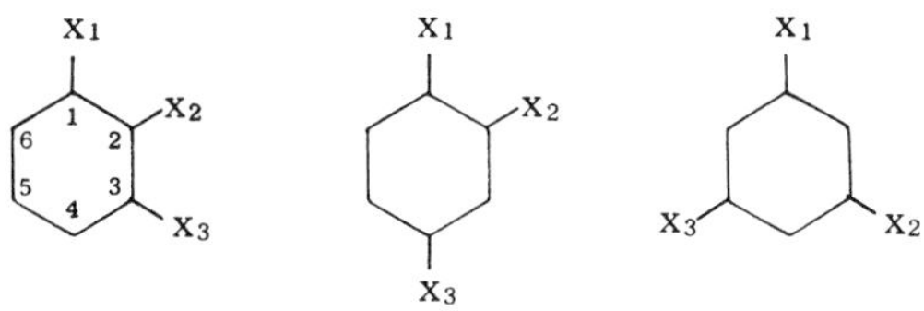

XXVIII XXIX XXX

For each positional number (1,2,3; 1,2,4; 1,3,5) the configurational as well as conformational isomers given in Table V are possible when X_1, X_2, and X_3 are different substituents:

---

[184] J. Sicher, F. Šipoš and J. Jonáš, *Collection Czech. Chem. Commun.* **26**, 262 (1961). Compare also: P. F. Sommer, C. Pascual, V. P. Arya, and W. Simon, *Helv. Chim. Acta* **46**, 1734 (1963).

[185] The isomers formed by the introduction of two substituents on one carbon atom (1,1,2; 1,1,3; 1,1,4) will be disregarded here.

TABLE V

ISOMERS OF TRISUBSTITUTEO CYCLOHEXANES

	Configuration of the substituents		Conformations
	X_1X_2	X_1X_3	
1,2,3 Compound	*trans*	*cis* (1)	eX_1, eX_2, $eX_3 \rightleftarrows aX_1$, aX_2, aX_3
	cis	*cis* (2)	eX_1, aX_2, $eX_3 \rightleftarrows aX_1$, eX_2, aX_3
	trans	*trans* (3)	eX_1, eX_2, $aX_3 \rightleftarrows aX_1$, aX_2, eX_3
	cis	*trans* (4)	eX_1, aX_2, $aX_3 \rightleftarrows aX_1$, eX_2, eX_3
1,2,4 Compound	*trans*	*trans* (1)	eX_1, eX_2, $eX_3 \rightleftarrows aX_1$, aX_2, aX_3
	cis	*trans* (2)	eX_1, aX_2, $eX_3 \rightleftarrows aX_1$, eX_2, aX_3
	trans	*cis* (3)	eX_1, eX_2, $aX_3 \rightleftarrows aX_1$, aX_2, eX_3
	cis	*cis* (4)	eX_1, aX_2, $aX_3 \rightleftarrows aX_1$, eX_2, eX_3
1,3,5 Compound	*cis*	*cis* (1)	eX_1, eX_2, $eX_3 \rightleftarrows aX_1$, aX_2, aX_3
	trans	*cis* (2)	eX_1, aX_2, $eX_3 \rightleftarrows aX_1$, eX_2, aX_3
	cis	*trans* (3)	eX_1, eX_2, $aX_3 \rightleftarrows aX_1$, aX_2, eX_3
	trans	*trans* (4)	eX_1, aX_2, $aX_3 \rightleftarrows aX_1$, eX_2, eX_3

The two conformations belonging to each configuration originate from the conversion of one chair form into the other. The substituents which were in the *equatorial* position at first, are now in the *axial* position, and *vice versa*.

b. Conformational Equilibria for Trisubstituted Cyclohexane Derivatives

In principle the same considerations can be applied to the trisubstituted cyclohexane derivatives as was the case with the disubstituted ones (see page 108). In this way we can arrive at a decision as to which of the two possible conformations of a certain configuration is preferred at equilibrium. The basic rule in conformational theory states that if the substituents are sterically equivalent, and if one can neglect dipole interactions, a molecule assumes that conformation with the greatest number of equatorial substituents.

If the substituents are no longer equal in size then a decision as to which conformation is preferred at equilibrium cannot be made with certainty, as will be shown below in detail for certain cases.

(*i*) *Menthols*

The trisubstituted cyclohexane compounds whose stereochemistry has been investigated in greatest detail are the isomeric menthols, as well as the 1,4-dimethylcyclohexan-2-ols.

Menthol (XXXI) has three asymmetric carbon atoms (*). The molecule exists in $2^3 = 8$ optically active forms, comprising four racemates. The four stereoisomeric forms are designated as *d,l*-menthol, *d,l*-neomenthol, *d,l*-isomenthol, and *d,l*-neoisomenthol. On the basis of extensive work by Read[186,187] and co-workers, their configurations are well

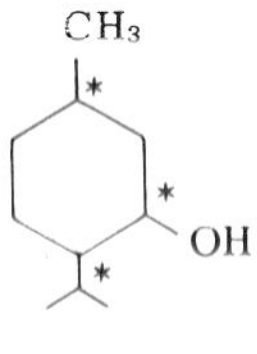

XXXI

established. Menthol, corresponding to the 1,2,4 compound in Table V, with isopropyl X_1, OH X_2, and CH_3 X_3, has the configuration (1) (*trans–trans*). Neomenthol is (2) (*cis-trans*), isomenthol is (3) (*trans-cis*), and neoisomenthol has the configuration (4) (*cis-cis*).

If the isopropyl group preferentially assumes the *equatorial* position, then conformations (XXXII) to (XXXV) result for the isomeric menthols:

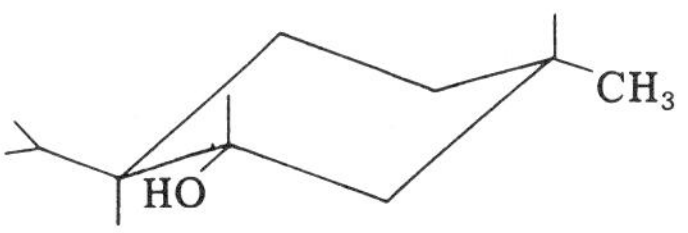

XXXII

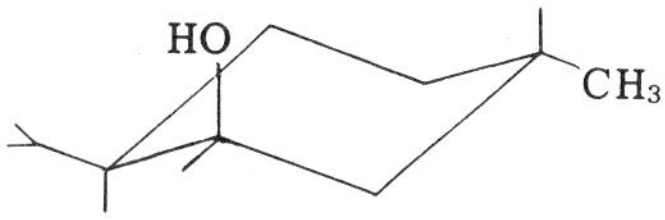

XXXIII

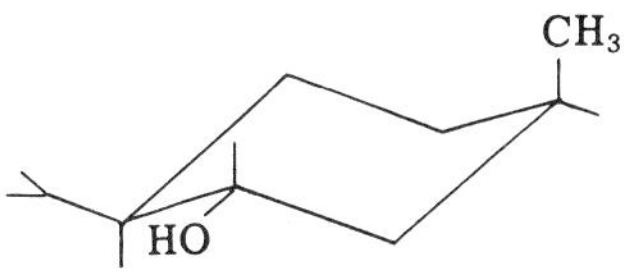

XXXIV

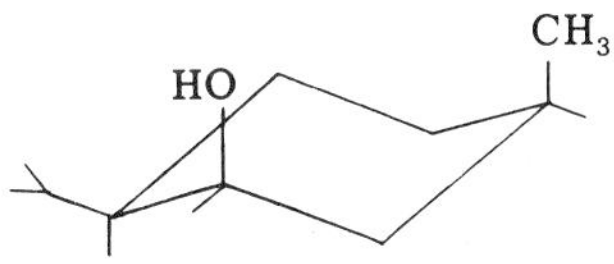

XXXV

These conformations have been verified extensively by various chemical and physical methods. It is not out of the question, however, if we disregard menthol which has all of its substituents in the *equatorial* position, to expect small amounts of the less stable conformation to existing in equilibrium. In the case of neoisomenthol, a distinct possibility exists for the formation of the second conformation (a-isopropyl,

---

186 J. Read, *Chem. Rev.* **7**, 1 (1930).

187 Cf. also J. L. Simonsen and L. N. Owen, "The Terpenes," Vol. I. Cambridge Univ. Press, London and New York, 1947.

e-OH, e-methyl). A rough estimate of the energy differences the substituents in the *equatorial* and *axial* positions confirms this. The difference in free energy between the *axial* and *equatorial* position of an isopropyl group amounts to ~2.3 kcal./mole (see p. 103). Assuming 2.4 kcal./mole (as estimated on page 116) for the 1,3-*diaxial* methyl–hydroxyl interaction, then both conformations must be present in equilibrium.

Various factors work in opposition in neoisomenthol; both conformations are energetically unfavorable. Eliel[188] favored the conformation with an equatorial position of the isopropyl group in his first discussion of the conformations of the isomeric menthols.[189] Cornubert[138] came to a different conclusion for the *n*-propyl analog of neoisomenthol (where the size of the *n*-propyl group is slightly different from that of the isopropyl group), placing the *n*-propyl group in an *axial* position.

Conformational analysis of the isomeric menthols at first was confined to chemical methods. As already mentioned, the first conformational analytical observations of the menthols were made by Eliel[188] who assigned to them the conformations with *equatorial* isopropyl groups. On the other hand, the concept of possible conformational equilibria was also indicated by Eliel. For analysis, he used values obtained by Read and Grubb[190] for the relative rates of esterification of the hydroxyl group of the isomeric menthols with *p*-nitrobenzoyl chloride in pyridine (25°). Read found the following values: Menthol 16.5, isomenthol 12.3, neoisomenthol 3.1, neomenthol 1.0. Since *equatorial* OH groups, as they are present in menthol (XXXII) and isomenthol (XXXIV), are esterified more rapidly than *axial* ones (see p. 259), menthol and isomenthol should be esterified more rapidly than neomenthol (XXXIII) and neoisomenthol (XXXV). On the other hand, neoisomenthol (XXXV) reacts more rapidly than neomenthol (XXXIII), which is hard to understand on the basis of an *axial* hydroxyl group for both compounds (see Sect. 2,b, *ii*). Because the methyl group in neoisomenthol is also *axial*, this compound should react more slowly than neomenthol. In the latter the methyl group is *equatorial* and does not hinder the OH group. The explanation proposed by Eliel lies in the fact that the esterification of the OH group takes place from the *equatorial* position. Neomenthol and neoisomenthol must be converted to the other chair forms before reaction. Since this is easier for neoisomenthol, (XXXV) as suggested above, than for neomenthol (XXXIII), for which two alkyl groups must then take up the unfavorable *axial* position, neoisomenthol reacts more rapi-

---

[188] E. L. Eliel, *Experientia* **9**, 91 (1953).

[189] Compare also (a) W. Hückel, H. Feltkamp, and S. Geiger, *Ann. Chem.* **624**, 142 (1959); (b) W. Hückel, ref. 40.

[190] J. Read and W. J. Grubb, *J. Chem. Soc.* p. 1779 (1934).

dly. Differences in the esterification and saponification rates of menthol and neomenthol had also been observed earlier by other investigators. Neomenthol reacted more slowly than menthol with acetic acid. The acid phthalate, as well as the acetate, of neomenthol are saponified with more difficulty than those of menthol.[191,192] These facts support the assigned conformations of these compounds.

Eliel's[188] conclusion that the isomeric menthols react only via the *equatorial* position of the hydroxyl, accounting in this way for the rate differences, was rejected by Winstein.[89] It was reasoned that conformations with an *equatorial* hydroxyl group could be disregarded in the case of neomenthol (XXXIII). Neomenthol exists mainly in the conformation with the *axial* hydroxyl group in equilibrium.

The solvolysis rates of the toluene sulfonates has also been tried as a means of conformational analysis[40] of the isomeric menthols. From the investigations of Hückel[40] *et al.*, however, it became apparent that this method is only useful to a limited extent for a quantitative conformational analysis of the isomeric menthols and other alkylated cyclohexanols.

Infrared spectroscopy has yielded further information. The differences between the *equatorial* and *axial* OH stretching vibrations have been used for the determination of the stable conformations of the isomeric menthols,[132] although the frequency differences are too small for the detection of less stable conformers (ν *axial* OH group $>$ ν *equatorial* OH group in $cm.^{-1}$). The maximum for the OH stretching vibration of menthol lies at 3628 $cm.^{-1}$, and of isomenthol at 3627 cm^{-1}, whereas neomenthol and neoisomenthol both absorb at 3632 $cm.^{-1}$ (measured in carbon tetrachloride solution). From these facts we can derive a conformation with an *equatorial* hydroxyl group for menthol and isomenthol and an *axial* hydroxyl group for neomenthol and neoisomenthol.

The isomeric menthols were investigated along with other *bis*-alkylated cyclohexanols[193] in the region of the C—(OH) stretching vibration, where frequency differences have been established between *equatorial* and *axial* hydroxyl groups (e.g., *equatorial* OH group $\sim$ 1060 $cm.^{-1}$, *axial* OH group $\sim$ 955 $cm.^{-1}$, measurements in carbon disulfide). These data suggest that menthol (XXXII) and neomenthol (XXXIII) exist in the conformations with *equatorial* and *axial* OH groups, respectively. For isomenthol (XXXIV) a homogeneous conformation with *equatorial* OH group is also derived, while a conformational equilibrium

[191] G. Vavon and A. Couderc, *Compt. Rend. Acad. Sci.* **179**, 405 (1924), *Bull. Soc. Chim. France* **39**, 666 (1926).

[192] Cf. also O. Zeitschel and H. Schmidt, *Ber.* **59**, 2298 (1926).

[193] W. Hückel and J. Kurz, *Ann. Chem.* **645**, 194 (1961).

is not excluded for neoisomenthol (XXXV). Quantitative investigations of the conformational equilibria are not available.

Investigations of equilibrium position gave information concerning the relative stability of the isomeric menthols.[189] If a mixture of the four isomeric menthols, obtained by the catalytic hydrogenation of thymol, is treated with Raney nickel at 250 atmospheres and 200°, the following equilibrium composition is obtained: 57% menthol, 29% neomenthol, 14% isomenthol, and 0% neoisomenthol. These values agree well with the stabilities expected from conformational considerations. The energetically quite unfavorable neoisomenthol disappears entirely at the attainment of equilibrium.

(ii) Dimethylcyclohexanols

The conformations of the 2,5-dimethylcyclohexanols, corresponding in their configuration to the menthols, also has been investigated by solvolysis of the toluenesulfonates, saponification of their carboxylic esters, infrared spectra, and equilibria.[189,193] The three dimethylcyclohexanols whose configurations are analogous to menthol, neomenthol, and isomenthol possess the same conformations as these compounds. The isomer corresponding to neoisomenthol behaves differently. The saponification rate of the phthalate, in comparison to that of neoisomenthol, is larger by about a factor of 7, whereas the solvolysis of the toluenesulfonate proceeds more slowly. It does not show the high reaction rate of an *axial* tosylate. Examination of the C—(OH) stretching vibration shows a band of strong intensity in the *equatorial* region.

From these results it follow that, in contrast to neoisomenthol, the dimethylcyclohexanol of corresponding configuration, has an equilibrium conformation with the hydroxyl group preferentially in an *equatorial* position (XXXVI). The equilibrium investigations carried out under the same conditions as for the isomeric menthols also agree with this conclusion, i.e., 14% of this dimethylcyclohexanol was found to be present.[189] A preponderance of the conformation with the *equatorial* hydroxyl group is energetically preferred in contrast to neoisomenthol which has an *axial* OH. In this form only one substituent, a methyl group, exists in an *axial* position; the other two (OH and CH_3) are *equatorial* (XXXVI).

(iii) Carvomenthols

The conformational observations made for the menthol series[194] can also be carried out for the isomeric carvomenthols (XXXVII).

[194] (a) Concerning the conformations of the so-called *n*-menthols: (these are similar in configuration to the menthols; but, in place of the isopropyl group, they have an *n*-propyl

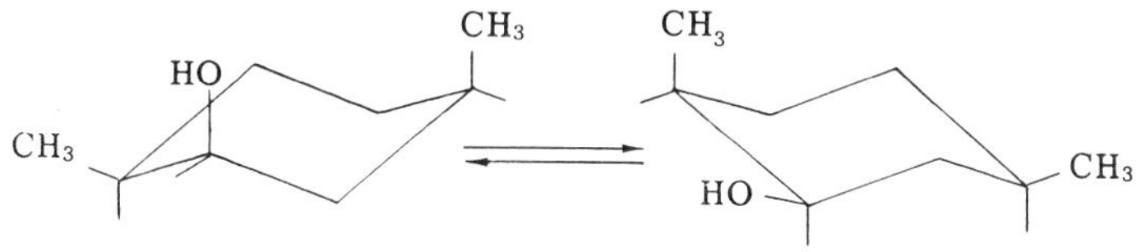

XXXVI

(Carvomenthol: e-isopropyl, e-OH, e-methyl; neocarvomenthol: e-isopropyl, a-OH, e-methyl; isocarvomenthol: e-isopropyl, a-OH, a-methyl; neoisocarvomenthol: e-isopropyl, e-OH, a-methyl.) Except for the interpretation of various chemical reactions, more detailed experimental investigations are not yet available.[195] For the *trans* relationship of the

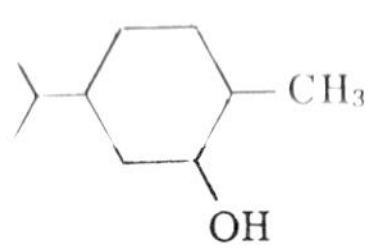

XXXVII

isopropyl and methyl groups, the normally expected conformations with an *equatorial* isopropyl group would result.[188,195] In the *cis* series, exceptions again occur for the isomers which correspond in configuration to isomenthol and neoisomenthol. The hydroxyl group in the *cis* series comes close to the *axial* methyl group. There arises a conformation for isocarvomenthol,[195c] with hydroxyl and methyl groups *trans* to each other, having two substituents in *axial* positions. This conformation should not be very stable, even though it is not as unfavorable energetically as neoisomenthol. In this case the *axial* substituents are located on different sides of the ring (XXXVIII).

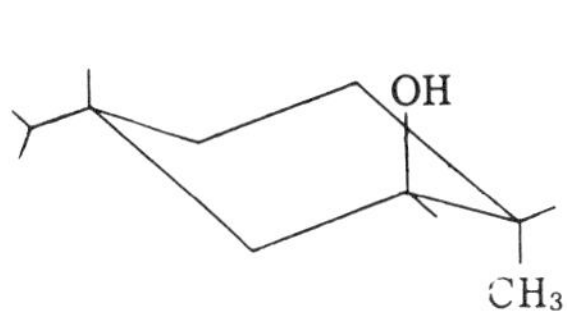

XXXVIII

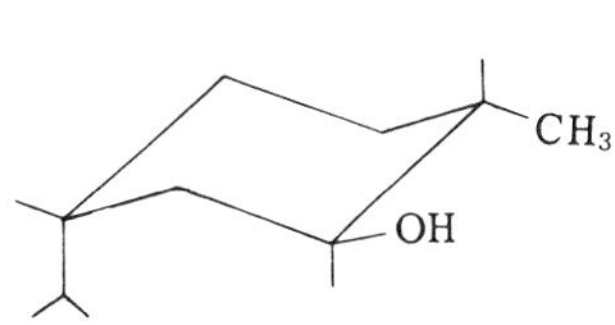

XXXIX

group) cf. Ref. 138; (b) R. Cornubert, F. Lainé, and H. Lemoine, *Bull. Soc. Chim. France* p. 1340 (1958); (c) F. Lainé, H. Lemoine, and R. Cornubert, *ibid.* p. 1346 (1958); R. Cornubert and H. Lemoine, *ibid.* p. 1747 (1960).

[195] (a) A. K. Bose, *Experientia* **8**, 458 (1952); (b) compare also: Y.R. Naves, *Helv. Chim. Acta*, **67**, 308 (1964); (c) for the correct configurations of the carvomenthols see E. E. Royals and J. C. Leffingwell, *J. Am. Chem. Soc.* **86**, 2067 (1964).

It remains to be seen how much of the conformation shown as (XXXIX) is present at equilibrium.[196] If the hydroxyl group is introduced into a position *cis* to the methyl group, no extraordinary situation arises, since both the hydroxyl and isopropyl groups are *equatorial*.[197]

Similar considerations can be used for the isomeric menthylamines and carvomenthylamines,[195] which can be assigned the same conformations as the alcohols—menthylamine: e-isopropyl, e-NH_2, e-methyl; neomenthylamine: e-isopropyl, a-NH_2, e-methyl; isomenthylamine: e-isopropyl, e-NH_2, a-methyl; neoisomenthylamine: e-isopropyl, a-NH_2, a-methyl.

Correspondingly, the same situation should hold for carvomenthylamines (see conformations of the carvomenthols, p. 135). More detailed investigations concerning the positions of the conformational equilibria are not available.[196b]

(iv) Halocyclohexanes

There is only a small amount of experimental data available concerning trihalogenated cyclohexanes. 1,2,3-Tribromocyclohexane which originally was assigned to the e,e,e conformation[198] (m.p. 51°) was investigated with the help of electron diffraction methods.[199] It could be determined that a mixture exists in the gas phase consisting of ~80% e,e,e and ~20% a,a,a conformation of this compound. An energy difference of 1.2 kcal./mole between a,a,a and e,e,e-1,2,3-tribromocyclohexane was determined.[198]

5. Polysubstituted Cyclohexane Compounds[200]

The introduction of more than three substituents into the cyclohexane ring leads to a further increase in the number of possible configurational isomers (cf. the trisubstituted compounds p. 129).[201] If we disregard,

[196] (a) Investigations concerning the conformations of the isomeric carvomenthols are in progress. W. Hückel *et al.* unpublished; (b) cf. S. Schröter and E. L. Eliel, *J. Am. Chem. Soc.* **86**, 2066 (1964).

[197] Further investigations of trisubstituted cyclohexanes: cf. for example (a) O. Hassel and P. Andersen, *Acta Chem. Scand.* **2**, 527 (1948); (b) D. H. R. Barton and G. A. Schmeidler, *J. Chem. Soc.* p. 1197 (1948); (c) M. I. Batuev, A. A. Akrem, A. V. Kamernitzki, and A. D. Matveeva, Izvest. Akad. Nauk SSSR p. 1665 and 1668 (1959); (d) E. G. Peppiatt and R. J. Wicker, *J. Chem. Soc.* p. 3122 (1955).

[198] (a) R. Cornubert, A. Rio, and P. Senechal, *Bull. Soc. Chim. France* p. 46 (1955); (b) R. Cornubert, *ibid.* p. 996 (1956).

[199] V. A. Atkinson and K. Thurmann-Moe, *Acta Chem. Scand.* **14**, 497 (1960).

[200] See also the detailed description of H. D. Orloff, *Chem. Rev.* **54**, 347 (1954).

[201] A description of the configuration will be ignored in the following. In principal it can be done, as for the trisubstituted compounds, by assigning the relative configuration between a pair of substituents when the substituents are numbered.

for simplification, those position isomers which arise by the introduction of two substituents on a carbon atom, i.e., substitute each ring carbon with only one substituent, and finally consider only those situations where all substituents are identical, then there would be three position isomers for a tetrasubstituted cyclohexane. They are the 1,2,3,4-, 1,2,3,5-, and 1,2,4,5 compounds. Each of these position isomers can exist in various configurations or conformations. The number of theoretically possible conformations for the 1,2,3,4 compound is ten, for the 1,2,3,5 compound twelve, for the 1,2,4,5 compound seven.[202] As a consequence of the possibility of converting two conformations into each other by ring conversion, the number of separable forms (configurational isomers) is reduced to six in the case of the 1,2,3,4 and 1,2,3,5 compounds and for the 1,2,4,5 compound to five.

For example the conformations given in Table VI result for 1,2,3,4-tetrasubstituted cyclohexanes.

TABLE VI

POSSIBLE CONFORMATIONS OF 1,2,3,4-TETRASUBSTITUTED CYCLOHEXANES

1	eeee	⇄	aaaa
2	eeea	⇄	aaae
3	eeae	⇄	aaea
4	eaae	⇄	aeea
5	eeaa	⇄	(aaee)
6	eaea	⇄	(aeae)

The conformation in the right-hand column of Table VI forms from that in the left-hand column by chair–chair interconversions and *vice versa*. The conformations in parentheses are not counted because, as is apparent from a model, they are identical with the starting conformation.

[202] The determination of the number of position isomers as well as the number of theoretically possible conformations of substituted cyclohexanes with from one to twelve identical substituents ($C_6H_{11}X$ to C_6X_{12}) was accomplished by R. Riemschneider and P. Geschke, *Angew. Chem.* **65**, 390 (1953); R. Riemschneider, *Österr. Chem. Ztg.* **55**, 102, 161 (1954); R. Riemschneider and I. Geschke, *Sci. Insect Control* (*Japan*) **20**, 31 (1955); compare also R. Riemschneider, *Österr. Chem. Ztg.* **56**, 133 (1955). It was done with the help of formulae developed by the authors as well as counting all nonidentical conformations and summarizing in so-called configurational tables. Taking into account the possibility that two substituents can be on the same carbon atom, there are altogether 91 position isomers from $C_6H_{11}X$ to C_6X_{12}, which can be summarized as follows: $C_6H_{11}X$ 1, $C_6H_{10}X_2$ 4, $C_6H_9X_3$ 6, $C_6H_8X_4$ 12, $C_6H_7X_5$ 13, $C_6H_6X_6$ 18, $C_6H_5X_7$ 13, $C_6H_4X_8$ 12, $C_6H_3X_9$ 6, $C_6H_2X_{10}$ 4, C_6HX_{11} 1, C_6X_{12} 1. For $C_6H_{10}X_2$ the following would be possible: 1,1; 1,2; 1,3; 1,4 = 4. For $C_6H_{10}X_3$: 1,1,2; 1,1,3; 1,1,4; 1,2,3; 1,2,4; 1,3,5 = 6, etc.

There are ten conformations theoretically possible which, as has been mentioned, are reduced to six configurational isomers due to the inability to separate the conformers. Two of them (Nos. 1 and 4 in Table VI) are optically inactive (*meso* forms) because they possess a plane of symmetry. Nos. 2, 3, 5, and 6, consist of resolvable *d,l* forms. Which of the two interchangeable conformations of the various pairs is more stable depends on the type and size of the substituents. Because of the fact that the presence of a number of *axial* substituents results in 1,3 interactions, it can be assumed that the rule that the conformation with the most *equatorial* substituents is the more stable one is also valid here.

For the 1,2,3,5 compound there are six separable configurational isomers, as mentioned above, due to the same considerations. For the 1,2,4,5 compound there are five separable configurational isomers.

With five identical substituents, using the above-mentioned provisions, it is possible to have the substituents in only the 1,2,3,4,5 positions. Twenty conformations are theoretically possible, which again can be reduced to ten separable configurational isomers due to the interchangeability of conformations. The resulting conformations are given in Table VII. Of these, Nos. 1, 6, 8 and 9 (Table VII) are optically inactive (*meso* forms), because they possess a plane of symmetry. The isomers 2, 3, 4, 5, 7, and 10 (Table VII) consist of resolvable *d,l* forms.

TABLE VII

POSSIBLE CONFORMATIONS OF PENTASUBSTITUTED CYCLOHEXANES

1	e e e e e ⇄ a a a a a
2	a e e e e ⇄ e a a a a
3	a a e e e ⇄ e e a a a
4	a e a e e ⇄ e a e a a
5	a e e a e ⇄ e a a e a
6	a e e e a ⇄ e a a a e
7	e a e e e ⇄ a e a a a
8	e a e a e ⇄ a e a e a
9	e e a e e ⇄ a a e a a
10	e e a a e ⇄ a a e e a

For six identical substituents in positions 1, 2, 3, 4, 5, 6 on the cyclohexane ring, thirteen conformations are possible as Hassel[74,202a] and Bijvöet[203] were able to show with hexachlorocyclohexane (Table VIII).

[202a] O. Hassel, *Quart. Rev.* (*London*) **7**, 221 (1953).
[203] J. M. Bijvöet, *Rec. Trav. Chim.* **67**, 777 (1948).

TABLE VIII

POSSIBLE CONFORMATIONS OF HEXASUBSTITUTED CYCLOHEXANES

		1a–3a Interactions			1a–3a Interactions	Isomeric hexachloro-cyclohexane[204]
1	e e e e e e	(0)	⇄	a a a a a a	(6)	β
2	a e e e e e	(0)	⇄	e a a a a a	(4)	δ
3	a a e e e e	(0)	⇄	e e a a a a	(2)	α
4	a e e a e e	(0)	⇄	e a a e a a	(2)	ϵ
5	a e a e e e	(1)	⇄	e a e a a a	(3)	η
6	a a e a e e	(1)	⇄	(e e a e a a)	(1)	ζ
7	a e a e a e	(3)	⇄	(e a e a e a)	(3)	ι *
8	a a a e e e	(1)	⇄	(e e e a a a)	(1)	γ

* Unknown.

The sixteen theoretically possible conformations according to Table VIII are reduced, as indicated, to thirteen, because conformers Nos. 7 and 8 are identical with the starting conformation and No. 6 has interchangeable *d* and *l* forms through conversion of the ring (in Table VIII set in parentheses). Because of ring interconversion, there are necessarily only eight separable isomers present. Except for No. 3 all are optically inactive. In Table VIII an attempt at clarification of the stabilities of the conformers is added by indicating the number of 1,3 *diaxial* chlorine interactions. (These numbers are in parentheses.) The more chlorine atoms found in an *axial* position the less stable the conformation becomes. Two chlorine atoms in 1,3 *diaxial* positions are still possible without giving rise to a decided distortion of the molecule. If three chlorine atoms are *axial*, as for γ-hexachlorocyclohexane (No. 8 in Table VIII), then, due to spatial proximity of the substituents, a distortion of the tetrahedral angles of the molecule[205] occurs.

With the introduction of seven substituents into the cyclohexane ring, thirteen position isomers are possible[202]. Of the 1,1,2,3,4,5,6-compounds investigated to date, the heptachloro cyclohexanes are

[204] The Greek letters α, β, γ ... have no connection with the steric structure of hexachlorocyclohexane, but are trivial designations representing the order of discovery. Cf. R. Riemschneider, *Z. Naturforsch.* **10b**, 177 (1955). The method of notation proposed by Riemschneider was used: R. Riemschneider, *Z. Naturforsch.* **10b**, 605 (1955); R. Riemschneider, M. Spät, W. Rausch, and E. Böttger, *Monatsh. Chem.* **84**, 1068 (1953).

[205] G. W. van Vloten, C. A. Kruissink, B. Strijk, and J. M. Bijvöet, *Acta. Cryst.* **3**, 139 (1950); compare also R. Riemschneider, *Z. Angew. Entomol.* **48**, 423 (1961).

especially mentioned.[200] Of twenty theoretically possible conformations, ten separable isomers exist.

In the case of polysubstituted cyclohexanes with different substituents, the possible configurations and conformations can also be ascertained according to the methods of Riemschneider.[205a] The number of possibilities is much greater in these cases. Investigations of the conformations of polysubstituted cyclohexanes, notably the halogen-substituted cyclohexanes, have been undertaken with the help of physical methods such as electron diffraction, X-ray diffraction, infrared spectroscopic and nuclear magnetic resonance investigations and measurements of the dipole moments. The chemical methods for conformational elucidation, employed with previously mentioned compounds, play a secondary role.

A very detailed account of polysubstituted cyclohexanes is beyond the scope of this review. We shall restrict ourselves to a short tabulation of the results: For halogenated cyclohexanes, the tetra-, penta-, hexa-, hepta-, and octachlorocyclohexane configurations and conformations have been partially elucidated by the methods mentioned above, as they also have been for other cyclohexanes with mixed substituents.[206]

The hexachlorocyclohexanes will be given special treatment; their conformations (Table VIII) were defined by the basic investigations of Hassel[207] and his co-workers by means of electron diffraction in the gas phase. The conformations of the α, β, γ, δ, and ϵ isomers could be

[205a] R. Riemschneider, *Z. Naturforsch.* **10b**, 183, 605 (1955).

[206] Ref. 200 with a review of the literature to 1953. (a) R. Riemschneider, H. Karl, and R. Bös, *Ann. Chem.* **580**, 192 (1953) with further literature; (b) O. Hassel, Chr. Römming, and E. Hadler-Vihovde, *Acta. Chem. Scand.* **8**, 788 (1954), (1,1,4,4-tetrachlorocyclohexane); (c) E. Wang Lund, *ibid.* **8**, 1378 (1954), (1,2-dichloro-4,5-dibromocyclohexane); (d) O. Bastiansen, O. Hassel, and E. Hadler-Vihovde, *ibid.* **8**, 1951 (1954), (1,2,3,4-tetrahalogencyclohexanes); (e) R. Riemschneider and S. Bäker, *Z. Naturforsch.* **9b**, 751 (1954), (dibromo-tetrachlorocylcohexanes); (f) R. Riemschneider, *Chem. Ber.* **88**, 1437 (1955), (dibromotetrachloro- and tetrabromodichlorocyclohexanes); (g) R. Riemschneider and W. Triebel, *ibid.* **88**, 1442 (1955), (bromotrichloro- and bromopentachlorocyclohexane); (h) R. Riemschneider, *ibid.* **88**, 1694 (1955), (octachlorocyclohexane); (i) K. Lunde, *Acta. Chem. Scand.* **10**, 690 (1956), (1,2-dichloro-4,5-dibromocyclohexane), see ref. 199b; (j) R. Riemschneider and G. Mau, *Chem. Ber.* **90**, 2713 (1957), (1-iodo-2,3,4,5,6-pentachlorocyclohexane, 1-iodo-2,4,5-trihalocyclohexane); Concerning the polyfluorocyclohexanes see: (k) R. P. Smith and J. C. Tatlow, *J. Chem. Soc.* p. 2505 (1957) with further references; (l) R. Stephens and J. C. Tatlow, *Chem. & Ind. (London)* p. 821 (1957); (m) D. E. M. Evans, J. A. Godsell, R. Stephens, J. C. Tatlow, and E. H. Wiseman, *Tetrahedron* **2**, 183 (1958); (n) J. A. Godsell, M. Stacey, and J. C. Tatlow, *ibid.* **2**, 193 (1958); (o) D. Steele and D. H. Whiffen, *ibid.* **3**, 181 (1958); (p) R. Stephens, J. C. Tatlow, and E. H. Wiseman, *J. Chem. Soc.* p. 148 (1959); (g) E. Nield, R. Stephens, and J. C. Tatlow, *ibid.* p. 159 (1959); (r) J. Feeney and L. H. Sutcliffe, *Trans. Faraday Soc.* **56**, 1559 (1960); (s) G. van Dyke Tiers, *Proc. Chem. Soc.* p. 389 (1960).

[207] O. Bastiansen, O. Ellefsen, and O. Hassel, *Research (London)* **2**, 248 (1949); *Acta. Chem. Scand.* **3**, 918 (1949); Cf. also N. Norman, *ibid.* **4**, 251 (1950) and ref. 205.

determined, and resulted in the observation that the conformational equilibria lie practically entirely on the side of the more stable conformer[208] (Table VIII). The indicated conformations of the hexachlorocyclohexanes were also verified by measurements of the dipole moments.[209]

A series of investigations is also available which describes the elucidation of the conformations of other polysubstituted cyclohexanes.[210] Included are the polyhydroxycyclohexanes (cyclitols), of which the hexahydroxycyclohexanes (inositols) have been very thoroughly investigated because of their wide occurrence in nature.[210,211] All of the eight separable isomers are known, and thus they form the only series of the hexasubstituted cyclohexane for which all isomers have been isolated. Although the conformations of the inositols have not been subjected to a direct analysis using physical methods, an assignment of the conformations has been possible (Table IX). The methods

TABLE IX

CONFORMATION OF THE INOSITOLS[a]

		Prefix[b]
1	e e e e e e	scyllo
2	a e e e e e	myo (meso)
3	a a e e e e	dextro (or levo)
4	a e e a e e	neo
5	a e a e e e	epi
6	a e a e a e	cis
7	a a a e e e	muco
8	a e a a e e	allo

[a] Only the stable conformers are listed.
[b] See ref. 212.

[208] Cf. also R. Riemschneider, *Chem. Ber.* **89**, 2713 (1956); H. Elias, K. H. Lieser, and H. W. Kohlschütter, *Chem. Ber.* **93**, 2128 (1960).

[209] Cf. E. L. Lind, M. E. Hobbs, and P. N. Gross, *J. Am. Chem. Soc.* **72**, 4474 (1950); E. Hetland, *Acta. Chem. Scand.* **2**, 678 (1948).

[210] Cf. ref. 200, E. G. Peppiatt and R. J. Wicker, *J. Chem. Soc.* p. 3122 (1955); R. J. Wicker, *ibid.* p. 2165 (1956); R. J. Jeffries and B. Milligan, *ibid.* p. 4384 (1956); R. J. Wicker, *ibid.* p. 3299 (1957); J. Alkonyi, *Chem. Ber.* **92**, 1130 (1959), *ibid.* **93**, 1449 (1960); R. Wicker, *ibid.* **93**, 1448 (1960).

[211] Summary: S. J. Angyal, *Quart. Rev.* (*London*) **11**, 212 (1957), where further literature is cited; See also S. J. Angyal and D. S. McHugh, *J. Chem. Soc.* p. 3682 (1957); S. J. Angyal, M. Tate, and S. D. Gero, *ibid.* p. 4116 (1961); S. J. Angyal and R. M. Hoskinson, *ibid.* p. 2985, 2991 (1962); cf. also G. R. Barker, *ibid.* p. 624 (1960).

[212] For the nomenclature of the inositols see ref. 211 as well as H. G. Fletcher, Jr., L. Anderson, and H. A. Lardy, *J. Org. Chem.* **16**, 1238 (1951); S. J. Angyal and P. I. Gilham, *J. Chem. Soc.* p. 3691 (1957).

consisted of a comparison with other polysubstituted cyclohexanes and with sugars, (p. 324) as well as studies of chemical reactions such as biological dehydrogenation,[211] formation of isopropylidene derivatives[211] and formation of borate complexes.

Except for No. 3 (*d* or *l*), which is optically active, all others are *meso* forms. Scylloinositol (No. 1) is the most stable isomer (Table IX). An estimate of the free energy differences of the inositols was undertaken by Angyal on the basis of the equilibrium constants of the complex formation with boric acid.[174] According to this data, myoinositol (one *axial* hydroxyl group) is ~0.9 kcal./mole higher in energy than scyllo inositol; epiinositol (two *axial* hydroxyl groups on the same side of the ring) is ~2.8 kcal./mole higher; and *cis*-inositol (three *axial* hydroxyl groups on the same side of the ring) is ~5.7 kcal./mole higher.

6. Conformation and Physical Properties

Aside from chemical and physical investigations, as well as energy considerations, there have been reference to various other physical properties of molecules, which have been used to draw conclusions concerning the configuration and conformation of a molecule. Among these parameters are—boiling point, density and refractive index. These were summarized in the so-called Auwers-Skita rule (see p. 86) which stated that the *cis* configuration of a compound has the higher boiling point,[212a] density, and refractive index. For alcohols, measurements of intermolecular association and viscosity can be used for configurational and conformational elucidation within certain limits.

We may once again consider some physical constants of the isomeric dimethylcyclohexanes,[213] whose configurations (see page 110) are established, so as to test the Auwers-Skita rule (Table X).

A comparison of the listed constants with the empirical Auwers-Skita rule shows that the rule only holds for the 1,2 and 1,4 compounds. The opposite is true for the 1,3 compounds. For other 1,3-disubstituted cyclohexane derivatives as well, the Auwers-Skita rule does not hold (cf. p. 110). The use of this rule led, in many cases, to the assignment of

[212a] Concerning the boiling point see: H. van Bekhum, A. van Veen, P. E. Verkaade, and B. M. Wepster, *Rec. trav. Chim.* **80**, 1310 (1961), who suggested that the Auwers-Skita rule or the conformational rule (see below) should be applied only to the density and refractive index, since the rule frequently fails for the boiling point.

[213] "Selected Values of Physical and Thermodynamic Properties of Hydrocarbons and Related Compounds" American Petroleum Institute Research Project **44**, Carnegie Press, 1953.

TABLE X

PHYSICAL CONSTANTS OF DIMETHYLCYCLOHEXANES

	Boiling point	n_D^{25}	d_4^{25}
cis-1,2	129.7	1.4336	0.7922
trans-1,2	123.4	1.4247	0.7720
cis-1,3	120.1	1.4206	0.7620
trans-1,3	124.5	1.4284	0.7806
cis-1,4	124.3	1.4273	0.7787
trans-1,4	119.4	1.4185	0.7584

the wrong configurations before the correct configurations were determined by other methods (see p. 110)[214]

On the basis of the true geometry of the cyclohexane ring, whose isomer relationship is based on the various possible conformations, the Auwers-Skita rule has been revised by Allinger[215] as follows: for alicyclic epimers, which do not differ in their dipole moments, "that isomer which has the highest boiling point,[212a] highest index of refraction, and highest density is the isomer which possesses the least stable configuration," or, expressed in another way, which possesses the highest heat content (enthalpy). In other words this so-called conformational rule indicates that those isomers which are capable of assuming a total *equatorial* conformation, in comparison with isomers having axial substituents, possess the lower boiling point, refractive index, and density.

Assuming a nearly ideal behavior of two isomeric forms of a molecule, the larger density and higher refractive index of a configuration can be related to a smaller molecular volume.[215]

The higher boiling point can be related to the smaller molecular volume under the same assumptions, even though the relation is less clear here. The smaller molecular volume leads to an increase in enthalpy because of the closer packing of the atoms and the resulting increase in repulsive forces. At the same time, this decreases the stability of the compound and also decreases the distance between the various molecules. The van der Waals attraction is increased, which leads to an increase in boiling point.

[214] Compare also: B. A. Kazanskii, A. L. Liberman, and N. J. Tyun'kin, *Dokl. Akad. Nauk, SSSR* **134**, 93 (1960); G. J. Nikishín, V. D. Vorobev, and E. D. Lubuzh, *Zh. Obshih. Khim.* **30**, 3548 (1960).

[215] (a) N. L. Allinger, *Experientia* **10**, 328 (1954); cf. also (b) N. L. Allinger, *J. Org. Chem.* **21**, 915 (1956); (c) N. L. Allinger, *J. Am. Chem. Soc.* **79**, 3443 (1957); (d) N. L. Allinger, *ibid.* **80**, 1953 (1958); (e) N. L. Allinger, *ibid.* **81**, 232 (1959); (f) N. L. Allinger and R. J. Curby, Jr., *J. Org. Chem.* **26**, 933 (1961).

The conformational rule is no longer valid for isomers which differ greatly in their dipole moment; instead, the so-called van Arkel or dipole rule is applied.[215c,216] The latter rule was designed initially for *cis–trans* isomeric olefins[216] and states the following: "The isomer with the higher dipole moment possesses the higher boiling point, refractive index, and density." In the cases investigated, especially by Allinger,[215c,217] it was shown that the three stated physical properties follow this rules.

If an alkyl and a polar group such as hydroxy or amino are on the ring, a generally valid rule can no longer be assumed at least for the boiling point, according to available material. Eliel[218,219] was able to show this with the isomeric 2-, 3-, and 4-methyl cyclohexanols. By using the conformational rule,[220] the *diequatorial* compounds (*trans*-2, *cis*-3, and *trans*-4-methylcyclohexanol) should have the lower boiling point, refractive index, and density compared to their partially *axial* epimers (*cis*-2, *trans*-3, and *cis*-4-methylcyclohexanol). From careful measurements of these physical constants[218,221] it is apparent that the refractive

TABLE XI

Boiling Points of Methylcyclohexanols

Methylcyclohexanol		Boiling Point (°C)
cis-2	(e, a)	165°/760 mm
trans-2	(e, e)	166,5/760 mm
cis-3	(e, e)	173/745 mm
trans-3	(e, a)	168–169°/745 mm
cis-4	(e, a)	168–169°/745 mm
trans-4	(e, e)	172–173°/745 mm

index and density do follow the conformational rule, but the boiling point does not, as shown for the boiling point in Table XI. The *diequatorial*

[216] A. E. van Arkel, *Rec. Trav. Chim.* **51**, 1081 (1931); **53**, 246 (1934).

[217] A detailed discussion of the relation between conformation and physical properties appears under R. B. Kelly, *Can. J. Chem.* **35**, 149 (1957).

[218] E. L. Eliel and R. G. Haber, *J. Org. Chem.* **23**, 2041 (1958).

[219] Cf. also W. Hückel and A. Hubele, *Ann. Chem.* **613**, 27 (1958); W. Hückel, M. Maier, E. Jordan, and W. Seeger, *ibid.* **616**, 46 (1958); compare also H. van Bekhum, A. van Veen, P. E. Verkade, and B. M. Wepster, ref. 212a.

[220] Thereby is assumed that the dipole moments of the isomers do not vary greatly from each other, cf. ref. 221.

[221] Cf. also A. Skita and W. Faust, *Ber.* **64**, 2878 (1931).

isomer has the higher boiling point in all cases shown. Therefore, the conformational rule is reversed in the case of the boiling point. The reasons for this were given by Eliel as the formation of intermolecular hydrogen bonds, which leads to higher association and a resulting rise in the boiling point (see below). The formation of intermolecular hydrogen bonds[222] is favored in the isomers with *equatorial* hydroxyl groups (*trans*-2, *cis*-3, *trans*-4) compared with those with an *axial* hydroxyl group, because the *equatorial* hydroxyl is more accessible than the *axial* one. However, it is not certain if the boiling point is determined solely by the formation of hydrogen bonds. Examples are known[219] in which ethers corresponding to the isomeric alcohols show the same deviations from the conformational rule.[215]

The same considerations are valid for the dialkylated cyclohexanols, such as the 2,5-dimethylcyclohexanols, the menthols, and the carvomenthols. The conformational rule cannot be applied here to the boiling points. The compounds with an *equatorial* hydroxyl group (menthol, isomenthol, carvomenthol, isocarvomenthol,) boil higher than their isomers with *axial* hydroxyls.

The symmetry properties of the molecule also determine the boiling point within a series of dialkylated cyclohexanols, in agreement with the conformational rule. The 2,5-dimethylcyclohexanols belonging to the *cis*-series (the methyl groups are *cis*, one of them *axial*) boil higher than the corresponding alcohols in the *trans* series. For the isomeric menthols, the same effect is decisive for the boiling point differences (menthol, 216°; isomenthol, 219°). They vary only slightly in their association. For the same reason the boiling point of neoisomenthol (215°) is higher than that of neomenthol (212°). The same is valid for the corresponding menthyl- and carvomenthylamines,[187] although less definitive. The conformational rule is approximately followed[217] in the case of the density and refractive index of the carvomenthols, whereas the menthols are exceptions[217] here as well.

As mentioned above, the particular ability of alcohols to associate, i.e. formation of intermolecular hydrogen bonds, has been referred to for a conformational analysis of substituted cyclohexanols.[133a,134,222c] This can be carried out by a study of the OH stretching vibrations in the region of 3350 to 3650 cm.$^{-1}$,[133a,134] as well as cryoscopic molecular weight determinations and calculations of the degree of association.[222c] From available data, all alkylated cyclohexanols show bands for free and bound hydroxyl in the region of 3350–3650 cm.$^{-1}$ The frequencies of the

---

[222] Cf. (a) H. L. Goering, R. L. Reeves, and H. H. Espy, *J. Am. Chem. Soc.* **78**, 4926 (1956); (b) E. G. Peppiatt and R. J. Wicker, *J. Chem. Soc.* p. 3122 (1955); R. J. Wicker, *ibid.* p. 2165 (1956).

[222c] W. Hückel, H. Feltkamp, and S. Geiger, *Ann. Chem.* **637**, 1 (1960).

free hydroxyl of the *cis–trans* isomeric alcohols are only slightly different (see p. 113), but the values for the bound hydroxyl are spread out much more. A qualitative comparison of the spectra of the isomeric alcohols shows that the intensity of the bonded band for isomers with *axial* hydroxyl is comparatively smaller than for those with an *equatorial* hydroxyl group. Of the cases studied (2- and 3-alkylated cyclohexanols, with alkyl=methyl, ethyl, isopropyl, cyclohexyl; and 2,5-dimethylcyclohexanols, menthols) the following rule is in agreement with the above results: *axial*, or preponderantly *axial*, alcohols bond less strongly than *equatorial* ones, although the difference for the *cis*- and *trans*-3-alkylcyclohexanols is small. A neighboring group in the 2-*trans* position also decreases the bonding effect of *equatorial* hydroxyls greatly. At the same time, the difference in bonding between the *equatorial trans* and the *axial cis* isomers becomes greater. With increasing size of the neighboring group, the strength of the bonding diminishes.

The molecular weights determined cryoscopically in benzene, and the degree of bonding[222c] calculated from this, give the same sequence as the boiling points do for the isomeric menthols. Neomenthol is bonded the least, followed closely by neoisomenthol. Menthol and isomenthol are bonded decidedly more, with stronger dependence of the molecular weight on concentration. The measurements in cyclohexane show a stronger bonding for neomenthol than for neoisomenthol, with a simultaneously stronger concentration dependence of the molecular weight for the former. In cyclohexane solution too, menthol and isomenthol are also more strongly bonded than their epimers. The results are in good agreement with the assigned conformations of the isomeric menthols. If the hydroxyl group is *equatorial* (menthol and isomenthol), then a stronger bonding takes place than when the hydroxyl is in an *axial* position (neomenthol, neoisomenthol). The same results were obtained for the isomeric 2,5-dimethylcyclohexanols[222c]. The isomers corresponding to menthol and isomenthol are more strongly bonded than the isomers corresponding to neomenthol. The 2,5-dimethylcyclohexanol (corresponding to neoisomenthol), of all the four isomeric alcohols, is bonded the most and shows a great dependence of concentration on molecular weight. In agreement with the above-derived (p. 134) conformation, this alcohol has an *equatorial* hydroxyl group.

7. Cyclohexene

Investigations of the conformations which result from the introduction of a double bond into the cyclohexane ring are smaller in number. As a result, we cannot give the same certainty to conformational considerations of cyclohexene as was possible with cyclohexane.

From the investigations and energy considerations carried out so far, the following can be stated concerning the structure of cyclohexene: Cyclohexene can also exist in two conformations which correspond to the chair and the boat form of cyclohexane.[223]

If we designate the two trigonal carbon atoms of the double bond as 1 and 2, then the neighboring carbon atoms 3 and 6 lie in the same plane, while carbon atom 5 lies above and carbon atom 4 below the plane of the ring. The resultant conformation is called "half-chair."[224]

If we look in the direction of the plane of the ring, then the interchangeable conformations can be depicted as projections (XL) or perspectives (XLI), as in the case of cylcohexane:

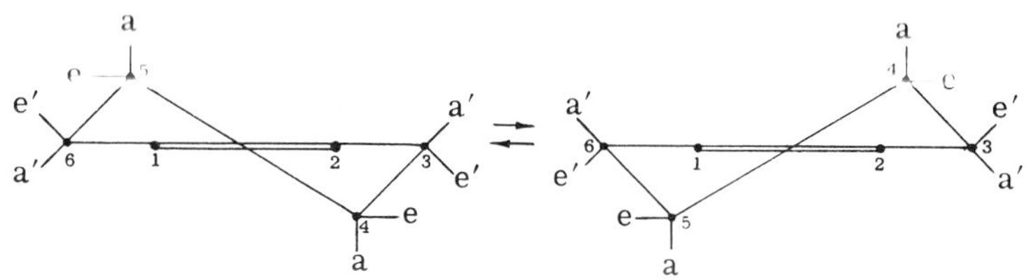

XL

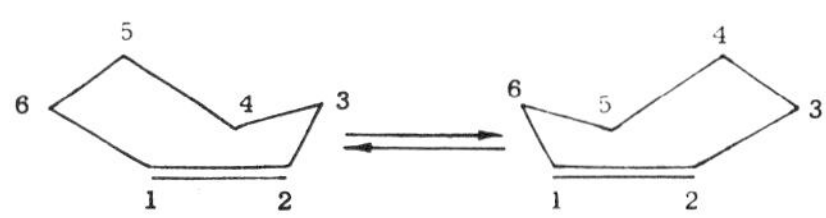

XLI

A "half-boat" conformation is also possible for cyclohexene, with the carbon atoms 1, 2, 3, and 6 in a plane, while the carbon atoms 4 and 5 both lie above this plane. (XLII—projection; XLIII—perspective).

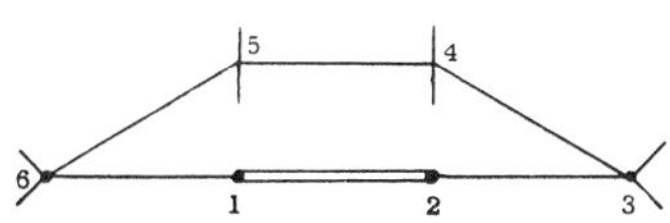

XLII

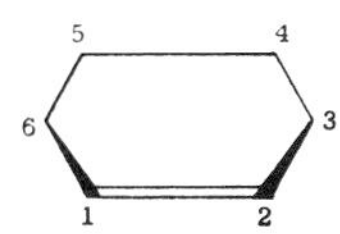

XLIII

[223] (a) R. A. Raphael and J. B. Stenlake, *Chem. & Ind.* (*London*) p. 1286 (1953); (b) D. H. R. Barton, R. C. Cookson, W. Klyne, and C. W. Shoppee, *ibid.* p. 21 (1954); (c) E. J. Corey and R. A. Sneen, *J. Am. Chem. Soc.* **77**, 2505 (1955); (d) R. Bucourt, *Bull. Soc. Chim. France* p. 1262 (1963); R. Bucourt and D. Hainaut, *Compt. Rend. Acad. Sci.* **258**, 3305 (1964).

[224] M. W. Lister, *J. Am. Chem. Soc.* **63**, 143 (1941).

Of the two conformations, half-chair and half-boat, the first is the more stable. We can convince ourselves of this by a study of a model and consideration of nonbonded hydrogen interactions. The energy difference between the two conformations was estimated at 2.7 kcal./mole by Pitzer *et al.*[36] The measurement of the infrared and Raman spectra of cyclohexene in the gaseous, liquid, and solid phase, yielded bands for the skeletal and CH vibrations which indicate a half-chair conformation.[225] The half-chair conformation was further demonstrated with more certainty by X-ray and electron diffraction studies of derivatives such as 3,4,5,6-tetrachlorocyclohexene[226] and pentachlorocyclohexene.[227]

A study of the model (XL) demonstrates that the C—H bonds in the half-chair conformation of cyclohexene are not geometrically equivalent. Relative to each other, the C—H bond on C-3, C-4, C-5, and C-6. assume the *staggered* conformation. The two carbon atoms C-4 and C-5, which do not lie in the plane, have one *axial* and one *equatorial* C—H bond, as does cyclohexane. On carbon atoms C-3 and C-6, lying in the plane, the direction of the C—H bonds varies from a true *axial* or *equatorial* orientation. They are therefore called "*quasi-axial*" (a′) and "*quasi-equatorial*" (e′).[223]

The few data available in the literature permit no sure conclusions about the differences in stability of substituents in the *equatorial* or *quasi-equatorial* position and *axial* or *quasi-axial* positions on the cyclohexene ring.[222a] Sakashita[228] has concluded recently from infrared and Raman spectra that in the liquid state a halogen is more stable in the *axial* 3- and 4-position than in the *equatorial* position. Though investigations are available showing that *trans*-1,2-disubstituted cyclohexenes are more stable than the corresponding *cis* compounds[229] and thus correspond to the situation with cyclohexanes, the example of a 3,5-disubstituted cyclohexene with the *trans* isomer more stable than the *cis* isomer no longer corresponds to the cyclohexane analog.[217,230,231]

---

[225] K. Sakashita, *J. Chem. Soc. Japan* **77**, 1094 (1956); cf. also K. Sakashita, *Nippon Kagaku Zasshi* **79**, 329 (1958).

[226] O. Bastiansen and J. Markali, *Acta. Chem. Scand.* **6**, 442 (1952); O. Bastiansen, *ibid.* **6**, 875 (1952); cf. also H. D. Orloff, A. J. Kolka, G. Calingaert, M. E. Griffing, and E. R. Kerr, *J. Am. Chem. Soc.* **75**, 4243 (1953); A. J. Kolka, H. D. Orloff, and M. E. Griffing, *ibid.* **76**, 1244 (1954); (pentachloro- and hexachlorocyclohexene) cf. ref. 200; R. Riemschneider, *Angew. Chem.* **65**, 543 (1953).

[227] R. S. Pasternak, *Acta Cryst.* **4**, 316 (1951); cf. also J. M. Lindsey and W. H. Barnes, *ibid.* **8**, 227 (1955).

[228] K. Sakashita, *Nippon Kagaku Zasshi* **81**, 49 (1960); **80**, 972 (1959), *Chem. Abstr.* **54**, 2008, 12015 (1960); compare also E. W. Garbisch, Jr., *J. Org. Chem.* **27**, 4249 (1962).

[229] See ref. 223, including its further references; cf. also G. Stork and W. N. White, *J. Am. Chem. Soc.* **78**, 4604 (1956).

[230] H. L. Goering, J. P. Blanchard, and E. F. Silversmith, *J. Am. Chem. Soc.* **76**, 5409

The half-chair conformation described for cyclohexene is also present for cyclohexene oxide (1,2-epoxycyclohexane), as was determined by X-ray diffraction[232] in the gas phase.

The four carbon atoms 1, 2, 3, and 6 lying closest to the oxygen lie approximately in a plane, while carbon atom 5 lies above this plane and carbon-atom 4 below it (XLIV).

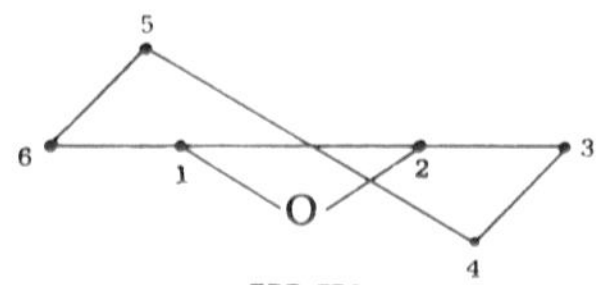

XLIV

Viewed from the perspective of the cyclohexane ring, the oxygen atom is attached in such a way that one bond is *equatorial* and the other *axial*. Thereby, the ring system of cyclohexene oxide retains a certain mobility.

8. Cyclohexanones

a. Cyclohexanone

The introduction of a trigonal atom into the cyclohexane ring has little effect on the form of the cyclohexanone molecule, if we disregard small deformations of the valence angles. Therefore, cyclohexanone differs very little in its geometry from cyclohexane and preferably assumes the chair conformation[233] (XLV). The possible twist conformations of substituted cyclohexanones will be dealt with in Chapter 6, page 284. The energy difference between the chair form and the

(1954). The equilibration of *trans*-5-methyl-2-cyclohexenol acid phthalate in acetonitrile yields 77% of the *trans* isomer at 80°. Cf. also H. L. Goering and E. J. Silversmith *ibid.* **79**, 348 (1957).

[231] For the conformation of cyclohexadiene-1,4 cf. F. H. Herbstein, *J. Chem. Soc.* p. 2292 (1959).

[232] B. Ottar, *Acta. Chem. Scand.* **1**, 283 (1947).

[233] Compare the detailed discussion of W. Moffitt, R. B. Woodward, A. Moscowitz, W. Klyne, and C. Djerassi, *J. Am. Chem. Soc.* **83**, 4013 (1961). The inner angle of the carbonyl group is given a value of 116°, while the C—C bonds of the trigonal carbon atom of the carbonyl group, in comparison to a normal C—C bond (1.545 Å) are assumed to be shortened to 1.50 Å. [Cf. C. C. Costain and B. P. Stoicheff, *J. Chem. Phys.* **30**, 777 (1959); L. C. Krisher and E. B. Wilson, Jr., *ibid.* **31**, 882 (1959); and also R. D. Stolow, *J. Am. Chem. Soc.* **84**, 686 (1962).] In older models, no shortening of the C—C bond of the trigonal carbon atom was assumed and the inner angle of the carbonyl group was set at ~120°. As a result, a substantial deformation of cyclohexanone was apparent in comparison with cyclohexane. [Cf. E. J. Corey and R. A. Sneen, *J. Am. Chem. Soc.* **77**, 2505 (1955); R. C. Cookson, *J. Chem. Soc.* p. 282 (1954)].

most favorable of the twist forms of cyclohexanone was estimated at 2.7 kcal./mole.[234] As a result, it is only about half as great as the corresponding energy differences for cyclohexane (see page 45f.). A possible reason for this could be that the *eclipsed* interactions between the carbonyl oxygen and the neighboring *equatorial* hydrogens in the chair form (XLV), are partially abolished in the *twist*

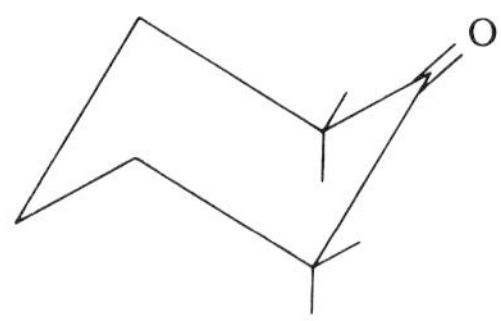

(XLV)

form of cyclohexanone. In addition, the *eclipsed* interactions between the neighboring hydrogen atoms in the *twist* form of cyclohexanone are partially abolished by the presence of a keto group, in comparison to the corresponding form of cyclohexane.[234]

[234] N. L. Allinger, *J. Am. Chem. Soc.* **81**, 5727 (1959). Allinger considered three flexible forms of cyclohexanone in his calculations (XLVI, XLVII, XLVIII).

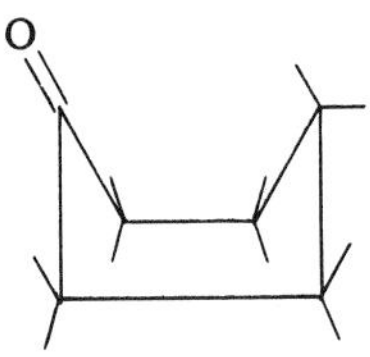

XLVI

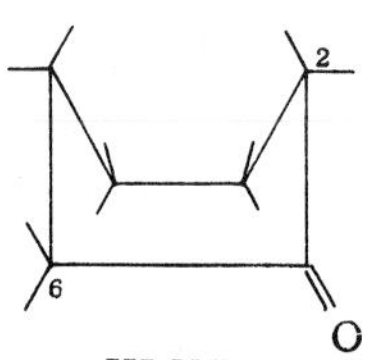

XLVII

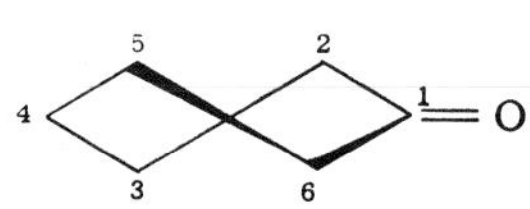

XLVIII

In the chair form (XLV), the two neighboring *equatorial* hydrogen atoms are approximately *eclipsed* with respect to the oxygen of the carbonyl group. If 0.8 kcal./mole for the energy per CH–oxygen interaction is introduced, then the total energy of the chair form (XLV) and of conformation (XLVI) is around $2 \times 0.8 = 1.6$ kcal./mole greater than that of cyclohexane, all of whose C—H bonds are *staggered*. The energy of conformation (XLVII) is 0.8 kcal./mole larger due to the *eclipsed* hydrogen on carbon atom 2. On the other hand, the energy is decreased by 1.6 kcal./mole due to the omission of the

As already mentioned, the two adjacent *equatorial* hydrogens in cyclohexanone (XLV) are in practically an *eclipsed* position with respect to the carbonyl group. This is also true of a substituent in position 2 to the keto group. If the substituent assumes the *axial* position on the same carbon atom, then it is in a *staggered* position with respect to the carbonyl oxygen. It was thus concluded in older investigations[235] that an alkyl group adjacent to the keto group was subject to an interaction which was generalized as the "2-alkyl ketone effect".[236] A value of ~1.0 kcal./mole was calculated[236] for the interaction energy between an *equatorial* methyl group and the neighboring carbonyl group, based on older measurements of the location of the equilibrium between carvomenthone and isocarvomenthone.[237,238]

In more recent investigations, it has been possible to measure the 2-alkyl ketone effect directly by equilibration of 2-alkyl-4-*tert*-butyl cyclohexanones[239] and 2,6-dialkyl cyclohexanones.[240] It has been ascertained[239,240] that no 2-alkyl ketone effect can be observed for the 2-methyl ketone.[241,242] The energy difference between an *axial* and an *equatorial* methyl group in the examples studied was found to be between 1.6 and 1.8 kcal./mole, which is identical with the value found for the methyl group in cyclohexane (see Table I, page 103). On the other hand, the 2-alkyl ketone effect becomes noticeable for the ethyl, isopropyl, and *tert*-butyl groups. It reduces the energy difference between the

---

eclipsed position on carbon atom 6. This means that conformation (XLVII) has an energy of about 0.8 kcal./mole lower than the chair form, as does the regular boat form of cyclohexane. With the aid of the calculated energy value of 5.1 kcal./mole for cyclohexane in the twist form, the energy difference between (XLV) and (XLVII) becomes $5.1 - 1.6 - 0.8 = 2.7$ kcal./mole. (XLVII) is then converted into (XLVIII) and the preferred *twist* conformation attained. The energy of (XLVIII), relative to (XLV), probably does not vary much from the estimated value of 2.7 kcal./mole.

A more accurate description of the preferred *twist* form of cyclohexanone is given by C. Djerassi and W. Klyne, *Proc. Natl. Acad. Sci. U.S.* **48**, 1093 (1962). Accordingly, in (XLVIII) C-1, C-2, C-6 and C-4 lie in a plane, while C-5 lies above this plane and C-3 below it.

[235] P. A. Robins and J. Walker, *Chem. & Ind. (London)* p. 772 (1955); *J. Chem. Soc.* p. 1789 (1955).

[236] W. Klyne, *Experientia* **12**, 119 (1956).

[237] R. G. Johnston and J. Read, *J. Chem. Soc.* p. 1138 (1935).

[238] Cf. also C. Beard, C. Djerassi, T. Elliott, and R. C. C. Tao, *J. Am. Chem. Soc.* **84**, 874 (1962).

[239] N. L. Allinger and H. M. Blatter, *J. Am. Chem. Soc.* **83**, 994 (1961).

[240] B. Rickborn, *J. Am. Chem. Soc.* **84**, 2414 (1962).

[241] Cf. also N. L. Allinger, M. A. DaRooge, and R. B. Herman, *J. Org. Chem.* **26**, 3626 (1961).

[242] M. J. T. Robinson and W. D. Cotterill, private communication. The measurements of Robinson and Cotterill even indicate a reverse 2-alkylketone effect. Compare also W. D. Cotterill and M. J. T. Robinson, *Tetrahedron Letters* p. 1833 (1963).

axial and *equatorial* ethyl group, in comparison to that in cyclohexane (see Table I), to 1.1–1.2 kcal./mole.[239,240] For the isopropyl group, the difference dropped to 0.4–0.5 kcal./mole[239,240,243] and for *tert*-butyl to ~1.5 kcal./mole.[240] Before these measurements became available, the absence of a 2-alkyl ketone effect had been predicted in 2-methylcyclohexanone by Fieser and Fieser.[244] The distance of the *equatorial* methyl group from the oxygen of the carbonyl group is too great for a van der Waals' repulsion effect.

Aside from the 2-alkyl ketone effect, the 3-alkyl ketone effect must also be considered in a cyclohexanone molecule.[235,236,245] As can be seen from (L), in comparison to the saturated system (XLIX), an *axial* substituent in positions 3 or 5 of cyclohexanone is exposed to only one *axial hydrogen* interaction. This leads to a lowering of the energy of the *axial* substituent; which would, for example in the case of the methyl group, amount to 0.9 kcal./mole less than a usual CH_3—H *diaxial* interaction.

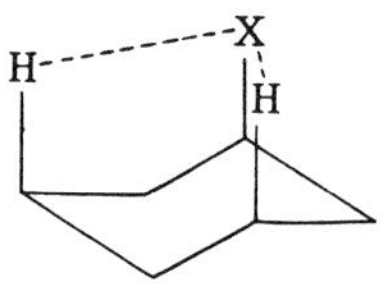

XLIX

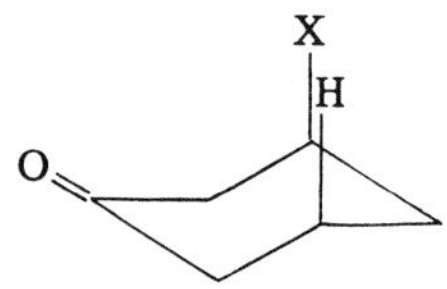

L

Experimental proof for the 3-alkyl ketone effect came from the equilibration studies of *cis*- and *trans*-2,5- and 3,5-dimethylcyclohexanone[242,246] which showed a lowering of the energy of a 3-*axial* methyl group in cyclohexanone of 0.5–0.6 kcal./mole in comparison to the saturated system.[247]

b. 2-Halocyclohexanones

The conformations of the 2-halocyclohexanones have been quite thoroughly investigated by physical methods, such as infrared and

[243] Cf. also the investigation of the conformation of isomenthone with the help of optical rotatory dispersion: C. Djerassi, *Tetrahedron* **13**, 13 (1961). According to this, the isopropyl group adjacent to the ketogroup preferably assumes the *axial* position in isomenthone.

[244] L. F. Fieser and M. Fieser, "Steroids." p. 213. Reinhold, New York, 1959.

[245] A. S. Dreiding, *Chem. & Ind. (London)* p. 1419 (1954); cf. also O. H. Wheeler and J. Z. Zabicky, *Can. J. Chem.* **36**, 656 (1958).

[246] N. L. Allinger and L. A. Freiberg, *J. Am. Chem. Soc.* **84**, 2201 (1962).

[247] However, see B. Rickborn[240] who concluded from the equilibration between menthone and isomenthone that the energy of an *axial* methyl group in cyclohexanone is practically unchanged. That is to say, the 3-alkylketone effect is hardly noticeable; cf. also R. L. Augustine and J. A. Caputo, *J. Am. Chem. Soc.* **86**, 2751 (1964).

ultraviolet spectroscopy, dipole moment measurements and measurements of the optical rotatory dispersion,[248a] as well as polarographic reduction,[248b] with special attention to the bonds next to the C = O group.

For α-halogenated steroid ketones in which the position of the halogen is fixed because of their rigid structure, it was shown first that a characteristic difference exists in the absorption of the C = O stretching vibration, depending on whether the halogen is *equatorial* or *axial*.[249] In the first case, in which the halogen lies practically in the same plane as the C=O group, a shift of the absorption maximum to higher frequencies takes place relative to the unsubstituted ketone. An *axial* halogen has no influence as to the position of the C=O absorption frequency (see further, page 233). Characteristic differences with respect to the carbonyl absorption are also found in the ultraviolet spectra, depending on an *equatorial* or *axial* position of the α-halogen. In Chapter 5, page 233 a more detailed study of this point will be made.

Corey[250,251] applied these results to simple, 2-halocyclohexanones which undergo conformational equilibration, and attempted to determine the conformations by investigation of the infrared spectra. Because of the possibility of conversion of the molecule, the halogen atom (X in LI) can assume the *equatorial* as well as the *axial* position.

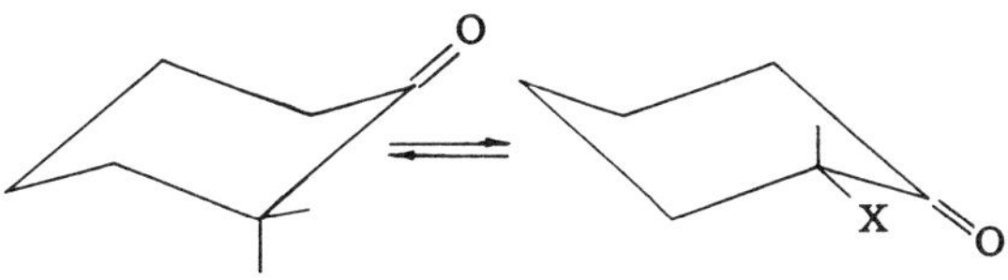

LI

From investigations of the infrared spectra as well as from calculations, Corey[250] came to the conclusion that in 2-chloro- and 2-bromocyclohexanone, the halogen is situated almost exclusively (~97%) in an *axial* position. When an additional *axial* substituent is introduced in the 3 position relative to the halogen, the resulting 1,3-*diaxial* interaction, can force the halogen substituent into an *equatorial* position. Such was said to be the case for *cis*-2,6-dibromo- and 2-bromo-4,4′-dimethylcyclohexanone.[252]

---

[248] (a) Compare: C. Djerassi, "Optical Rotatory Dispersion," p. 115. McGraw-Hill, New York, 1960. (b) A. M. Wilson and N. L. Allinger, *J. Am. Chem. Soc.* **83**, 1999 (1961).

[249] R. N. Jones, D. A. Ramsay, F. Herling, and K. Dobriner, *J. Am. Chem. Soc.* **74**, 2828 (1952).

[250] See ref. 31e, E. J. Corey, *J. Am. Chem. Soc.* **75**, 3297 (1953); E. J. Corey, J. H. Topie, and W. A. Wozniak, *ibid.* **77**, 5415 (1955); E. J. Corey and H. J. Burke, *ibid.* **77**, 5418 (1955).

[251] (a) Cf. also R. C. Cookson, *J. Chem. Soc.* p. 282 (1954); (b) N. J. Leonard and F. H. Owens, *J. Am. Chem. Soc.* **80**, 6039 (1958).

[252] E. J. Corey, *J. Am. Chem. Soc.* **75**, 3297 (1953); *ibid.* **76**, 175 (1954).

On the other hand, Kumler and Huitric [252a] came to the conclusion that the halogen atom does not assume the *axial* position, but is found mainly in the *equatorial* position. These workers arrived at this conclusion by measurements of the dipole moment of 2-chloro- and 2-bromocyclohexanone in comparison with the theoretical value for an *axial* placed halogen. A large part of the cyclohexanone molecule was said to exist in a flexible form, no longer corresponding to a true chair. This would at least explain the strong solvent dependence of the dipole moment.

Newer investigations of the conformations of 2-halocyclohexanones by Allinger[253] showed that this contradictory interpretation can be avoided by the assumption of a conformational equilibrium between *axial* and *equatorial* 2-halocyclohexanones. By investigation of the infrared and ultraviolet spectra of 2-bromocyclohexanone in various solvents, it could be shown that 2-bromocyclohexanone exhibits a conformational equilibrium between the *axial* and *equatorial* bromine. The *axial* conformer predominates and the direction of the equilibrium depends mainly on the solvent. The higher the polarity of the solvent, the higher the proportion of *equatorial* conformer. On the basis of older measurements of the dipole moments,[252a] the energy difference between *axial* and *equatorial* bromine in 2-bromocyclohexanone was calculated as 0.45 kcal./mole in heptane, 0.24 kcal./mole in benzene, and 0.02 kcal./mole in dioxane.[253] Parallel studies by other investigators[254] produced different results. They also determined, on the basis of infrared investigations and measurements of dipole moments, a conformational equilibrium for 2-chlorocyclohexanone. In benzene solution and in the liquid phase, the conformer with *equatorial* halogen predominates, while the solid is assumed to be exclusively in the *equatorial* conformation.

The interpretation of the investigations of the conformational equilibria of the 2-halocyclohexanones is based only on theoretical calculations of the dipole moments for both conformers (and comparison with the measured dipole moments). Thus a more accurate determination of the position of the equilibrium was not possible.

As in the case of cyclohexane, the conformation of the cyclohexanone molecule can be fixed by the introduction of a large substituent. Thus, the possibility of obtaining a standard substance is achieved and permits the determination of the conformational equilibria of analogous mole-

---

[252a] W. D. Kumler and A. C. Huitric, *J. Am. Chem. Soc.* **78**, 3369 (1956).

[253] J. Allinger and N. L. Allinger, *Tetrahedron* **2**, 64 (1958).

[254] K. Kozima and Y. Yamanouchi, *J. Am. Chem. Soc.* **81**, 4159 (1959); cf. also K. Kozima and E. Hirano, *ibid.* **83**, 4300 (1961); S. Yaroslavsky and E. D. Bergmann, *J. Chem. Phys.* **33**, 635 (1960); cf. also M. L. Josien and C. Castinel, *Bull. Soc. Chim. France* p. 801 (1958).

cules which, however, are capable of ring conversion. As standard compounds in which the position of the halogen atom is fixed in an axial or equatorial position by the *tert*-butyl group, *cis*- and *trans*-2-bromo-4-*tert*-butylcyclohexanone were first chosen by Allinger[255] (LII and LIII, X = Br).

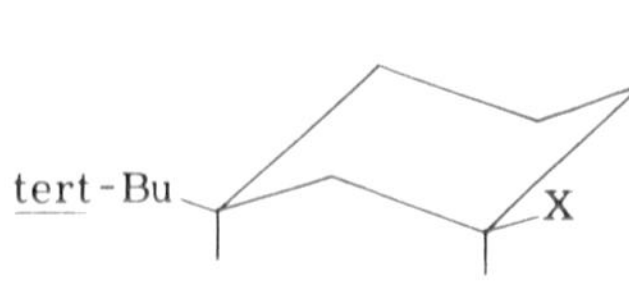

LII

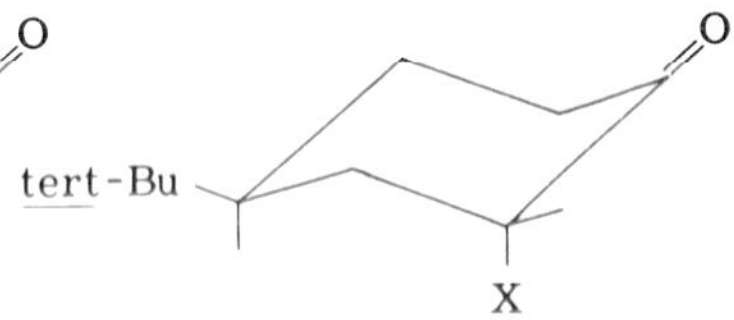

LIII

Studies of the infrared and ultraviolet spectra of these compounds and especially measurement of the dipole moments,[255,256] permitted a determination of the conformational equilibrium of 2-bromocyclohexanone, without additional theoretical assumptions. It is only assumed that the *tert*-butyl group has no measurable influence on the C=O group (see page 96). For *equatorial* 2-bromo-4-*tert*-butylcyclohexanone (LII, X=Br) with both bond moments (C—Br and C=O) in nearly the same plane, a large dipole moment (4.27 D) was found, in good agreement with the theoretically calculated values. The *axial* epimer (LIII, X=Br) had a value of 3.20 D in benzene.[256] The measurement of the dipole moment of 2-bromocyclohexanone under the same conditions gave a value of 3.49 D, from which the fraction of conformer with *axial* bromine can be calculated as 76% in benzene solution. In heptane ~85% of the conformer with *axial* bromine is present.[256a]

Comparative investigations of *cis*- and *trans*-2-chloro-4-*tert*-butyl cyclohexanone (LII and LIII, X=Cl) with 2-chlorocyclohexanone gave a preponderance of *axial* conformer for the latter in nonpolar solvents (~75% in octane). In polar solvents, the *equatorial* conformer predominates (~65% in dioxane).[257a] The *equatorial* position is more preferred by chlorine than by bromine.

The point of conformational equilibrium of 2-fluorocyclohexanone was also determined by comparison of its dipole moment with that of *cis*- and *trans*-2-fluoro-4-*tert*-butylcyclohexanone (LII and LIII,

[255] N. L. Allinger and J. Allinger, *J. Am. Chem. Soc.* **80**, 5476 (1958).

[256] N. L. Allinger, J. Allinger, and N. A. LeBel, *J. Am. Chem. Soc.* **82**, 2926 (1960); cf. also R. Cetina and J. L. Mateos, *J. Org. Chem.* **25**, 704 (1960).

[256a] Compare also: E. W. Garbisch, Jr., *J. Am. Chem. Soc.* **86**, 1780 (1964) for conformational analysis of 2-bromocyclohexanone by nuclear magnetic resonance spectroscopy.

[257a] N. L. Allinger, J. Allinger, L. A. Freiberg, R. F. Czaja, and N. A. LeBel, *J. Am. Chem. Soc.* **82**, 5876 (1960).

X=F).[257b] In heptane, 2-fluorocyclohexanone exists with 52% of the *equatorial* conformation, but in dioxane, the value found for the *equatorial* conformer is 85%.

Measurement of optical rotatory dispersion has also been successfully applied to the determination of the absolute configuration and the conformations of the 2-haloketones, particularly by Djerassi[258] and co-workers. By preparation of the optically active *trans*-2-bromo-5-methylcyclohexanone[258e] (LIV, X=Br), they succeeded, with the help of optical rotatory dispersion, infrared, ultraviolet[258e,f] and dipole moment measurements, in determining quantitatively the conformational equilibrium of this compound in various solvents.

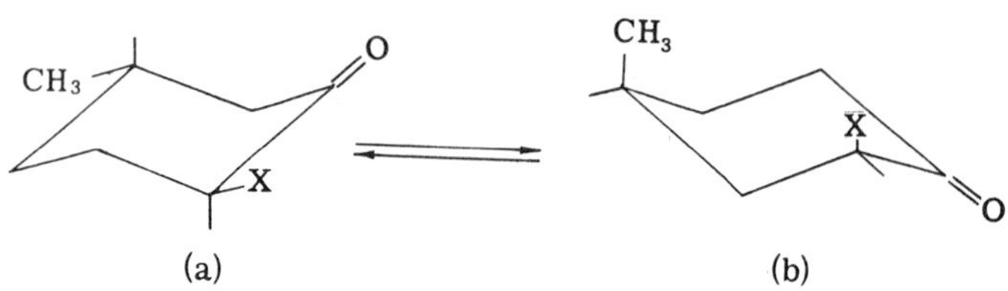

LIV

From known values for the free energy differences between *axial* and *equatorial* methyl and bromine in cyclohexanone, the energy differences, and from these the conformational equilibrium, can be estimated for the above compound. These calculations can be compared with the results of the measurement. A relatively good agreement between the calculated and found values was obtained. The free energy difference between *axial* and *equatorial* conformers was found

---

[257b] N. L. Allinger and H. M. Blatter, *J. Org. Chem.* **27**, 1523 (1962); cf. also A. S. Kende, *Tetrahedron Letters* p. 13 (1959); N. L. Allinger, H. M. Blatter, M. A. DaRooge, and L. A. Freiberg, *J. Org. Chem.* **26**, 2550 (1960); H. M. Kissman, A. M. Small, and M. J. Weiss, *J. Am. Chem. Soc.* **82**, 2312 (1960); N. L. Allinger, M. A. DaRooge, and C. L. Neumann, *J. Org. Chem.* **27**, 1082 (1962).

[258] (a) C. Djerassi and W. Klyne, *J. Am. Chem. Soc.* **79**, 1506 (1957); (b) C. Djerassi, J. Osiecki, R. Riniker, and B. Riniker, *ibid.* **80**, 1216 (1958); (c) C. Djerassi, I. Fornaguera, and O. Mancera, *ibid.* **81**, 2383 (1959); (d) C. Djerassi, N. Finch, and R. Mauli, *ibid.* **81**, 4997 (1959); (e) C. Djerassi, L. E. Geller, and E. J. Eisenbraun, *J. Org. Chem.* **25**, 1 (1960); (f) N. L. Allinger, J. Allinger, L. E. Geller, and C. Djerassi, *ibid.* **25**, 6 (1960); (g) C. Djerassi, E. J. Warawa, R. E. Wolff, E. J. Eisenbraun, *ibid.* **25**, 917 (1960); (h) J. Allinger, N. L. Allinger, L. E. Geller, and C. Djerassi, *ibid.* **26**, 3521 (1961); (i) Cf. also C. Djerassi, *Record of Chem. Progr.* (*Kresge-Hooker Sci. Lib.*) **20**, 101 (1959); (k) W. Klyne, *Bull. Soc. Chim. France* p. 1396 (1960); (l) K. Wellman and C. Djerassi, *J. Am. Chem. Soc.* **85**, 3516 (1963).

at 0.35 (heptane), 0.56 (benzene), and 0.77 (dioxane) kcal./mole, corresponding to ~40%, ~30%, and ~20% *axial* conformer (LIVb; X = Br), i.e., in all solvents studied, the conformer with the *equatorial* substituents (LIVa) predominated. However, a noticeable solvent dependence of the conformational equilibrium between the two chair conformations is present here too. Analogous investigations of *trans*-2-chloro-5-methylcyclohexanone (LIV, X=Cl)[258f] also gave a preponderance of that conformation with *equatorial* chlorine (LIVa, X=Cl). In this case, the *equatorial* position is even more preferred than in the corresponding bromo compound. Djerassi and co-workers[258g] were further able to show, in an elegant investigation, that the introduction of a *tert*-butyl group fixes the conformation. Measurements of the rotatory dispersion, as well as infrared and ultraviolet spectra, of *trans*-2-bromo-5-*tert*-butylcyclohexanone in various solvents, showed, contrary to *trans*-2-bromo-5-methylcyclohexanone, the complete absence of a conformational equilibrium.[259]

c. 1,4-Cyclohexanedione

The conformation of 1,4-cyclohexanedione will now be discussed briefly. From older measurements of the dipole moment, LeFèvre[260] concluded that this compound no longer exists exclusively in the chair form, but that over 10% must be in the boat conformation (cf. p. 275).

Aside from the chair conformation (LV), two boat conformations have been considered (LVI and LVII), as well as a twist conformation (LVIII).[261]

Of the two structures (LVI) and (LVII), conformation (LVI) with two pairs of eclipsed C—H bonds and a 1,4-dipolar repulsion of the keto groups is energetically the least favorable. Thus, that proportion existing in the boat form should be present in conformation (LVII), or in a conformation between (LVII) and (LVIII). The energy difference between (LVII) and (LV) has been estimated by Allinger[261] as no more than 2 kcal./mole, but its exact magnitude and direction are uncertain. Later it was experimentally determined[262] that 1,4-cyclohexanedione is a mixture of the conformations (LV) and (LVII) in a ratio of 80:20, by measurements of the molecular polarizability and comparison with the dipole moment measurements. However, by a comparison of the

---

[259] For the conformation of 2,4-dibromomenthone see J. A. Wunderlich and W. N. Lipscomb, *Tetrahedron* 11, 219 (1960).

[260] C. G. LeFèvre and R. J. W. LeFèvre, *J. Chem. Soc.* p. 1696 (1935).

[261] N. L. Allinger, *J. Am. Chem. Soc.* 81, 5727 (1959); compare also G. A. Bottomley and P. R. Jeffries, *Australian J. Chem.* 14, 657 (1961).

measured dipole moment[262] with the one calculated for the conformations (LVII) or (LVIII), Allinger and Freiberg[263], in a recent investigation, came to the conclusion that 1,4-cyclohexanedione exists preferentially in the conformations (LVII) or (LVIII) in both the liquid state and in solution at room temperature.[263a] Thus, this represents

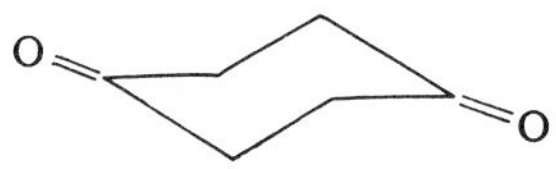

LV

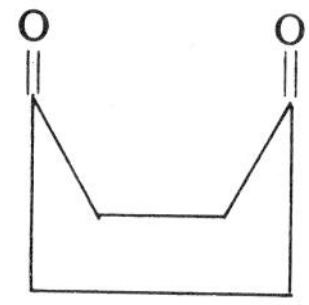

LVI

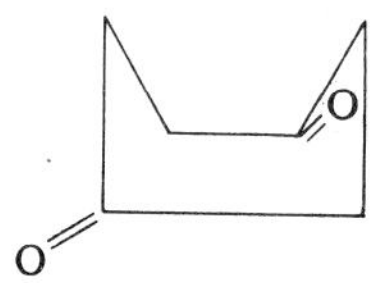

LVII

LVIII

the first example in which a twist conformation is not forced upon a molecule by steric interactions due to substituents on the ring. This is the case, for example, for *cis*-2,5-di-*tert*-butylcyclohexane-1,4-dione, which also preferentially assumes the twist conformation[264] (see also page 275).[265]

III. CONFORMATION OF CYCLOHEPTANES

1. Cycloheptane

Although Mohr[266] had concerned himself with the spatial structure of cycloheptane, less information has appearaed concerning the confor-

---

[262] C. G. LeFèvre and R. J. W. LeFèvre, *Rev. Pure Appl. Chem.* **5**, 261 (1955) and Ref. 25; (b) M. T. Rogers and J. M. Cannon, *J. Phys. Chem.* **65**, 1417 (1961); (c) compare also: C. Y. Chen and R. J. W. LeFèvre, *Australian J. Chem.* **16**, 917 (1963).

[263] N. L. Allinger and L. A. Freiberg, *J. Am. Chem. Soc.* **83**, 5028 (1961); compare also: A. Mossel, C. Romers, and E. Havinga, *Tetrahedron Letters* p. 1247 (1963).

[263a] For the solid state see: P. Groth and O. Hassel, *Proc. Chem. Soc.* p. 218 (1963).

[264] R. D. Stolow and C. B. Boyce, *J. Am. Chem. Soc.* **83**, 3722 (1961).

[265] For the conformation of cyclohexenone, see W. B. Whalley, *Chem. & Ind.* (*London*) p. 1024 (1962).

mations of cycloheptane and its derivatives in comparison to the detailed studies of cyclohexanes.

As Mohr[266] had first shown, two forms of cycloheptane can be constructed. Like the chair and boat form of cyclohexane, they can be transformed into each other by conversion. In contrast to cyclohexane, both forms of cycloheptane are flexible, so that a considerable difference may exist in the relative position of the substituents to each other. The *trans* valences of two neighboring atoms can approach each other up to 23°, almost lying in one plane. The *cis* valences can diverge from each other up to ~100° (according to the recent calculations of Hendrickson, see below).

The problem of the conformations of the cycloheptane ring has been reinvestigated only recently,[261,267,269a,269b,269c] without resulting in an unequivocal answer. Hendrickson[268] has calculated, with the aid of a computer, the energies of the possible conformations of cycloheptane, depending on the various geometric parameters. This has presented a detailed representation of the conformations of cycloheptane.

The two types of conformations to be considered, are called chair and boat form[261,268,269b,269c] (LIX and LX), as with cyclohexane.

As mentioned above, in contrast to cyclohexane, both forms of cycloheptane are flexible and capable of pseudorotation. However, in the chair (LIX) as well as in the boat (LX) conformation, an interaction which cannot be neglected appears between the hydrogens on carbon atoms 3 and 6. To avoid this, (LIX) and (LX) go into the conformation where the interaction energy is at a minimum, thus assuming the *twist*-chair (LXI) conformation and a corresponding *twist*-boat conformation. In this way, the 3,6-H—H interactions are not entirely avoided but are reduced, so that an additional valence angle distortion appears. Hendrickson's[268] calculations have shown that the energetically most favorable conformation of cycloheptane is the *twist*-chair conformation (LXI), in which the valence angles are spread to 112°. This form differs from the chair conformation (LIX) by 2.1 kcal./mole, while the *twist*-boat

---

[266] E. Mohr, *J. Prakt. Chem.* [2] **103**, 316 (1922).

[267] Cf. also: J. Sicher, J. Jonáš, M. Svoboda, and O. Knessl, *Collection Czech. Chem. Commun.* **23**, 2141 (1958).

[268] J. B. Hendrickson, *J. Am. Chem. Soc.* **83**, 4537 (1961); *ibid.* **84**, 3355 (1962); *Tetrahedron* **19**, 1387 (1963).

[269a] V. Prelog, *Colloq. Intern. Chim. Org. Probl. Stereochim. Montpellier*, 1959, in *Bull. Soc. Chim. France* p. 1433 (1960).

[269b] Compare also: E. Billeter, Th. Bürer and H. H. Günthard, *Helv. Chim. Acta* **40**, 2046 (1957).

[269c] R. Pauncz and D. Ginsburg, *Tetrahedron* **9**, 40 (1959).

conformation is less stable than the *twist*-chair conformation by 2.5 kcal./mole.

The conformational analysis of substituted cycloheptanes also turned out to be more difficult than had been the case for substituted cyclohexanes. As can be seen from (LXI) (as well as LIX and LX), the positions

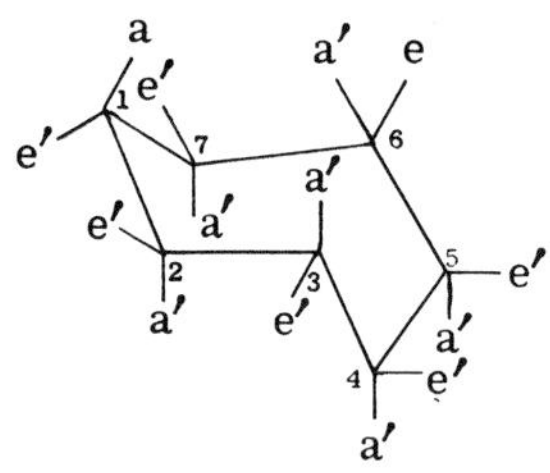

LIX

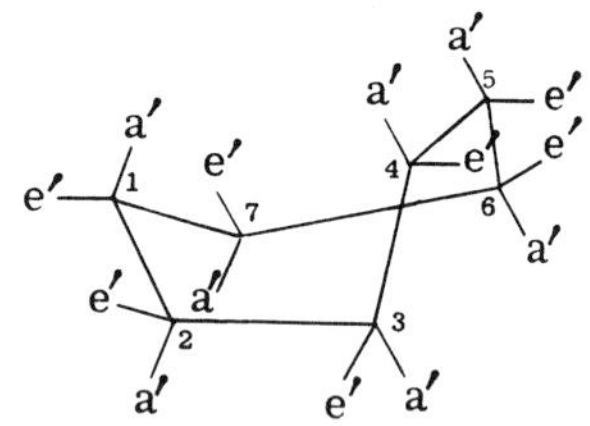

LX

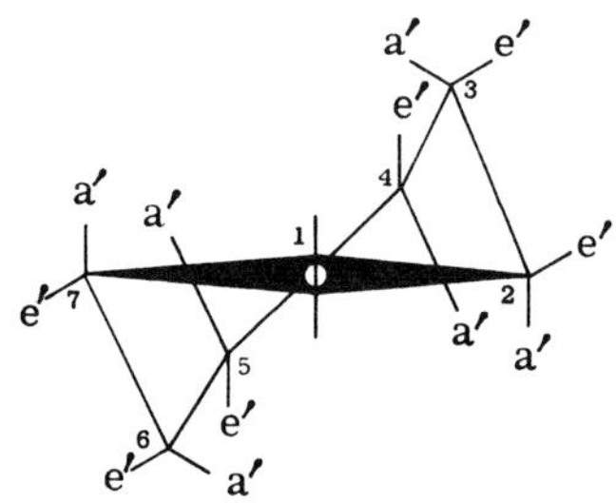

LXI

of the hydrogen atoms correspond approximately to the *axial* and *equatorial* positions of cyclohexane. Therefore, they can be designated[270] as *quasi-axial* (a′) and *quasi-equatorial* (e′). A monosubstituted cycloheptane would preferably assume the *twist*-chair conformation with the substituent *quasi-equatorial* (see LXI). A given substituent, such as methyl, moves through all possible *quasi-equatorial* positions as a result of pseudorotation, traversing the chair (LIX) and *twist*-chair (LXI) conformations. While the dihedral angles between the bonds in cycloheptane are not equal (in contrast to cyclohexane, in which they are 60°, four different bonds are differentiated in cycloheptane as a result of their different dihedral angles), methylcycloheptane preferably assumes the various *twist*-chair conformations with a *quasi-equatorial* methyl group.

Carbon-atom 1 of the *twist* conformation (LXI) possesses two equivalent positions and in this way differs characteristically from the other C atoms. Both positions on C-1 practically correspond energetically to the *quasi-equatorial* positions of the other carbon atoms. Therefore,

[270] N. L. Allinger, *J. Am. Chem. Soc.* **81**, 232 (1959).

1,1-disubstituted cycloheptanes preferably would assume only *one* *twist–chair* conformation with the substituents on C-1 (see LXI).

While only the e,a conformation is possible for the *cis* isomer of a 1,2-disubstituted cyclohexane in the chair form and e,e and a,a conformations for the *trans* isomer (see page 105), two *cis* and two *trans* conformations are possible in cycloheptane. If the two substituents are in positions 2,3 of the *twist*-chair conformation (LXI), for example, then the *cis*-2e′, 3a′ and 2a′, 3e′ positions are no longer identical. The same is true for the *trans*-2,3 position (2e′, 3e′ and 2a′, 3a′). The various conformations of the *cis* isomer can be converted into each other either by pseudorotation of the *twist*-chair conformation or by flipping into the boat conformation and back again. This is also true of the *trans* isomer.

Hendrickson[268] determined the preferred conformations of di- and polymethylcycloheptanes more accurately by calculating the energies of a *quasi-axial* methyl group in the various positions of the *twist-chair* conformation of methylcycloheptane and determining the dihedral angles. For example, *cis*-1,2-dimethylcycloheptane preferably assumes the *twist*-chair conformation in which one methyl group is on C-1 (see LXI), while the second one takes up the e′ position on the adjacent carbon. In *trans*-1,2-dimethylcycloheptane, every e′, e′ position in the *twist*-chair conformation, where the various positions are interconvertible by pseudorotation, is of equal energy. The same is true for the *cis*-1,3 and *trans*-1,4 compound, while the 1,3e′ or 1,6e′ and the 2e′, 7e′ conformations are preferred to the others in *trans*-1,3-dimethylcycloheptane (see LXI). For *cis*-1,4-dimethylcycloheptane, the 1,4e′ or 1,5e′ and the 2e′, 6e′ conformations are preferred over the others (see LXI). In addition, the calculations revealed that the preferred conformations of all the dimethylcycloheptanes hardly varied in their energy content.

Experimental criteria concerning the energy difference between *cis*- and *trans*-dimethylcycloheptanes are not available except for an investigation by Allinger.[270] By equilibration of *cis*- and *trans*-3,5-dimethylcycloheptanone, the energy difference between the two isomers was determined at 0.8 kcal./mole. Because of the presence of a carbonyl group, this result cannot be applied directly to the hydrocarbons, i.e. this does not conflict with the calculations of Hendrickson. However, Sicher *et al.* were able to show[271] that *cis*-1,2-cycloheptane dicarboxylic acid is converted to the *trans*-dicarboxylic acid by way of the ester and equilibrium conditions. This indicates an energy difference between the various conformers.

Of the monosubstituted cycloheptanes, Chiurdoglu[272] and co-workers

[271] J. Sicher, F. Šipoš, and J. Jonáš, *Collection Czech. Chem. Commun.* **26**, 262 (1961).

[272] G. Chiurdoglu, L. Kleiner, W. Masschelein, and J. Reisse, *Bull. Soc. Chim. Belges* **69**, 143 (1960).

have investigated the conformational equilibria of bromo and chlorocycloheptane with the help of infrared spectroscopy. By making use of the C–halogen stretching vibration of the halocycloheptanes in the solid and liquid states, it could be shown that an equilibrium exists just as in the case of the halogen-substituted cyclohexanes (see page 100). Whether or not this is an equilibrium between *quasi-equatorial* and *quasi-axial* positions of the halogen atom has not been clarified. Chiurdoglu[272] also discussed the possibility of two different *quasi-equatorial* positions of the halogen, which could be possible as mentioned above due to the nonequality of the methylene groups.[273,274]

From the limited chemical data available, we can only conclude that the cycloheptane skeleton has a relatively large flexibility; a detailed conformational analysis is not possible. Ayres and Raphael[275] prepared *trans*-1,2-cycloheptane dicarboxylic acid and found that it readily forms a *trans* anhydride in contrast to *trans*-1,2-cyclohexane dicarboxylic acid. *Cis*- and *trans*-1,2-cycloheptane diol form cyclic ketals. Both also increase the conductance of a boric acid solution,[276] unlike the cyclohexanediols. The reaction of *cis*- and *trans*-2-aminocycloheptanol with nitrous acid also proceeds differently, than for the corresponding cyclohexane compound.[277] These briefly mentioned observations permit no further definite conclusions concerning the conformations of the cycloheptane ring, discussed above but indicate a greater flexibility of this ring system.

2. Cycloheptanone

A discussion of the possible conformations of cycloheptanone was undertaken by Allinger[215e,261] based on a study of models and measure-

[273] B. Tursch, Thèse, Université Libre de Bruxelles, 1959, and ref. 102 in Chap. 2.

[274] For intramolecular hydrogen bonds of *cis*- and *trans*-cycloheptane-1,2-diol as well as *cis*- and *trans*-2-aminocycloheptanol, see L. P. Kuhn, *J. Am. Chem. Soc.* **76**, 4323 (1954); J. Sicher, M. Horak, and M. Svoboda, *Collection Czech. Chem. Commun.* **24**, 950 (1959).

[275] D. C. Ayres and R. A. Raphael, *J. Chem. Soc.* p. 1779 (1958).

[276] P. Hermans and C. J. Maan, *Rec. Trav. Chim.* **57**, 643 (1938); (b) J. Böeseken, *ibid.* **58**, 856 (1939).

[277] (a) J. W. Huffman and J. E. Engle, *J. Org. Chem.* **24**, 1844 (1959); cf. also J. W. Huffman and J. E. Engle, *ibid.* **26**, 3116 (1961); (b) cf. for other reactions: R. Criegee, E. Höger, G. Huber, P. Kruck, F. Marktscheffel, and H. Schellenberg, *Ann. Chem.* **599**, 1181 (1956); (c) J. Weinstock, S. N. Lewis, and F. G. Bordwell, *J. Am. Chem. Soc.* **78**, 6072 (1956); (d) R. Heck and V. Prelog, *Helv. Chim. Acta* **38**, 1541 (1955); (e) H. C. Brown and G. Ham, *J. Am. Chem. Soc.* **78**, 2735 (1956); (f) H. J. E. Loewenthal and P. Rona, *J. Chem. Soc.* p. 1429 (1961); W. Hückel and J. Wächter, *Ann. Chem.* **672**, 62 (1964).

ments of the dihedral angles. Allinger[215e,261] considered the following conformations: (LXII) to (LXVI). Conformation (LXII) is unfavorable, compared to the hydrocarbon, as its calculated energy would be higher. If the keto group is introduced into the *twist* form[278] of the chair (LXIII), then the calculated energy drops by 5 kcal.

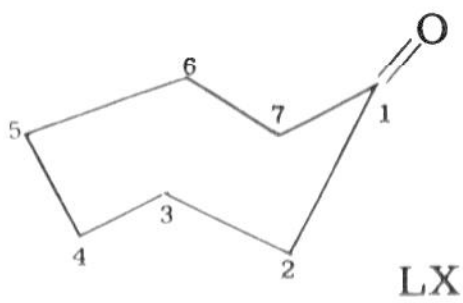

LXII

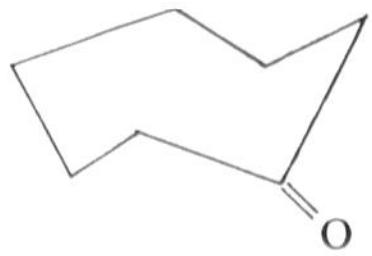

LXIII

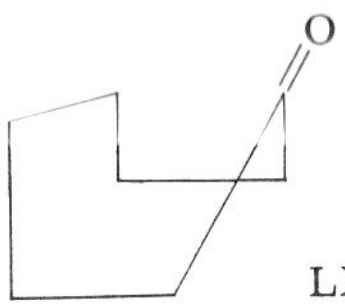

LXIV

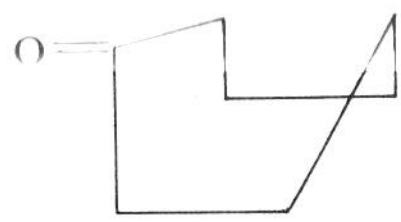

LXV

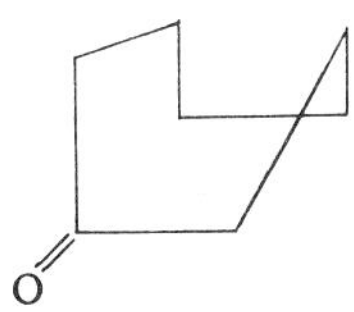

LXVI

Allinger[215e,261] calculated energies of 5.5, 3.4, and 3.5 kcal. for the conformations (LXIV), (LXV) and (LXVI), disregarding the transannular interaction mentioned for cycloheptane itself (page 159), the magnitude of which is uncertain. If this amount is large, then conformations (LXV) and (LXVI) will not exist in significant amounts. Conformation (LXIII) would then be the favored conformation, but it does not seem impossible that (LXIV) would then also exist in small amounts.

Measurements of molecular polarizability[33a] do not lead to the

[278] To clarify the difference between the chair form (LXII) and the *twist* form (LXIII) which is not readily apparent from the diagrams, it is expedient to build a ball and stick model of the seven-membered ring. It immediately becomes clear that conformation (LXII) is less favorable than (LXIII) because the C—H valences on C-4 and C-5 in (LXII) assume an eclipsed position. In the twist form (LXIII), which is obtained by twisting C-4 and C-5 counterclockwise a certain amount, the positions of the C—H valences on C-4 and C-5 become more favorable.

definite conclusion that cycloheptanone is present in a conformation corresponding to (LXIII).[279,280]

IV. MEDIUM-SIZED RINGS

The conformations of medium-sized rings, have already been treated in Chapter 2. Therefore, we shall deal only briefly in the following with some derivatives of medium-sized rings.

1. Cyclooctanone

Again on the basis of theoretical considerations, similar to those used for cyclooctane (see p. 52), Allinger[261] also derived the conformation of cyclooctanone.[281,282] Cyclooctanone can assume two conformations which are differentiated as crown (LXVII) and boat conformation (LXVIII). The conformations shown in (LXVII) and (LXVIII) are

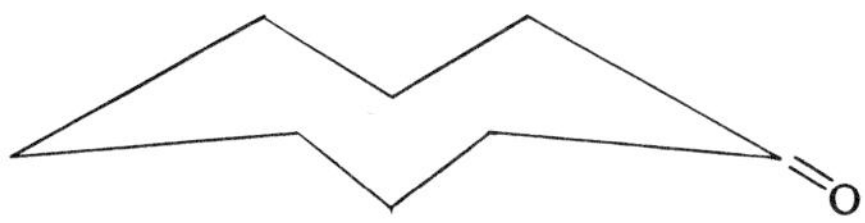

LXVII

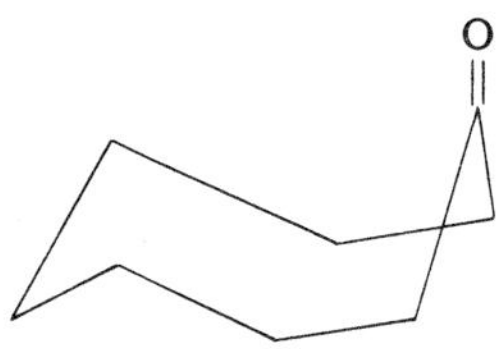

LXVIII

also denoted as "O-outside" and "O-inside"[283] conformations. In one of the conformations, O-outside (LXVII), the oxygen lies approximately in the imagined plane of the ring. In the other conformation,

[279] For the conformation of 4-cyclohepten-1-one: N. L. Allinger and W. Szkrybalo, *J. Org. Chem.* **27**, 722 (1962).

[280] For attempts to obtain information concerning the conformation of cyclanones by infrared spectroscopic investigations see references 269b and 31j and page 169. For measurements of the optical rotatory dispersion of 3-methyl-cycloheptanone see C. Djerassi and G. W. Krakower, *J. Am. Chem. Soc.* **81**, 237 (1959); ref. 248a.

[281] For *cis*- and *trans*-5-*tert* butylcyclooctanol and a discussion of their reactivities from the viewpoint of the conformation of the cyclooctane ring, see: N. L. Allinger and S. Greenberg, *J. Am. Chem. Soc.* **84**, 2394 (1962).

[282] Cf. also N. J. Leonard, T. W. Milligan and T. L. Brown, *J. Am. Chem. Soc.* **82**, 4075 (1960).

[283] V. Prelog, *J. Chem. Soc.* p. 420 (1950).

O-inside (LXVIII), the oxygen is approximately perpendicular to the plane. On the basis of theoretical considerations, it was originally assumed[261] that conformations (LXVII) and (LXVIII) did not differ greatly in their stability. Allinger[284] attempted to solve the question of the stability of the two conformers experimentally by measurement of the dipole moment of a cyclooctanone derivative. The 5-(*p*-chlorophenyl) cyclooctanone gave a dipole moment of 3.39 D in benzene. The large *p*-chlorophenyl group cannot go into an *axial* position, yet its polarity is not so large as to force the molecule into another conformation by interaction with the C=O group. The dipole moments of the O-outside and O-inside conformations were calculated with the help of the known values of the C=O group in cyclooctanone, of the *p*-chlorophenyl group dipole, and direct measurement of the dipole angles on a model.

Because conformation (LXVII) is flexible in contrast to (LXVIII), the calculated dipole moment for 5-(*p*-chlorophenyl)cyclooctanone in conformation (LXVII) fluctuates between 3.52 and 1.58 D, corresponding to the two possible extremes of the *twist* conformation, while a dipole moment of 4.17 D was calculated for (LXVIII).[285] These results were the basis of a preference for conformation (LXVII) of cyclooctanone.

Infrared investigations of cyclooctanone[286] and 5-(*p*-chlorophenyl)-cyclooctanone[285] also led to the conclusion of one conformation, probably corresponding to the crown form (LXVII). Only at higher temperatures is the existence of a further conformation noted by the appearance of a new C=O band.

Among the derivatives of cyclooctanone, 2-bromocyclooctanone has been investigated.[287] From theoretical considerations, as well as measurements of the dipole moments, ultraviolet, and infrared studies, Allinger and Allinger[287a,b] came to the conclusion that 2-bromocyclooctanone is capable of assuming various conformations, partly crown and partly boat conformations.

---

[284] N. L. Allinger and S. Greenberg, *J. Am. Chem. Soc.* **81**, 5733 (1959); compare also N. L. Allinger, S. P. Jindal, and M. A. DaRooge, *J. Org. Chem.* **27**, 4290 (1962).

[285] N. L. Allinger and S. E. Hu, *J. Am. Chem. Soc.* **83**, 1664 (1961).

[286] See ref. 102a in Chap. 2. (a) G. Chiurdoglu, Th. Doehaerd, and C. Goldenberg, *Compt. Rend. Acad. Sci.* **250**, 3495 (1960); (b) G. Chiurdoglu, Th. Doehaerd, and B. Tursch, *Colloq. Intern. Chim. Org. Probl. Stereochim.* in *Bull. Soc. Chim. France* p. 1322 (1960); (c) E. Billeter, T. Bürer, and H. H. Günthard, *Helv. Chim. Acta* **40**, 2046 (1951); (d) T. Bürer and H. H. Günthard, *ibid.* **40**, 2054 (1957); (e) T. Bürer and H. H. Günthard, *ibid.* **43**, 1487 (1960); cf. also: N. J. Leonard and F. H. Owens, *J. Am. Chem. Soc.* **80**, 6039 (1958).

[287] (a) J. Allinger and N. L. Allinger, *J. Am. Chem. Soc.* **81**, 5736 (1959); (b) J. Allinger and N. L. Allinger, *J. Org. Chem.* **25**, 262 (1960).

2. Cyclononanone and Cyclodecanone[288a]

If we assume that the cyclodecanone ring possesses the same conformation as *trans*-1,6-diaminocyclodecane dihydrochloride shown on page 55, three conformations (LXIX, LXX, LXXI) are possible for cyclodecanone: In (LXIX) and (LXX), one of the so-called intra-annular

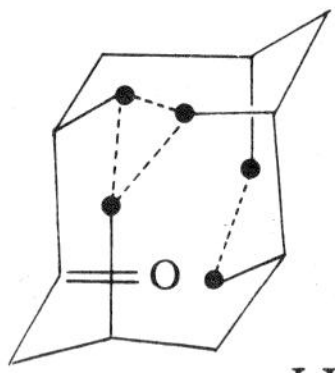

LXIX

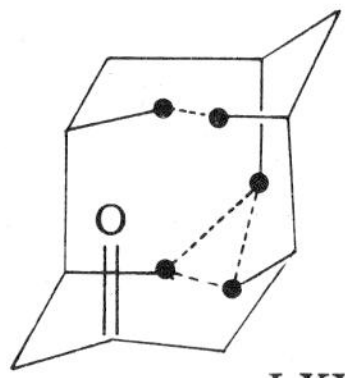

LXX

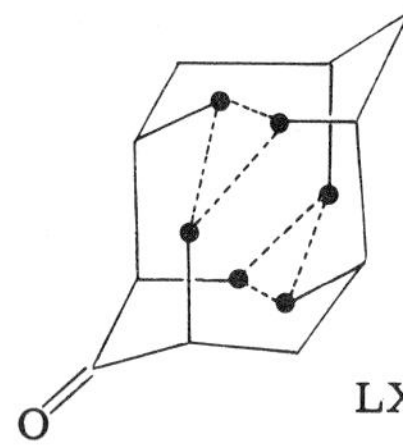

LXXI

hydrogen atoms (concerning the designation "intra-annular," see page 171) is removed, i.e. two of the six transannular hydrogen interactions present in cyclodecane are missing, stabilizing this conformation designated as "O-inside."[288b] There is no such destraining present in (LXXI) which is also designated the "O-outside "conformation.

Similar considerations may also be made for cyclononanone, in which it is assumed that the carbonyl group removes two of the six transannular hydrogen interactions present in the corresponding hydrocarbon, and thus stabilizes the ketone.

3. General Considerations Concerning Medium-Sized Rings[289]

With increasing ring size, conformational analysis becomes more difficult. The properties of the medium-sized rings show a noticeable dependence on ring size. In comparison to acyclic compounds, in which a steady increase in the density with respect to molecular weight is observed in a homologous series, cyclic compounds usually show a density maximum or a depression in molecular refraction in the range of the medium-sized rings.[283]

Chemical reactions, as well, show peculiarities in the range of the medium-sized rings. Only two examples will be mentioned (for a detailed discussion of the reactivity of medium-sized rings, see ref. 289 b, c and d):

[288a] V. Prelog, private communication; cf. ref. 289d.

[288b] Compare V. Prelog, ref. 283.

[289] Compare: (a) J. D. Dunitz and V. Prelog, *Angew. Chem.* **72**, 896 (1960); (b) J. Sicher, *in* P. B. D. de la Mare and W. Klyne, eds., *Progr. Stereochem.* **3**, 202 (1962); (c) R. A. Raphael, *Proc. Chem. Soc.* p. 97 (1962); (d) V. Prelog, *in* "Pure and Applied Chemistry." Butterworths, London, Vol. 6, p. 545 (1963).

In the acetolysis of cycloalkyl tosylates (C_5-C_{14}), the medium-sized rings react more rapidly than the cyclohexyl tosylates, with cyclodecyl tosylate exhibiting a reaction rate maximum.[289e]

The dependence of the equilibrium constant on the ring size becomes evident in the reaction between cyclanones and hydrogen cyanide in alcoholic solution.[283,290]

If we compare the measured dissociation constants or the pK values ($= -\log K$) of the ring-homologous cyclanone cyanohydrins with ring size C_5 to C_{18} in their dependence on the number of ring members, then we obtain a distinct maximum in the range of the medium-sized rings. The influence of ring size on the dissociation constants can be shown in this way: cyclohexanone reacts almost quantitatively with hydrogen cyanide, whereas no reaction at all can be detected with cyclodecanone. The dissociation constants of the cyclanone cyanhydrins of rings with more than thirteen members, lie between that of cyclohexanone and that of cyclodecanone. They are of about the same magnitude as those of aliphatic ketones of similar molecular weight.

The reaction rate is controlled mainly by the ring size, with other examples available in addition to the ones cited.[289b,289c,289d] In order to explain the reactivity of the medium-sized ring compounds, the I-strain theory proposed by H. C. Brown in general, currently is referred to (see page 82). With this theory, the differences in reaction rates (or in equilibria) observed for various cyclic compounds depend on the relief or increase of strain (Pitzer and Baeyer strain) during the reaction. Accordingly, the high reaction rate of a certain reaction, denotes a relief of ring strain by a change in coordination number (see page 82). If the reaction slows up, then a change in coordination number is resisted by a special geometry of the ring. The S_N1 solvolysis of cycloalkyl tosylates mentioned before, in which a carbonium ion is formed in the rate-determining step, is an indication of the effect of ring size on the ability to form a carbonium ion, i.e. a trigonal carbon atom is formed (transition $sp^3 \rightarrow sp^2$). While strain-free cyclohexane resists such a transition, the comparatively higher reaction rate of the medium-sized rings indicates that a relief of strain takes place,[291] through the formation of a carbonium ion, i.e. attained through the loss of a bond and the formation of a 120° angle.

[289e] R. Heck and V. Prelog, *Helv. Chim. Acta* **38**, 1541 (1955); H. C. Brown and G. Ham, *J. Am. Chem. Soc.* **78**, 2735 (1956).

[290] (a) L. Ruzicka, Pl. A. Plattner, and H. Wild, *Helv. Chim. Acta* **28**, 613 (1945); (b) V. Prelog and M. Kobelt, *ibid.* **32**, 1187 (1949).

[291] Compare also L. Schotsmans, P. J. C. Fierens, and T. Verlie, *Bull. Soc. Chim. Belges* **68**, 580 (1959); H. C. Brown and M. Borowski, *J. Am. Chem. Soc.* **74**, 1894 (1952); compare also: P. v. R. Schleyer, *J. Am. Chem. Soc.* **86**, 1854 (1964).

The reverse process takes place in the reaction of cyclanones with hydrogen cyanide (transition $sp^2 \rightarrow sp^3$). The medium-sized ring ketones react more slowly with hydrogen cyanide in agreement with the I-strain theory, because the removal of the trigonal carbon of the carbonyl group in the addition reaction, leads to an increase in strain.[292]

It is pertinent to refer, at this point, to one of the most important effects which results from the special geometry of medium-sized rings, i.e. the occurrence of transannular substitution and elimination reactions, which have been studied mainly by Prelog and co-workers on the cyclodecane ring,[293] and also by Cope[294] and co-workers on the cyclooctane ring.[289b,289c]

A number of other methods were used for the medium-sized rings, to obtain answers concerning their conformations. The use of infrared spectroscopy is especially cited in this regard.

Kuhn[167b] made a thorough investigation of the *cis*- and *trans*-1,2-diols of the medium-sized rings with the aid of infrared spectroscopy. From the measured $\Delta\nu$ values between free and bound OH groups (see page 124), certain conclusions could be drawn concerning the conformation of the rings. In rings smaller than cyclodecane, the OH groups of the *cis* diols are closer together than the OH groups of the *trans* diols. In rings larger than cyclodecane, the *trans* OH groups are closer together than the *cis* OH groups. Sicher *et al.*,[294a] carried out the same investigations on *cis*- and *trans*-2-aminocyclanols. However, as a result of the differences in hydrogen bond strengths the results vary from those of the diols.

The normal and medium-sized cycloalkanes and derivatives were investigated by Günthard and co-workers[295] as well as Chiurdoglu and co-workers.[296]

Although the above-named investigators did not always arrive at the

---

[292] Compare also H. C. Brown and K. Ichikawa, *Tetrahedron* 1, 221 (1957); G. Chiurdoglu, T. Doehaerd, and B. Tursch, *Colloq. Intern. Chim. Org. Probl. Stereochim., Montpellier*, 1959, in *Bull. Soc. Chim. France* p. 1322 (1960).

[293] (a) V. Prelog and A. R. Todd, "Perspectives in Organic Chemistry." p. 96. Wiley (Interscience), New York, 1956; (b) V. Prelog, *Angew. Chem.* **70**, 145 (1958); (c) V. Prelog, *Colloq. Intern. Chim. Org. Probl. Stereochim.* in *Bull. Soc. Chim. France* p. 1322 (1960); (d) cf. also J. Sicher, M. Svoboda. J. Jonas, and F. Mares, *ibid.* p. 1438 (1960).

[294] See A. C. Cope and P. E. Peterson, *J. Am. Chem. Soc.* **81**, 1643 (1959), with further references.

[294a] J. Sicher, M. Horák, and M. Svoboda, *Collection Czech. Chem. Commun.* **24**, 950 (1959).

[295] (a) Th. Bürer and H. H. Günthard, *Helv. Chim. Acta* **43**, 1487 (1960); references 31j, 269b, 297; (b) Th. Bürer and H. H. Günthard, *ibid.* **39**, 356 (1956); (c) cf. also Th. Bürer and H. H. Günthard, *Chimia (Aarau)* **11**, 96 (1957); (d) Measurements of the dipole moments of cyclanones see H. H. Günthard and T. Gäumann, *Helv. Chim. Acta* **34**, 39 (1951).

[296] References 102a in Chap. 2 and 286b; (a) G. Chiurdoglu, R. Fuks, and B. Tursch, *Bull. Soc. Chim. Belges* **67**, 191 (1958); (b) G. Chiurdoglu, R. DeProost, and B. Tursch, *ibid.* **67**, 198 (1958).

same conclusions in the interpretation of their investigations, the most important results are outlined briefly:

The infrared spectra of medium-sized ring hydrocarbons possess the same shape over a large temperature range in the solid and the liquid states.[297] This was investigated particularly thoroughly in the case of cyclodecane. The spectra of the cyclanones are not sufficiently clear cut to answer the question of whether or not the conformations of the medium ring compounds are the same in the solid and liquid state.

The spectra of the ring ketones were taken at various temperatures (+ 60° to − 170°); the even-numbered and odd-numbered ketones showed a different behavior.[31j,269b] For even-numbered ring ketones, the spectrum changes suddenly at the melting point, while in the rest of the temperature range only slight changes were observed. For odd-numbered ring ketones, the spectrum also changes, not at the melting point but at a transition point which lies lower than the melting point. The change in spectrum may be interpreted as the possible disappearance of one or more conformations at the given point.

Aside from the infrared spectra of the cyclanes C_9 to C_{17} and the corresponding ring ketones, Chiurdoglu[296] also investigated the cyclanols and halogen cyclanes in the region of the $\delta(CH_2)$ vibrations. An analysis of the spectra led Chiurdoglu to the hypothesis that all ring systems studied exist in only one preferred conformation. Probably only at higher temperatures can indications be found for the appearance of a further unstable conformation.[286a] With the exception of cyclopentane and cyclohexane, there were found for all other rings at least two bands in the region of 1450 cm.$^{-1}$. The largest difference of the $\delta(CH_2)$ bands is designated with $(\Delta\delta)_{max}$, for which in the series C_5 to C_{17} the following values in cm.$^{-1}$ were found:

C_5	C_6	C_7	C_8	C_9	C_{10}	C_{11}	C_{12}	C_{13}	C_{14}	C_{15}	C_{16}	C_{17}
0	0	11	27	43	38	30	23	14	17	11	13	10

Because this effect is independent of concentration and also is found in the vapor phase, it is probably intramolecular and not intermolecular in origin. By a comparison of molecules which have identical CH_2 groups, such as cyclopentane, cyclohexane, adamantane, or the *n*-paraffins, and which show only one $\delta(CH_2)$ vibration, with the geometry of the medium-sized rings whose CH_2 groups are not equivalent, the following rule is set up: A molecule which shows *n* different $\delta(CH_2)$ bands has at least *n* types of CH_2 groups which differ in their environment.[298]

[297] E. Billeter and H. H. Günthard, *Helv. Chim. Acta* **41**, 338, 686 (1958).

[298] However, see Günthard *et al.*, ref. 295a, who could derive no connection in their investigations between the number of observed $\delta(CH_2)$-bands and the number of non-equivalent CH_2-groups.

This rule holds for the hydrocarbons, ketones, alcohols, and halogen derivatives of the medium-sized rings. For all cyclanones, the $\delta(CH_2^{\alpha})$ band is uniform. The activated CH_2 groups in α and α' position to the keto group are equivalent. They are arranged similarly, relative to their environment. Chiurdoglu derived from these data crown forms of the symmetry groups C_2, C_s and C_{2v} for the eight,- nine-, and ten-membered rings. A noticeable deviation of the valence angle can take place due to transannular interaction of the hydrogens.

In conclusion, considering the results portrayed in Chapter 2, page 49f., we refer once more to special model studies of the medium-sized rings.

The hypothetical models used for medium-sized rings were "crown" forms. The dihedral angles in regular crown forms can be calculated. For the eight-, nine-, and ten-membered rings they amount to 96°, 115°, and 126°,[299] respectively. These dihedral angles correspond to the unfavorable *eclipsed* conformations. In this fashion it was believed possible to explain the "nonclassical" strain of these rings. As was shown on page 52f., this view could be verified only for the eight-membered ring on the basis of X-ray investigations.[300] It was no longer valid for higher ring homologs. In these compounds no *eclipsed* conformations were found. The models can be constructed mainly from *staggered* conformations.

The type of tetrahedral ring member in the model of medium-sized rings with more than eight carbon atoms is best described by one of the "partial conformations" depicted in (LXXII) to (LXXV). The partial conformations embrace five ring members.[300] The partial conformations (LXXII) and (LXXV) need not be explained any further. They occur as structural elements in many organic compounds. The crown form of cyclooctane, for example, consists of ring elements of type (LXXII). However, the dihedral angles deviate strongly from the energetically most favored ones. For rings with 9, 10, and 12 carbon atoms, the partial conformations (LXIII) and (LXXIV) are characteristic.[300]

The position of the substituents has already been discussed briefly for cycloheptane (page 160f.).

For cyclononane and cyclodecane, the circumstances are much more difficult to describe. In completely asymmetric cyclononane, no hydrogen is equivalent to any other one.[300] Even in the highly symmetrical cyclodecane there are still six nonequivalent positions. According to a proposal by Dunitz and Prelog,[300] the classification of the hydrogens is simplified for practical purposes by separation into two groups:

[299] L. Pauling, *Proc. Nat. Acad. Sci. US* **35**, 495 (1949).

[300] Cf. J. D. Dunitz and V. Prelog, *Angew. Chem.* **72**, 898 (1960).

In cyclononane and cyclodecane six hydrogens point towards the inside of the ring and are characterized by having especially small transannular distances (see page 55). These hydrogens are designated as "*intra-annular*," all the rest as "*extra-annular*."[301]

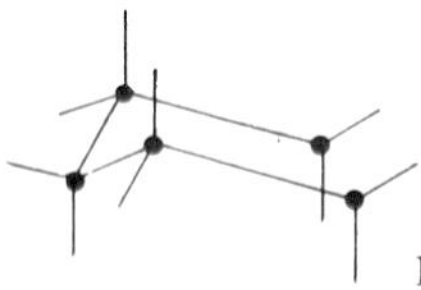

LXXII

LXXIII

LXXIV

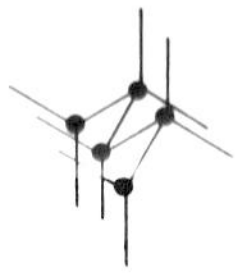

LXXV

The following was predicted for substituted medium-sized rings concerning the position of the substituent[300]:

As with cyclohexane, cycloheptane, and cyclooctane the larger substituents assume the energetically favored *equatorial* or *quasi-equatorial* position, in cyclononane and cyclodecane the *extra-annular* position of a substituent is preferred. *Intra-annular* hydrogens cannot be replaced by larger substituents, as a study of the model shows, without greatly changing the energetically favorable conformation of the ring. If the ring is substituted in the 1,1-position, then the substituents can only be on a carbon atom which has no *intra-annular* hydrogens, i.e. on ring members of the partial conformation shown in (LXXIII).[300]

If a tetrahedral carbon atom is replaced by a trigonal one (carbonium ion, carbonyl carbon), then a strong reduction of the strain takes place if it concerns a ring member which possesses an *intra-annular* hydrogen. Probably such trigonal ring members will assume the position in the ring where the resulting strain is minimal (compare cyclononanone and cyclodecanone, pp. 54, 55).

[301] See also: V. Prelog, W. Küng, and T. Tomljenović, *Helv. Chim. Acta* **45**, 1352 (1962).

CHAPTER 4

Bicyclic Compounds[1]

I. HYDRINDANES[2]

Hydrindane is a ring system in which a six-membered ring is joined to a five-membered ring in the 1,2-position. As with decalin, the five-membered ring can be connected to the six-membered ring in a *cis* or *trans* arrangement (I and II).

I II

The existence of this ring system in many natural products, whose stereochemistry usually can be related to that of the simple system, confers a special interest to the conformations of hydrindane and its derivatives.

Conformational analysis of the hydrindanes and various derivatives, as well as compounds in which a hydrindane moiety is present, is relatively difficult. It will be shown subsequently that no common denominator for all compounds can be found. This is partly due to the fact that, in contrast to other bicyclic compounds such as decalin, the energy difference between *cis*- and *trans*-hydrindane is very small. The influence of substituents, or structural changes such as the introduction of a keto group, can bring about considerable effects.

[1] For the stereochemistry of bicyclo[3.3.0]octanes see (a) P. F. G. Nau, "L'influence de la chaine pentagonale en série bicyclique." Thèse Montpellier, 1958; (b) R. Granger, P. F. G. Nau, and J. Nau, *Bull. Soc. Chim. France* p. 1807, 1811 (1959); (c) R. Granger, P. F. G. Nau, and J. Nau, *ibid.* p. 217, 1225 (1960); (d) R. Granger, P. F. G. Nau and J. Nau, *Colloq. Intern. Chim. Org. Probl. Stereochim., Montpellier,* 1959 in *Bull. Soc. Chim. France* p. 1350 (1960); (e) for the stereochemistry of bicyclo[5.3.0]decanes (perhydroazulenes), see: N. L. Allinger and V. B. Zalkow, *J. Am. Chem. Soc.* **83**, 1144 (1961).

[2] Cf. descriptive review: G. Quinkert, *Experientia* **13**, 381 (1957); E. P. Serebryakov and V. F. Kucherov, *Usp. Khim.* **32**, 1177 (1963) [*Russ. Chem. Rev.* (*Engl. Transl.*), p. 523 (1963)].

1. Hydrindane and Substituted Hydrindanes

The fusing of a six-membered and a five-membered ring can take place in *cis* or *trans* orientations of the 1,2 valences. Considering a chair form of the six-membered ring, certain bond orientations are required to do this: The *cis* form requires an *equatorial* and an *axial* bond (III), the *trans* form two *equatorial* bonds (IV). In the case of the isomeric

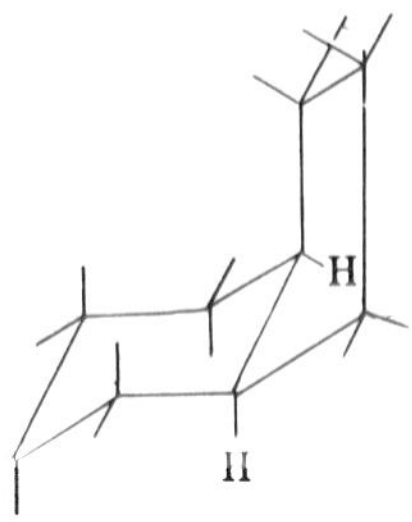

III

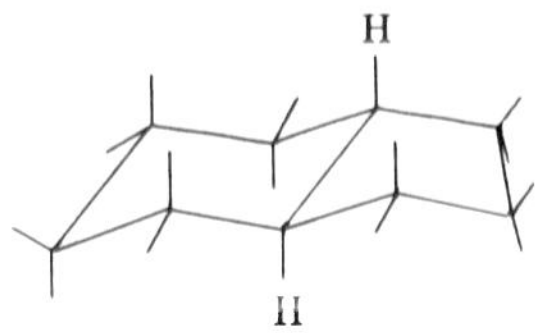

IV

decalins, each isomer can exist in a strain-free arrangement (page 56f.). However, with hydrindane, the more planar five-membered ring (in comparison to the six-membered ring) is fused to the six-membered ring, leading to a strong influence of the geometry and strain conditions. In every case, in order to attach the rigid five-membered ring, the former C—H bonds must be twisted out of their normal position. The following results become obvious from a study of models of the isomeric hydrindanes.

For *cis*-hydrindane, an *equatorial* and an *axial* bond must be drawn closer to each other in order to join a six-membered ring with a five-membered ring. The dihedral angle of the C—H bonds must be decreased (V). The result is a deformation of the chair form in such a way that the 1,3-*axial* hydrogens avoid each other. A deformation approaching the twist form[3] occurs to a certain extent. The over-all energy level is not influenced very much by this. Similar to *cis*-decalin (see page 57), *cis*-hydrindane also is capable of flipping from one conformation into the other. An approximate value of 6.4 kcal./mole for the energy barrier of the chair–chair inter-conversion in *cis*-hydrindane was ascertained from the change in the nuclear magnetic resonance spectrum at low temperatures.[3a]

The *equatorial* bonds of the two neighboring carbon atoms must

[3] W. G. Dauben and K. S. Pitzer, *in* "Steric Effects in Organic Chemistry" (M. S. Newman, ed.), Chap. 1. Wiley, New York, 1956.

[3a] W. B. Moniz and J. A. Dixon, *J. Am. Chem. Soc.* **83**, 1671 (1961).

approach each other in *trans*-hydrindane (VI). As a result, the six-membered ring becomes more puckered. This deformation leads to a closer approach of the 1,3-*axial* hydrogens, as well as to a strained bond angle[4a] although the distortion cannot be large.[4b] The result is an increase in the energy in comparison to a strain-free fused bicyclic compound, for which two normal *equatorial* bonds are used (*trans*-decalin). In other words, the energy difference between *cis* and *trans*-hydrindane (of which *trans*-hydrindane, nevertheless, is the more stable) should become small in comparison to that of the isomeric decalins.

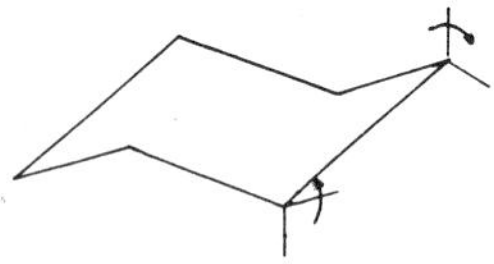

V

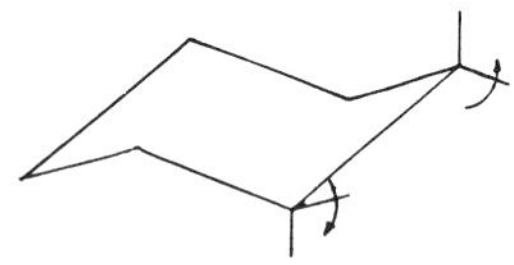

VI

In actual fact, the energy (enthalpy) difference between *cis*- and *trans*- hydrindane is small in comparison to that of the isomeric decalins. The difference amounts to 2.7 kcal./mole for the decalins (see page 59), but was observed to be only 0.74 ± 0.52 kcal./mole (liquid, 298° K) for the hydrindanes.[5,6] ($\Delta F = 0.24 \pm 0.52$ kcal./mole)

The relative stabilities of the hydrindane isomers were investigated by Allinger[7,7a] by establishing the equilibrium *cis* ⇄ *trans* over palladium-charcoal catalyst in the liquid phase. Below 466° K, as mentioned above, *trans*-hydrindane predominates and is the more stable isomer. Above this temperature, the conditions are reversed. *Cis*-hydrindane is more stable than the *trans* isomer in the temperature range of 466° to 638° K. The reason for the greater stability of the *cis* isomer in this range lies

[4] (a) S. J. Angyal and C. G. Macdonald, *J. Chem. Soc.* p. 686 (1952), used this idea for the first time in order to explain the differences of the ketal formation between *cis*- and *trans*-1,2-cyclohexanediol. (b) Cf. C. Djerassi, D. Marshall, and T. Nakano, *J. Am. Chem. Soc.* **80**, 4853 (1958).

[5] C. C. Browne and F. D. Rossini, *J. Phys. Chem.* **64**, 927 (1960).

[6] Cf. also (a) W. Hückel, *Ann. Chem.* **533**, 1 (1938); (b) W. Hückel, M. Sachs, J. Yantschulewitsch, and F. Nerdel, *ibid.* **518**, 155 (1935). Here the *trans* form was reported as 1.8 kcal./mole more stable than the *cis* form. The measured heats of combustion (*cis*-hydrindane 1347.5 kcal./mole, *trans*-hydrindane 1345.5 kcal./mole) were close to the limit of error of this method, because the substances had to be burned, "under cover." (Cf. ref. 10 in Chap. 1, page 82, footnote 5).

[7] N. L. Allinger and J. L. Coke, *J. Am. Chem. Soc.* **82**, 2553 (1960); cf. also N. L. Allinger, R. B. Hermann, and C. Djerassi, *J. Org. Chem.* **25**, 922 (1960).

[7a] Compare also K. R. Blanchard and P. v. R. Schleyer, *J. Org. Chem.* **28**, 247 (1963), ($\Delta H = 0.58 \pm 0.05$ kcal./mole by equilibration of *cis*- and *trans*-hydrindane with $AlBr_3$).

in its higher entropy in comparison to that of the *trans* isomer. Because the energy content of the isomeric forms differs only slightly, the entropy makes itself effective at higher temperatures to the extent that the *cis* isomer predominates. The higher entropy of *cis*-hydrindane is attributed by Allinger[7] to its spatial structure. In *trans*-hydrindane, the five-membered ring is more rigid than the five-membered ring in *cis*-hydrindane, as can be seen from a model (cf. also III and IV). The five membered ring is thus more capable of pseudorotation in the latter (see page 14), which leads to a higher entropy. On the basis of an entropy higher by 2.3 cal./degree mole for *cis*-hydrindane ΔH was calculated at 1.07 ± 0.09 kcal./mole (at 552° K).

No definitive experimental data are as yet available concerning the conformation of the five-membered ring in *cis*-hydrindane. However, on the basis of infrared investigations with *trans*-hydrindanol-2, the five-membered ring in *trans*-hydrindane probably exists in the envelope conformation, in agreement with a model (IV).[8]

Whether the six-membered ring in *cis*-hydrindane exists in the twist form, caused by the twisting imparted by the fused five-membered ring, has been discussed intensively.[9] In this way, the five-membered ring would lose the additional strain. A study of models lessens the probability that such an assumption is valid, particularly because of the energetically unfavorable boat form. Eliel[9h] was able to show that the oxidation rates of the *cis*- and *trans*-2-oxa-*cis*-hydrindane-5,6-diols with lead tetraacetate corresponded with the oxidation rates of the *cis*- or *trans*-cyclohexanediols but not with the cyclopentanediols. This result, as well as infrared investigations, was used as an argument against the twist form.

Granger and Nau[1a,9k,9l] synthesized the elusive *cis*-hydrindane-5,6-diol. Its oxidation with lead tetraacetate was repeated. A boat form of the six-membered ring in *cis*-hydrindane-5,6-diol was not consistent with the measured reaction rate. Nevertheles, the reaction rate, is a good deal higher than that of *cis*-cyclohexanediol (*cis*-1,2-cyclohexane-diol, $k_{25} = 8.36$, $k_{cis/trans} = 21.9$; and *cis*-5,6-hydrindanediol, $k_{25} = 107$, $k_{cis/trans} = 151$). This indicates a closer proximity of the hydroxyl

[8] W. Hückel and J. Kurz, *Ann. Chem.* **645**, 194 (1961).

[9] (a) A. S. Dreiding, *Chem. & Ind. (London)* p. 992 (1954); (b) H. G. Derx, *Rec. Trav. Chim.* **41**, 318 (1922); (c) W. Hückel and H. Friedrich, *Ann. Chem.* **451**, 132 (1926); (d) A. Kandiak, *J. Chem. Soc.* p. 922 (1931); (e) R. S. Thakur, *ibid.* p. 2147 (1932); (f) R. P. Linstead, *Ann. Rep. Progr. Chem. (Chem. Soc. London)* **32**, 306 (1935); (g) W. G. Dauben and J. Jiu, *J. Am. Chem. Soc.* **76**, 4426 (1954); (h) E. L. Eliel and C. Pillar, *ibid.* **77**, 3600 (1955); (i) L. F. Fieser and M. Fieser, "Organic Chemistry," p. 308, Reinhold, New York, 1956; (j) R. A. Raphael, *in* "Chemistry of Carbon Compounds" (E. H. Rodd, ed.), Vol. II, Part A, p. 311. Elsevier, Amsterdam, 1953; (k) R. Granger, P. F. G. Nau, J. Nau, and C. François, *Bull. Soc. Chim. France* p. 496 (1962); (l) R. Granger, P. F. G. Nau, and C. François, *ibid.* p. 1902 (1962).

groups to each other. A conformation with a deformed chair form of the six-membered ring was obtained for *trans*-hydrindane-5,6-diol, but no precise conformation for *cis*-hydrindane-5,6-diol was obtained, even though an analysis of the infrared spectra had been made according to Kuhn's procedure (calculation of the $\Delta\nu$ values between free and bonded OH groups, see page 124).

Let us consider also the influence of an angular methyl group on the stability of the hydrindane system. In the case of the 8-methylhydrindane system, we can say in general that the *cis* isomer is the more stable one. In the structurally more complicated compounds of the steroid (C/D rings or A-nor/B rings) and triterpene series (cf. page 223)[10], the same rule applies, although there are exceptions.

The effect of an angular methyl group is such that its introduction in *trans*-hydrindane causes more *gauche* interactions to arise than in *cis*-hydrindane. Contrary to the results for 9-methyl decalin, this brings about a greater stability of the *cis* compound.[11]

The following is known about the conformations and configurations of the hydrindanols:

The configurations of the monosubstituted hydrindanes are characterized by the same designations as for the substituted decalins, namely *cis-cis*, *cis-trans*, *trans-cis*, and *trans-trans* (see page 180). The first designation relates to the ring junctures; the second designation to the position of the substituent relative to the position of the C—C valence C-8–C-7 or C-8–C-1.

Of the hydrindanols with the OH group on the five-membered ring, configurations were assigned to *cis-cis*-1-hydrindanol and *cis-trans*-1-hydrindanol on the basis of their methods of preparation as well as the high rates of solvolysis of the toluenesulfonates of the *cis-cis* isomer.[12] A definite assignment for the two *cis*-2-hydrindanols is not possible from the saponification rates of the succinates or the solvolysis of the toluenesulfonates. These do not vary enough for the isomers[13]. Epimeric forms are not possible for *trans*-2-hydrindanol.

The *cis*-5-hydrindanols were investigated in regard to their stabilities.[9g,14] For each isomer, the *cis-cis* (VII and VIII) as well as the *cis-trans* compound (IX and X), two conformations are possible.

---

[10] Cf. Ref. 9a; W. E. Bachmann and A. S. Dreiding, *J. Am. Chem. Soc.* **72**, 1323 (1950); (b) K. Dimroth and H. Jonsson, *Ber.* **74**, 520 (1941).

[11] A diagram of the isomeric 8-methyl hydrindanes was omitted. The stability conditions become clear with a model, merely by counting the *gauche* interactions in *cis*- and *trans*-8-methylhydrindane.

[12] W. Hückel and M. Hanack, *Ann. Chem.* **610**, 106 (1957).

[13] W. Hückel and W. Egerer, *Ann. Chem.* **645**, 162 (1961).

[14] W. Hückel, *Ann. Chem.* **533**, 1 (1938); compare also R. Granger, H. Técher, and J. P. Girard, *Compt. Rend. Acad. Sci.* **251**, 2546 (1960).

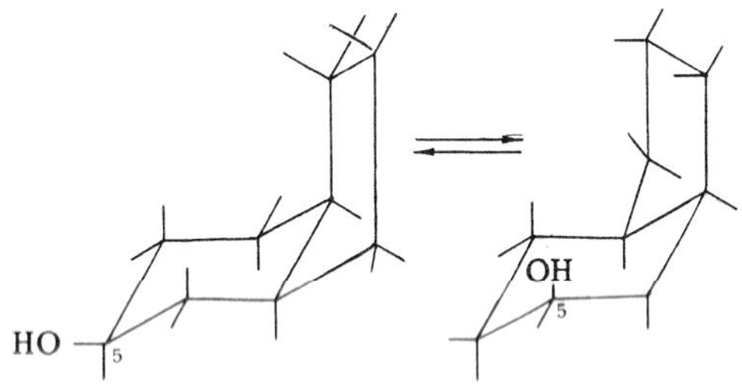

VII VIII

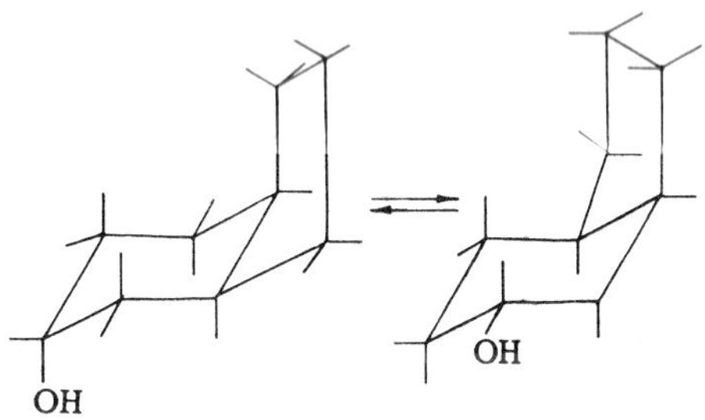

IX X

Equilibration with sodium in xylene or reduction of *cis*-5-hydrindanone with sodium in alcohol yielded a preponderant amount of the *cis-trans* compound (IX and X). If the hydroxyl group in the *cis-cis* compound is placed in the *axial* position (VIII), then an interaction between the OH group and the hydrogen at C-3 results[9g]. Such an interaction is not present in either conformation of the *cis-trans* isomer (IX and X). If we consider that the conformations (VII) and (X) have the same energy content, but (VIII) is of higher energy than (IX), then it can be expected that the *cis-trans* isomer is more stable thermodynamically.[9g,15]

The configuration of the OH group has been assigned for the *cis*-4-hydrindanols as well by means of equilibrium investigations.[16]

2. Hydrindanones

The presence of a keto group in the hydrindane system can lead to a shift in stability, as with the angularly substituted hydrindanes. The

[15] Concerning the corresponding hydrindanylamines, cf. ref. 9g.

[16] Ref. 14; (a) W. Hückel, R. Schlüter, W. Doll, and F. Reimer, *Ann. Chem.* **530**, 166 (1937); (b) cf. also W. Hückel and E. Goth, *Ber.* **67**, 2104 (1934).

reasons are not immediately apparent: *Trans*-2-hydrindanone possesses a lower heat of combustion than the *cis* isomer[17]; *cis*-1-hydrindanone[13,18,19] and *cis*-4-hydrindanone,[6b] however, predominate over the *trans* isomers at equilibrium.

Many factors have been brought out in attempting to explain this contrary behavior.[20] The alkyl ketone effects[2,21] (see page 151) and entropy effects[22] are noted. Also, the formation of a hydrogen bond between the keto group and the hydrogen atom on carbon atom 7, has been used as a possible explanation for the greater stability of *cis*-1-hydrindanone.[9i,j]

The application of alkyl ketone effects was employed thoroughly by Quinkert[2] in order to explain the stability of isomeric hydrindanones and methyl hydrindanones: In 1-hydrindanone, a 2-alkyl ketone effect appears in the *trans* form upon examining the atomic sequence O–C-1 C-8–C-7 (XI, X = H).

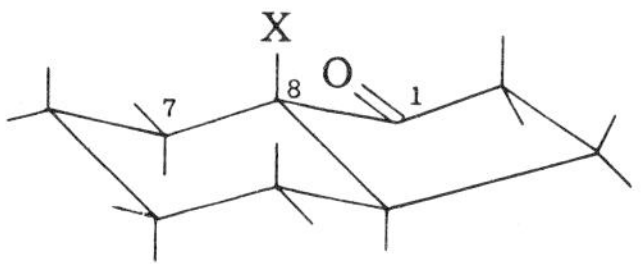

XI

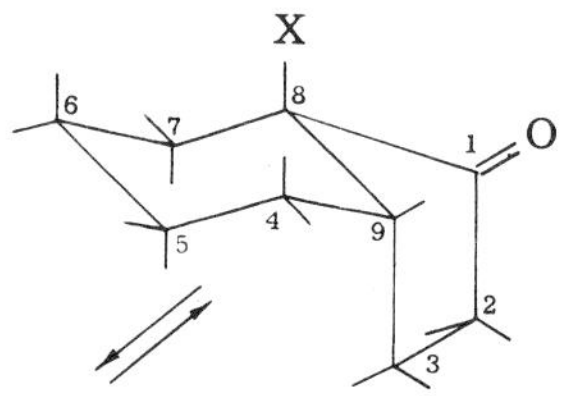

XII

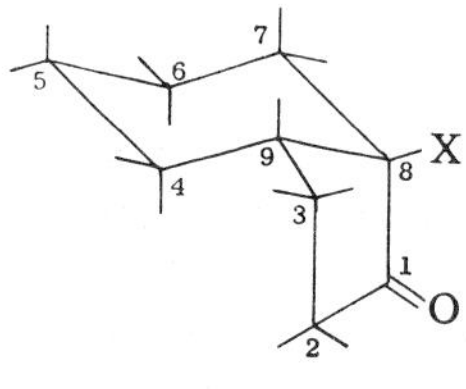

XIII

The *cis*-fused compounds can exist in two interconvertible conformations because the structure is not rigid (XII and XIII). The 2-alkyl ketone effect probably appears in the conformation shown in (XIII,

[17] (a) W. A. Roth and F. Müller, cited in W. Hückel and H. Friedrich, *Ann. Chem.* **451**, 160 (1926); (b) M. Sachs, Dissertation, Göttingen University, 1928.

[18] R. P. Linstead, *Ann. Rep. Progr. Chem.* (*Chem. Soc. London*) p. 305 (1935).

[19] H. O. House and G. A. Rasmussen, *J. Org. Chem.* **28**, 31 (1963).

[20] Compare: W. G. Dauben, *Colloq. Intern. Chim. Org. Probl. Stereochim. Montpellier*, 1959 in *Bull. Soc. Chim. France* p. 1338 (1960); G. Quinkert, ref. 2; W. G. Dauben, and G. J. Fonken, *J. Am. Chem. Soc.* **78**, 4736 (1956); C. Djerassi, T. T. Grossnickle, and L. B. High, *ibid.* **78**, 3166 (1956); A. S. Dreiding, *Chem. & Ind.* (*London*) p. 992 (1954).

[21] W. Klyne, *Experientia* **12**, 119 (1956).

[22] N. L. Allinger, *J. Org. Chem.* **21**, 915 (1956).

X=H) upon examination of the atomic chain O–C-1–C-8–C-7. This can best be seen with a model. Such an influence, however, does not exist in the conformation shown in (XII). It is now assumed that the small preference in energy of *trans*- over *cis*-hydrindane can be reversed by the stabilizing action of a 2-alkyl ketone effect. *Cis*-1-hydrindanone can assume a conformation not influenced by the 2-alkyl ketone effect.

The 2-alkyl ketone effect does not appear in 2-hydrindanone. The *trans* isomer is the more stable one in this case, as in the case of hydrindane itself. In 4-hydrindanone there exists a 2-alkyl ketone effect in the *trans* form [O–C-4–C-9–C-3, compare with (XI)]. Such an effect is also present in one of the conformations of the *cis* isomer [O–C-4–C-9–C-3, compare with (XII)]. No such effect can be detected in the second possible conformation (cf. XIII). The *cis*-fused isomer of 4-hydrindanone is therefore more stable.

In the case of 5-hydrindanone, the 3-alkyl ketone effect takes place in only one of the conformations of the *cis* form (cf. XIII). This can be seen on a model representing the conformation (XIII), O–C-5 to C-9 and C-3. This should lead to a stabilization of the *cis*-fused isomer.[23] The application of this method to the isomeric 8-methylhydrindanones always results in a preference for the *cis*-fused form,[2] on the basis of stability relations between *cis*- and *trans*-methylhydrindane.

Aside from the alkyl ketone effects, entropy effects can be used to explain the greater stability of the *cis*-hydrindanone system,[24] as mentioned previously. Although experimental data are lacking, except for 2-hydrindanone, entropy becomes a determining factor for those compounds which do not differ greatly in energy content. The *cis* form of hydrindane possesses a larger positive entropy due to its greater flexibility, as has already been discussed (see page 175). As a result, this form predominates at equilibrium at higher temperatures. The same considerations can also be applied to 8-methylhydrindane.[22] Although, here as well, no experimental data are available, it can be determined from other parameters such as density and refractive index that the *cis* and *trans* forms have the same energy content. As a result of the higher entropy of the *cis* form, it should predominate at equilibrium.

A conformational analysis has been attempted[25] for structurally more

[23] Cf. the physical constants given by R. Granger, P. Nau, and J. Nau, *Bull. Soc. Chim. France* p. 531 (1958); also R. Granger, H. Técher, and J. P. Girard, *Compt. Rend. Acad. Sci.* **251**, 2546 (1960).

[24] See ref. 21. For 8-methyl-5-hydrindanone see C. Djerassi, D. Marshall, and T. Nakano, *J. Am. Chem. Soc.* **80**, 4853 (1958).

[25] Cf. especially (a) N. L. Allinger, R. B. Hermann, and C. Djerassi, *J. Org. Chem.* **25**, 922 (1960), with further references. Cf. also ref. 20; (b) J. F. Biellmann, D. Francetic, and G. Ourisson, *Tetrahedron Letters* **18**, 4 (1960).

complicated compounds containing a hydrindanone moiety, such as steroids, using similar considerations. The *cis* isomers were determined to be the more stable ones by equilibrium measurements. The stability differences between two isomers, however, are sometimes very small.

We especially call attention to the fact that all the methods considered for explaining stability relationships in the hydrindane and hydrindanone series are qualitative in nature. This is especially true for the alkyl ketone effects, which, as shown on p. 151 are doubtful, due to recent measurements on substituted cyclohexanones. A quantitative treatment has so far not been possible because of the number of various effects[26] to be considered and because of the lack of numerical data on the magnitude of these effects.

II. DECALINS

The decalins have been considered in Chapter 2 (page 56f.). Therefore, only the substituted decalins will be treated in the following.

As in the case of the hydrindanes, two different linkages are considered for *cis*- and *trans*-fused decalin, as viewed from one of the six-membered rings. In *cis*-decalin, there is an *equatorial* and an *axial* bond; in *trans*-decalin there are two *equatorial* bonds. Both linkages, as a study of a model shows, are possible without a deformation of the chair conformation of the six-membered ring, thus differing from hydrindane. *Cis*-decalin is flexible. One conformation can transform into the other by flipping. The *trans*-fused form leads to a rigid model by inclusion of both *equatorial* bonds (cf. Chapter 2, page 57).

1. Monosubstituted Decalins

Each of the 1- and 2-substituted decalins exists in four stereoisomeric forms. They are differentiated as *cis-cis*, *cis-trans*, *trans-cis*, and *trans-trans* forms. As mentioned before (cf. page 17), the first term relates to the type of ring fusion (*cis* or *trans*). The second term relates to the relative position of the substituent to the C-9–C-8 bond. The configurations of the 1- and 2-decalols, as well as those of the corresponding

[26] Cf. N. L. Allinger and S. Greenberg, *J. Org. Chem.* **25**, 1399 (1960). Allinger considered in his investigations of the stability of the hydrindanone system in steroids, influences produced by strained ring systems through the so-called conformational transmission. (Compare p. 225.) Compare also: W. G. Dauben, ref. 20.

amines, have been elucidated mainly by W. Hückel[27] and collaborators. Recently, the conformations of the decalols were thoroughly investigated.[28]

a. Decalols

(i) 1-Decalols

The configurations of the four stereoisomeric, racemic 1-decalols are definitely established. They were based chiefly on their method of

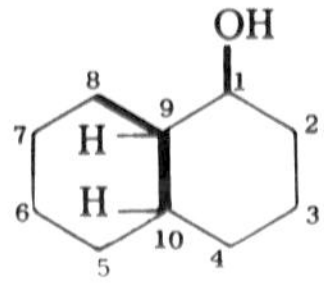

1. *cis-cis*-1-decalol
m.p. 93°

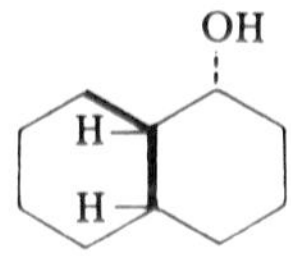

2. *cis-trans*-1-decalol
m.p. 55°

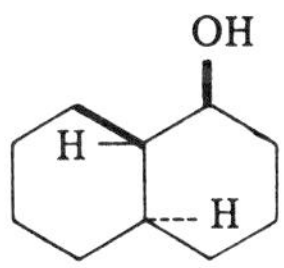

3. *trans-cis*-1-decalol
m.p. 49°

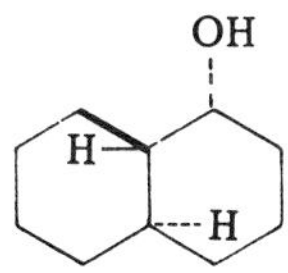

4. *trans-trans*-1-decalol
m.p. 63°

---

[27] (a) Ref. 23a in Chap. 1; (b) W. Hückel, *Ber.* **58**, 1449 (1925); (c) W. Hückel, *Angew. Chem.* **39**, 842 (1926); (d) Ref. 23b in Chap. 1; (e) W. Hückel and E. Frank, *Ann. Chem.* **477**, 137 (1930); (f) W. Hückel, R. Danneel, A. Gross, and H. Naab, *ibid.* **502**, 99 (1933); (g) W. Hückel and H. Naab, *ibid.* **502**, 136 (1933); (h) W. Hückel, Ber. **67A**, 129 (1934); (i) W. Hückel, *Ann. Chem.* **533**, 1 (1938); (k) W. Hückel and Ch. Kühn, *Ber.* **70**, 2479 (1937); (l) W. Hückel, H. Havekoss, K. Kumetat, D. Ullmann, and W. Doll, *Ann. Chem.* **533**, 128 (1938); (m) W. Hückel, W. Tappe, and G. Legutke, *ibid.* **543**, 191 (1940); (n) W. Hückel, *Ber.* **77**, 805 (1944); (o) W. Hückel and G. Stelzer, *Chem. Ber.* **88**, 984 (1955).

[28] (a) W. Hückel and R. Schwen, *Ann. Chem.* **604**, 97 (1957); (b) Ref. 39 in Chap. 3; (c) W. Hückel, *Acta Chim. Acad. Sci. Hung,* **18**, 28 (1958); (d) Ref. 57c in Chap. 3; (e) W. Hückel, M. Maier, E. Jordan, and W. Seeger, *Ann. Chem.* **616**, 48 (1958); (f) W. Hückel and W. Kraus, *Chem. Ber.* **92**, 1158 (1959); (g) W. Hückel and R. B. Rashingkar, *Ann. Chem.* **637**, 20 (1960). Ref. 32; (i) W. Hückel and J. Kurz, *Ann. Chem.* **645**, 194 (1961); (k) W. Hückel, *Colloq. Intern. Chim. Org. Probl. Stereochim., Montpellier,* 1959 in *Bull. Soc. Chim. France* p. 1369 (1960); (l) W. Hückel, D. Maucher, O. Fechtig, J. Kurz, M. Heinzel, and A. Hubele, *Ann. Chem.* **645**, 115 (1961); (m) W. Hückel and H. Feltkamp, *Ann. Chem.* **649**, 21 (1961); (n) W. Hückel and D. Rücker, *Ann. Chem.* **666**, 30 (1963).

formation—which will not be detailed here[29]—as well as on their physical properties (Auwers-Skita rule). While no doubt exists[30] in the assignment of configurations of the isomeric 1-decalols, the attempts to determine the preferred conformations of these compounds have not been as fruitful.

Attempts have been made to use the saponification rates of the carboxylic esters,[27h,27l,30a] as well as the solvolysis rates of the toluenesulfonates[31] of the isomeric 1-decalols for a conformational analysis. Infrared spectroscopy has also been utilized for conformational analysis.[28i,32]

Let us first compare the two chemical methods, the saponification of the carboxylic esters and the solvolysis of the toluenesulfonates (cf. p. 190f.).

We recall the already mentioned general rule concerning the saponification rates of carboxylic esters as well as toluenesulfonates, depending upon their position on the six-membered ring (page 115, ref. 140). *Axial* carboxylic ester groups are saponified more slowly than *equatorial* ones[33]; *axial* toluenesulfonates are solvolyzed more rapidly than *equatorial* ones. (Cf. page 259.)

The reason for this can be found in the difference in reaction mechanism between the solvolysis of carboxylic esters and sulfonic esters (cf. page 259). During saponification of carboxylic esters, the attack of the reagent takes place on the carbonyl group, if we disregard exceptions. The C—O bond, whose direction is determined by conformational analysis, remains undisturbed. During the solvolysis of toluenesulfonic esters, the solvent directly attacks, in two ways, the carbon atom to which

---

[29] Ref. 23a in Chap. 1; ref. 27b, 27f in this chapter; cf. also A. Skita, *Ann. Chem.* **427**, 267 (1922).

[30] Compare also W. G. Dauben, R. C. Tweit, and C. Mannerkantz, *J. Am. Chem. Soc.* **76**, 4420 (1954).

[30a] H. D. Orloff, *Chem. Rev.* **54**, 347 (1954).

[31] See references 27h, 28a, 28b, 28c, 28e; also (a) I. Moritani, S. Nishida, and M. Murakami, *J. Am. Chem. Soc.* **81**, 3420 (1959); (b) Cf. also S. Nishida, *ibid.* **82**, 4290 (1960); (c) Cf. also H. L. Goering, H. H. Espy, and W. D. Closson, *J. Am. Chem. Soc.* **81**, 329 (1959).

[32] W. Hückel and Y. Riad, *Ann. Chem.* **637**, 33 (1960).

[33] However, it may be mentioned at this point that there are exceptions to this rule. In the case of the 3-acetoxycholestan-5-ols and 3-acetoxycoprostan-5-ols, the *axial* acetate is saponified more rapidly than the *equatorial* one. [H. B. Henbest and B. J. Lovell, *J. Chem. Soc.* p. 1965 (1957)]. The reversal of the saponification rates, in the case of monoesters of 1,3-diols, can result from the possibility of a hydrogen bond from the *axial* OH to the *axial* ester group. The hydrolysis is facilitated by the hydrogen bond. Compare also: R. West, J. J. Korst, and W. S. Johnson, *J. Org. Chem.* **25**, 1976 (1960).

the bond is attached[34]: (1) The sulfonic acid residue must be solvated from one side; (2) concurrently, the solvent must attach itself to the carbon from the other side. The conditions for the possibility of this attack are different for the e- and a-position of the sulfonic acid residue. The backside attack of the carbon atom with an e-directed substituent is quite difficult in contrast to attack on the substituent itself. A cursory study of a model makes this apparent. It follows that the solvolysis of an *equatorial* sulfonate residue is more difficult than that of an *axial* one, in which case the rear attack of the solvent on the carbon atom carrying the substituent can proceed almost freely. From this simple explanation, the more rapid solvolysis of an *axial* sulfonic acid group becomes more obvious. However, we can not predict how much more rapidly the reaction proceeds from the a-position than from the e-position (cf. page 260).

A more simplified picture results[35] if we attribute the increased reaction rate to steric strain. This strain arises from the interaction of the tosyl group with the 3,5-situated *axial* hydrogens, whereby the ionization is facilitated.

In the case of the carboxylic ester, the reaction rates reverse themselves. An *equatorial* carboxylic ester group has a more open location than an *axial* one. The attack of the solvent on the carbonyl is hindered by the 1,3-*axial* hydrogens. *Equatorial* carboxylic ester groups are therefore saponified more rapidly.

The reaction rate constants which were found are given in Tables I and II.

Consider first the *trans-cis* and the *trans-trans*-1-decalols for an interpretation of the listed constants.

A rigid ring system exists in the case of a *trans*-fused decalin. A conversion of the two-chair form is not possible. If we ignore a possible deformation of the ring system, then the two isomeric *trans*-1-decalols should be homogeneous in conformation. From the assigned configurations it follows, as is apparent from a model, that the OH group can assume only the *axial* position in the *trans-cis*-1-decalol and the *equatorial* position in *trans-trans*-1-decalol (see XVII and XVIII page 188).

The measured reaction rates for the carboxylic esters and toluenesulonates agree well with the theory if we assume that the reaction rate

---

[34] Cf. W. Hückel and K. Tomopulos, *Ann. Chem.* **610**, 78 (1957), with further references; ref. 39, Chap. 3, and 28e.

[35] S. Winstein and N. J. Holness, *J. Am. Chem. Soc.* **77**, 5562 (1955).

TABLE I

SAPONIFICATION OF CARBOXYLIC ESTERS[a]

1-Decalol	Rate constant k in aqueous sodium hydroxide (liters $mole^{-1}$ min^{-1})	
	Acid succinate 40°	Acid phthalate 60°
cis-cis	0.452	0.064
cis-trans	0.130	0.021
trans-cis	0.012	Very small
trans-trans	0.244	0.034

[a] See refs. 27h, 27l.

TABLE II

SOLVOLYSIS OF TOLUENESULFONATES

1-Decalol	Rate constant k (sec^{-1}) at			Activation energy
	30°	40°	50°	
	Ethanolysis ($k \times 10^6$)			
cis-cis[a,b]	3.0	10.7	36.5	24.2
trans-cis[a,b]	1.39	5.4	18.6	25.1
cis-trans[c]	—	—	0.81	—
trans-trans[d]	—	—	0.28	27.0
	Acetolysis ($k \times 10^5$)			
	80°	75.5°		
cis-cis	118[e]	—		—
cis-trans	6.81[e]	—		—
trans-cis	86[e]	59.3[f]		—
trans-trans	1.63[e]	1.10[f]		—

[a] See ref. 28a. [b] See ref. 57e in Chap. 3. [c] See ref. 28m. [d] See ref. 28l. [e] See ref. 31c. [f] See ref. 31a.

depends only on the *axial* or *equatorial* position of the ester group. The acid succinate of *trans-cis*-1-decalol is saponified 20 times more slowly than that of *trans-trans*-1-decalol. The rate constant of the acid phthalate of *trans-cis*-1-decalol is so small that it could not be determined accurately.[271] The reaction rates for the ethanolysis and acetolysis of the tosylates of the two *trans*-fused 1-decalols agree well with the above-stated rule: *Trans-cis*-decalyl toluenesulfonate with an *axial* tosyl group is solvolyzed ~65 times more rapidly by ethanolysis and ~50 times more rapidly by acetolysis than the *trans-trans* isomer with an equatorial tosyl group.

While there are no contradictions for the *trans*-fused 1-decalols in a kinetic conformational analysis, the situation is not clear for the corresponding *cis*-1-decalols.

The position of the OH group can no longer be derived from a study of a model. We can see from this study that *cis*-decalin possesses two types of 1-positions, which have a *peri*-relationship to each other:

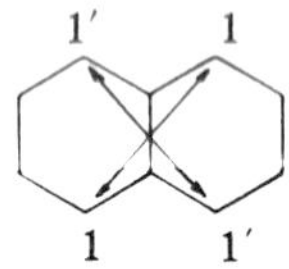

In the one species (1), a substituent which is *cis* to the ring fusion assumes an e-position, in the other species (1′) an a-position. A *trans* substituent follows an *axial* (1) position in the first case; an *equatorial* position (1′) in the second case.[36] In using molecular models for the purpose of illustration, the rearrangement of substituents from 1 to 1′ can be achieved by the flipping of one conformation into the other. *Cis*-decalin, contrary to rigid *trans*-decalin, is capable of this.[37] (XIV, cf. also page 57).

If the reaction rate of a substituent depended only on whether it were in an e- or a-position, then we would find the same reaction rate for

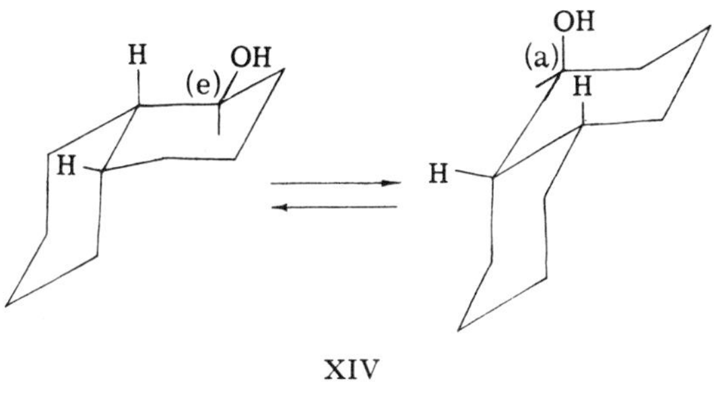

XIV

[36] It is, therefore, always possible to give a substituent in position 1 the preferred *equatorial* conformation.

[37] Compare also ref. 6 in Chap. 2, page 41; (b) J. A. Mills, *J. Chem. Soc.* p. 260 (1953).

derivatives of the isomeric *cis*-1-decalols. We can assign the preferred *equatorial* position to the substituent in both cases. However, this is not the case.

Let us consider now the saponification rates of the carboxylic esters. The higher saponification rate of *cis-cis*-1-decalol carboxylic ester in comparison to that of the *cis-trans* isomer, indicates an *equatorial* position of the carboxylic ester residue.[38] Yet the saponification proceeds twice as fast as that of *trans-trans*-1-decalol carboxylic ester, for which an *equatorial* position of this group has been determined.

The carboxylic esters of *cis-trans*-1-decalol react about three times more slowly than those of *cis-cis*-1-decalol. The following explanation is given for this behavior[30a]. The a-position of the hydroxyl of *cis-cis*-1-decalol is especially unfavorable, because not only is there an interaction with the *axial* hydrogen on C-3, but also with the *peri*hydrogen on C-5. In the case of the *cis-trans*-1-decalol, as can be seen on the model, the influence of the hydrogen on C-5 is of no consequence because it is further removed. The lower rate of saponification of the carboxylic ester of *cis-trans*-1-decalol, in contrast to the *cis-cis*-1-decalol, is explained by the presence of an equilibrium mixture between *equatorial* and *axial* conformers. Because of the presence of conformers with an *axial* carboxylic ester group, a lower saponification rate results.

The disturbing fact in this explanation lies therein, namely that we include a conformation with an *axial* ester group for a *trans* alcohol. We generally do not consider these for *trans*-2-substituted cyclohexanols[38,39] (cf. page 113). The conformational analysis of the *cis*-1-decalols on the basis of saponification rates of carboxylic esters does not lead to an unequivocal result, nor does it lead to an actual contradiction.

Contradictions arise when we analyze the solvolysis of the tosylates and compare them with the saponification rates of the carboxylic esters.[39] The tosylate of *cis-trans*-1-decalol is stable. It exhibits a low solvolysis rate, just as do the tosylates of *trans*-2-alkylated cyclohexanols. This agrees with an *equatorial* position of the esterified hydroxyl, but no longer with the assumption of an equilibrium between the conformers with an *axial* and *equatorial* tosyl group. The toluenesulfonate of *cis-cis*-1-decalol, on the other hand, exhibits a very high reaction rate. It even reacts much faster than the toluenesulfonate of *trans-cis*-1-decalol (Table II). The value determined for the ethanolysis corresponds approximately to the values found for the *cis*-2-alkylcyclohexanols. Because we relate

[38] Compare S. Winstein and N. J. Holness, *J. Am. Chem. Soc.* **77**, 5574 (1955).
[39] Compare also: W. Hückel and M. Hanack, *Ann. Chem.* **616**, 18 (1958).

high solvolysis rates with an *axial* position of the esterified hydroxyl, we have to assume[39] this position for *cis-cis*-1-decalol. This is done despite the unfavorable interaction with the hydrogen on C-5 and the conformational analysis conclusions from the saponification of the carboxylic ester, which favor the *equatorial* position of the hydroxyl.

The dubious value of a conformational analysis based on reaction kinetics, especially applied to the tosylates, can be seen from this detailed example.[39,39a] To explain the reasons for this, even as far as they are known, would involve even greater detail. We should point out once more the different mechanisms involved in the saponification of carboxylic esters and toluene sulfonic esters. In the latter case, the shape of the molecule may be different from normal, especially with the possibility of an elimination reaction taking place at the same time. Further, the idea may be discussed, that a kinetic conformational analysis may lead to contradictory results because the positions of the hydroxyl group, carboxylic ester, and toluene sulfonic ester groups are not identical.

Infrared spectroscopy was, therefore, also called upon for a conformational analysis of the decalols.[28i,32,40,40a]

All four stereoisomeric 1-decalols were investigated in carbon disulfide. As was demonstrated for the alkylated cyclohexanols (page 113), the *equatorial* and *axial* hydroxyl groups differ characteristically in the absorption of the C—(OH) stretching vibrations [ν-C—(OH)]. The difficulty of making a positive statement concerning the conformation or possible conformational equilibrium in the decalols, just as was the case with the alkylated cyclohexanols (page 113), is attributed to assigning definite bands to the C—(OH) stretching vibration. The range of the C—(OH) stretching vibration ($\sim$940 to 1070 cm^{-1}) is also the range of the skeletal vibrations. The bands considered up to now as C—(OH) stretching vibrations and the conformations derived from these for the 1-decalols must be viewed with reservations.[32]

A possibility in assigning the C—(OH) stretching vibration was found by Hückel and Kurz[28i] in the following: If carefully dried hydrogen chloride is passed into a carbon disulfide solution of the alcohol to be investigated, then a shift or splitting of the C—(OH) stretching vibration band takes place due to the presumed formation of the oxonium ion while the skeletal vibrations are not affected. It thus becomes possible to assign the C—(OH) vibrations of the decalols with greater certainty and to determine their conformations. The values listed in Table III were obtained for the C—(OH) stretching vibrations.

[39a] W. Hückel *et al.*, *Ann. Chem.* **624**, 142 (1959).
[40] Cf. M. Hanack, *Forsch. Fortschr.* 34, 259 (1960).
[40a] For NMR investigations of decalols cf. Ref. 19 in Chap. 2.

TABLE III

C—OH STRETCHING VIBRATIONS OF 1-DECALOLS

1-Decalol	ν - C—(OH) (cm^{-1})	Position of the OH group
cis-cis	1027	e
cis-trans	1041	e
trans-cis	942	a
trans-trans	1020	e

According to Table III, *cis-cis* as well as *cis-trans*-1-decalol are in the conformation with an *equatorial* hydroxyl group (XV and XVI). *Trans-cis* and *trans-trans*-1-decalol exist in the only possible conformation according to model studies with an *axial* and an *equatorial* hydroxyl group (XVII) and (XVIII), respectively.

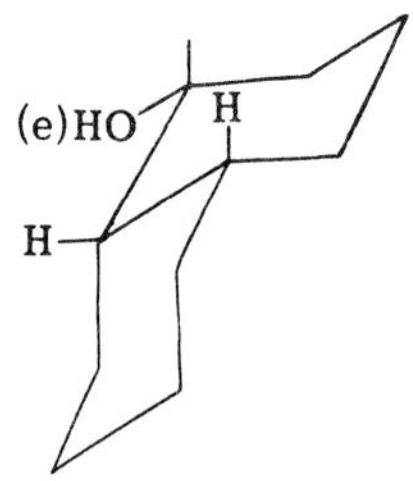

XV

cis-cis-1-decalol

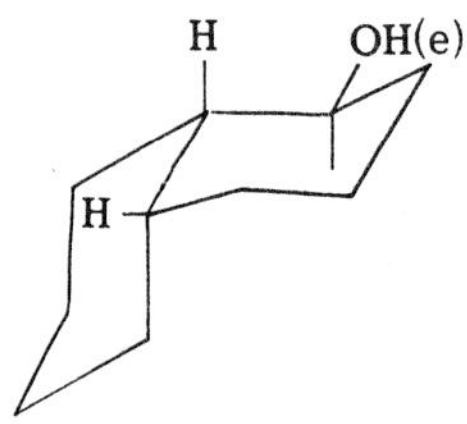

XVI

cis-trans-1-decalol

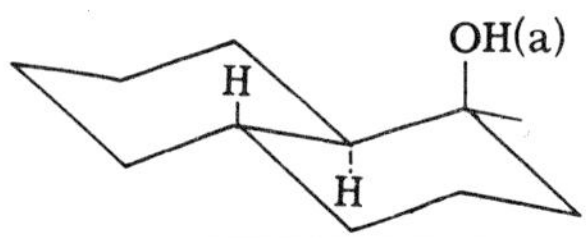

XVII

trans-cis-1-decalol

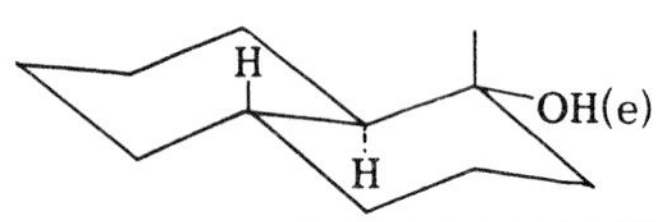

XVIII

trans-trans-1-decalol

(ii) 2-Decalols

The 2-decalols exist in four stereoisomeric racemic forms, as do the 1-decalols. As a result of three asymmetric carbon atoms, there are eight optically active forms possible.

The configurational assignment followed, as for the 1-decalols, from their mode of formation, their chemical properties (such as thermal decomposition of the methyl xanthates, relationships between the *cis*-2-

decahydronaphthoic acids and the 2-decalols on the basis of stereospecific reactions) and their physical properties (Auwers-Skita rule).[41]

Although there no longer is any doubt concerning the configuration of the 2-decalols, their assignment was much more difficult to arrive at than had been the case for the 1-decalols.[28g] In the 1-decalol series, which corresponds to the 1,2-substituted cyclohexanes, the classical rules of configuration[42] (Auwers-Skita rule, conditions of formation) are very reliable. This was demonstrated later for the 1,2-disubstituted cyclohexane compounds in greatly expanded data. 2-Substituted decalins can be compared with the 1,3- (or 1,4-) substituted cyclohexane compounds. The classical rules for configurational determinations can be applied to these only with reservations.

For this reason Hückel[43] dispensed, at first, with a configurational assignment after the discovery of the four isomers of 2-decalols, and 2-decalyl amines. These were decided on later, after more extensive chemical observations were available (see the accompanying formulas).

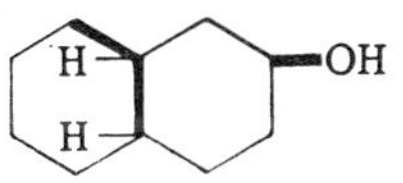

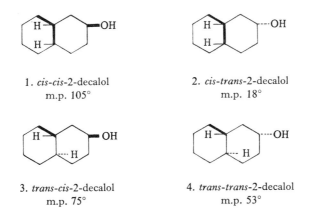

1. *cis-cis*-2-decalol
m.p. 105°

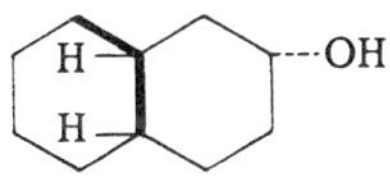

2. *cis-trans*-2-decalol
m.p. 18°

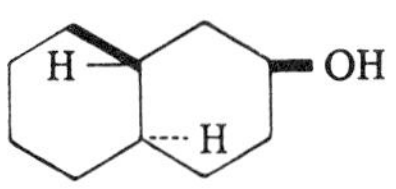

3. *trans-cis*-2-decalol
m.p. 75°

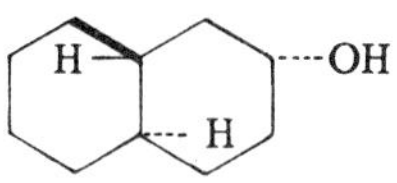

4. *trans-trans*-2-decalol
m.p. 53°

The conformational analysis is again mainly based on kinetic data of the carboxylate saponifications (Table IV), and the solvolysis of the tosylates (Tables V and VI), as well as the evaluation of the ν-C—(OH) vibrations in the infrared spectra.

The conformations of the two-*trans*-fused 2-decalols can be derived from a study of models because of the rigidity of the ring system. The hydroxyl group can only assume the *equatorial* position in *trans-cis*-2-

[41] References 23a, 23b, in Chap. 1; 27e, 27g, 27m, 27n; W. G. Dauben, and E. Hoerger, *J. Am. Chem. Soc.* **73**, 1504 (1951); cf. also ref. 30.

[42] A. Skita, Ref. 59b in Chap. 3, *Ann. Chem.* **427**, 255 (1922).

[43] W. Hückel, *Ann. Chem.* **451**, 109 (1927).

TABLE IV

SAPONIFICATION OF CARBOXYLIC ESTERS[a]

2-Decalol	Rate constant k in aqueous sodium hydroxide (liters $mole^{-1}$ min^{-1})	
	Acid succinate 40°	Acid phthalate 60°
cis-cis	1.16	Not measured
cis-trans	0.70	0.078
trans-cis[b]	1.22	0.134
trans-trans[b]	0.15	0.022

[a] See ref. 27h, 27l. [b] See also ref. 27n.

TABLE V

SOLVOLYSIS OF TOLUENESULFONATES

2-Decalol	Rate constant k (sec^{-1}) at		
	40°	50°	60°
	Ethanolysis ($k \times 10^6$)		
cis-cis[a,b,c]	—	2.9	10.4
cis-trans[b,c]	—	2.1	7.1
trans-cis[c]	—	0.55	2.0
trans-trans[c]	6.8	2.6	8.4
	Acetolysis[d] ($k \times 10^5$)		
		75.5°	
trans-cis		2.14	
trans-trans		6.61	

[a] See ref. 27e. [b] See ref. 27g. [c] See ref. 27n. [d] See refs. 31a and b.

decalol and the *axial* position in *trans-trans*-2-decalol[44] (XXV and XXVI). The great differences in the saponification rates of the carboxylic esters are in accord with these findings: The acid succinate of *trans-cis*-2-decalol is saponified about nine times more rapidly than the correspond-

[44] Cf. ref. 38. Winstein even employed *trans-trans*-2-decalol as a standard example of a conformationally homogeneous alcohol in addition to *trans*-3-and-*trans*-4-*tert*-butylcyclohexanol. Cf. also W. Hückel and M. Hanack, ref. 39.

ing ester of *trans-trans*-2-decalol. The acid phthalate is saponified about six times faster (Table IV).

The same applies to the ethanolysis and acetolysis of the tosylates (Table V). The toluenesulfonate of *trans-cis*-2-decalol with an *equatorial* hydroxyl group reacts much more slowly (it is one of the slowest reacting toluenesulfonates) than that of *trans-trans*-2-decalol with an *axial* hydroxyl group.

Briefly, we shall touch upon the differences in reaction rates of the *trans*-fused *axial* and *equatorial* 1- and 2-decalyl toluenesulfonates.[31a,b] The relative rates decrease in the following order: 1-*axial* $\gg$ 2-*axial* $>$ 2-*equatorial* $>$ 1-*equatorial* (Tables II and V). The reasons for the high reaction rate of the *trans-cis*-1-decalyl toluenesulfonate (1-*axial*) in comparison to the *trans-trans*-2 isomer (2-*axial*) are related to the difference in magnitude of the 1 3-*diaxial* hydrogen interaction. In the *trans-trans*-2 compound, the tosylate group is exposed to two *axial* hydrogen interactions; in the *trans-cis*-1 compound to three. This can be seen on a model (XXVI and XVII). The ionization is thus facilitated in the latter case.

In the series of the *cis*-fused 2-decalols, a kinetic conformational analysis leads to contradictory results (as seen in Tables IV and V) as it did for the corresponding 1-decalols. *Cis-cis*-2-decalol was assigned a conformation with a predominantly *equatorial* hydroxyl.[45] This conformation is in agreement with the saponification rate of the acid succinate. It is saponified as rapidly as the acid succinate of *trans-cis*-2-decalol (*equatorial* hydroxyl) and 2/3 as rapidly as the acid succinate of *cis-trans*-2-decalol. The latter was therefore assigned a conformation with an axial hydroxyl group,[45] or assumes an equilibrium between the conformation with *axial* and *equatorial* hydroxyl groups.[46] However, if we now consider the reaction rates of the toluenesulfonates (Table V), we must assign *cis-cis*-2-decalol an *axial* hydroxyl.

This example again demonstrates that a kinetic conformational analysis need not even qualitatively lead to the same results from studies of various reactions. As a result of the flexibility of the *cis*-decalin structure, a different e:a relation of the individual derivatives need not be solely responsible for this. As a possible explanation, a change in conformation of the carbon skeleton as a result of various effects of distant atoms should be considered. The reaction rate could be influenced in this way.

A study of the model by Dauben and Pitzer,[3] independent of the kinetic conformational analysis, indicated an *equatorial* position of the

[45] H. D. Orloff, *Chem. Rev.* **54**, 419, 423 (1954).

[46] Cf. Chap. 2, ref. 3, p. 28, and refs. 15 and 37b.

hydroxyl group in *cis-cis*-2-decalol. If we look at the two interchangeable conformations of *cis-cis*-2-decalol (XIX and XX) and of *cis-trans*-2-decalol (XXI and XXII), then there exists a characteristic difference between the isomers, in regard to a conformational equilibrium.

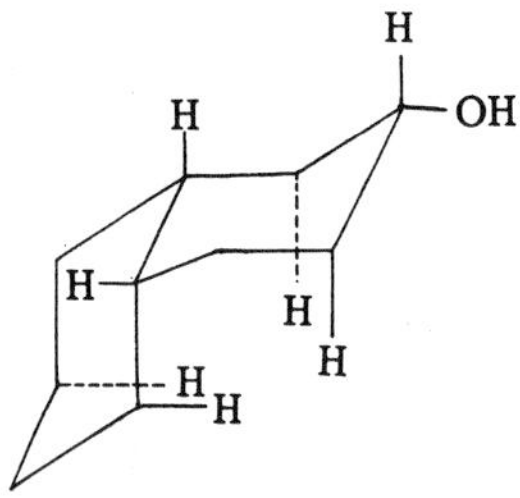

XIX

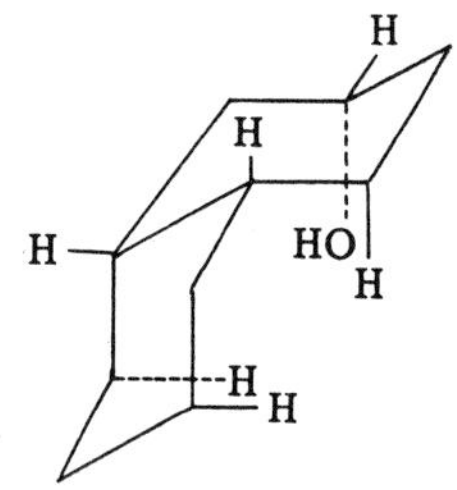

XX

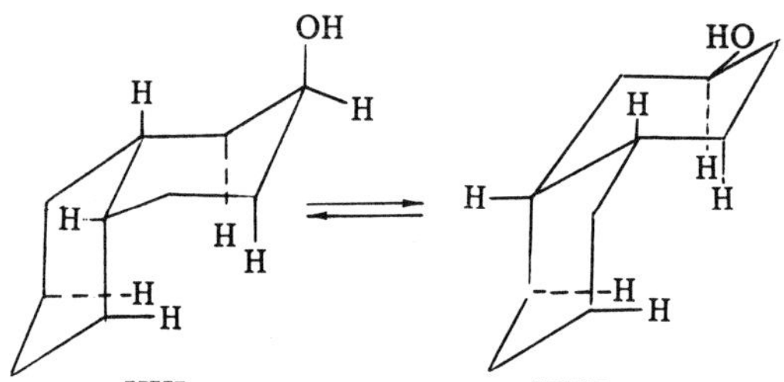

XXI XXII

In the conformation of the *cis-cis*-2-decalol, with an *axial* hydroxyl group (XX), the OH group assumes a position in which it is not only exposed to the normal 1,3-hydrogen interaction, but also to an additional *gauche* butane interaction (C-8–C-9) which is much larger. Only the normal hydrogen interactions apply for *cis-trans*-2-decalol with an *axial* hydroxyl (XXI). The larger *equatorial* character of *cis-cis*-2-decalol can, therefore, be explained in this way as well. There should then be more of the *axial* hydroxyl conformation in *cis-trans*-2-decalol.

However, such use of a model to observe the long range effect of atoms should be viewed with care, as has been noted frequently. The following experimental result is presented as an illustration: According to the considerations by Dauben and Pitzer[3] *cis-trans*-2-decalol must predominate at equilibrium between *cis-cis*- and *cis-trans*-2-decalol. This was based on the premise that conformations (XIX) and (XXII) contain the same energy. [On the basis of these assumptions, conformation (XXI) is present in larger concentration with a lower energy than (XX). The sum of the concentrations of (XX) and (XXII), therefore, should be greater than that of (XIX) and (XX).]

Past experimental results seemed to have contradicted these views entirely. The establishment of an equilibrium between the two isomeric *cis*-2-decalols in the presence of an excess of sodium in refluxing decalin provided 80% *cis-cis*-2-decalol.[47] However, this value is only a rough approximation. A correction of this value was presented by Hückel and Stelzer[27o] in the reduction of *cis*-2-decalone with sodium in alcohol. This resulted in a mixture of alcohols, as expected, approaching the alcoholate equilibrium (see page 98). In this way 60% *cis-cis*-2-decalol was obtained. Recently the equilibrium between the two *cis*-2-decalols has been reinvestigated, using both an excess and less than the required amount of sodium to establish the equilibrium.[48] With an excess of sodium, 66% *cis-cis*-2-decalol was obtained; with less than the required amount of sodium, 52% *cis-cis*-2-decalol was obtained. The direction of the equilibrium of this case, as also found in others,[48a] depends on the amount of sodium present.

These latter results agree somewhat better with the above theoretical considerations. With less than the required amount of sodium, the alcohols exist mainly in the free form, i.e., the isomeric alcohols possess about the same amount of free energy under these conditions, a fact which cannot be quite understood according to the above statements. Possibly this may indicate[48] that the conformations (XX) and (XXI) are present in such small amounts that they have no influence on the equilibrium, or that their free energies are comparable.

Barton's rule[49] (cf. p. 269) stands in contradiction to the *equatorial* conformation of the hydroxyl group in *cis-cis*-2-decalol. It was set up for the cyclohexane ring and is usually valid. According to the rule the alcohol with an *axial* hydroxyl is formed predominantly during catalytic hydrogenation of the ketone in acidic solution. *Cis-cis*-2-decalol is obtained from the ketone under these conditions, but in ~75% yield.[27c,o] The reduction of the ketone with aluminum isopropoxide, which also should produce the *axial* alcohol, yields *cis-cis*-2-decalol in 75% yield as well.[28g]

An investigation of the infrared spectra of the isomeric 2-decalols had already been carried out by Dauben[50] and co-workers in 1952. However, no relation to the conformation can be obtained from the published spectra.

A repeated measurement of the spectra under the conditions described

---

[47] W. Hückel and H. Naab, *Ber.* **64**, 2137 (1931).

[48] W. Masschelein, paper presented at the *International Symposium of Organic Chemistry, Brussels*, 1962.

[48a] O. R. Rodig and L. C. Ellis, *J. Org. Chem.* **26**, 2197 (1960).

[49] D. H. R. Barton, *J. Chem. Soc.* p. 1027 (1953).

[50] W. G. Dauben, E. Hoerger, and N. K. Freeman, *J. Am. Chem. Soc.* **74**, 5206 (1952).

for the 1-decalols allowed a definite assignment of the *equatorial* and *axial* C—(OH) stretching vibrations[28i,32,40] and led to the following bands (Table VI):

TABLE VI

C—OH STRETCHING VIBRATIONS OF 2-DECALOLS

2-Decalols	ν - C—(OH) (cm⁻¹)	Position of the OH group
cis-cis	1028	e
cis-trans	1050, 1008 (946)	Mainly e
trans-cis	1028	e
trans-trans	947	a

According to this, *cis-cis*-2-decalol has a conformation with an *equatorial* hydroxyl group (XXIII). There is no sign of a conformational equilibrium. The spectrum of *cis-trans*-2-decalol is not unequivocal. Although the two bands at 1050 and 1008 cm.$^{-1}$ indicate a preponderance of *equatorial* hydroxyl, a weak band at 946 cm.$^{-1}$ indicates the presence of the conformer with *axial* hydroxyl group. The equilibrium is shifted, however, in favor of the *equatorial* conformer (XXIV). *Trans-cis* and

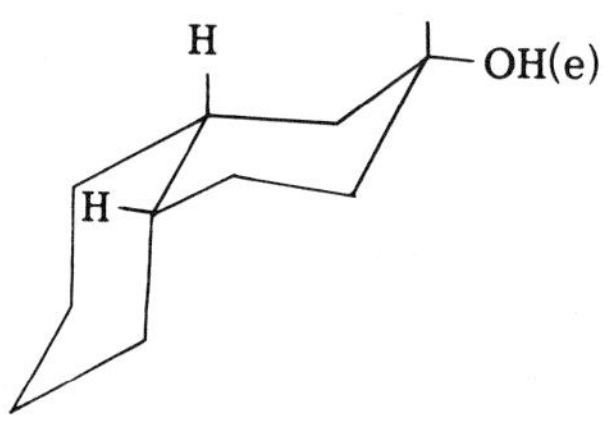

XXIII

cis-cis-2-decalol

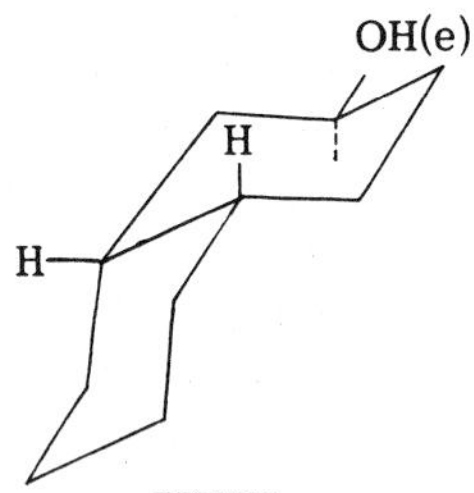

XXIV

cis-trans-2-decalol

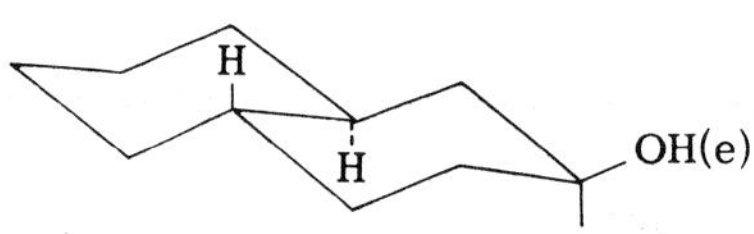

XXV

trans-cis-2-decalol

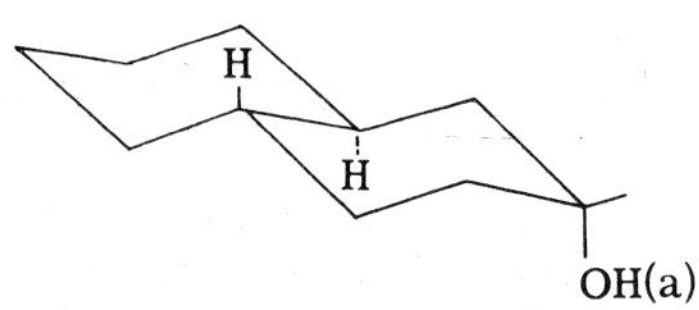

XXVI

trans-trans-2-decalol

trans-trans-2-decalol have hydroxyl groups that uniformly occupy the *equatorial* and *axial* position (XXV and XXVI) as required from a study of the models.[51]

b. Decalylamines

We shall deal only briefly with the conformation of the isomeric 1- and 2-decalylamines. In comparison to the numerous physical and chemical methods available to clarify the conformations of alicyclic alcohols, experimental results concerning the corresponding amines are but few. Methods such as infrared spectroscopy, which gave an insight into the conformation of the alcohols, have not yet been applied successfully to the decalyl amines. Conformational assignments up to now rest exclusively on chemical reactions.[46] As has been reiterated on numerous occasions, a conformational analysis based solely on chemical methods is fraught with uncertainties.

The four stereoisomeric forms of 1-decalylamine and those of 2-decalylamine have been prepared by Hückel and co-workers.[27f,43,52] The configurations, assigned on the basis of their preparation (catalytic hydrogenation of the corresponding *cis*- and *trans*-decalone oximes to give an amino group in a *cis* position, alkaline reduction = amino group in a *trans* position) are defined for the isomeric 1-decalylamines as well as the *trans*-2-decalylamines. Doubt has been expressed[53] concerning the configurational assignments (see the tabulation) originally put forth by Hückel for the *cis*-2-decalylamines.

The conformations can again be obtained from the study of a model of the *trans* amines, the rigidity of the system being due to the *trans* juncture of the rings. We obtain the conformations corresponding to the decalols [cf. (XVII) and (XVIII); (XXV) and (XXVI)]: *trans-cis*-1- and *trans-trans*-2-decalylamine carry the amino group in an *axial* position; the amino group of *trans-trans*-1- and *trans-cis*-2-decalylamine lies in an *equatorial* position.

Because of the possibility of ring conversion, the same limitations for the corresponding decalols are valid for the isomeric *cis*-1- and 2-decalylamines.

Only one reaction, the deamination of the decalylamines with nitrous acid, studied by Hückel and co-workers,[27i] has been used[46] for a conformational analysis. Conformational analyses carried out later were based

---

[51] Mass spectrometry of isomeric 2-decalols: K. Biemann and J. Seibl, *J. Am. Chem. Soc.* **81**, 3149 (1959).

[52] See also ref. 23a, 23b in Chap. 1.

[53] W. G. Dauben and E. Hoerger, *J. Am. Chem. Soc.* **73**, 1504 (1951); ref. 30; cf. also ref. 27o.

1-Decalylamines	Configuration	M.p. of the corresponding decalols
Cis-cis-1-decalylamine m.p. 8°	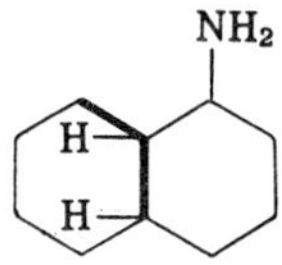	93°
Cis-trans-1-decalylamine m.p. −2°	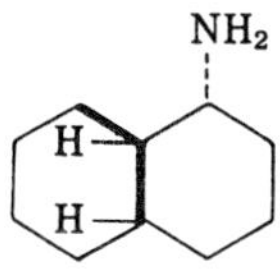	55°
Trans-cis-1-decalylamine m.p. −18°	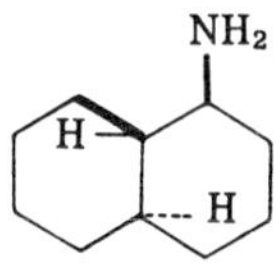	49°
Trans-trans-2-decalylamine m.p. −1°	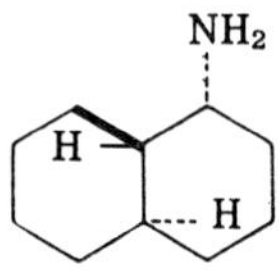	63°

2-Decalylamines	Configuration	M.p. of the corresponding decalols
Cis-cis-2-decalylamine m.p. 20°	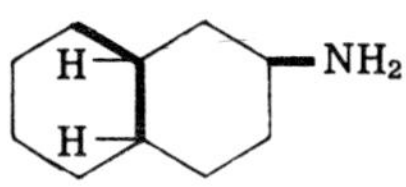	105°
Cis-trans-2-decalylamine m.p. 14°	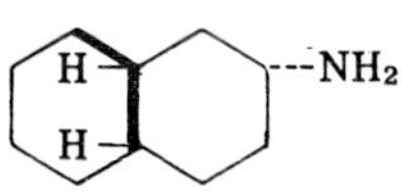	18°
Trans-cis-2-decalylamine m.p. 15°	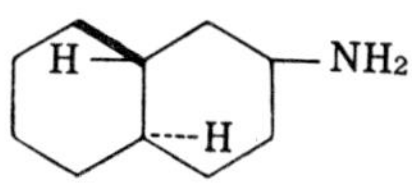	75°
Trans-trans-2-decalylamine m.p. −47°	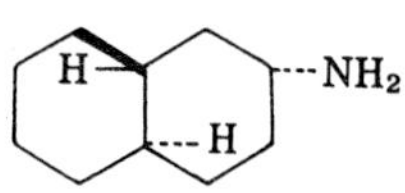	53°

on these results (Table VII), and the following rule, set up by Bose and Mills,[54] was used for the reaction of alicyclic amines with nitrous acid: If the amino group is in the stable *equatorial* position, then the nitrous acid deamination proceeds with retention of configuration, i.e. the *equatorial* alcohol is formed. If the amino group is in the *axial* position, then the reaction proceeds by way of a Walden inversion and in addition, an unsaturated hydrocarbon is formed. Bose and Mills[54] based their hypothesis mainly on investigations of the menthane, decalin, and hydrindane systems.[55]

More recent investigations of alkylated cyclohexylamines[56,57] cast some doubt on the Bose-Mills rule (cf. p. 265). It can be shown,[57] that the rule was not valid for some amines with an *axial* amino group. However, the rule checked relatively well for *equatorial* amines. As an example, the epimeric 2- and 3-methylcyclohexylamines were studied[57,58]: In *cis*-2-methylcyclohexylamine, in which the amino group should be located in the *axial* position, corresponding to its configurationally analogous alcohol, up to 90% retention is observed; in *trans*-3-methylcyclohexylamine there is up to 70% retention. At the same time about 50% elimination takes place. The *equatorial* amines, *trans*-2- and *cis*-3- again follow the rule. For both, the configuration is largely retained, the hydrocarbon formation being very small.[58] However, it is not out of the question that the investigated methyl cyclohexylamines are not conformationally uniform. The deamination of the conformationally pure *cis*-4-*tert*-butylcyclohexylamine, which possesses an *axial* amino group, proceeds preponderantly with the formation of Δ^3-*tert*-butylcyclohexene (77%). In addition, 10% alcohol of the same configuration (*cis*) and 13% alcohol with the opposite configuration (*trans*) are formed.[59]

The one rule not subject to error, on the other hand, has been the one concerning hydrocarbon formation during the reaction of nitrous acid with amines: Amines with *equatorial* NH_2 groups form practically none. Amines with *axial* NH_2 groups usually form much unsaturated hydrocarbon during deamination.

---

[54] Ref. 37b; A. K. Bose, *Experientia* **9**, 256 (1953); ref. 53.

[55] J. Read, A. M. R. Cook, and M. I. Shannon, *J. Chem. Soc.* p. 2223 (1926); (b) J. Read and G. J. Robertson, p. 2168 (1927); (c) J. Read and G. Walker, p. 308 (1934); R. G. Johnston and J. Read, p. 1138 (1935); ref. 27i.

[56] Cf. W. Hückel and G. Ude, *Chem. Ber.* **94**, 1026 (1961).

[57] W. Hückel and K. D. Thomas, *Ann. Chem.* **645**, 177 (1961).

[58] For further examples among the steroids see (a) C. W. Shopee, D. E. Evans, and G. H. R. Summers, *J. Chem. Soc.* p. 97 (1957); (b) C. W. Shoppee, R. J. W. Cremlyn, D. E. Evans, and G. H. R. Summers, *ibid*, p. 4364 (1957); (c) also M. M. Claudon, *Bull. Soc. Chim. France* p. 627 (1950).

[59] W. Hückel and K. Heyder, *Chem. Ber.* **96**, 220 (1963).

Table VII surveys the reaction of the decalylamines with nitrous acid.

TABLE VII

DEAMINATION OF DECALYLAMINES[a]

Decalylamine	Reaction products from deamination: Decalol: Total (%)	Decalol: Inversion (%)	Decalol: Rentention (%)	Octalines (%) ($\Delta^{1,2}$, $\Delta^{1,9}$, and $\Delta^{1,10}$)	Con-formations
cis-cis-1	100	0	100	0	*e*
cis-trans-1	75	10	90	25	e/a
trans-cis-1	30	90	10	70	a
trans-trans-1	100	0	100	0	e
cis-cis-2	100	0	100	0	e
cis-trans-2	70	10	90	30	e/a
trans-cis-2	100	0	100	0	e
trans-trans-2	30	90	10	70	a

[a] See ref. 27i.

The following conclusions[60] can be drawn from the results listed in Table VII. Let us disregard the Bose-Mill's rule, because of previously mentioned reasons, and only look at the formation of the hydrocarbons. If the amino group is in the *axial* position, then the elimination reaction of the *axial* hydrogen *trans* to the NH_2 group is strongly favored. The amount of hydrocarbon, therefore, is large during the reaction with nitrous acid. If the amino group is *equatorial*, then there is no opportunity for *trans* coplanar elimination. The amount of hydrocarbon formed is quite small.

The positions assigned to the NH_2 group in the *trans*-fused decalylamines, on the basis of models, agree with these facts. *Trans-cis*-1- and *trans-trans*-2-decalylamine give a high yield of hydrocarbon (Table VII), i.e. the NH_2 group is *axial*. *Trans-trans*-1- and *trans-cis*-2-decalylamine form practically no hydrocarbon. Therefore, they possess an *equatorial* NH_2 group.

Further, we assign the conformation with an *equatorial* amino group to

[60] Attention is drawn to the fact that the values listed in Table VII were based on older investigations by Hückel and co-workers (ref. 27i). Since the methods of analysis of stereoisomeric compounds have made notable advances through the introduction of gas chromatography and infrared spectroscopy, it would be desirable to check one or more of the results in order to obtain still more exact values.

cis-cis-1- and *cis-cis*-2-decalylamine for the same reasons (Table VII). Conclusions as to conformation are more difficult for *cis-trans*-1- and *cis-trans*-2-decalylamine because of the amount of hydrocarbon formed. Both form decidedly less than the *axial trans*-amines (Table VII), so that we are inclined to assume a mixture of conformers with *axial* and *equatorial* amino groups, with the *equatorial* conformer predominating. The fact that 90% decalol of the same configuration was found in both cases (Table VII) does not contradict this result. We have seen previously, that, contrary to Bose-Mills rule, it is also possible to react *axial* amines with nitrous acid and observe retention of configuration. Further, we can count on the possibility of tertiary alcohol formation[56,57] during the nitrous acid reaction of amines having neighboring substituents. The analytical results must undergo a reappraisal with respect to the alcohols formed.

If we compare the results of this conformational analysis with the results we have obtained for the decalols (see pages 188 and 194), we can say that the conformations of the corresponding alcohols and amines are generally about the same. We would not be wrong if we assume that this is also valid for other amines not yet investigated, i.e. that the hydroxyl and amino groups behave alike in respect to the assumption of a certain conformation.[60a]

In conclusion let us return once more to the results shown in Table VII for *trans-cis*-1- and *trans-trans*-2-decalylamine. These results do not agree with those for various other amines in respect to retention of configuration of the *axial* amino group as established from the configuration of the decalols formed. In both cases, a conformation with an axial amino group was derived from studies of models and from hydrocarbon formation in the deamination reaction. The Bose-Mills rule for an *axial* amino group was, therefore, met in this case. An experimental re-examination of these results would be desirable.[60]

c. 9-Substituted Decalins

The replacement of a bridgehead hydrogen in decalin by another substituent leads to a change in the energy difference between the *cis*- and *trans*-decalins, relative to the nonsubstituted isomeric pair. This has also been shown for hydrindane.

Let us consider 9-methyldecalin. As Turner[61] was able to demonstrate, the following is obtained by introducing a methyl group in position 9 of *trans*- and *cis*-decalin: Substitution of a methyl group in position 9 of

---

[60a] Compare: H. Feltkamp and N. C. Franklin, *Tetrahedron* (in press).

[61] R. B. Turner, *J. Am. Chem. Soc.* **74**, 2118 (1952).

trans-decalin leads to the appearance of four additional *gauche* interactions (XXVII). However, substitution of a methyl group in position 9 of *cis*-decalin leads to only two additional *gauche* interactions (XXVIII).

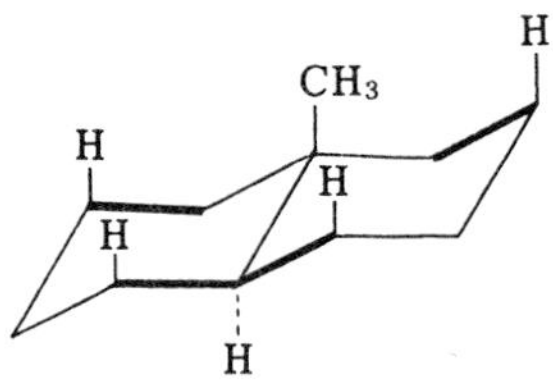

XXVII

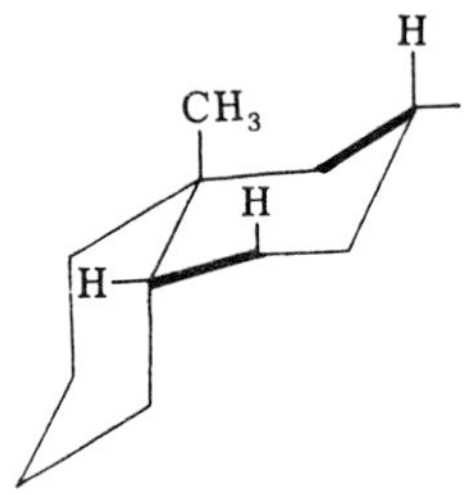

XXVIII

As has been shown (page 59), nonsubstituted *cis*-decalin differs from the *trans*- isomer by three *gauche* interactions. With these considerations, we now have five *gauche* interactions in *cis*-decalin and four in *trans*-decalin. This indicates that the 9-methyldecalins differ from each other by only one *gauche* interaction and that *trans*-9-methyl decalin is only lower in energy by 0.9 kcal./mole. The experimental data available to date agree with this conclusion.[62] The heats of combustion of the two isomeric 9-methyl decalins have been measured[63] and the enthalpy of the *trans* isomer was found to be about 1.39 ± 0.64 kcal./mole (liquid, 298° K) lower than that of the *cis*-isomer. The establishment of the equilibrium over a palladium-charcoal catalyst at various temperatures, led to the determination of $\Delta H = -0.55 \pm 0.28$ kcal./mole, in good agreement with the value obtained from the heat of combustion.[64] The composition of the equilibrium mixture of the isomeric 9-methyl decalins lies at 55% of the *trans*-isomer and is extensively temperature independent.

d. Decalones

As in the case of the hydrindanones (p. 178f.) the factors determining the stabilities and conformations of the decalones are not very clear.

---

[62] (a) W. G. Dauben, J. B. Rogan, and E. J. Blanz, Jr., *J. Am. Chem. Soc.* **76**, 6384 (1954); (b) W. E. Bachmann, A. Ross, A. S. Dreiding, and P. A. Smith, *J. Org. Chem.* **19**, 235 (1954); (c) A. Ross, P. A. Smith, and A. S. Dreiding, *ibid.* **20**, 905 (1955); (d) F. Sondheimer and D. Rosenthal, *J. Am. Chem. Soc.* **80**, 3995 (1958); cf. also ref. 21. For stereochemistry of 9-hydroxy- and 9-aminodecalin cf. W. G. Dauben, R. C. Tweit, and R. L. MacLean, *ibid.* **77**, 48 (1955).

[63] W. G. Dauben, O. Rohr, A. Labbauf, and F. D. Rossini, *J. Phys. Chem.* **64**, 283 (1960).

[64] N. L. Allinger and J. L. Coke, *J. Org. Chem.* **26**, 2096 (1961).

According to Klyne[21] the alkyl ketone effects (p. 151) play an important part. A few examples will now be given.

We may recall (page 59) that the difference in energy between *cis*- and *trans*-decalin depends essentially on the three *gauche*-butane interactions which are present in *cis*-decalin. In *cis*-2-decalone of the conformation (XXIX, R = H), one of the normal *gauche* interactions is replaced by an interaction of the 3-alkyl ketone type between the methylene group at C-8 and the keto group at C-2.[65]

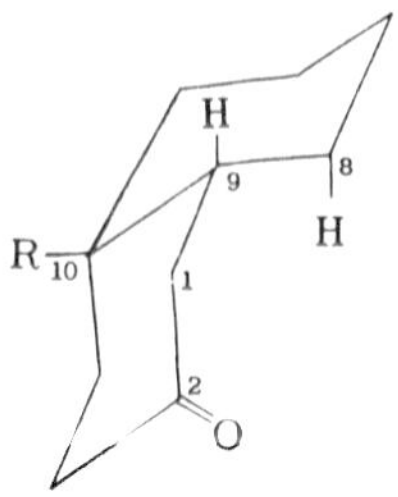

XXIX

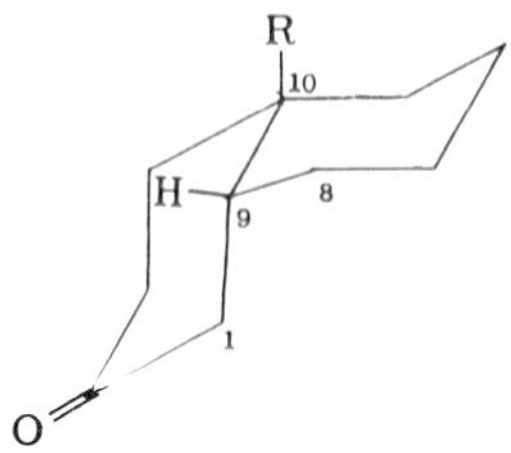

XXX

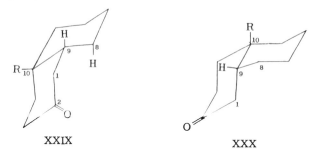

The second possible conformation of *cis*-2-decalone (XXX, R = H) is less favorable. The stabilizing influence of the 3-alkyl ketone effect of conformation (XXIX, R = H) is not present. This conformation has three normal *gauche* interactions.

Trans-2-decalone reveals no *gauche* interactions (see page 59). The energy difference between *cis*- and *trans*-2-decalone must, therefore, be smaller than that between *cis*- and *trans*-decalin (2.7 kcal./mole). The difference in the heat of combustion between *cis*- and *trans*-2-decalone according to earlier measurements amounts to 2.2 kcal./mole.[66]

These strictly theoretical concepts of the conformations of *cis*-2-decalone have no experimental support. In the case of *cis*-10-methyl-2-decalone, the experimental results even contradict the conformation given above as the preferred one. According to the theory, the preferred conformation should be (XXIX, R = CH_3; "nonsteroid" conformation). Djerassi and co-workers[67] compared the rotatory dispersion curve of *cis*-10-methyl-2-decalone with that of a rigid 3-keto-5β-steroid as in (XXX). They came to the conclusion that in *cis*-10-methyl-2-decalone the "steroid" conformation (XXX) predominates.[68]

[65] Cf. also (a) A. S. Dreiding, *Chem. & Ind. (London)* p. 1419 (1954); (b) D. A. H. Taylor, *ibid.* p. 250 (1954).

[66] W. Hückel, *Ann, Chem.* **451**, 109 (1927).

[67] (a) C. Djerassi and D. Marshall, *J. Am. Chem. Soc.* **80**, 3986 (1958); (b) C. Djerassi, "Optical Rotary Dispersion" p. 185 ff. McGraw-Hill, New York, 1960.

[68] (a) W. Klyne, *Colloq. Inter. Chim. Org. Probl. Stereochim., Montpellier*, 1959, in *Bull. Soc. Chim. France*, p. 1396 (1960); (b) W. Moffitt, R. B. Woodword, A. Moscowitz,

On the basis of values obtained by Johnson[69] and Robins and Walker[70] for the 2-alkyl ketone effect (1.0 kcal./mole) and the 3-alkyl ketone effect (0.9 kcal./mole), (cf. p. 151) Klyne[21,71] calculated approximate energy values for *cis*- and *trans*-1-decalone. *Cis*-1-decalone exists in two conformations (XXXI and XXXII, R=H). *Trans*-1-decalone possesses no *gauche* interactions; only the 2-alkyl ketone effect must be considered (XXXIII,

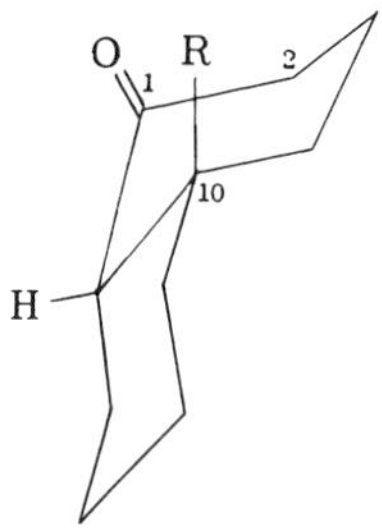

XXXI

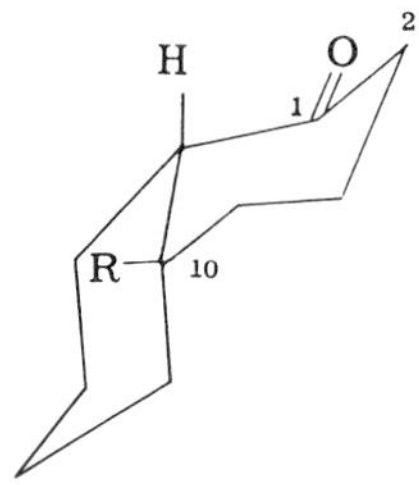

XXXII

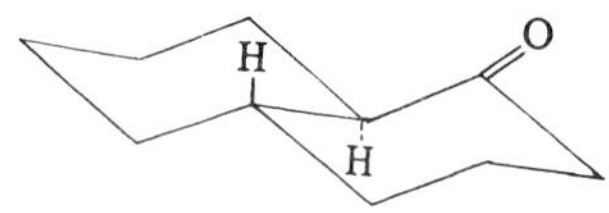

XXXIII

1-C=O/C-8). The estimated increase in energy amounts to ~1 kcal./mole. In *cis*-1-decalone of conformation (XXXI, R = H), in addition to the destabilizing 2-alkyl ketone effect, a 3-alkyl ketone effect (1-C=O/C-7) comes into play. The estimated energy amounts to 2.5 kcal./mole. No alkyl ketone effect appears for conformation (XXXII, R = H). The estimated energy amounts to ~2.4 kcal./mole. Which of these two conformations is preferred by *cis*-1-decalone cannot be determined from the estimated energy values because of the small difference. In addition, the energy values used for the alkyl ketone effects may be viewed with reservations[72] (see page 151). Measurements of the optical rotatory dispersion of optically active *cis*-1-decalone lead to the conclusion that *cis*-1-decalone preponderantly assumes the "steroid" conformation (XXXII), although conformation (XXXI) cannot be excluded.[72a]

It could be determined experimentally that *trans*-1-decalone is more

W. Klyne, and C. Djerassi, *J. Am. Chem. Soc.* **83**, 4013 (1961); (c) C. Djerassi and W. Klyne, *J. Chem. Soc.* p. 4929 (1962); (d) C. Djerassi and W. Klyne, *ibid.* p. 2390 (1963); compare also C. Djerassi, J. Burakevich, J. W. Chamberlin, D. Elad, T. Toda, and G. Stork, *J. Am. Chem. Soc.* **86**, 465 (1964).

[69] W. S. Johnson (a) *Experientia* **7**, 315 (1951); (b) *J. Am. Chem. Soc.* **75**, 1498 (1953).

[70] P. A. Robins and J. Walker, *J. Chem. Soc.* p. 1789 (1955); *Chem. & Ind.* (*London*) p. 722 (1955).

[71] Cf. also H. E. Zimmerman and A. Mais, *J. Am. Chem. Soc.* **81**, 3644 (1959).

[72] Cf. N. L. Allinger and H. M. Blatter, *J. Am. Chem. Soc.* **83**, 994 (1961) and further references in Chap. 3 concerning the alkyl ketone effects.

[72a] C. Djerassi and J. Staunton, *J. Am. Chem. Soc.* **83**, 736 (1961).

stable than *cis*-1-decalone.[73] The equilibrium lies 95% on the side of *trans*-1-decalone at 220°.

Similar considerations were used by Klyne[21] for the 9-methyl-1-decalones. *Cis*-9-methyl-1-decalone, in agreement with experimental results,[74] proved to be more stable.

Cis-10-methyl-1-decalone has been investigated by Djerassi[67] with the help of optical rotatory dispersion. In this case, the corresponding "nonsteroid" type of conformation is preferred (XXXI) in agreement with the predictions.[68a]

2. Disubstituted Decalins

Of the disubstituted decalins, we shall consider only briefly the decalindiols. Studies of conformations have been carried out on various decalindiols. Some of these are decalin-2,3-diol,[75] decalin-2,6-diol,[76] decalin-1,4-diol,[77] and decalin-1,5-diol.[78]

A detailed explanation of the methods used for elucidating the configurations and conformations of the decalindiols mentioned would go beyond the scope of this text. The assignment of the configurations and conformations came about through chemical methods—measurement of the rates of oxidation with lead tetraacetate, sodium periodate, or chromic acid—as well as physical investigations. These latter included evaluation of the infrared spectra and measurement of the optical rotatory dispersion of the corresponding optically active diols. Other properties were also considered successfully for the assignment, such as their behavior on paper chromatography.[76,77]. The more polar compounds with *equatorial* hydroxyl groups move more slowly than the less polar compounds with *axial* hydroxyl groups.[79] The elution rates in column chromatographic separation as well as the retention times in gas chromatographic separations have also been used.[27f,77b]

In order to clarify the circumstances, the frequently studied decalin-1,4-diol is singled out and its configurational and conformational pos-

[73] W. Hückel, *Ann. Chem.* **441**, 1 (1925).

[74] A. Ross, P. A. Smith, and A. S. Dreiding, *J. Org. Chem.* **20**, 905 (1955).

[75] Md. Erfan Ali and L. N. Owen, *J. Chem. Soc.* p. 2119 (1958) with further references; cf. also H. B. Henbest, M. Smith, and A. Thomas, *ibid.* p. 3293 (1958).

[76] R. L. Clarke and C. M. Martini, *J. Am. Chem. Soc.* **81**, 5716 (1959).

[77] (a) P. Baumann and V. Prelog, *Helv. Chim. Acta* **41**, 2362 (1958); ref. 27f; (b) W. Hückel and W. Kraus, *Chem. Ber.* **95**, 233 (1962); (c) H. Feltkamp and W. Kraus, *Ann. Chem.* **651**, 11 (1962).

[78] P. Baumann and V. Prelog, *Helv. Chim. Acta* **41**, 2379 (1958).

[79] Cf. D. H. R. Barton, *J. Chem. Soc.* p. 1027 (1953); K. Savard, *J. Biol. Chem.* **202**, 457 (1953).

sibilities will be given in greater detail. Six diastereoisomeric forms are possible for decalin-1,4-diol (XXXIV), as for all decalindiols mentioned, and they are either *racemic* or *meso* forms. Let us start

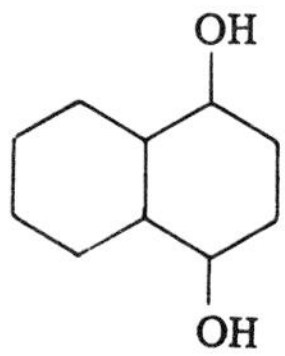

XXXIV

with the *trans*-fused 1,4-decalin diols. The following forms can be constructed with models:

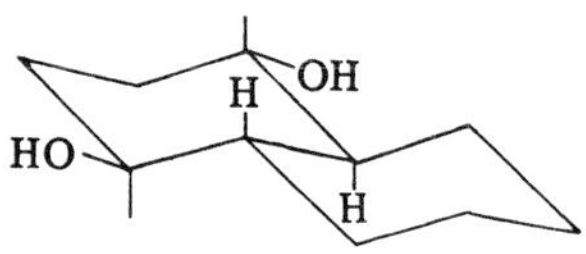

XXXV

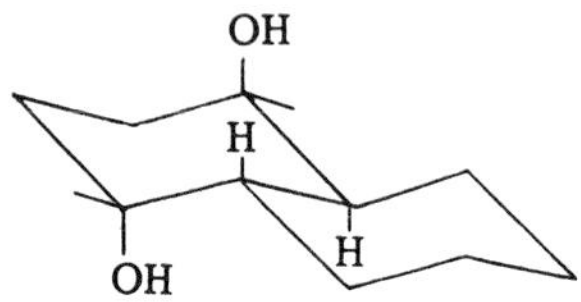

XXXVI

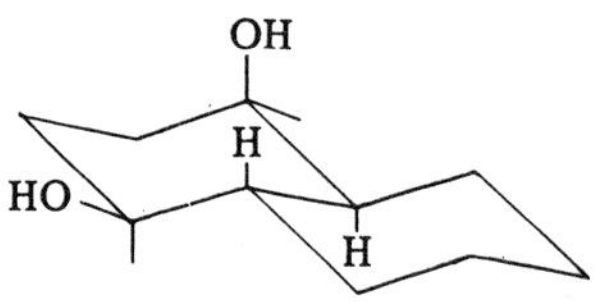

XXXVII

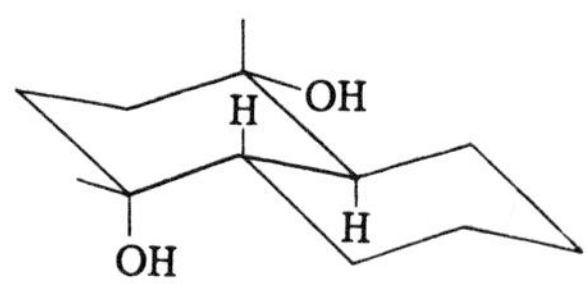

XXXVIII

Forms (XXXV) and (XXXVI) are not interchangeable because of the *trans*-fusion of the rings and the resulting rigidity of the molecules (see page 57). (XXXV) and (XXXVI) are configurational isomers, with *equatorial* OH groups in one case (XXXV) and *axial* OH groups in the other (XXXVI).[80]

The forms (XXXVII) and (XXXVIII) are identical, as can be seen with a model. One substituent is *axial*, the other is *equatorial*. The corresponding mirror images can be constructed for the forms (XXXV),

[80] The situation can be clarified in another way. If we replace the nonsubstituted ring at the juncture with methyl groups, for example, we obtain 2,3-dimethyl cyclohexane 1,4-diol. The compounds corresponding to forms (XXXV) and (XXXVI) are not interchangeable, as can be seen with a model. Only when the two methyl groups are replaced by hydrogen are the resulting 1,4-cyclohexanediols identical.

(XXXVI) and (XXXVII) (= XXXVIII) because of the asymmetric carbon atoms in position 1 and 4, i.e. all three diols form racemates.

The accompanying models can be constructed for the *cis*-fused decalin-1,4-diols (XXXIX to XLIIa).

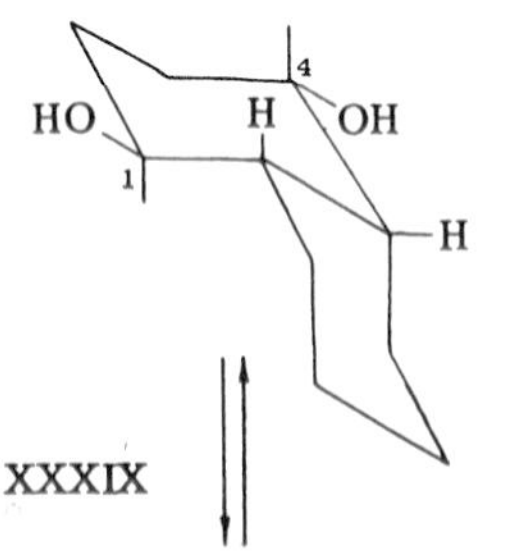

XXXIX

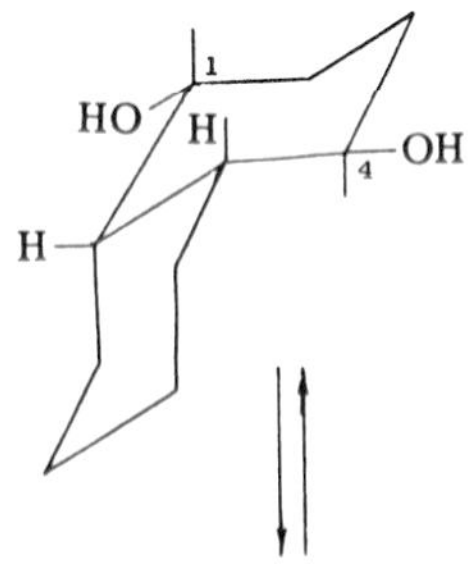

XXXIXa

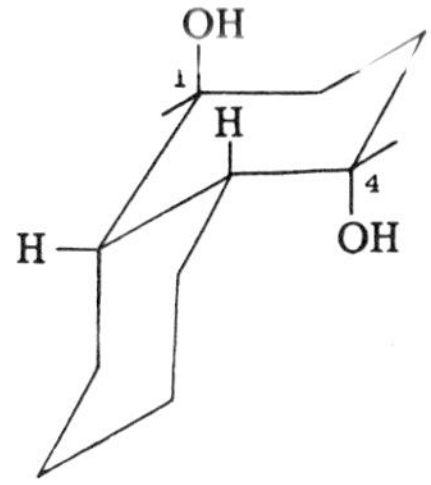

XL

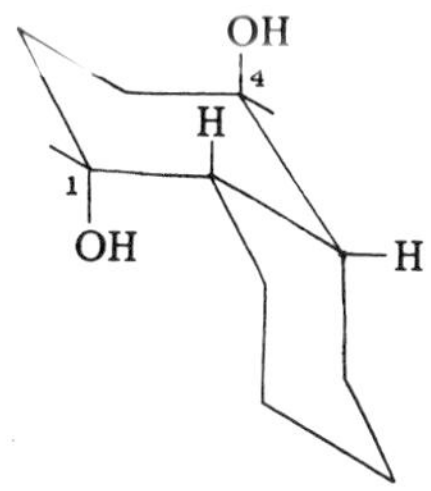

XLa

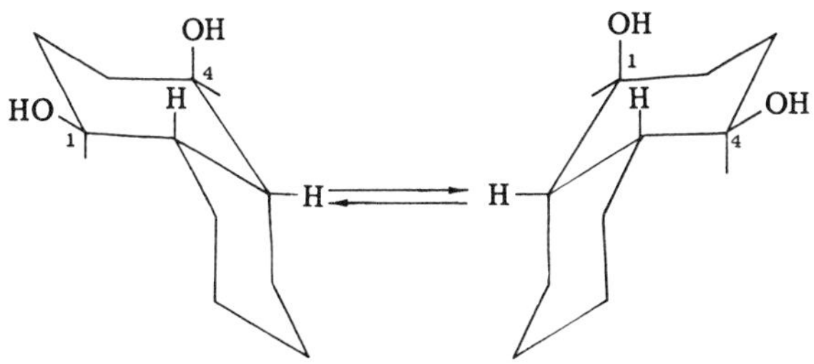

XLI

XLIa

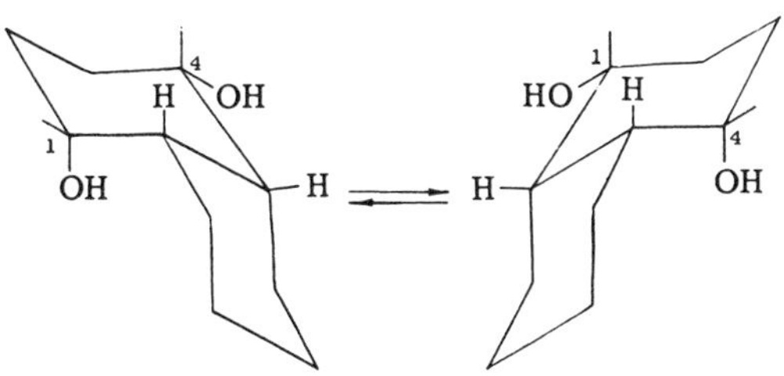

XLII

XLIIa

Form (XXXIX), can be changed into form (XL) by conversion of both rings. (XL) is the mirror image of (XLa). (XLa) can be converted to (XXXIXa), as well, which is the mirror image of (XXXIX). This indicates that (XXXIX) and (XLa) will exist together as a racemate. Form (XLI) can be converted to its mirror image (XLIa) by a conversion of both rings. The same is true of forms (XLII) and (XLIIa). Both compounds, therefore, represent inseparable *meso* forms.

As mentioned above, six separable stereoisomeric forms of decalin-1,4-diol result, four of which are racemates and two *meso* forms. Similar considerations can be applied as well to the other position isomers of the decalindiols.

All six stereoisomers of the decalin-1,4-diols have been prepared in pure form.[77] The conformations of the *trans*-fused compounds, have been determined by the methods mentioned above. The conformational assignment in the *cis*-fused series has not yet been possible with the same certainty for the compounds isolated. The racemic diol (XXXIX) probably exists in the conformation with *equatorial* hydroxyl groups.

Investigations of the conformations of other disubstituted decalins, such as the amines, are few. The separation of the stereoisomeric forms is very difficult. Extensive investigations of the decalin dicarboxylic acids were carried out by Nasarov *et al.*[81] Of the eight stereoisomers theoretically possible, seven decalin-1,2-dicarboxylic acids were prepared and their conformations tentatively assigned.

3. Unsaturated Compounds

Little is known concerning the conformations of unsaturated systems corresponding to the decalins. Except for tetralin, experimental support is lacking, due mainly to the fact that it is so difficult to obtain pure isomers.

The introduction of one double bond in decalin leads to octalin, of which a number of structural isomers are possible ($\Delta^{9,10}$, $\Delta^{1,9}$, $\Delta^{1,2}$, $\Delta^{2,3}$-octalin). From studies of models, the following can be seen: The ring in which a $\Delta^{1,2}$ or $\Delta^{2,3}$ double bond is introduced (ring B), is in each case, forced into the half-chair conformation described under cyclohexene (page 146). This is true of *cis*- and *trans*-$\Delta^{1,2}$- or $\Delta^{2,3}$-octalin. An accurate calculation of the molecular geometry of trans-$\Delta^{1,2}$- and trans-$\Delta^{2,3}$-octalin was carried out by Corey and Sneen[81a] by vector

[81] I. N. Nasarov, V. F. Kucherov, and V. M. Andreev, *Proc. Acad. Sci. USSR* **102**, 1127 (1955); *Bull. Acad. Sci. USSR Div. Chem. Sci.* (*Eng. Trans.*) p. 951 (1956); *ibid.* p. 1091 (1956); *ibid.* p. 471 (1957), with further references.

[81a] E. J. Corey and R. A. Sneen, *J. Am. Chem. Soc.* **77**, 2505 (1955); see also R. Bucourt, *Bull. Soc. Chim. France.* p. 1262 (1963).

analysis. It developed that the introduction of a double bond in ring B deformed not only this ring, but also, to a small extent, ring A. The deformation of the valence angles calculated for ring A is only slightly larger in *trans*-$\Delta^{1,2}$-octalin than in *trans*-$\Delta^{2,3}$-octalin, in which the deformation is in the opposite direction. It was postulated further, that the distance between the *axial* hydrogen or substituent on C-8 and the hydrogen or substituent on C-10 varies in the two structures. From this it was concluded that *trans*-$\Delta^{2,3}$-octalin is more stable than *trans*-$\Delta^{1,2}$-octalin. The same is valid for the corresponding 10-methyl octalins.

Both rings of $\Delta^{9,10}$-octalin (XLIII) in which the double bond is common to both rings, exist in the half-chair conformation.[82] In the case of $\Delta^{1,9}$-octalin, ring B exists in a deformed half-chair form, ring A

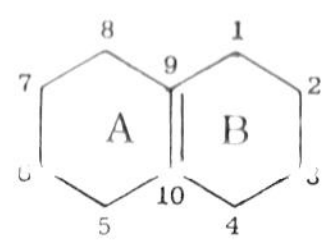

XLIII

in a deformed chair form. The relative stability of the *cis-trans* isomeric $\Delta^{1,2}$- and $\Delta^{2,3}$-octalins has also been predicted by Taylor.[65b] The introduction of the double bond at $\Delta^{1,2}$ in *cis*-decalin decreases the hydrogen interactions existing in *cis*-decalin at C-1–C-5, C-1–C-7, and C-3–C-6 to one, namely the C-3–C-7 interaction. If the double bond is introduced at $\Delta^{2,3}$, then the interactions are decreased by only one (C-3–C-7). From this, it was concluded that *cis*-$\Delta^{1,2}$-octalin is more stable than *cis*-$\Delta^{2,3}$-octalin. From similar considerations, Taylor[65b] came to the same conclusion for the corresponding *trans*-octalins: *Trans*-$\Delta^{2,3}$-octalin is more stable than *trans*-$\Delta^{1,2}$-octalin.

The steric structure of tetralin (XLIV) has been quite thoroughly investigated by Drehfahl *et al.*[83] As a result of the aromatic moiety in tetralin, the carbon atoms 1 and 4 lie in the same plane as the aromatic

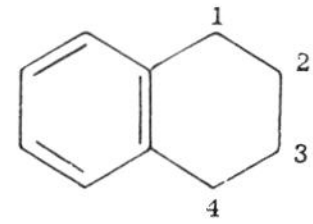

XLIV

[82] M. Mousseron, M. Mousseron, and M. Granier, *Bull. Soc. Chim. France*, p. 1418 (1960).

[83] (a) G. Drehfahl and K. Ponsold, *Chem. Ber.* **91**, 266 (1958); (b) G. Drehfahl and D. Martin, *ibid.* **93**, 2497 (1960).

ring. In this way a half-chair conformation of the partially hydrogenated ring is produced which is only slightly favored energetically over the half-boat form. Drehfahl[83] could show through his investigations of the acyl migration of corresponding derivatives of the isomeric 1-amino-2-hydroxytetralins, that, differing from the 2,3-disubstituted compounds,[84] a deviation in the reaction behavior was detected. From this it was concluded that there are changes in location of the substituents at positions 1 and 4, as compared to the same on cyclohexane (*quasiaxial* and *quasiequatorial* position, see page 148). The substituents at positions 2 and 3 have a location similar to those on cyclohexane.

[84] J. Kiss and J. Fodor, *Acta. Chim. Acad. Sci. Hung.* **5**, 365 (1955).

CHAPTER 5

Polycyclic Compounds

I. PERHYDROPHENANTHRENES AND PERHYDROANTHRACENES

1. Perhydrophenanthrenes

Application of the principles of conformational analysis to the perhydrophenanthrene structural nucleus which occurs in steroids and triterpenes has led to a better understanding of the stabilities of the individual possible isomers.[1]

Perhydrophenanthrene (I) possesses four asymmetric carbon atoms, located at C-11, -12, -13, and -14. There are six stereoisomeric forms, of which four are racemic and two are *meso*.

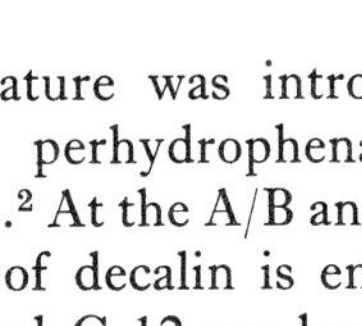

(I)

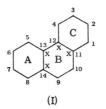

The following nomenclature was introduced for representation of the configurations of the perhydrophenanthrenes. It is general for condensed polyclic systems.[2] At the A/B and B/C ring junctures the same notation used in the case of decalin is employed. If the hydrogens at C-13 and C-14 or C-11 and C-12 are located on the same side of the ring, a *cis* configuration is present otherwise, a *trans* configuration. The configuration at C-12–C-13 is characterized by the expressions "*syn*," when the hydrogens on these carbon atoms are on the same side, or "*anti*" when they are on opposite sides. In the planar representation, the

[1] Cf. references 6 in Chap. 2, Vol. I, p. 50; 30a in Chap. 4; 3 in Chap. 4; R. Riemschneider, *Öster. Chemiker Ztg.* **57**, 38 (1956).

[2] R. P. Linstead, *Chem. & Ind. (London)* **56**, 510 (1937).

trans-anti-trans configuration, for example, would appear as shown in (II) and the *cis-syn-cis* configuration as shown in (III):

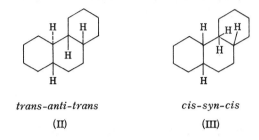

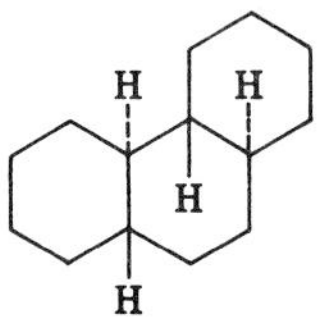

trans-anti-trans

(II)

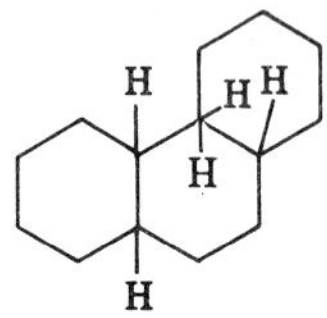

cis-syn-cis

(III)

The hydrogens at the ring fusions can also be indicated in the planar representation by a black dot if they lie above the plane, according to the suggestion of Linstead.[2,3] The structures (II) and (III) would then be shown as follows (IV and V):

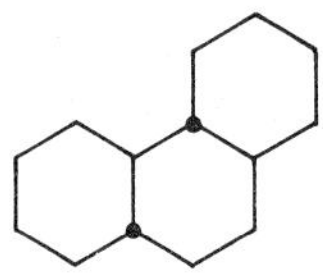

trans-anti-trans

(IV)

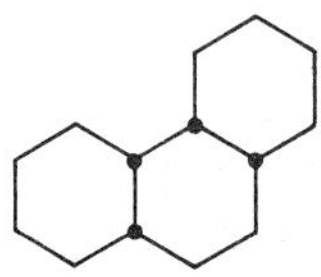

cis-syn-cis

(V)

The configurations of the six stereoisomeric perhydrophenanthrenes [*cis-syn-cis*(*meso*); *cis-syn-trans* (rac.); *trans-syn-trans*(*meso*); *cis-anti-cis* (rac.) *cis-anti-trans* (rac.); and *trans-anti-trans* (rac.)] have been established and defined by Linstead[4] and co-workers on the basis of chemical methods.

Before we discuss the conformational analysis of the various isomeric perhydrophenanthrenes, some chemical observations will be described. These will permit conclusions concerning the relative stabilities of the isomers, as well as the peculiarities inherent to polycyclic systems.

Linstead[4] was able to show that *cis-syn-trans*-9-ketoperhydrophenanthrene (VI)[5] is stable under the conditions of a catalytic hydrogenation. The alcohol formed from this hydrogenation also has the *cis-syn-trans* structure. No rearrangement to the *trans-syn-trans* form (VII) had taken place.

---

[3] (a) R. P. Linstead and A. L. Walpole, *J. Chem. Soc.* p. 842 (1939); (b) R. P. Linstead, R. R. Whetstone, and P. Levine, *J. Am. Chem. Soc.* **64**, 2014 (1942).

[4] R. P. Linstead and R. R. Whetstone, *J. Chem. Soc.* p. 1428 (1950); ref. 3b and earlier investigations.

[5] The first prefix is used for the ring juncture nearest to the functional group.

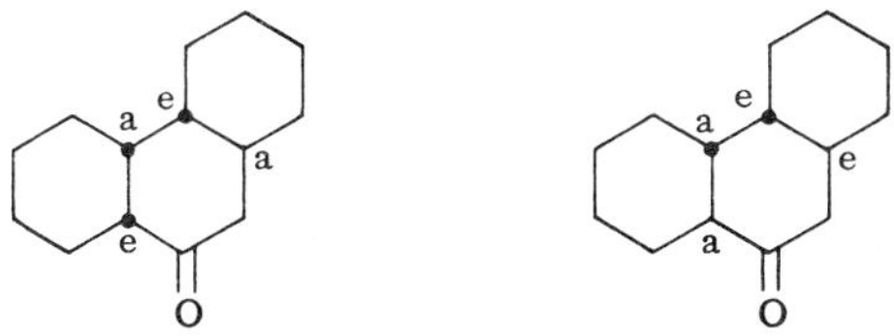

(VI) (VII)

In contrast to this, Linstead[4] showed that *cis-syn-cis*-9-ketoperhydrophenanthrene (VIII) could be isomerized to the *trans-syn-cis* ketone (IX). (IX) is more stable than (VIII). The *cis-anti-trans* ketone (X) is con-

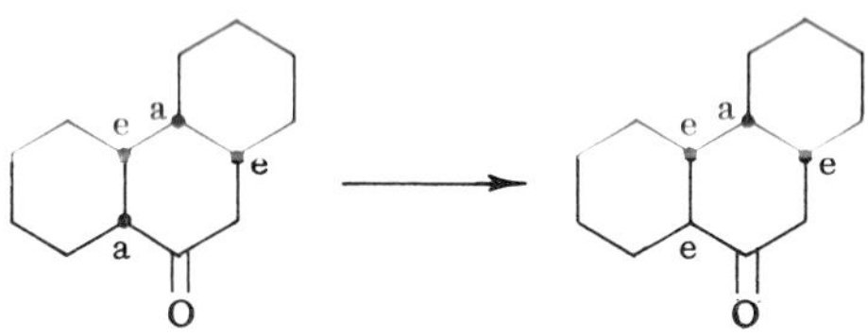

(VIII) (IX)

verted to the *trans-anti-trans* isomer (XI) in the same way.

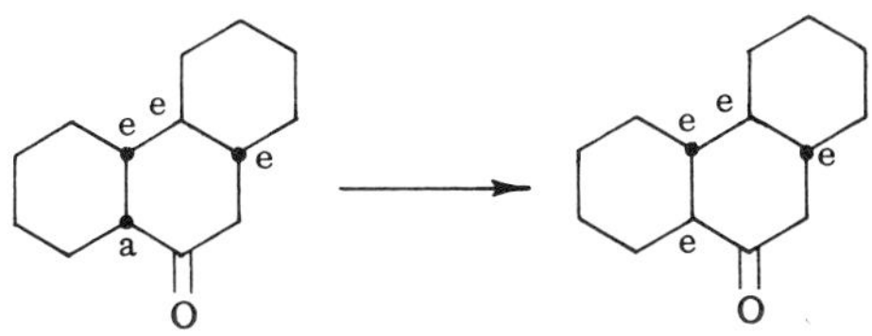

(X) (XI)

The explanation for the difference in behavior of the compounds described can be found in the following: The conversion of (VIII) into (IX) or (X) into (XI) represents the normal behavior in the isomerization of a ketone. The keto group is located adjacent to a *cis*-fused ring. As an example, *cis*-1-decalone is easily converted to *trans*-1-decalone (see page 203).

Ring B (cf. formula I) in conformation (VIII) has two *axial* and two *equatorial* C—C bonds.[6] This follows from the assigned configuration and the rules of conformation derived in Chapter 3. There are one *axial* and three *equatorial* bonds in configuration (IX). As was shown in

[6] W. S. Johnson, *Experientia* **7**, 315 (1951).

Chapter 3, the conformation with the greatest number of *equatorial* substituents is the more stable one, with the exceptions noted. The same is valid for the pair of isomers (X) → (XI). In the *cis-anti-trans* ketone (X), one *axial* and three *equatorial* valences are present, as viewed from ring B. Four *equatorial* valences are necessary in the *trans-anti-trans* isomer (XI). (XI) is, therefore, more stable than (X).

Let us now consider the pair of isomers (VI) and (VII). In the *cis-syn-trans* ketone (VI), rings A and B (cf. formula I) have a *cis* juncture. Viewed from ring B, this involves an *equatorial* and an *axial* bond. However, since rings B and C have a *trans* juncture, involving two equatorial bonds of ring B, the isomeric *trans-syn-trans* ketone (VII) can only be formed by relinquishing the chair conformation of ring B. As there already is an *axial* bond at carbon 13 in the *cis-syn-trans* isomer (VI), the ring juncture in (VII) would have to be formed by two *axial* bonds. This is not possible with the retention of the chair conformation of ring B. A juncture of ring A and B in the *trans* configuration for (VII) is only possible if ring B flips into the boat form (XII). The explanation for the greater stability of (VI) rests in the well-known (cf. Chapter 2, p. 44f.) energetic preference of a chair over a boat.

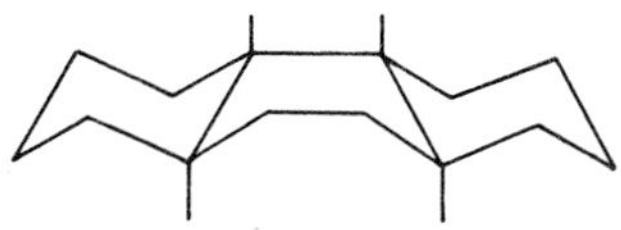

(XII)

As shown on p. 59 the difference in energy between *cis-* and *trans-*decalin had been estimated by Johnson.[6,6a] *Trans*-decalin was assigned the energy value of zero according to his method. The higher energy of *cis*-decalin was estimated on the basis of the three *gauche*-interactions. Johnson[6a] applied the same method to the isomeric perhydrophenanthrenes and thereby attempted to obtain an order of stabilities in a more quantitative way.

The bonds extending from the central ring B in the formulae (XIII) to (XVIII) are designated as e and a, or as be (*boat-equatorial*) and ba (*boat-axial*).[7]

Only *equatorial* bonds from ring B are involved in *trans-anti-trans*-perhydrophenanthrene (XIII), in which all rings are *trans* fused. The *trans-anti-trans* isomer is, therefore, the most stable of the perhydrophenanthrenes. Nevertheless, it possesses a *gauche* interaction (C-4–C-5),[8] which amounts to between 0.8 and 0.9 kcal./mole, as noted previously

[6a] See ref. 126 in Chap. 2.

[7] Compare S. J. Angyal and J. A. Mills, *Rev. Pure Appl. Chem.* **2**, 185 (1952).

[8] The *gauche* interactions are indicated by arrows in formulae (XIII) to (XVIII).

(see page 31). The value of 0.8 kcal./mole is assigned to *trans-anti-trans*-perhydrophenanthrene as the isomer of lowest energy.

The *trans-anti-cis* (XIV) and *cis-syn-trans* (XV) isomers differ from the *trans-anti-trans* isomer by the same amount. One ring of each of these isomers is fused in *cis* position to the central ring B; thus three

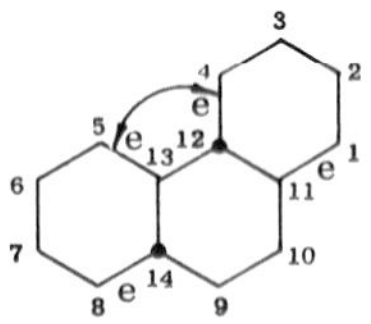

~0.8 kcal./mole
trans-anti-trans
(XIII)

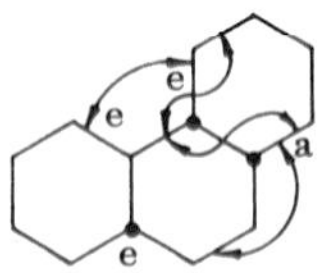

~3.2 kcal./mole
trans-anti-cis
(XIV)

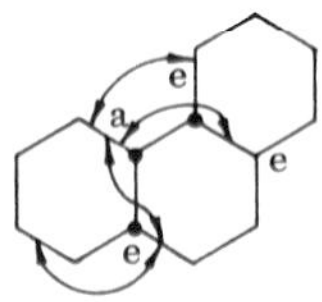

~3.2 kcal./mole
cis-syn-trans
(XV)

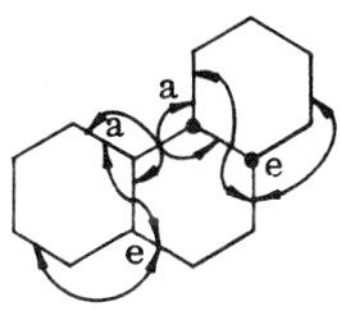

~4.8 kcal./mole
(XVIa)

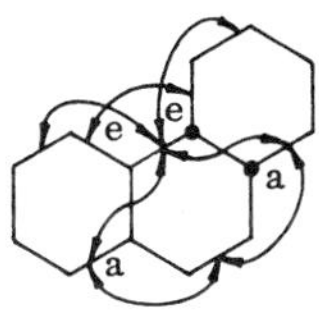

~5.6 kcal./mole
cis-anti-cis
(XVIb)

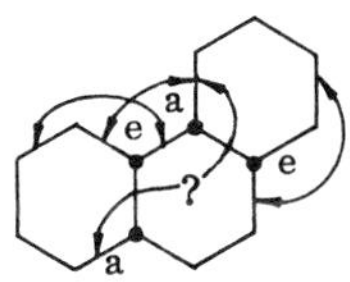

cis-syn-cis
(XVII)

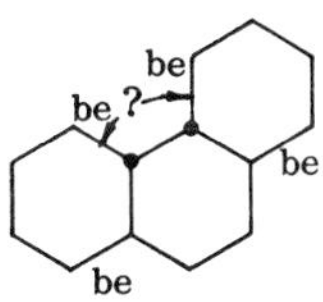

trans-syn-trans
(XVIII)

equatorial bonds and one *axial* bond are involved. Four *gauche* interactions are indicated for (XIV) and (XV); they can easily be counted by means of a model. Both isomers have been assigned $4 \times 0.8 = 3.2$ kcal./mole, or differ by 2.4 kcal./mole from the *trans-anti-trans* isomer (XIII).

The *cis-anti-cis* compound (XVI) cannot be treated as easily.[9] It

[9] Compare: J. Castells, G. A. Fletcher, E. R. H. Jones, G. D. Meakins, and R. Swindells, *J. Chem. Soc.* p. 2627 (1960).

involves two *axial* and two *equatorial* bonds that can be arranged in various ways. They can be arranged as in (XVIa) with both *axial* bonds in a 1,2-position of the central ring, or as in (XVIb) with the two *axial* bonds in a 1,4-position. There are six *gauche* interactions (or two "*cis*-decalin" interactions) present in (XVIa). The value of 4.8 kcal./mole is assigned to (XVIa). In addition an angular *gauche* interaction appears in (XVIb), so that the value would be 5.6 kcal./mole in this case. Conformation (XVIa) is therefore preferred.

In the case of the *cis-syn-cis* isomer (XVII), an estimate of the energy is practically impossible. We can make no statement concerning the magnitude of the interactions because two *axial* bonds on the central ring are in a 1,3 arrangement on the same side of the ring. The estimate of the energy is only conjecture.

Johnson[6a] compared this interaction with the one that occurs in 1,3-dimethylcyclohexane. This was estimated at about 5.4 kcal./mole.[10,10a] Because methylene groups and not methyl groups account for the interaction in the previous case, only 8/9 of this value, or 4.8 kcal./mole, is used. In contrast, Dauben and Pitzer[11] arrived at a higher value; they postulate a value of 8 to 9 kcal./mole because of the greater rigidity of the methylene groups. In addition, the *gauche* interactions indicated in formula (XVII) must also be considered.

In *trans-syn-trans*-perhydrophenanthrene (XVIII), the central ring assumes the boat conformation as mentioned above (see XII). Even though the energy difference between boat and chair form is quite accurately known from various methods of determination (~6.0 kcal./mole, cf. page 45f.), this value alone cannot be used for an estimate of the energy. We have to consider, in addition, a *cis*-butane interaction between C-4 and C-5 (formula XVIII). The size of this interaction is doubtful.[11a]

2. Perhydroanthracenes

Five stereoisomeric forms are possible for perhydroanthracene, and these were treated by Johnson[6a] in the same manner as were the perhydrophenanthrenes. Here too, the primary purpose was to establish an order of stability.

---

[10] C. W. Beckett, K. S. Pitzer, and R. Spitzer, *J. Am. Chem. Soc.* **69**, 2488 (1947).

[10a] See page 109.

[11] W. G. Dauben and K. S. Pitzer, *in* "Steric Effects in Organic Chemistry" (M.S. Newman, ed.), Chap. 1. Wiley, New York, 1956.

[11a] For octahydrophenanthrenes, cf. ref. 83b in Chap. 4.

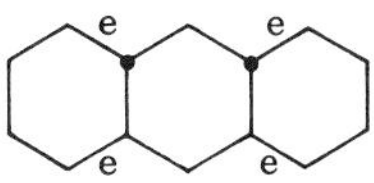

0 kcal./mole
trans-syn-trans
(XIX)

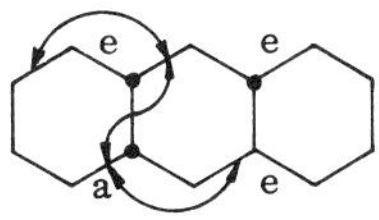

~ 2.4 kcal./mole
cis-syn-trans
(XX)

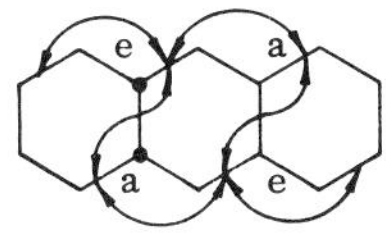

~ 4.8 kcal./mole
cis-anti-cis
(XXI)

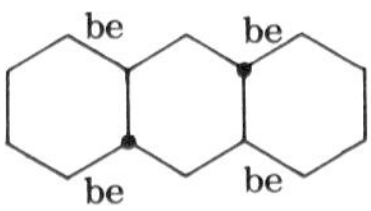

> 6.0 kcal./mole
trans-anti-trans
(XXII)

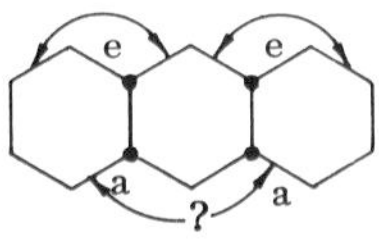

cis-syn-cis
(XXIII)

The *cis-syn-cis* (XXIII), *cis-syn-trans* (XX), and *trans-syn-trans* (XIX) isomers[12] have been known for quite some time. The *cis-anti-cis* (XXI)[13] and *trans-anti-trans* (XXII)[14] perhydroanthracenes, however, have only been synthesized recently and their configurations established.

The application of the method described under perhydrophenanthrene for the determination of the relative stabilities of the perhydroanthracenes produced the following result[6a]: The most stable isomer is *trans-syn-trans*-perhydroanthracene (XIX), as was also shown by equilibrating (XIX), (XX), and (XXII) with aluminum bromide; in which (XIX) predominates to the extent of 96% of equilibrium.[13b] All bonds leading from the central ring are *equatorial*, and no *gauche* interactions are involved. A zero energy value is therefore arbitrarily assigned to it. The *cis-syn-trans* isomer (XX) possesses one *axial* and three *equatorial* bonds from the central ring. One of the rings has a *cis* juncture. Three *gauche* interactions result as with *cis*-decalin, and 2.4 kcal./mole is assigned to (XX) which is in good agreement with the value of 1.9 kcal./mole obtained from the equilibration.[13b]

In the *cis-anti-cis* isomer (XXI) there are two *axial* and two *equatorial* bonds present. Two rings form a *cis* juncture with the central ring. Therefore, two "*cis*-decalin" interactions are involved, a total of six *gauche* interactions. The assigned value is 4.8 kcal./mole.

---

[12] J. W. Cook, N. A. McGinnis, and S. Mitchell, *J. Chem. Soc.* p. 286 (1944).

[13] (a) R. K. Hill and J. G. Martin, *Proc. Chem. Soc.* p. 390 (1959); (b) R. K. Hill, J. G. Martin, and W. H. Stouch, *J. Am. Chem. Soc.* **83**, 4006 (1961); (c) N. S. Crossley and H. B. Henbest, *J. Chem. Soc.* p. 4413 (1960); (d) R. L. Clarke and W. S. Johnson, *J. Am. Chem. Soc.* **81**, 5706 (1959).

[14] R. L. Clarke, *J. Am. Chem. Soc.* **83**, 965 (1961).

The central ring must again assume the boat or twist conformation in the *trans-anti-trans* isomer (XXII). From calorimetric measurements it has been shown,[14a] that (XXII) is 5.39 ± 0.86 kcal./mole higher in energy than the all-chair form (XIX).

It is not possible to assign an energy value to the *cis-syn-cis* isomer (XXIII). Just as in the case of *cis-syn-cis*- perhydrophenanthrene, two *axial* bonds are located in a 1,3-position to each other. It was not possible to estimate the amount of this interaction.[15]

II. STEROIDS

1. General Introduction

A discussion of the individual steroids on the basis of stereochemistry or conformation would go beyond the scope of this book. The amount of data accumulated in the last few years has been tremendous. A series of recent and excellent descriptions are available for further information on specific compounds.[16]

The reason for treating the steroids in greater detail can be briefly outlined as follows: Apart from Chapters 3 and 4, energetics or physical methods have been intentionally stressed for a determination of the conformations of the compounds treated up to now. The reasons for this have been stated frequently, and result from the fact that chemical conformational analysis of compounds of simple structure, with certain exceptions, does not yield unequivocal results. This latter is due to the flexible conformations common to such compounds. If a chemical reaction with an activation energy normally larger than the energy barrier between the individual conformations, is used for a conformational analysis, then there is no guarantee that the molecule will react in the conformation of its ground state.

The steroids (and triterpenes), in contrast, which superficially may

---

[14a] J. L. Margrave, M. A. Frisch, R. G. Bautista, R. L. Clarke, and W. S. Johnson, *J. Am. Chem. Soc.* **85**, 546 (1963).

[15] For the conformations of the dodecahydrophenanthrenes see S. K. Balasubramanian, *Tetrahedron* **12**, 196 (1961).

[16] (a) L. F. Fieser and M. Fieser, "Steroids." Reinhold, New York, 1959. (b) D. H. R, Barton in Lettré-Inhoffen-Tschesche, "Über Sterine, Gallensäuren und verwandte Naturstoffe," Vol. 2, 2nd ed. Enke, Stuttgart, 1959; (c) Cf. also W. Klyne, "The Chemistry of Steroids." Methuen, London, 1957; (d) Cf. also D. H. R. Barton and R. C. Cookson, *Quart. Rev.* (*London*) **10**, 44 (1956); (e) Cf. also D. H. R. Barton and G. A. Morrison, Conformational analysis of steroids and related natural products. *In* "Progress in the Chemistry of Organic Natural Products." (L. Zechmeister, ed.), Vol. 19, p. 165. Springer, Berlin, 1961.

appear very complicated structurally, are easier compounds to work with from the view-point of conformational analysis. As shall be illustrated, they are no longer capable of forming various conformations. The conformations are "frozen" because of the special nature of the ring junctures. If this is the case, i.e. if the possibility for the formation of energetically less favorable conformations in the transition states of the reactions is not available or rather, improbable, then a chemical conformational analysis is easier. Relationships between the conformation of a molecule and its reactivity can be derived. The fundamental observations leading to the development of conformational analysis were, in fact, made by Barton[17] with compounds with rigid conformations, viz. mostly with steroids. It was not observed until later that simple compounds are more difficult to treat. The regularity of generalizations made regarding the pertinent reactivity of steroid systems could only be applied tenuously toward the reactivities of systems, which are capable of forming different conformations. It should be possible, for this reason, to find a relationship between the conformation and reactivity of a steroid. It should also be possible to find reactions which can be applied to chemical conformational analysis without calling upon many auxiliary hypotheses.

Before considering these problems individually, some generalities about the conformations of steroids may be summarized. Let us briefly consider some important rules of nomenclature and recall the classical stereochemistry of the steroids.[18]

The numbering of the steroid skeleton (or perhydrocyclopentanophenanthrene nucleus) is shown in structure (XXIV). The position of the substituent is designated by the number of the carbon atom on which the substituent is located.

The steroid nucleus has a planar representation with the methyl groups at carbons 10 and 13 above the plane. Nuclear substituents which are situated above the plane (or on the same side as the angular methyl groups at C-10 and C-13), are β oriented. They are drawn with a solid line. Substituents situated below the plane of the nucleus are α-oriented. They are symbolized by a dotted (or broken) line. If the configuration of the substituent is unknown, the prefix ζ is used and the bond shown as a wave line.

---

[17] D. H. R. Barton, *Experientia* **51**, 316 (1950); D. H. R. Barton and W. J. Rosenfelder, *J. Chem. Soc.* p. 1048 (1951); D. H. R. Barton, *ibid.* p. 1027 (1953).

[18] Cf. International Union of Pure and Applied Chemistry, "Nomenclature of Organic Chemistry." Butterworths, London, 1958; *J. Am. Chem. Soc.* **82**, 5577 (1960). The IUPAC rules are based on the ones which were formulated at the Conference of the Ciba-Foundation in London, 1950. Cf. *J. Chem. Soc.* p. 3526 (1951); *Helv. Chim. Acta* **34**, 1680 (1951); cf. also L. F. Fieser and M. Fieser, *Tetrahedron* **8**, 360 (1960).

The perhydrocyclopentanophenanthrene nucleus (XXIV) contains six asymmetric carbon atoms at the ring junctures A/B, B/C, and C/D. This would indicate that 64 stereoisomeric forms are possible. As far as is known to date, only certain types of ring fusions exist in all naturally occurring steroids. The number of stereoisomers is thereby greatly reduced.

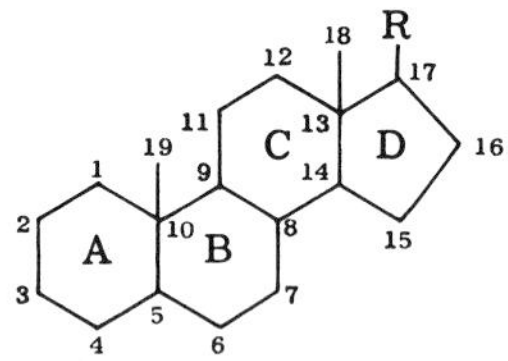

(XXIV)

Rings A and B may be joined *trans* as well as in the *cis* position. Rings B and C are always fused with a *trans* arrangement. Rings C and D, with some exceptions,[19] are generally fused in a *trans* orientation.

In all naturally occurring steroids, the substituents on C-9 and C-10 exist in an *anti* arrangement (see page 209). The same is true, with some exceptions,[19] for the substituents on C-8 and C-14. In summary, the largest number of steroids are distinguished by a preponderance of *trans* configurations on adjoining ring junctures. In the case of a *trans* juncture of rings A and B (5α-steroids), almost planar structure results (see XXV).[20]

As has already been seen in the discussion of the perhydrophenanthrenes, that conformation of a system of condensed cyclohexane rings which contains the largest number of chair forms possesses the greatest stability.[11,21] With some exceptions, practically all steroids can be constructed from chair forms. The simplicity of the conformational analysis lies in the fact that each of the steroid conformations constructed only from chair forms represents the only conformation. (Concerning boat or twist conformations of steroids, see Chapter 7.) An interconversion of the chair forms is not possible because of the rigidity of the system. The assignment of conformations, therefore, follows from the configuration.

---

[19] Cardiac aglycones, toad poisons.

[20] Cf. detailed summary of literature references by R. B. Turner, *in* "Natural Products Related to Phenathrene" (L. F. Fieser and M. Fieser, eds.), 3rd ed., p. 620 ff. Reinhold, New York, 1949; Cf. also A. Heusner, *Angew. Chem.* **63**, 59 (1951).

[21] H. R. Nace and R. B. Turner, *J. Am. Chem. Soc.* **75**, 4063 (1953).

A few examples are given here:

(XXV) shows the conformation of a 5α-steroid (XXVI) (rings A/B *trans*).

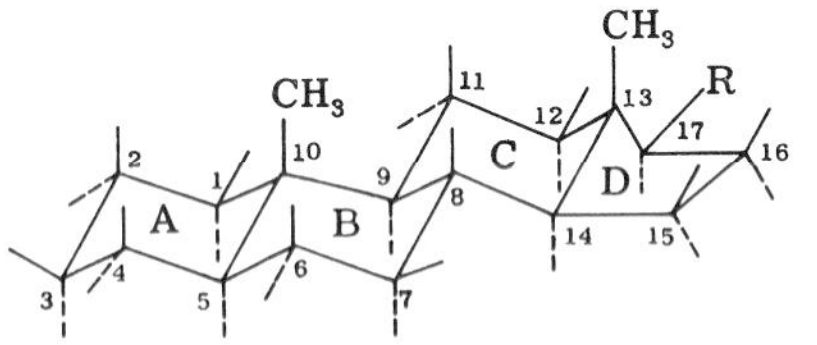

(XXV)

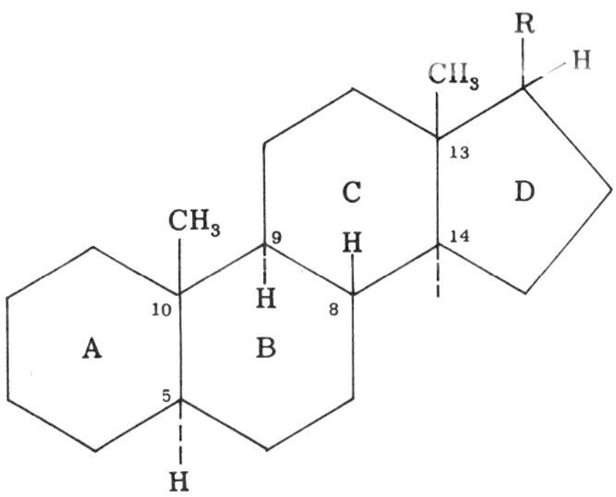

(XXVI)

For example:

R = H, 5α-androstane

R = $-C_2H_5$, 5α-pregnane

R = $-CH(CH_3)CH_2CH_2CH_3$, 5α-cholane

R = $-CH(CH_3)CH_2CH_2CH_2CH(CH_3)_2$, 5α-cholestane.

As can clearly be seen, this conformation is derived from the *trans*-(5,10)-*anti*(10,9)-*trans*(9,8)-*anti*(8,14)-*trans*(14,13)-configuration.

(XXVII) shows the conformation of a 5β-steroid (XXVIII) in which rings A and B are fused in *cis* position [*cis*(5,10)-*anti*(10,9)-*trans*(9,8)-*anti*(8,14)-*trans*(14,13)]. As in *cis*-decalin (see page 57), ring A retains the chair conformation, and rings B, C, and D are no different from those in the 5α-steroids.

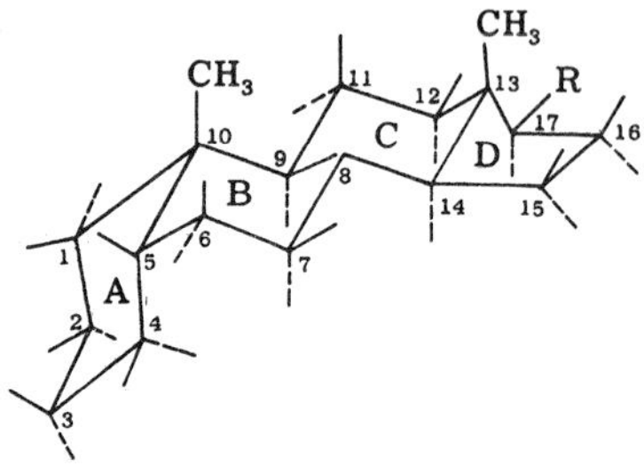

(XXVII)

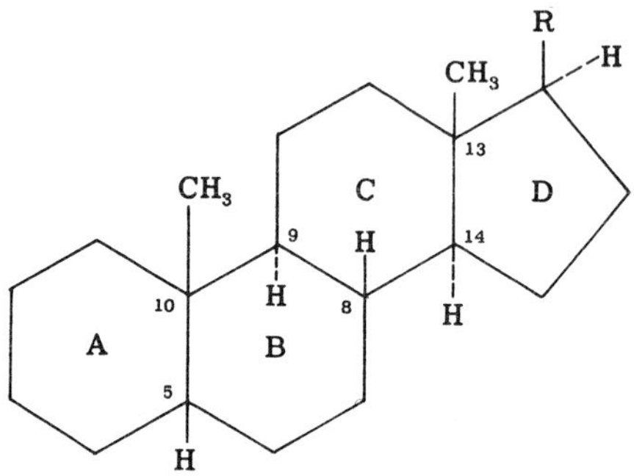

(XXVIII)

For example:

R = H, 5β-androstane
R = $-C_2H_5$, 5β-pregnane
R = $-CH(CH_3)CH_2CH_2CH_3$, 5β-cholane
R = $-CH(CH_3)CH_2CH_2CH_2CH(CH_3)_2$, 5β-cholestane.

Because the conformations of the rings are fixed the same must also be true of the substituents. The *axial* or *equatorial* position of a substituent is unequivocally determined for every carbon atom, in contrast to those simple cyclohexane derivatives which are subject to ring interconversion. However, it must be borne in mind that ring *A* can, nevertheless, still exist in a twist or boat conformation in special cases (see Chapter 7).

The all-chair conformation of the steroid nucleus has been confirmed with X-ray analysis on cholesteryl iodide.[21a]

[21a] C. H. Carlisle and D. Crawfoot, *Proc. Roy. Soc.* **A184**, 64 (1945); compare also: J. Fridrichson and A. McL. Mathieson, *J. Chem. Soc.* p. 2159 (1953); J. M. Ohrt, B. A. Haner, and D. A. Norton, *Nature*, **199**, 1180 (1963).

There exists a definite relationship between the α and β configuration of each substituent on the steroid nucleus and the *axial* or *equatorial* position. This is summarized in Table I. [Cf. (XXV) and (XXVII).[21a]]

TABLE I

CONFORMATIONS OF STEROIDS

Position [see formula (I)]	5α-series (A/B *trans*) configuration		5β-series (A/B *cis*) configuration	
	α	β	α	β
1	a	e	e	a
2	e	a	a	e
3	a	e	e	a
4	e	a	a	e
5	a	—	—	a(A)e(B)[a]
6	e	a	e	a
7	a	e	a	e
8	—	a	—	a
9	a	—	a	—
10	—	a	—	e(A)a(B)[a]
11	e	a	e	a
12	a	e	a	e
13	—	a	—	a
14	a	—	a	—
15	—[b]	—[b]	—[b]	—[b]
16	—[b]	—[b]	—[b]	—[b]
17	—[b]	—[b]	—[b]	—[b]

[a] a(A) or e(B) indicates that the substituent is located in *axial* position in relation to ring A or in *equatorial* position in relation to ring B.
[b] For position of substituents on ring D refer to text below.

Whereas the conformations of the substituents on the six-membered rings may be found in Table I, the question of the preferred conformations of substituents on the five-membered ring (ring D) remains open. As has been discussed in Chapter 3 (page 75) the five-membered ring is not planar, but assumes puckered conformations which have been designated as the envelope and half-chair forms. In the originally expressed and, later, widely applied representation of the steroid molecule (rings C/D trans) by Barton,[22] ring D was shown in the envelope conformation (XXVIIIa). Here C-14 lies below the plane of C-13,

[21a] Cf. ref. 16a, b, c.
[22] D. H. R. Barton, *J. Chem. Soc.* p. 1027 (1953).

C-15, C-16 and C-17. The half-chair conformation (XXVIIIb) in which C-13 is above and C-14 the same distance below the plane formed by C-15, C-16, and C-17 has also been used at various times.[23] Brutcher and Bauer have recently shown[24] that there is another, not previously considered, envelope conformation (XXIIIc) possible, in which C-13

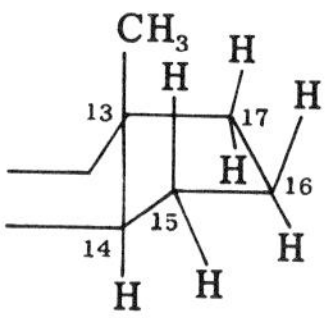

(XXVIII a)

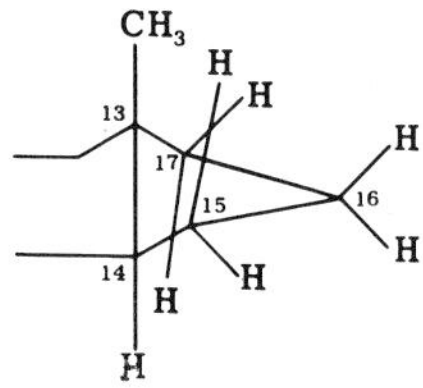

(XXVIII b)

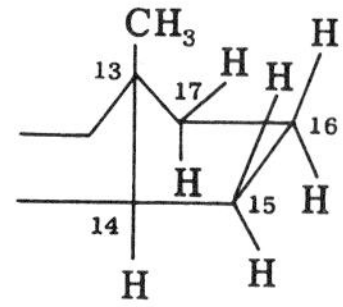

(XXVIII c)

lies above the plane formed by C-14, C-15, C-16, and C-17. Through a careful analysis of angle strain, dihedral angles, and 1,3 interactions which arise between the *axial* methyl group at C-13 and the β-hydrogens at C-15 and C-16 in the three conformations, these conformations were examined with various substituents on ring D. As a result, conformation (XXVIIIc) is preferred for a non-substituted ring D (androstane) because only in this conformation does the angular methyl group at C-13 assume a symmetrical position with respect to the β-hydrogens at C-15 and C-16.

If ring D is substituted at position 17β, as is the case for most of the steroids, then the envelope conformation (XXVIIIc) is also the preferred conformation with lowest energy. The 17β substituent assumes an *equatorial* position in this conformation (see XXVIIIc). For 17α-substituted steroids there exists a large interaction between the substituent in the 17α position and the hydrogen at C-14. As a result, conformation (XXVIIIa) would probably be the more stable one in this case. The 17β epimer is more stable by $\sim$3 kcal./mole compared to the 17α epimer.

For 17-ketosteroids, ring D also exists preferentially in conformation (XXVIIIa). The infrared spectra of 16α- and 16β-bromo-17-ketosteroids showed[24] that both bromine atoms produce an identical shift of 12 cm.$^{-1}$ in the carbonyl stretching frequency[25] (cf. page 233). This is the result of an identical angle between the 17-keto group and the

[23] See ref. 3f in Chap. 3. Cf. also C. W. Shoppee and J. C. P. Sly, *J. Chem. Soc.* p. 345 (1959).

[24] F. V. Brutcher, Jr. and W. Bauer, Jr., *J. Am. Chem. Soc.* **84**, 2233, 2236, (1962).

[25] Cf. also J. Fishman and C. Djerassi, *Experientia* **16**, 138 (1960); J. Fishman and W. R. Bigerstaff, *J. Org. Chem.* **23**, 1190 (1958); J. Fajkos and J. Joska, *Chem. & Ind. (London)* p. 872 (1960); *ibid.* p. 1162 (1960); Collection Czech. Chem. Commun. **25**, 2863 (1960); *ibid.* **26**, 1118 (1961); J. Fishman, *J. Org. Chem.* **27**, 1745 (1962).

bromine, which is possible only in conformation (XXVIIIa) (cf. also page 291). That the envelope conformation (XXVIIIc) is not present in this case, although the 1,3 interactions are at a minimum in this conformation, was explained by Brutcher.[24] He attributes it not only to the *eclipsed* position of the hydrogens at C-15 and C-16, but also to the inability to attain a trigonal hybridization at C-17. For 16-ketosteroids, however, conformation (XXVIIIc) again seems to be the more stable one.[24,26,27]

Nevertheless, the fusion of the five-membered ring with the six-membered ring still presents some difficulties with respect to the stability.

We had noted (page 176) for the 8-methylhydrindane system that as a result of a greater "*gauche*" character of the *trans* compound in contrast to the parent hydrindane, the *cis* compound had to be the more stable one. Otherwise, the *trans* fusion of a six-membered ring with a five-membered ring does not lead to any essential change in the conformation of the former. Djerassi was able to show this with optical rotatory dispersion studies of 8-methylhydrindanone[28] as an example.

This rule, that the *cis* compound is the more stable one, can be applied to the steroid systems as well. Dreiding[29] summarized this on the basis of examples that were supplied by other investigators. A *trans* to *cis* isomerization of rings C/D for stereochemically established steroid compounds was observed. Most of the compounds studied (11-oxosteroids) had a double bond[30] at the 8,9 position. Barton[31] and co-workers, however, were able to demonstrate that the *trans* juncture of rings C and D in a series of 15-ketostanols was more stable than the *cis* juncture. This was later verified by measurements of the optical rotatory dispersion.[32] Further, Djerassi and co-workers were readily able to isomerize 15-ketosapogenins with a 14α-(C/D-*trans*) juncture into the 14β-(C/D-*cis*) epimer. This result was verified by optical rotatory dispersion.[32]

---

[26] Cf. however, J. Fishman, *Chem. & Ind. (London)* p. 1078 (1961); J. Fajkos and J. Joska, *ibid.* p. 872 (1960).

[27] For an investigation of the conformation of the side chain of steroids see N. L. Allinger and M. A. DaRooge, *J. Am. Chem. Soc.* **83**, 4256 (1961).

[28] C. Djerassi, D. Marshall, and T. Nakano, *J. Am. Chem. Soc.* **80**, 4853 (1958).

[29] (a) A. S. Dreiding, *Chem. & Ind. (London)* p. 992 (1954) with further references; (b) Cf. also E. L. Eliel and C. Pillar, *J. Am. Chem. Soc.* **77**, 3600 (1955); (c) C. Djerassi, W. Frick, G. Rosenkranz, and F. Sondheimer, *ibid.* **75**, 3496 (1953).

[30] In the case of the A-nor steroids (ring A is a 5-membered ring), both the 3-keto and the 2-keto compounds are more stable in the *cis* configuration of rings A and B. See F. L. Weisenborn and H. E. Applegate, *J. Am. Chem. Soc.* **81**, 1960 (1959).

[31] C. S. Barnes, D. H. R. Barton, and G. F. Laws, *Chem. & Ind. (London)* p. 616 (1953); D. H. R. Barton and G. F. Laws, *J. Chem. Soc.* p. 52 (1954).

[32] C. Djerassi, T. T. Grossnickle, and L. B. High, *J. Am. Chem. Soc.* **78**, 3166 (1956); C. Djerassi, R. Riniker, and B. Riniker, *ibid.* **78**, 6362 (1956); cf. also C. Djerassi, J. Fishman, and T. Nambara, *Experientia* **17**, 566 (1961).

It has not been possible thus far to present a satisfactory and uniform explanation for the variable stability of the C/D ring juncture in steroids. A whole series of explanations has been offered[33] but have not clarified the situation because of the inadequate quantitative experimental data available.

A series of equilibrium measurements was carried out on steroids by Allinger and Djerassi and the difference in the free energy was determined.[33] Allinger[33,34] attempted to find a starting point for the calculation of energy differences in structurally complicated hydrindane and hydrindanone systems by consideration of the 2- and 3-alkyl ketone effects described for the hydrindanones, the *gauche* interactions, and the entropy difference between the isomeric hydrindanes. By equilibration in methanol, the relative stability of the *cis* and *trans* juncture of steroidal 4- and 6-ketones has also been measured. If the keto group is in position 4, as in cholestan-4-one, then 99% of the equilibrium mixture has rings A and B *trans*-fused and 1% consists of the A/B *cis* compound. For cholestan-6-one, however, the equilibrium mixture contains 88% A/B *trans* and 12% A/B *cis* compounds.[35]

Dauben[36] also called attention to the difficult problem of determining the relative stabilities of various steroids. The stability of a molecule can be greatly affected by small changes in structure. The variation in stability[36] between 15-ketocholesterol (XXIX)[31] and 14β-artebufogenin (XXX)[37] can be cited as an example. (XXIX) is more stable with a *trans* juncture of rings C and D, but (XXX) is more stable with a *cis* juncture of rings C and D. The compounds differ only at the A/B juncture,

---

[33] Cf. the literature summary of N. L. Allinger, R. B. Hermann, and C. Djerassi, *J. Org. Chem.* **25**, 922 (1960). Cf. also J. Fishman, *J. Am. Chem. Soc.* **82**, 6143 (1960).

[34] Cf. N. L. Allinger and S. Greenberg, *J. Org. Chem.* **25**, 1399 (1960). Allinger in his investigations of the stability of the hydrindanone system in steroids, considered influences produced by strained ring systems through the so-called conformational transmission. (Compare page 225).

[35] N. L. Allinger, M. A. Darooge, and R. B. Herman, *J. Org. Chem.* **26**, 3626 (1961). The occurrence of the so-called reflex interaction has also been used to explain the relative stabilities of *cis*- and *trans*-steroid hydrindanones: J. F. Biellmann, D. Francetic and G. Ourisson, *Tetrahedron Letters* **18**, 4 (1960); G. Ourisson, *Ing. Chimiste* **43**, 95 (1961). The term "reflex effect" (reflex interaction) has been introduced by G. Ourisson to describe an enhancement of an 1,3-interaction by bond angle distortion, due to non-bonded interactions of bulky groups. C. Sandris and G. Ourisson, *Bull. Soc. Chim. France*, p. 1524 (1958); R. Hanna, C. Sandris, and G. Ourisson, *ibid.* p. 1454 (1959); J. F. Biellmann, R. Hanna, G. Ourisson, C. Sandris, and B. Waegell, *ibid.* p. 1429 (1960); B. Waegell and G. Ourisson, *ibid.* p. 495 (1963); *ibid.* p. 496 (1963); *ibid.* p. 503 (1963); compare also: L. C. G. Goaman and D. F. Grant, *Tetrahedron* **19**, 1531 (1963).

[36] W. G. Dauben, *Colloq. Intern. Chim. Org. Probl. Stereochim. Montpellier*, 1959 in *Bull. Soc. Chim. France* p. 1338 (1960).

[37] H. Linde and K. Meyer, *Helv. Chim. Acta* **42**, 807 (1959).

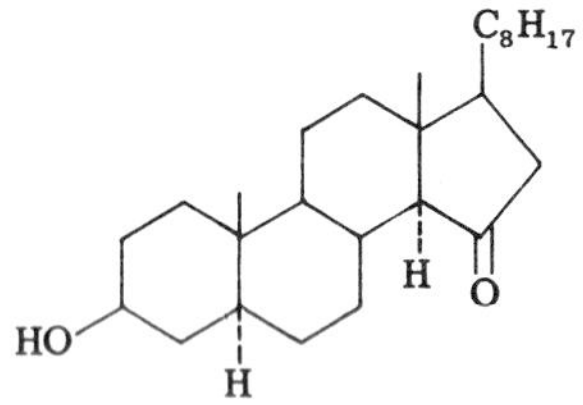

(XXIX)

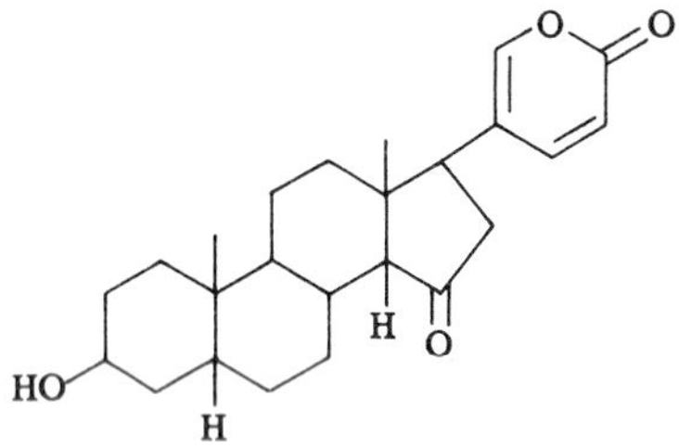

(XXX)

as well as the moieties at C-17. From this one example if follows that even small changes in the molecule may have a decided influence on the stability. These effects demonstrated by Dauben[36] are comparable to the "conformational transmission" effects.

2. Conformational Transmission

The so-called "conformational transmission" effect discovered by Barton[38] can be defined briefly as follows: Unsaturated substituents (and to a small extent saturated ones) bring about a local distortion of the molecule. This distortion is transmitted in a saturated molecule by a slight deformation of the valence angles and a slight change in the atomic coordinates.

This effect was discovered in the base-catalyzed condensations of benzaldehyde with triterpenoid ketones, which react as follows: (XXXI; XXXII):

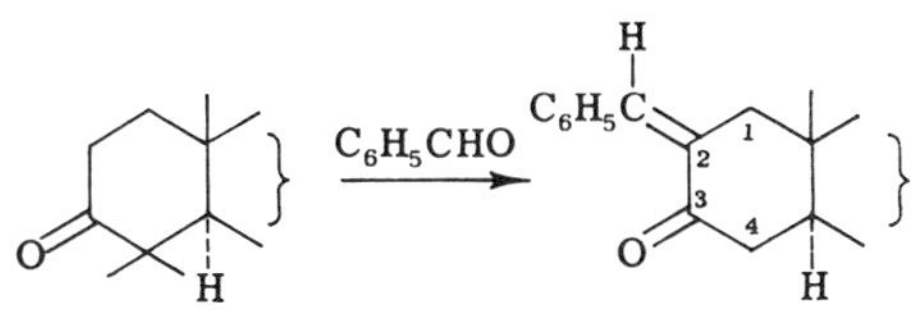

(XXXI) (XXXII)

Some illustrative examples follow. The rate of the condensations of the following triterpenoid-3-ketones with benzaldehyde was measured and based on lanonst-8-enone (XXXIII) = 100: lanostanone (XXXIV) = 55; lanost-7-enone (XXXV) = 17.[38d]

[38] See ref. 56; also (a) D. H. R. Barton, A. J. Head, and P. J. May, *J. Chem. Soc.* p. 935 (1957); (b) D. H. R. Barton, *in* "Kekule Symposium on Theoretical Organic Chemistry," p. 127. Butterworths, London, 1958; (c) D. H. R. Barton, F. McCapra, P. J. May, and F. Thudium, *J. Chem. Soc.* p. 1297 (1960).

[38d] For further examples, see ref. 38a.

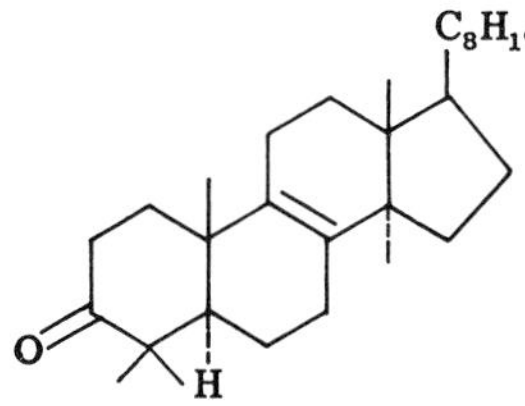

(100)
(XXXIII)

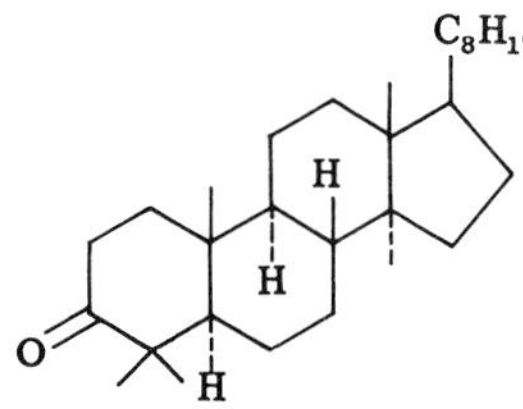

(55)
(XXXIV)

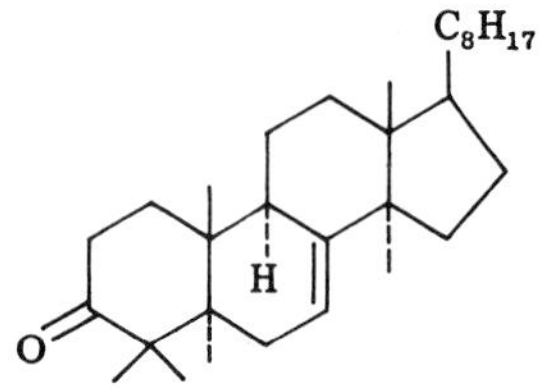

(17)
(XXXV)

The double bond in (XXXIII) causes the rate to increase over that of (XXXIV), an effect also noted with other examples.[38a] A methyl group at C-8 also has an influence on the condensation rate with benzaldehyde, as shown in other examples.[38a]

The investigations were also extended to the 3-keto steroids.[38b,c] The condensation rates were: cholestanone (XXXVI, R = C_7H_{18}) = 182; stigmastanone (XXXVI, R = $C_{10}H_{21}$) = 180; 17β-hydroxyandrostan-3-one (XXXVI, R = OH) = 188. These react at the about same rate with benzaldehyde and form exclusively the 2-benzylidene compound. Bisbenzylidene compounds, formed by reaction at C-4 as well were never detected. The rates are higher than those for the corresponding saturated triterpenoid compounds (XXXIV) because the interaction between the C—C bond which is formed at C-2 (XXXII), and methyl groups at C-4 is absent.

Cholest-6-en-3-one (XXXVII) has a markedly higher rate due to the "conformational transmission" effect (= 645). However, if the double

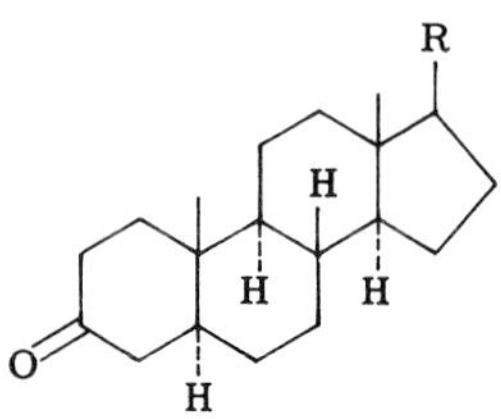

(XXXVI)

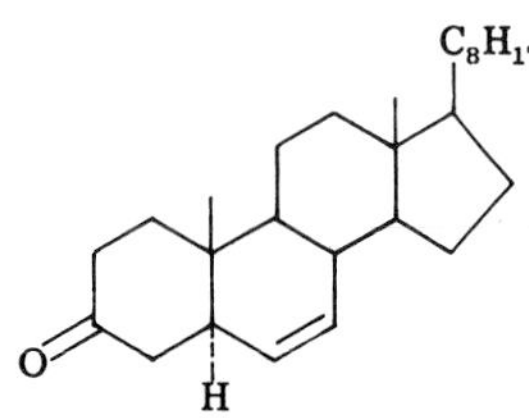

(XXXVII)

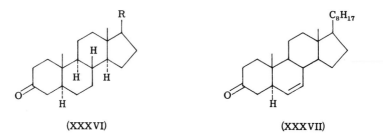

bond is moved to position 7 (ergosta-7,22-dienone, XXXVIII), then the rate of condensation with benzaldehyde drops to the surprisingly low value of 43. The obvious thought that a hyperconjugation effect is involved can be excluded.[38b,c] Additional keto groups, at positions other than C-3, also cause an increase in the rate. 7-Ketocholestanone XXXIX, X = O) shows a relative reactivity of 615. Here also, this cannot result only from a polar effect of the keto group, but results in part from its influence in producing angle distortion (conformational transmission).

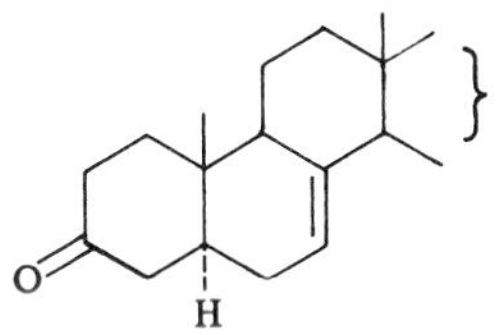

(43)

(XXXVIII)

If the 7-keto group is replaced by a methylene group (XXXIX, X = CH_2) a relative rate of 365 is retained even though the polar effect of the keto group has been removed.

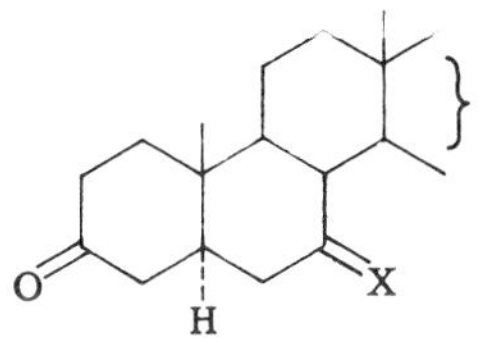

(X = O, 615; X = CH_2, 365)

(XXXIX)

These effects described briefly have also been expanded in a quantitative sense by Barton[38c] to structurally analogous steroids. The rates are expressed as the rates of a saturated ketone (as the standard) multiplied by a series of group rate factors. The latter depends on the type and position of the group considered.[39]

Mention must also be made of the optical rotatory dispersion studies by Djerassi and co-workers[40] on the problem of "conformational transmission." If the double bond in triterpenoid-3-ketones is shifted from the 7-8 to the 8-9 position, then a reversal of the Cotton effect curve[41]

[39] For additional investigations into the problem of conformational transmission see: (a) O. H. Wheeler and J. L. Mateos, *Can. J. Chem.* **36**, 712 1958); (b) O. H. Wheeler and J. Z. Zabicky, *ibid.* **36**, 656 (1958); (c) O. H. Wheeler and V. S. Gaind, *ibid.* **36**, 1735 (1958); (d) J. Mathieu, M. Legrand, and J. Valls, *Bull. Soc. Chim. France* p. 549 (1960); (e) M. Legrand, V. Delaroff, and J. Mathieu, *ibid.* p. 1346 (1961); (f) R. Bucourt, *ibid.* p. 1262 (1963); (g) M. J. T. Robinson and W. B. Whalley, *Tetrahedron* **19**, 2123 (1963).

[40] C. Djerassi, O. Halpern, V. Halpern, and B. Riniker, *J. Am. Chem. Soc.* **80**, 4001 (1958); C. Djerassi, J. Osiecki, and W. Closson, *ibid.* **81**, 4587 (1959).

[41] C. Djerassi, "Optical Rotary Dispersion," pp. 185 ff. McGraw-Hill, New York, 1960.

from negative to positive takes place. This is in agreement with the results of Barton and can be related to a conformational transmission effect. However, such an effect could not be observed for the steroid-3-ketones. A better understanding of the phenomenon of conformational transmission must await further investigations.[39g]

The effect of conformational transmission must, nevertheless, be taken into consideration in investigations of the relative stabilities of steroids. However, investigations[34] along these lines with a few compounds have indicated that any strain introduced by a C/D *trans* ring juncture is not of a decisive nature in determining the stability of the A/B ring juncture.

3. Stability of the Substituents

In regard to the stability of substituents on the steroid system, the well-known rule of simple cyclohexane derivatives is valid. *Equatorial* substituents are more stable than *axial* ones. It has been shown for sterols that the *equatorial* alcohol was more stable than the *axial* epimer in every case studied.[42]

Trans-1,2-halogens can cause a shift of the preferred position of the halogens in simple cyclohexanes. It was shown (page 121f.) that *diaxial* conformations of *trans*-1,2-dihalocyclohexanes predominate. Barton and co-workers could show that corresponding *trans*-dihalo derivatives of steroids are more stable in the *diequatorial* conformations.[17,43] The reasons for this must be found in the fact that the opposing electrostatic effects, which in simple 1,2-halogenated cyclohexanes leads to a predominance of the *diaxial* conformation, are compensated by another factor. The *axial* methyl groups at C-10 and C-13 exert a steric interaction on an *axial* halogen, which counteracts the electrostatic interaction of halogen atoms in the *diequatorial* position. The latter then becomes more stable.

The phenomenon in which *diaxial* dibromides rearrange into the corresponding *diequatorial* compounds was discovered by Barton[43,44] and will be explained in more detail with specific examples.

---

[42] See references 17, 31 and cf. in addition: (a) W. Klyne and W. M. Stokes, *J. Chem. Soc.* p. 1979 (1954); (b) Pl. A. Plattner, A. Fürst, and H. Els, *Helv. Chim. Acta* **37**, 1399 (1954).

[43] See for example ref. 101; (a) G. H. Alt and D. H. R. Barton, *J. Chem. Soc.* p. 4284 (1954); Ref. 56; (b) D. H. R. Barton, *Bull. Soc. Chim. France* p. 973 (1956); Cf. also ref. 38b; (c) D. H. R. Barton, *Svensk Kem. Tidskr.* **71**, 256 (1959); cf. also J. F. King, R. G. Pecos and R. A. Simmons, *Can. J. Chem.* **41**, 2187 (1963).

[44] D. H. R. Barton and J. F. King, *J. Chem. Soc.* p. 4398 (1958).

When *diaxial* 2β, 3α-dibromocholestane (XL[45], X = Y = Br) is heated, it rearranges to the *diequatorial* 2α, 3β-dibromide (XLI, X = Y = Br). The equilibrium mixture contains 93% of the *diequatorial* epimer.[17,43,44] The first example of such a rearrangement was known much

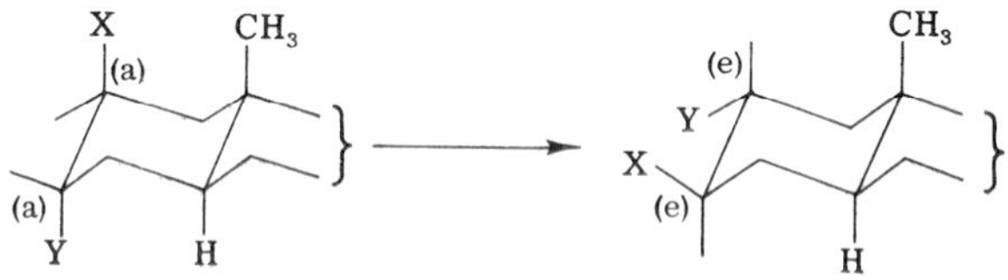

(XL) (XLI)

earlier,[46] and consists of the spontaneous mutarotation of "normal" cholestene dibromide (5α, 6β-dibromocholestane) to the 5β, 6α isomer.

Further examples are 3α, 4β-dibromocholestane, which rearranges to the *diequatorial* 3β, 4α epimer[43a] upon heating; 2β-bromo-3α-chlorocholestane (XL, X = Br, Y = Cl), which rearranges to the *diequatorial* 2α-chloro-3β-bromocholestane (XLI, Y = Cl, X = Br); and 3α-bromo-2β-chlorocholestane (XL, Y = Br, X = Cl), which rearranges to the *diequatorial* 2α-bromo-3β-chloroisomer[43a] (XLI, Y = Br, X = Cl). The equilibrium position attained during rearrangement can be reached from both sides.[47]

The rearrangement, as mentioned, takes place because the steric interaction of an *axial* halogen is eliminated in the *diequatorial* form. It is intramolecular and proceeds by way of a polar transition state[48] (XLII). The bromide ion remains bound to the rest of the molecule during the rearrangement.

If different substituents prone to rearrangement are chosen such that

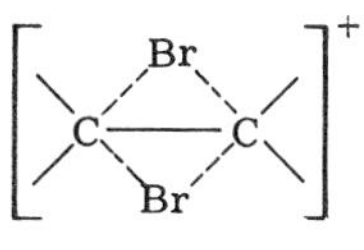

(XLII)

---

[45] For simplification, only Ring A is drawn (cf. Formula XXV, page 219).

[46] J. Mauthner and W. Suida, *Monatsh. Chem.* **15**, 85 (1894); J. Mauthner, *ibid.* **27**, 421 (1906); cf. also H. Bretschneider, Z. Főldi, F. Galinovski, and G. v. Fodor, *Ber.* **74**, 1451 (1941).

[47] D. H. R. Barton and A. J. Head, *J. Chem. Soc.* p. 932 (1956).

[48] C. A. Grob and S. Winstein, *Helv. Chim. Acta* **35**, 782 (1952); Cf. H. Kwart and L. B. Weisfeld, *J. Am. Chem. Soc.* **78**, 635 (1956).

one of them is relatively electronegative and the other relatively electropositive, then the rearrangement should take place more easily.

Barton and King[44] were able to show that this does occur. 2β-Bromo-3α-acetoxycholestane (XL, X = Br, Y = OAc) is rearranged by heat to 2α-acetoxy-3β-bromocholestane (XLI, Y = OAc, X = Br); 96% of the *diequatorial* epimer is formed. The 2β-chloro-(XL, X = Cl, Y = OAc) and the 2β-iodo-(XL, X = I, Y = OAc) 3α-acetoxycolestanes were also isomerized to the corresponding 3β-chloro-(XLI, X = Cl, Y = OAc) and 3β-iodo-(XLI, X = I, Y = OAc) 2α-acetoxycholestanes. The rate showed first order kineties and varied according to the halogen substituent. The rates of rearrangement depend on the relative polarizability of the halogen atom and the ease with which the carbon-halogen bond can be broken (I > Br > Cl). In addition, the benzoate of 2β-bromocholestan-3α-ol (XL, X = Br, Y = OBz) was rearranged to the *diequatorial* epimer (XLI, X = Br, Y = OBz). The rate was greater than that of the corresponding acetate because of the phenyl group. Benzoates substituted in the *para* position were also investigated. The rate of rearrangement was greater or smaller than the unsubstituted benzoate, depending on the electron-releasing ability (*p*-methoxy group)

TABLE II

STABILITIES OF α-BROMOKETOSTEROIDS

Location of the keto group	Predicted stable configuration for bromine in α position	
	Ring A/B *trans*	Ring A/B *cis*
1	C-2 α (e)[a]	C-2 β (e)[a]
2	C-1 α (a) C-3 α (a)	C-1 β (a) C-3 β (a)
3	C-2 α (e) C-4 α (e)	C-2 β (e) C-4 β (e)
4	C-3 α (a) C-5 α (a)	C-3 β (a) C-6 α (a)
6	C-5 α (a) C-7 α (a)	C-6 α (a) C-7 β (e)
7	C-6 α (e)	C-6 α (e)
11	C-9 α (a) C-12 α (a)	C-9 α (a) C-12 α (a)
12	C-11 α (e)	C-11 α (e)

[a] The isomer with *axial* bromine differs only slightly in its stability.

or the electron-withdrawing ability (*p*-nitro group) of the substituent, respectively.

The relatively complicated steric and electropolar effects involved in the stability of α-halo ketones in the steroid series have already been considered in the treatment of simpler α-halo ketones (see page 152f.). Corey[49] derived the preferred position of a bromine atom in keto-steroids on the basis of these effects. The keto group may be located in rings A, B, or C, and rings A and B may be joined either *cis* or *trans*. It was assumed that bromination was under thermodynamic control (see also page 276). The predicted relative stabilities of the epimeric α-bromoketo steroids are summarized in Table II.

4. Physical Methods for Conformational Analysis of Steroids

In addition to the chemical studies of the stabilities of steroid compounds, physical methods for the determination of configurations and conformations were, of course, employed. Several of these methods are familiar from discussions in previous chapters; namely infrared and ultraviolet spectra, optical rotatory dispersion, and occasionally Raman spectra and nuclear magnetic resonance. Recently molecular polarisability has been used,[50] in measurements. The rigid structure of steroids and the lack of conformational isomers greatly simplifies an interpretation of the results.

As has been shown for conformationally homogeneous *tert*-butylcyclohexanols (see page 98), the *axial* and *equatorial* hydroxyl groups differ characteristically in the position of the C—(OH)-stretching bands: This had been established previously for steroid alcohols. For secondary steroid alcohols, the C—(OH) stretching vibration occurs at about 1040 cm.$^{-1}$, while it is somewhat lower, at about 1000 cm.$^{-1}$,[51,52] for

[49] E. J. Corey, *Experientia* **9**, 329 (1953).

[50] J. M. Eckert and R. J. W. LeFèvre, *J. Chem. Soc.* p. 1081 (1962).

[51] Cf. E. A. Braude, and E. S. Waight *in Progr. Stereochem.* (W. Klyne, ed.), Vol. 1, loc. cit. ref. 6 in Chap. 2.

[52] See for example: (a) R. N. Jones, P. Humphries, F. Herling, and K. Dobriner *J. Am. Chem. Soc.* **73**, 3215 (1951); (b) A. R. H. Cole, R. N. Jones, and K. Dobriner, *ibid.* **74**, 5571 (1952); (c) A. Fürst, H. H. Kuhn, R. Scotoni, Jr., and H. H. Günthard, *Helv. Chim. Acta* **35**, 951 (1952); (d) A. R. H. Cole, *J. Chem. Soc.* p. 4969 (1952) (e) R. N. Jones, E. Katzenellenbogen, and K. Dobriner, *J. Am. Chem. Soc.* **75**, 158 (1953); (f) H. Rosenkrantz and L. Zablow, *ibid.* **75**, 903 (1953); (g) J. E. Page, *J. Chem. Soc.* p. 2017 (1955); (h) A. R. H. Cole, *Fortschr. Chem. Org. Naturstoffe* **13**, 1 (1956); (i) I. L. Allsop, A. R. H Cole, D. E. White, and R. L. S. Willix, *J. Chem. Soc.* p. 4868 (1956); (j) R. N. Jones and G. Roberts, *J. Am. Chem. Soc.* **80**, 6121 (1958); (k) cf.

axial alcohols. An evaluation of the O—H stretching vibration as a function of its location has also been carried out with triterpenoid alcohols.[52h,53] Primary, secondary, *axial*, and *equatorial*, as well as tertiary alcohols show characteristic absorption frequencies. The absorption for *axial* alcohols lies between 3635 and 3638 cm.$^{-1}$, for *equatorial* alcohols between 3628 and 3630 cm^{-1}. when the hydroxyl is at C-3, and between 3623 and 3635 cm.$^{-1}$ when the hydroxyl is at C-11. Tertiary alcohols show OH stretching vibrations in the range between 3613 and 3618 cm.$^{-1}$, but a distinction between an *axial* and an *equatorial* position is not yet possible with certainty.

As has been indicated (page 113), the difference in frequencies between *axial* and *equatorial* OH stretching vibrations is decidedly smaller than the differences in the C-(OH) stretching vibrations. Nevertheless, the former is frequently used in conformational elucidation because the assignment is much simpler, as noted (page 113) in reference to the difficulties in assigning the C—(OH) stretching vibration for the structurally simpler decalols.

The presence of an intramolecular hydrogen bond can be determined from the position of the OH stretching vibration (cf. the cyclohexanediols, page 123). Conclusions as to the preferred conformations are possible from such measurements. It could be shown from a number of steroid and triterpene examples that neighboring *diaxial* groups (OH or COOR) are too far apart to form hydrogen bonds. However, *diequatorial* or *equatorial-axial* substituents form hydrogen bonds whose strength depends on the type of group and its position.[53b,54]

Various halohydrins of steroids have also been studied.[55] In the molecules where an intramolecular hydrogen bond is possible due to the relative positions of the halogen and hydroxyl groups, the OH stretching vibration is noticeably lowered ($\sim$25–48 cm.$^{-1}$). If no hydrogen bond is possible between the two neighboring groups, then the observed shift is negligible.

The infrared spectra of halogenated steroids also show a clear relation-

---

also K. Dobriner, E. R. Katzenellenbogen, and R. N. Jones, "Infrared Absorption Spectra of Steroids—An Atlas." Wiley (Interscience), New York, 1953. G. Roberts, B. F. Gallagher, and R. N. Jones, "Infrared Absorption Spectra of Steroids—An Atlas." Vol. II. Wiley (Interscience), New York, 1958; (l) cf. also F. Dalton, G. D. Meakins, J. H. Robinson, and W. Zaharia, *J. Chem. Soc.* p. 1566 (1962).

[53] (a) A. R. H. Cole, G. T. A. Müller, D. W. Thornton, and R. L. S. Willix, *J. Chem. Soc.* p. 1218 (1959); (b) A. R. H. Cole and G. T. A. Müller, *ibid.* p. 1224 (1959).

[54] See also R. N. Jones, P. Humphries, F. Herling, and K. Dobriner, *J. Am. Chem. Soc.* **74**, 2820, 6319 (1952); H. B. Henbest and B. J. Lovell, *J. Chem. Soc.* p. 1965 (1957).

[55] A. Nickon, *J. Am. Chem. Soc.* **79**, 243 (1957); cf. also N. A. LeBel and R. F. Czaja, *J. Org. Chem.* **26**, 4768 (1961).

ship between *equatorial* and *axial* orientation of the halogen atom. The 2-, 3-, 5-, 6-, and 7-chloro- and bromosteroids of known configuration[56,57] were studied. The stretching vibration of an *equatorial* carbon-halogen bond was found to lie at higher frequencies than the corresponding *axial* carbon-halogen bond. The *equatorial* carbon-chlorine stretching vibration for 2-chlorosteroids occurs at 755 cm.$^{-1}$, for the 3-chlorosteroids at 782 cm.$^{-1}$, and for the 7-chlorosteroids at 749 cm.$^{-1}$. The corresponding *axial* isomers absorb at 693, 730–617, and 588 cm.$^{-1}$, respectively. The *equatorial* 2-bromosteroids absorb at 754–708 cm.$^{-1}$, 3-bromosteroids at 708–704 cm.$^{-1}$; their corresponding *axial* epimers show bonds at 662 and 692–591 cm.$^{-1}$.

A further relationship exists between the position of the carbonyl stretching bands and the *axial* or *equatorial* orientation of the halogen atoms (with the exception of fluorine) in α-halogenated steroid ketones. The carbonyl band is shifted to higher wave numbers by an *equatorial* halogen by 16–21 cm.$^{-1}$ (non-halogenated ketones ~ 1713–1719 cm.$^{-1}$). An *axial* halogen has practically no influence on the position of the carbonyl frequency.[58]

The situation is reversed for the ultraviolet spectra of α-halogenated steroid ketones. For the nonsubstituted carbonyl compounds, the carbonyl maximum lies at about 280–286 mμ. A slight hypsochromic effect (~5 mμ for bromine) is produced by an adjacent *equatorial* halogen substituent. However, an *axial* halogen shifts the ketone band to a distinctly longer wavelength (~28 mμ for bromine), with a simultaneous and noticeable increase in the extinction coefficient.[58e,59] The ultraviolet carbonyl maximum of steroid ketones is also distinctly influenced by a neighboring hydroxyl or acetoxyl group. An *axial* hydroxyl group brings about a bathochromic shift of 14 to 20 mμ. An *equatorial* hydroxyl group causes a hypsochromic shift of about 9–13 mμ. This has been investigated for several steroid ketols of the cholestane, bile acid, and cardiac aglycone series.[59b,60]

Axial acetoxy groups produce a bathochromic shift of 10 mμ; *equatorial*

---

[56] D. H. R. Barton, *Experientia* Suppl. II, p. 121 (1955).

[57] D. H. R. Barton, J. E. Page, and C. W. Shoppee, *J. Chem. Soc.* p. 331 (1956).

[58] (a) R. N. Jones, D. A. Ramsay, F. Herling, and K. Dobriner, *J. Am. Chem. Soc.* **74**, 2828 (1952); (b) R. N. Jones and F. Herling, *J. Org. Chem.* **19**, 1252 (1954); (c) E. G. Cummins and J. E. Page, *J. Chem. Soc.* p. 3847 (1957); Cf. also: (d) E. J. Corey, *J. Am. Chem. Soc.* **76**, 175 (1954); (e) D. N. Kirk and V. Petrov, *Tetrahedron* **12**, 95 (1961); (f) cf. also S. Inayama, *Pharm. Bull.* (*Tokyo*) **4**, 198 (1956).

[59] (a) R. C. Cookson, *J. Chem. Soc.* p. 282 (1954); (b) R. C. Cookson and S. H. Dandegaonker, *ibid.* p. 352 (1955); compare also R. C. Cookson and C. Hudec, *J. Chem. Soc.* p. 429 (1962).

[60] (a) G. Baumgartner and C. Tamm, *Helv. Chim. Acta.* **38**, 441 (1955); (b) J. Elks, G. H. Phillips, T. Walker, and L. J. Wyman, *J. Chem. Soc.* p. 4330 (1956).

acetoxy groups produce a hypsochromic shift of 5 mμ. Thus, the effect is in the same direction, though less pronounced.[59b,60,61]

Optical rotatory dispersion measurements represent a frequently used physical method for the solution of stereochemical problems, especially in steroids. The variation of rotatory power of optically active compounds with wavelength is measured. The characteristic shapes of rotatory dispersion curves, such as those of ketosteroids, are mainly determined by the spatial arrangements of substituents in the neighborhood of the carbonyl group. Empirical relations between the shapes of rotatory dispersion curves and the structural and conformational properties of the molecules were derived mainly by Djerassi and his group, by recording a large number of rotatory dispersion curves of various compounds. Further details of this elegant method lie beyond the scope of this book (see also page 280). A number of sources are available which go into greater detail.[62]

In comparison to other physical methods, nuclear magnetic resonance spectroscopy has been used only recently in stereochemical-analyses. As a result, only a limited amount of material concerning conformational analysis is available at present.[63]

5. Conformation and Reactivity

As has been stressed in previous chapters (see especially Chapter 3, page 117), a chemical conformational analysis of cyclohexane derivatives

[61] Cf. also C. Djerassi, O. Halpern, V. Halpern, O. Schindler, and C. Tamm, *Helv. Chim. Acta.* **41**, 250 (1958).

[62] See especially the comprehensive review by C. Djerassi, ref. 20 in Chap. 2; also there are C. Djerassi, *Record. Chem. Progr.* (*Kresge-Hooker Sci. Lib.*) **20**, 101 (1959) with further references; C. Djerassi, *Endeavour* **20**, 138 (1961); C. Djerassi, *Tetrahedron* **13**, 13 (1961); cf. also W. Klyne, ref. 68a in Chap. 4; W. Klyne, *in* "Advances in Organic Chemistry" (R. A. Raphael, E. C. Taylor, and A. Wynberg, eds.), Vol. 1, p. 239. Wiley (Interscience), New York, 1960; W. Klyne, *Tetrahedron* **13**, 29 (1961); G. G. Lyle and R. E. Lyle, *in* "Determination of Organic Structures by Physical Methods," (F. C. Nachod and W. D. Phillips, eds.), Vol. 2. Academic Press, New York, 1962.

[63] See for example J. N. Shoolery and M. T. Rogers, *J. Am. Chem. Soc.* **80**, 5121 (1958); G. Slomp, Jr. and B. R. McGarvey, *ibid.* **81**, 2200 (1959); G. Slomp, Jr. and F. McKellar, *ibid.* **82**, 999 (1960); J. S. G. Cox, E. O. Bishop, and R. E. Richards, *J. Chem. Soc.* p. 5118 (1960); R. F. Zürcher, *Helv. Chim. Acta* **44**, 1380 (1961); R. A. Y. Jones and A. R. Katritzky, *Chem. & Ind.* (*London*) p. 522 (1962); A. Nickon, M. A. Castle, R. Harada, C. E. Berkhoff, and R. O. Williams, *J. Am. Chem. Soc.* **85**, 2185 (1963); R. J. Abraham and J. S. E. Holker, *J. Chem. Soc.* p. 806 (1963); cf. A. D. Cross and I. T. Harrison, *J. Am. Chem. Soc.* **85**, 3223 (1963); cf. J. D. Roberts, "Nuclear Magnetic Resonance, Applications to Organic Chemistry." McGraw-Hill, New York, 1959; L. M. Jackman, "Applications of N. M. R. Spectroscopy in Organic Chemistry." Pergamon Press, London 1959; J. A. Pople, W. G. Schneider, and H. J. Bernstein, "High-Resolution Nuclear Magnetic Resonance." McGraw-Hill, New York, 1959.

with simple structure does not lead to unequivocal results. Hückel and co-workers had been able to show this in extensive investigations concerned with the solvolysis reaction of toluenesulfonates of variously substituted cyclohexanols and the resulting substitution and elimination reactions.[64] Straightforward relationships between conformation and reactivities of such compounds were difficult to observe[65] (see also Chapter 4, page 182f., conformation and reactivity of the five-membered ring). However, other reactions, such as the saponification rates of carboxylic acid esters of cyclohexanols of simple structure, at least showed qualitative agreement between the conformation and the reaction rate.

A series of other reactions, which as a result of their stereospecificity would seem to allow predictions concerning the conformation of the resulting compounds, gave uncertain results when applied to monocyclic compounds.[66] As discussed in Chapter 4, these restrictions are also required for the isomeric decalin derivatives. Still, derivatives of *trans*-decalin formed from two rigid six-membered rings, were more accessible to a conformational analysis than derivatives of *cis*-decalin.

The reasons for a cautious appraisal of conformational analyses stemming from chemical reactions, or the relationships between conformations and reactivities of such compounds, have been frequently reiterated. A conformational analysis by physical methods does not alter the conformation of the molecule. However, during a chemical reaction whose activation energy is larger than the energy barrier between the individual conformations, the possibility is introduced that the molecule reacts by way of a less stable conformation. Conclusions about the more exact steric structure of the molecules on the basis of chemical reactions also depend upon an exact knowledge of the reaction mechanism. In many cases this knowledge is not sufficient and conclusions about the conformation in the ground state cannot be made. A conformational analysis of simple compounds which are capable of assuming a number of conformations, even though such analysis is not impossible, must therefore be viewed critically.

At least one of the two difficulties encountered in a chemical conformational analysis can be mostly disregarded for steroids and the related triterpenes. As we have seen in Section 1 of this chapter, they are more easily treated from a conformational viewpoint. As a result of their rigid structure, the possibility of forming various conformations is greatly limited.

---

[64] See for example W. Hückel *et al.*, *Ann. Chem.* **624**, 142 (1959).

[65] Cf. also Ya. I. Goldfarb and L. I. Belenkii, *Usp. Khim.* p. 425 (1960); W. Masschelein, *Ind. Chim. Belge* **26**, 613 (1961).

[66] Compare for example: W. Hückel, M. Maier, E. Jordan, and W. Seeger, *Ann. Chem.* **616**, 46 (1958).

Hence it will be easier to recognize relationships between the conformation and the reactivity. In principle every chemical reaction may be used for a conformational analysis of the reaction center considered. This is carried out on the basis of an accurate knowledge of the reaction mechanism under the assumption mentioned above. Or, conversely, during the elucidation of the stereochemical feature of a reaction mechanism the conformation of the molecule must be considered. Even though the connection between conformation and reactivity has been considered at different times in the preceding chapters, this problem will be attacked once more in connection with the steroids.

In spite of the reasons mentioned above, we shall not limit ourselves only to examples of the steroids for explaining the reactions. Examples from simpler cyclic compounds will also be used. Characteristic examples will be cited in which conformation and reactivity considerations are valid.

Such relationships have been recognized by Barton and were set up as rules in his basic discussions of conformational analysis (see especially refs. 68, 69, 70). Despite the limitations indicated above for monocyclic compounds, insight into the stereochemistry of these compounds has been possible using these rules. If viewed with proper caution, they lead to an understanding of finer points of molecular structure.

a. Elimination Reactions

(*i*) *E2 Elimination*

As has already been shown during the consideration of the conformation of the five-membered ring (see page 81) bimolecular elimination reactions (type E2) proceed, with the most facility, when the four atoms involved in the reaction are located in one plane in the transition state.[67] The leaving groups are found in a coplanar position in cyclohexane compounds when they are *trans-diaxial.* If the substituents are *trans*, but *diequatorial*, or *cis* (a, e), then a transition state in which the four reaction centers lie in one plane is no longer possible. The elimination reactions then proceed relatively more slowly.

A few characteristic examples will be mentioned.

It could be shown[68,69,70,71] that during the debromination of steroid

[67] Cf. the literature summary on page 81 and also S. J. Cristol, J. Q. Weber, and M. C. Brindell, *J. Am. Chem. Soc.* **78**, 598 (1956); see also: C. Ingold, *Proc. Chem. Soc.* p. 265 (1962); J. F. Bunnet, *Angew. Chem.* **74**, 731 (1962).

[68] D. H. R. Barton, *Experientia* **6**, 316 (1950).

[69] D. H. R. Barton, *J. Chem. Soc.* p. 1027 (1953).

[70] D. H. R. Barton and R. C. Cookson, *Quart. Rev.* (*London*) **10**, 44 (1956).

[71] (a) D. H. R. Barton and E. Miller, *J. Am. Chem. Soc.* **72**, 1066 (1950); (b) D. H. R. Barton and J. Rosenfelder, *J. Chem. Soc.* p. 1048 (1951); (c) G. H. Alt and D. H. R. Barton, *ibid.* p. 4284 (1954).

dibromides with iodide ion, characteristic differences in the reaction rate exist: Of the two possible 5,6-dibromocholestan-3β-yl benzoates, one reacts very rapidly with potassium iodide and undergoes debromination. This is the "normal" cholesteryl dibromide benzoate which is initially formed during the bromination of cholesterol and subsequent benzoate formation. The isomer obtained by rearrangement in a suitable solvent reacts very slowly with potassium iodide.[71a]

The rapid reaction rate of the normal cholesteryl dibromide benzoate indicates a *diaxial* orientation of the bromide atoms on the basis of the above-mentioned transition state. Therefore, the normal cholesteryl dibromide benzoate is the 5α, 6β-dibromide XLIII whereas, the stable isomer has the bromine atoms in the 5β, 6α positions (XLIV) (cf. Table I, page 221).

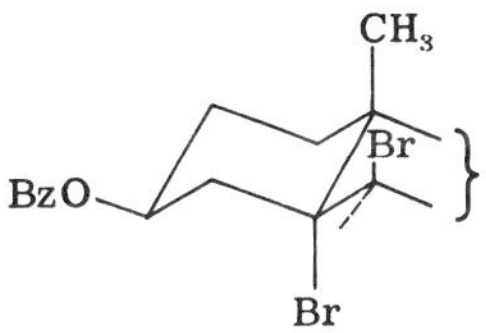

(XLIII)

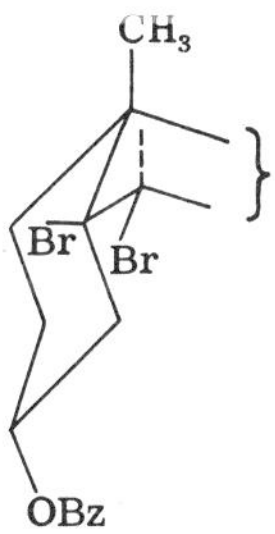

(XLIV)

This example can be supplemented by many additional examples in the steroid series. They demonstrate that the elimination of *diaxial* substituents is definitely preferred.

Table III summarizes the compounds, compiled by Barton,[68,69,70,71] which confirm the rule of preferred *diaxial* elimination.

This rule has also been applied to monocyclic compounds. (See page 81). From the rapid elimination rate of bromine with iodide ion in *trans*-1,2-dibromocyclopentane, a coplanar position of the bromine atoms was deduced and a conclusion was arrived at concerning the conformation of the five-membered ring.

The preferred E2 *trans* elimination has also been used for the determination of the configurations and conformations of simple cyclohexane derivatives. Only a few examples will be mentioned[72]:

The reactions of menthyl chloride (XLV) and neomenthyl chloride

[72] Cf. in more detail, D. S. Cram, *in* "Steric Effects in Organic Chemistry." (M. S. Newman, ed.), Chap. 6. Wiley, New York, 1956.

TABLE III[a]

Diaxial position of the substituents—elimination facilitated	*Diequatorial* position of the substituents—elimination difficult
Rings A/B *trans*	
2β,3α-dichloro-	2α,3β-dichloro-
2β,3α-dibromo-	2α,3β-dibromo-
3α-OTs and 2β or 4β-H	3β-OTs and 2α or 4α-H
3α,4β-dibromo-	3β,4α-dibromo-
5α,6β-dibromo-	5β,6α-dibromo-
Rings A/B *cis*	
3β-OTs and 2α-H	3α-OTs and 2β-H
11β,12α-dibromo-	11α,12β-dibromo
Δ^4-3-one series	
6β-H and 7α-OH	6α-H and 7β-OH

[a] Cf. Table I, page 221.

(XLVI) proceed in two different directions[73] under the conditions of an E2 reaction with sodium ethoxide. Neomenthyl chloride, with *axial* chlorine, yields a mixture of Δ^3-menthene and Δ^2-menthene in the ratio of 3:1. Menthyl chloride yields practically only Δ^2-menthene.

The formation of menthenes from menthyltrimethyl and neomenthyltrimethyl ammonium hydroxides via an E2 elimination gives practically the same result[73,74]: menthyl trimethyl ammonium hydroxide yields ~100% Δ^2-menthene. The isomeric neomenthyl trimethyl ammonium hydroxide yields ~30% Δ^2-menthene and ~60% Δ^3-menthene. A discussion of the mechanism of the reaction will not be included; only the stereochemical assumptions will be considered.[74,b,c]

For simplification consider again the isomeric menthyl chlorides. The facile formation of unsaturated hydrocarbon from neomenthyl chloride (XLVI) is readily understood without additional assumptions. Two *axial* hydrogens on carbons 2 and 4 are located opposite the *axial* chlorine at position 3. In both cases a coplanar *trans* elimination is possible and Δ^3-menthene predominates, in agreement with the Saytzeff rule.

[73] (a) W. Hückel, W. Tappe, and G. Legutke, *Ann. Chem.* **543**, 191 (1940); (b) E. D. Hughes, C. K. Ingold, and R. B. Rose, *J. Chem. Soc.* p. 3839 (1953).

[74] (a) N. L. McNiven and J. Read, *J. Chem. Soc.* p. 153 (1952); (b) A detailed discussion of the mechanism is found in E. D. Hughes and J. Wilby, *J. Chem. Soc.* p. 4094 (1960); T. H. Brownlee and W. H. Saunders, Jr., *Proc. Chem. Soc.* p. 314 (1961); (c) Cf. also D. Y. Curtin, R. D. Stolow, and W. Maya, *J. Am. Chem. Soc.* **81**, 3330 (1959).

The exclusive formation of Δ^{2}-menthene from menthyl chloride (XLV), cannot be explained without an additional assumption. Only on carbon 2 is there a hydrogen in *trans-equatorial* position. In order to assume the necessary coplanar position with the similarly *equatorial* chlorine in position 3, the molecule must flip into another conformation.

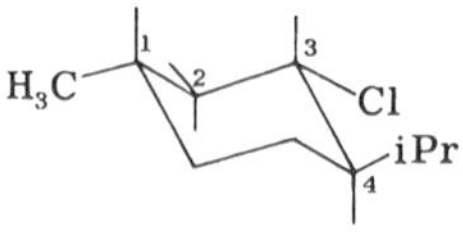

(XLV)

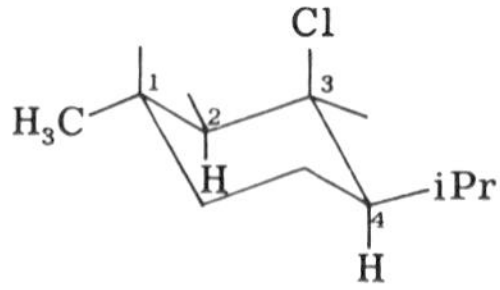

(XLVI)

The chlorine would then be in an *axial* position and coplanar to the axial hydrogen at position 2. The requirements for coplanarity of the reaction centers involved are thus met. The conversion of menthyl chloride from one conformation into the other has one consequence. The coplanar conformation represents an energetically unfavorable conformation. The isopropyl group and the methyl group must then be in *axial* positions, which results in an increase in energy, as set forth in Chapter 3 (page 131). A study of the model or an estimate of the necessary energy, however, makes this seem quite possible. Because of the high activation energy for reaction, the molecule may react by way of this conformation. Similar considerations can also be applied to other, reactions of menthyl and neomenthyl derivatives or similarly constructed monocyclic cyclohexane derivatives.

A few examples from the decalin series will also be considered briefly. Difficulties have arisen in the interpretation of the results of E2 elimination reactions. The toluenesulfonates of decalols were subjected to a elimination with sodium isopropoxide under the conditions of an E2 reaction by W. Hückel and co-workers.[75] The following results were obtained:

The toluenesulfonate of *cis-cis*-1-decalol, [*equatorial* OH-group, see (XV), page 188] yielded practically pure $\Delta^{1(9)}$ octalin.

The toluenesulfonate of *trans-cis*-1-decalol, [*axial* OH group, see (XVII), page 188] produced 90% unsaturated hydrocarbon, which was composed of 67% $\Delta^{1(2)}$-; 29% $\Delta^{1(9)}$-, and 4% $\Delta^{9(10)}$-octalin. The remaining was 10% *trans-cis*-1-decalol.

The toluenesulfonate of *trans-trans*-1-decalol, [*equatorial* OH group, see (XVIII), page 188] gave 58% unsaturated hydrocarbon, consisting

[75] W. Hückel, D. Maucher, O. Fechtig, J. Kurz, M. Heinzel, and A. Hubele, *Ann. Chem.* **645**, 115 (1961).

of 77% $\Delta^{1(2)}$-octalin, 19% $\Delta^{1(9)}$-octalin, and 4% $\Delta^{9(10)}$-octalin. The remaining 42% was *trans-trans*-1-decalol.

The interpretation of the results provides no difficulties in the case of *cis-cis*-1-decalyl toluenesulfonate and *trans-cis*-1-decalyl toluenesulfonate. The exclusive formation of $\Delta^{1(9)}$-octalin from the former can be easily understood for a reaction of the toluenesulfonate group in an *axial* position. (The E2 elimination produces the same results in this case as an El solvolysis of the tosylate).[73a] As has already been shown in the treatment of the decalols (page 185), the *equatorial* OH (or toluenesulfonate group) of *cis-cis*-1-decalol can easily be brought to the *axial* position by conversion of both rings. Then, as can be seen on a model, a coplanar *trans* arrangement arises with the tertiary hydrogen at C-9. The predominant formation of hydrocarbon from *trans-cis*-1-decalyl toluenesulfonate also results from the *axial* location of the toluenesulfonic ester group. An elimination according to coplanar conditions is possible with the tertiary hydrogen at C-9 as well as with the *axial* hydrogen at C-2. $\Delta^{1(2)}$ and $\Delta^{1(9)}$-octaline are formed. The orientation of the E2 eliminations ($\Delta^{1(2)}$-octalin and $\Delta^{1(9)}$-octalin are not formed in the same amounts) depends on other factors which will not be discussed here.[76]

A simple explanation for the hydrocarbon formation is not possible for *trans-trans*-1-decalyl toluenesulfonate according to the usual coplanar requirements of an E2 reaction. The OH group (or toluenesulfonate group) is *equatorial* (see XVIII, page 188) and an interconversion of the molecule is not possible because of the rigid *trans* juncture of both cyclohexane rings (cf. Chapter 4). A coplanar *trans* arrangement of the toluenesulfonic ester group with the neighboring hydrogens does not seem probable.

The relatively high amount of hydrocarbon formed can be explained by disregarding the coplanar *trans* position of the atoms involved as a necessary provision, or by using another explanation: At least one coplanar *trans* arrangement can be reached for *trans-trans*-1-decalyl toluenesulfonate if the molecule can assume another conformation in the transition state. The rigid model of the two-chair form (XVIII, page 188) of *trans*-decalin must be disregarded and one ring must assume a conformation not corresponding to the chair form. By "flip-

[76] The reason that some $\Delta^{9(10)}$-octalin is also produced from the toluenesulfonates of *cis-cis*-1-decalol and *trans-cis*-1-decalol is due to the fact that an E1 elimination by way of a carbonium ion takes place to a small extent as a side reaction. This side reaction cannot be surpressed completely. A rearrangement of $\Delta^{1(9)}$- to $\Delta^{9(10)}$-octalin is not expected from the action of an alkoxide. The formation of the decalol with the same configuration can be explained by the cleavage of the O—S bond of the toluenesulfonate through the action of alkoxide.

ping" up carbon atom 2, we can produce a true *trans* position for one of the hydrogens on this carbon and, thereby, bring the toluenesulfonic acid residue, C-1, C-2, and hydrogen into one plane. However, the formation of $\Delta^{1(9)}$-octalin cannot be explained by this process, because a coplanar *trans* arrangement with the tertiary hydrogen on C-9 is not possible under any conditions.

The example of *trans-trans*-1-decalol, described here in detail, represents one of the cases which cannot be incorporated into the scheme of a coplanar transition state for the E2 elimination without additional assumptions. Aside from this other examples have been found in which a coplanar *trans* arrangement of the atoms approaching elimination is not possible because of the rigid structure of the molecule: *endo-cis*-2,3-dichloronorbornane (XLVII) and *trans*-2,3-dichloronorbornane (XLVIII) were studied[77] with regard to the E2 elimination of HCl with sodium amylate.

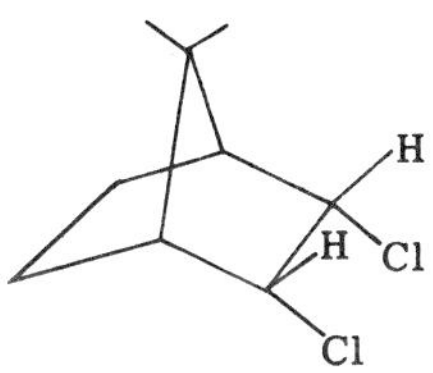

(XLVII)

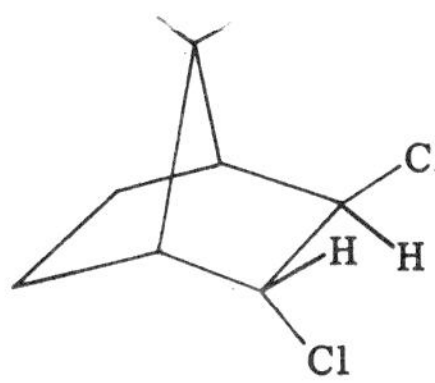

(XLVIII)

The flexibility of the cyclohexane system (existing here in the boat form) is restrained almost entirely by the CH_2 bridge (see also page 286). However, both chlorides are able to take part in a bimolecular elimination reaction, although the reaction rate is low. The *trans*-2,3 compound (XLVIII, hydrogen and chlorine in a *cis*-relationship) reacts a little faster than *cis*-2,3-dichloronorbornane (XLVII).

As a result of the limited flexibility of the system, it is not possible that an elimination can occur from the coplanar *trans* position. Nevertheless, because an E2 elimination is observed, steric factors seem to be less important in this case than the reaction mechanism. It is not relevant to go into this question at this time, but the possibility of a modified E2 elimination mechanism has been discussed[77] in which a carbanion of short lifetime is formed (carbanion mechanism, E1cb). The E2 elimination process would then not be the usual one-step concerted process, but rather the rate-determining step would be the formation of a carbanion by abstraction of the β-hydrogen with the base. Immediately

[77] S. J. Cristol and E. F. Hoegger, *J. Am. Chem. Soc.* **79**, 3438 (1957). For further examples see S. J. Cristol and N. L. Hause, *ibid.* **74**, 2193 (1952); S. J. Cristol and R. P. Arganbright, *ibid.* **79**, 3441 (1957).

thereafter, the chlorine would be eliminated as an anion. A mechanism of this type would no longer be bound to a *trans* coplanar transition state to the degree that it was in a one-step process (cf. the E1 elimination below). The provision for this possibility is that the concerted elimination must follow from a coplanar *trans* geometry. More recent investigations seem to indicate that a coplanar *cis* elimination is also possible. An E2 elimination would be especially preferred if the dihedral angle between the substituents to be eliminated equals 0° or 180°. (XLVIII) would therefore react more rapidly than (XLVII) because a coplanar *cis* position exists in (XLVIII).[78]

In this connection the possibility of *cis*-elimination (see page 246) can be recalled. With *trans*-2-*p*-toluenesulfonylcyclohexyl tosylate (XLIX), the β-hydrogen is made so acidic by the electron-withdrawing sulfone group that *cis*-elimination is preferred over the conformationally possible *trans* process. The conjugated sulfone (LI) is produced instead of the expected product (L).

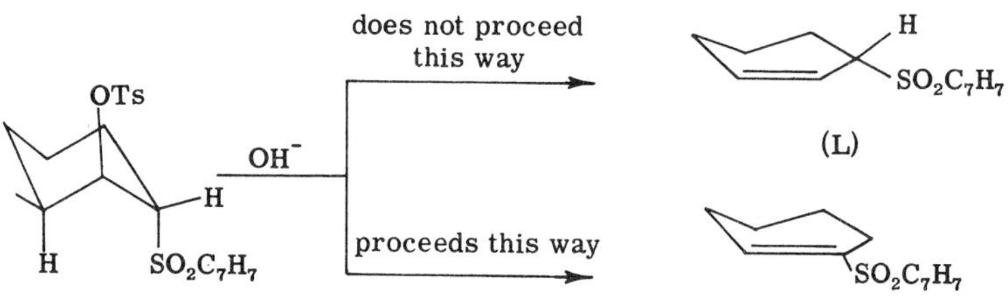

(XLIX) (L) (LI)

With these cited exceptions[80] to the normal E2 process, the problem stated earlier can now be recognized. In the examples reported, a planar transition state was regarded as a requirement for the elimination process. This was the basis of an assumption and can be circumvented involving a different mechanism of reaction brought about by other effects.

[78] C. H. DePuy, R. D. Thurn, and G. F. Morris, *J. Am. Chem. Soc.* **84**, 1314 (1962); compare also: N. A. LeBel, P. D. Beirne, E. R. Karger, J. C. Powers and P. M. Subramanian, *J. Am. Chem. Soc.* **85**, 3199 (1963).

[79] See for example F. G. Bordwell and R. J. Kern, *J. Am. Chem. Soc.* **77**, 1141 (1955); F. G. Bordwell and M. L. Peterson, *ibid.* **77**, 1145 (1955); J. Weinstock and F. G. Bordwell, *ibid.* **77**, 6706 (1955); J. Weinstock, R. G. Pearson, and F. G. Bordwell, *ibid.* **78**, 3468 (1956).

[80] For further examples, see ref. 72; cf. also S. J. Cristol and F. R. Stermitz, *J. Am. Chem. Soc.* **81**, 4692 (1960). Compare also S. Winstein, D. Darwish, and N. J. Holness, *ibid.* **78**, 2915 (1956) for the so-called "merged mechanism of substitution and elimination;" E. L. Eliel and R. Haber, *ibid.* **81**, 1249 (1959).

(ii) E1 Eliminations

In contrast to the E2 reaction, it is more difficult to find correlations of reactivity and conformation in elimination reactions proceeding via the E1 process. The more complicated mechanism and the ionic transition state involved make it more difficult to reach a conclusion as to the geometry of the molecule.

Although it was assumed, on the basis of a series of earlier investigations, that the E1 reaction depended on steric control[73a,81] in which a *trans* elimination by way of a coplanar transition state was favored, recent investigations with cyclohexane and decalin derivatives have shown that unequivocal relations with the stereochemistry of the molecule are harder to find.[64,75] Systematic investigations on rigid ring systems are not available; therefore, only examples from the classes of compounds named above will be considered.

A systematic investigation of the E1 hydrocarbon formation from solvolysis (ethanolysis)[82] of toluenesulfonates of stereoisomeric alkylcyclohexanols gave the following result[64,75]: The earlier point of view[69] (see above), that E1 eliminations are also predominantly *trans* oriented can be modified to a certain extent. *Trans* eliminations occur frequently during an E1 reaction, but they are not always required. Consequently the question arises as to whether a certain geometry of the molecule is necessary for the formation of an unsaturated hydrocarbon during an E1 solvolysis.[83] Even if the coplanar transition state (TsO—C—C—H) were a necessary assumption for an E1 elimination of the tosylate, this would only indicate that the toluenesulfonate group in *trans*-2-alkyl

[81] (a) Cf. also W. Hückel, O. Neunhoeffer, A. Gercke, and E. Frank, *Ann. Chem.* **477**, 131 (1930); (b) G. Vavon and M. Barbier, *Bull. Soc. Chim. France* [4] **49**, 567 (1931); (c) W. Hückel, *Ber.* **77**, 805 (1944); (d) M. L. Dhar, E. D. Hughes, C. K. Ingold, A. M. M. Mandour, G. A. Maw, and L. I. Woolf, *J. Chem. Soc.* p. 2093 (1948).

[82] Hückel and co-workers[64] were able to show that there were no fundamental differences between ethanolysis and acetolysis.

[83] This may be clarified by considering the different mechanisms of the two reactions (E1 and E2): The attack of alkoxide or another proton acceptor at the *trans* hydrogen of the β carbon atom initiates an E2 elimination. This corresponds to attack of iodide ion during the elimination of bromine from dibromides (see VIII, page 81). During the solvolysis of toluenesulfonates (E1), attack of the solvent takes place on the tosylate group as well as from the opposite side of the α carbon (on which the toluenesulfonate group, is attached). No special position seems to be required for the proton removed from the β carbon for this process. The idea of *axial* β-hydrogen participation developed by Winstein [cf. especially S. Winstein, B. K. Morse, E. Grunwald, H. W. Jones, J. Corse, D. Trifan, and H. Marshall, *J. Am. Chem. Soc.* **74**, 1127 (1952) and later investigations] is not considered a necessary requirement in the E1 elimination of an axial tosylate group by certain authors. [See refs. 64, 75; H. L. Goering and R. L. Reeves, *J. Am. Chem. Soc.* **78**, 4931 (1956).]

cyclohexyl toluenesulfonates would have to react by way of an *axial* position as in E2 eliminations.

This is theoretically possible because the molecule can convert *trans*-1,2 substituents from *diequatorial* (LII) into the *diaxial* conformation (LIII).

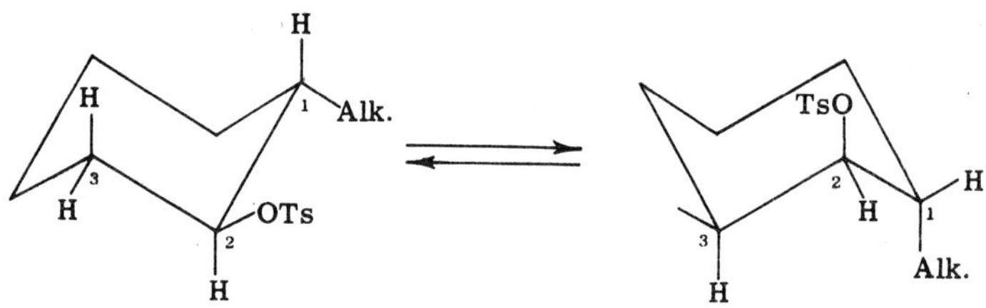

(LII) (LIII)

However, if the toluenesulfonate group is *axial* (LIII), then the *axial* valence at C-1 is occupied by an alkyl group and that at C-3 by a hydrogen which is available for elimination. If the elimination were to proceed in a *diaxial* manner, then the unsaturated hydrocarbons formed would be pure Δ^2-cycloolefins. This is true for E2 eliminations. However, Δ^1- and Δ^2-cycloolefins[64] are formed simultaneously in E1 reactions as shown in Table IV.

TABLE IV

CYCLOOLEFINS (%) FROM SOLVOLYSIS OF *Trans*-2-ALKYL TOLUENESULFONATES[a],[b]

Cycloolefin	Alkyl group					
	CH_3	$CH(CH_3)_2$	Menthol	Isomenthol	*tert*-Butyl	Cyclohexyl
Δ^1	32	48	—	—	53	51
Δ^2	68	52	32	94	47	49
Δ^3	—	—	68	6	—	—

[a] See ref. 64. [b] See ref. 84, below.

It can be seen from Table IV that only isomenthol yields practically homogeneous Δ^2-olefin.

For comparison, hydrocarbon formation by the E1 mechanism is also tabulated[75] for solvolyses (methanolysis) of three decalyl toluenesulfonates (Table V).

[84] The percentages give the relative composition of the cycloolefin mixtures. In addition to hydrocarbons, ethers are formed during the solvolysis by substitution (ethanolysis).

TABLE V

SOLVOLYSIS OF DECALYL TOLUENE SULFONATES[a]

Toluenesulfonate of	Conformation of alcohol	Total octalins (%)	Octalins (%)		
			$\Delta^{1(9)}$	$\Delta^{9(10)}$	$\Delta^{1(2)}$
cis-cis-1-decalol	e	90	91	9	—
trans-cis-1-decalol	a	76	75	24	1
trans-trans-1-decalol	e	19	29	10	61

[a] See ref. 85 below.

In all three cases a shift of the double bond to produce $\Delta^{9(10)}$-octalin takes place. *Trans-trans*-1-decalyl toluenesulfonate reacts the slowest due to its *equatorial* tosylate group, and yields only 19% hydrocarbon. The same conclusions regarding the necessity of maintaining a desired geometry of the reaction centers are reached for both E1 and E2 (see page 239) eliminations of this compound.

The question of the stereochemistry during the E1 elimination has been discussed by Winstein and Holness[86] for the conformationally pure *cis*- and *trans*-4-*tert*-butylcyclohexanyl toluene sulfonate, for which a definite geometry of the transition state was not disregarded. The solvolysis of these toluenesulfonates yields mainly Δ^3-*tert*-butylcyclohexene. Depending on the solvent, 83–87% is obtained from the *cis* isomer (*axial* tosyl group) and 67-76% from the *trans* isomer (*equatorial* tosyl group). Substitution reactions take place, concurrently, and proceed with practically complete Walden inversion for the *trans* isomer, and partial inversion for the *cis* isomer. Because of the possible *diaxial* relationship of the tosyl group and the β-hydrogen, olefin formation is favored with the *cis*-toluene sulfonate.

The *trans*-isomer (LIV) is a pure e form; it has no β-hydrogens in a suitable position. With the approach of a solvent molecule from the opposite side, the bond to the *equatorial* toluenesulfonate group is stretched to generate a carbonium ion intermediate.[87] This occurs to the extent that the C-1–OTs valence is almost parallel to the *axial* C—H

[85] In addition to hydrocarbon, decalyl methyl ethers are also formed. Varying amounts of the ether of the same configuration, the ether of opposite configuration, formed by a Walden inversion, and the tertiary ether occur.

[86] S. Winstein and N. J. Holness, *J. Am. Chem. Soc.* **77**, 5562 (1955).

[87] Cf. S. Winstein and D. Trifan, *J. Am. Chem. Soc.* **74**, 1147 (1952); (b) S. Winstein, E. Clippinger, A. H. Fainberg, and G. C. Robinson, *Chem. & Ind.* (*London*) p. 664 (1954).

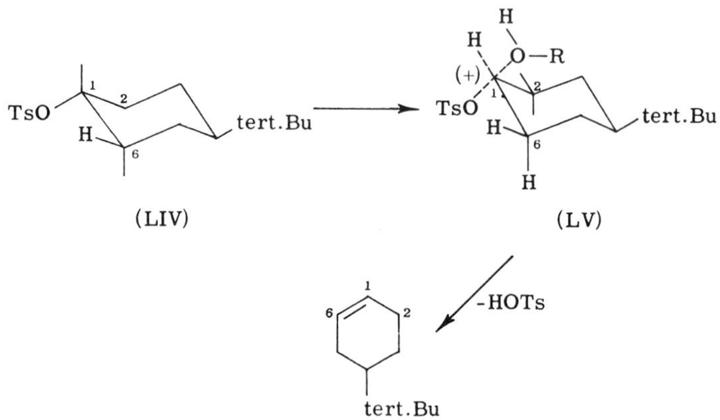

valences on C-2 and C-6 (LV). Elimination from the *equatorial* position in two directions can be followed from this illustration.

The few examples mentioned here demonstrate the difficulties in assuming a steric preference for an E1 elimination in cyclic compounds.

(*iii*) *Intramolecular Elimination*[88]

Further reactions which show a high degree of stereospecificity are the pyrolytic decompositions of xanthates (Tschugaeff reaction)[88c] and carboxylates. The "Cope" reaction,[88b] in which the oxides of tertiary amines are decomposed to olefins and a derivative of hydroxylamine, also proceeds unimolecularly, and, like the decomposition of the esters proceeds by way of a cyclic transition state (LVI, LVII, and LXIX).[89]

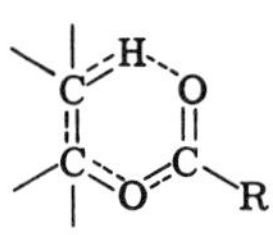

(LVI)

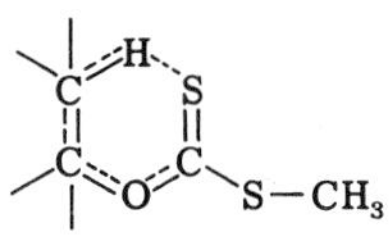

(LVII)

Hückel *et al.*[73a] first demonstrated that pyrolytic decomposition of menthyl and decalyl xanthates proceeds mainly in the direction of the

[88] (a) Cf. the summary by C. H. DePuy and R. W. King, *Chem. Rev.* **60**, 431 (1960); (b) A. C. Cope and E. R. Trumbull, *Org. Reactions* **11**, 317 (1960); (c) H. R. Nace, *ibid.* **12**, 57 (1962).

[89] C. D. Hurd and F. H. Blunk, *J. Am. Chem. Soc.* **60**, 2419 (1938); cf. also E. S. Gould, "Mechanism and Structure in Organic Chemistry." Holt, New York, 1959; A. C. Cope and A. S. Mehta, *J. Am. Chem. Soc.* **85**, 1949 (1963); C. H. De Puy and C. H. Bishop, *Tetrahedron Letters* p. 239 (1963).

cis-hydrogen. Menthyl xanthate (*equatorial* xanthate group, compare page 131) yields predominantly Δ^3-menthene. Neomenthyl xanthate (*axial* xanthate group compare page 131) yields mainly Δ^2-menthene. The xanthate group at C-3 and hydrogen at C-4 are *cis* in the first case, and *trans* in the latter case.

The following results were obtained from an investigation of two *trans*-decalyl-1-xanthates: The xanthate of *trans-cis*-1-decalol (LVIII) gave mainly $\Delta^{1(2)}$-octalin. The xanthate of the isomeric *trans-trans*-1-decalol (LIX) led to $\Delta^{1(9)}$-octalin (80%).

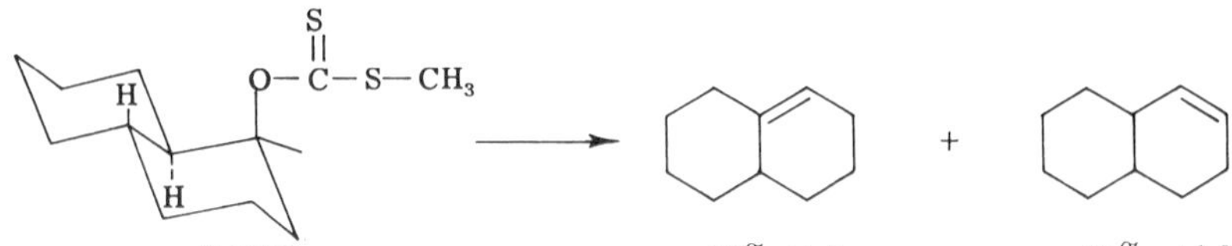

(LVIII) $\sim 20\%\ \Delta^{1,9}$ $\sim 80\%\ \Delta^{1,2}$

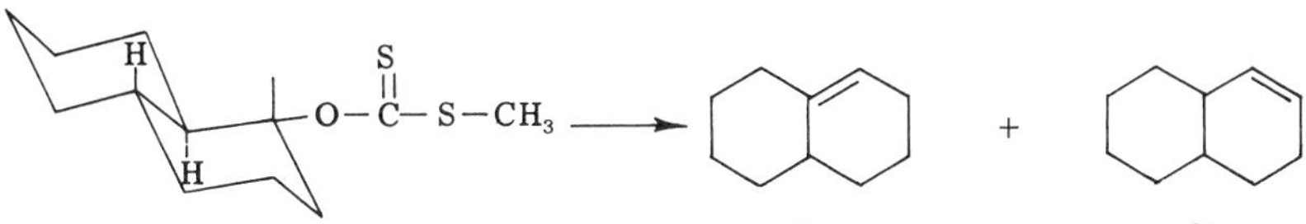

(LIX) $\sim 80\%\ \Delta^{1,9}$ $\sim 20\%\ \Delta^{1,2}$

If a *cis* hydrogen is available on both α-carbon atoms, as in (LIX), then the elimination proceeds preferentially in the direction of the ring juncture. In the other case (LVIII), the $\Delta^{1(2)}$-position is assumed by the double bond.

A recent example is *cis*-2-*tert*-butyl cyclohexyl xanthate (LX). Pyrolysis gave almost exclusively Δ^2-*tert*-butylcyclohexene (LXI), thereby representing a typical example of a *cis* elimination.[90]

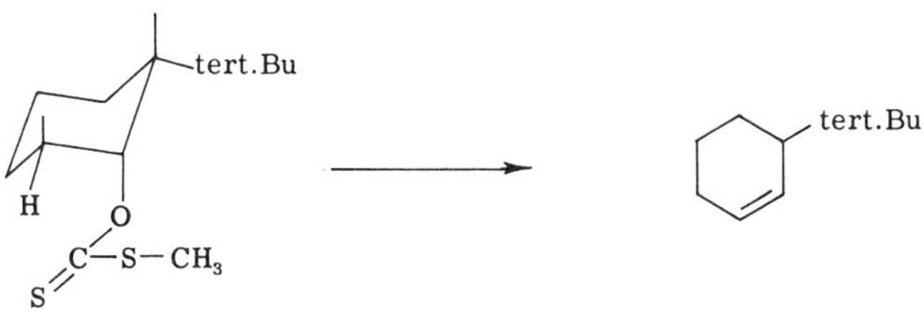

(LX) (LXI)

The elimination also goes preferentially in the direction of a *cis*-β-hydrogen in the pyrolysis of carboxylates, as shown by the following examples: *Cis*-2-methylcyclohexyl acetate (*axial* acetoxy group, LXII,

R=CH_3) gave mainly Δ^2-methylcyclohexene (LXIII, R=CH_3).[91] *Cis*-2-phenylcyclohexyl acetate (*axial* acetoxy group, LXII, R = phenyl) gave mainly Δ^2-phenylcyclohexene (LXIII, R=phenyl).[92] The corre-

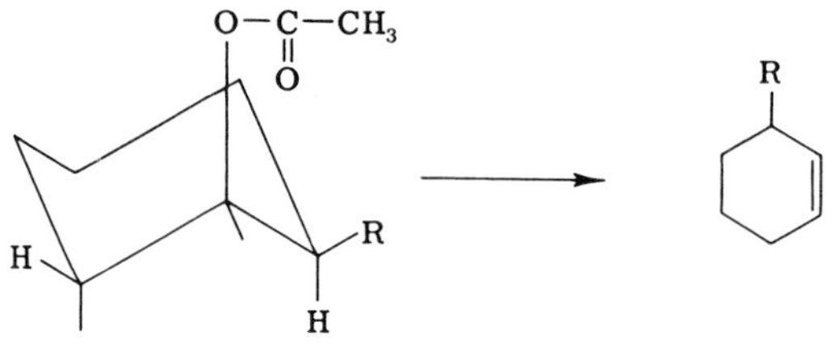

(LXII) (LXIII)

sponding *trans* isomers (*equatorial* acetoxy group) gave a mixture of Δ^1- and Δ^2-cyclohexenes.

The pyrolysis of the methyl carbonate of *trans-trans*-1-decalol, gave 59% of $\Delta^{1(9)}$-octalin and 41% of $\Delta^{1(2)}$-octalin.[93] The pyrolysis of the corresponding *cis-cis*-1-decalyl methyl carbonate still gave 25% of the $\Delta^{1(9)}$-octalin in addition to 75% $\Delta^{1(2)}$-octalin.[93]

A series of informative examples in which *cis* elimination occurs in steroids was reviewed by Barton.[71b,94] One example may be cited.[71b] The pyrolysis of cholestan-4β-yl benzoate (LXIV) gave pure cholest-3-ene (LXV), whereas the epimer, cholestan-4α-yl benzoate gave a mixture of 40% cholest-3-ene and 60% cholest-4-ene.[95]

The examples cited[95] indicate that the intramolecular elimination is

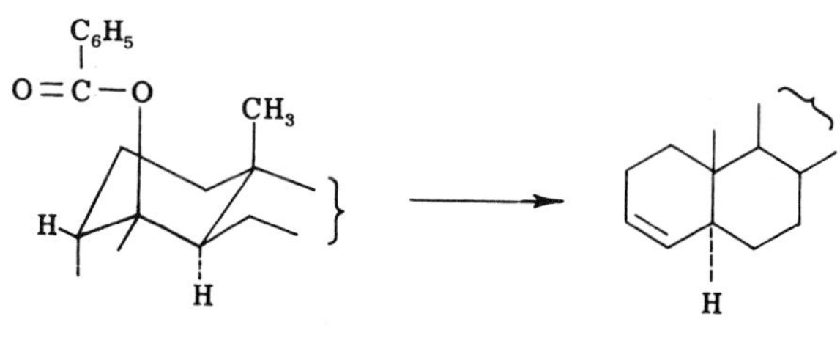

(LXIV) (LXV)

[90] F. G. Bordwell and P. S. Landis, *J. Am. Chem. Soc.* **80**, 6379 (1958).
[91] W. J. Bailey and L. Nicholas, *J. Org. Chem.* **21**, 854 (1956).
[92] E. R. Alexander and A. Mudrak, *J. Am. Chem. Soc.* **72**, 1810 (1950).
[93] J. W. Powell and M. C. Whiting, *Tetrahedron* **12**, 163 (1961).
[94] D. H. R. Barton, *J. Chem. Soc.* p. 2174 (1949).
[95] For further examples see ref. 88.

a process which does not proceed entirely in a stereospecific sense. Under special circumstances *trans* elimination can also occur, especially when the *trans-β-axial* hydrogen is made mobile by an electron-withdrawing group.[96]

The few examples considered here show once more that difficulties will be encountered in considering these still relatively stereospecific elimination reactions only from the point of view of stereochemistry.

A suggestion concerning the conformation of the cyclohexane ring in the transition state has been presented by DePuy and co-workers.[88,97] In contrast to other claims[98] DePuy and co-workers found that the pyrolysis of 1-methylcyclohexyl acetate (LXVI), led to 76% Δ^1-methylcyclohexene (LXVII) and only 24% methylenecyclohexane (LXVIII).[97]

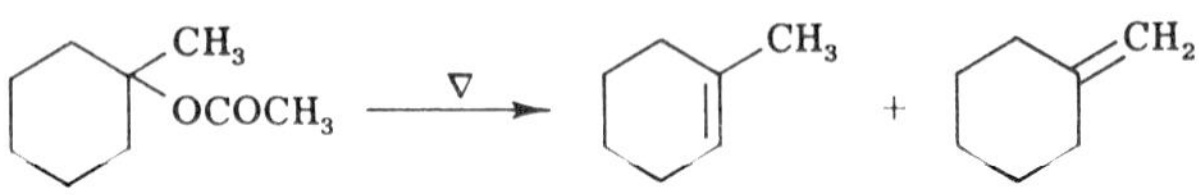

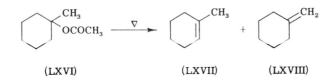

(LXVI) (LXVII) (LXVIII)

If absolute coplanarity were required for the formation of the cyclic transition state (LVI) (see page 246), then a deformation of the cyclohexane ring in the direction of the boat form would be required to produce (LXVII). This would lead to a true *cis* arrangement between the acetoxy group and the hydrogen involved in the reaction. No deformation of the ring would be required for the formation of methylenecyclohexane (LXVIII). A coplanar transition state between the acetoxy group and a hydrogen of the methyl group is always possible. However, because a relatively small amount of exocyclic olefin is formed, DePuy assumed that a coplanar arrangement is not necessary for the transition state of the ester pyrolysis. A true *cis* arrangement for the formation of an endocyclic double bond would require the boat form of higher energy, and it was assumed that no deformation of the chair form of the cyclohexane ring took place.

This view is strengthened by the observation that the corresponding "Cope" reaction, the pyrolysis of dimethyl(1-methylcyclohexyl)-amine oxide, yielded 97% methylene cyclohexane (LXVIII) and only 3%

[96] F. G. Bordwell and P. S. Landis, *J. Am. Chem. Soc.* **80**, 2450 (1958); see also D. G. Botteron and G. P. Shulman, *J. Org. Chem.* **27**, 2007 (1962) with further references.

[97] D. H. Froemsdorf, C. H. Collins, G. S. Hammond, and C. H. DePuy, *J. Am. Chem. Soc.* **81**, 643 (1959); Cf. also W. J. Bailey and W. F. Hale, *ibid.* **81**, 651 (1959); R. A. Benkeser and J. J. Hazdra, *ibid.* **81**, 228 (1959).

[98] Cf. for example T. D. Nevitt and G. S. Hammond, *J. Am. Chem. Soc.* **76**, 4124 (1954); J. G. Trynham and O. S. Pascual, *J. Org. Chem.* **21**, 1362 (1956).

Δ^1-methylcyclohexene (LXVII). In the ester pyrolysis, in contrast to the amine oxide example, there is one atom more present in the transition state (a six-membered ring is formed, cf. LVI). This leads to a greater flexibility and allows the carbonyl oxygen the possibility of reaching a neighboring "*gauche*" hydrogen without severe deformation of the ring. The elimination reaction of the amine oxide (LXIX) requires a coplanar arrangement because only five atoms are involved in the transition state. The energetically unfavorable boat form would have to be formed. The endocyclic elimination is therefore less favorable.

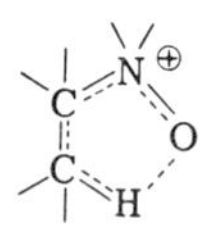

(LXIX)

The ratio of *exo* to *endo* olefin in the pyrolysis of 1-methylcyclohexyl acetate (LXVI) is independent of the conformation of the acetoxy group, whether *equatorial* or *axial*. In both cases the acetoxy group is equidistant from a *gauche* hydrogen. If the conformation of the acetoxy group is fixed by a *tert*-butyl group, as in *cis*-1-methyl-4-*tert*-butylcyclohexyl acetate (LXX) (*equatorial* acetoxy group) and *trans*-1-

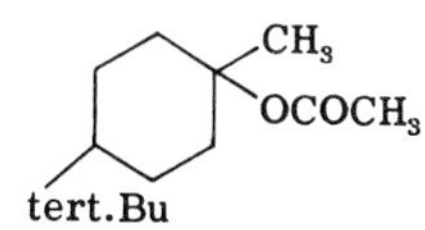

(LXX)

methyl-4-*tert*-butylcyclohexyl acetate (LXX) (*axial* acetoxy group), the *equatorial* and *axial* esters yield the same ratio of *exo* to *endo* olefin (23% *exo* and 77% *endo*). This ratio is the same as that obtained from the pyrolysis of 1-methylcyclohexyl acetate (LXVI). An influence of the conformation on the direction of the pyrolysis cannot be detected in this example.[99]

On the other hand, a difference does exist in the relative rates of pyrolysis between the *equatorial* and the *axial* acetates, as was shown for *cis*- and *trans*-4-*tert*-butylcyclohexyl acetates.[99] The rate for the *cis* acetate (*axial* acetoxy group) is 1.6 times greater than that of the *trans* acetate (*equatorial* acetoxy group), which makes one suspect that the rate difference is mainly due to a difference in ground state energies of the compounds, in which the *cis* isomer possesses the higher energy.

[99] C. H. DePuy and R. W. King, *J. Am. Chem. Soc.* **83**, 2743 (1961).

b. Addition Reactions

Another reaction which shows a high degree of stereospecificity is electrophilic addition to the C=C double bond.[99a] It has been known for some time that the additions of halogen or of hypohalous acids produce mainly *trans* compounds.[100] Barton[38a,38b,43a,43b,43c,56,69,70,101,102] was able to show that this configurational relationship could be extended by conformational considerations. The addition to cyclohexane systems proceeds in such a way that the *diaxial* addition product forms preferentially.

The mechanism generally considered as probable, at present, for the additions of halogen or hypohalous acid[103] to a double bond is the following: In both cases, the first step consists of the addition of a Cl^+ (or Br^+) to form a cyclic chloronium or bromonium ion (LXXI).

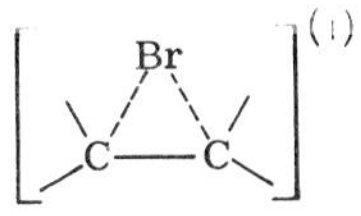

(LXXI)

The subsequent attack of an anion can proceed only from the side opposite to the transitional ring. Thus, in the case of an acyclic compound, a *trans* addition results.

Such an assumption would not necessarily produce the *diaxial* addition product for cyclic compounds, so the conformational stereospecificity requires a further explanation. In the rule of preferred *diaxial* elimination explained in Section 5,a, it was established that eliminations proceed readily because the participating centers possess a maximum of coplanarity in this arrangement. Barton[43] therefore, assumed that the *diaxial* conformation is preferred during an addition reaction because *diaxial* opening of the halonium ion (LXXI) in the transition state provides a maximum of coplanarity of the participating centers.

The following examples illustrate this point: The addition of bromine in carbon tetrachloride to cholest-2-ene (LXXII) proceeds by way of the bromonium ion (LXXIII, X=Br) and yields 88% of the *diaxial*,

---

[99a] For *cis*-additions of deuterium halides see: M. J. S. Dewar and R. C. Fahey, *J. Am. Chem. Soc.* **85**, 2245, 2248, 3645 (1963); *Angew. Chem.* **76**, 320 (1964).

[100] C. K. Ingold, "Structure and Mechanism in Organic Chemistry," pp. 685 ff. Cornell Univ. Press, Ithica, New York, 1953.

[101] D. H. R. Barton and E. Miller, *J. Am. Chem. Soc.* **72**, 1066 (1950).

[102] Cf. also P. B. D. de la Mare, *Quart. Rev. (London)* **3**, 126 (1949).

[103] Cf. for example H. B. Henbest and R. A. L. Wilson, *J. Chem. Soc.* p. 4136 (1959).

2β,3α-dibromocholestane (LXXIV, X=Br), and only 12 % of the *diequatorial*, 2α,3β-dibromocholestane (LXXV, X=Br).[43a]

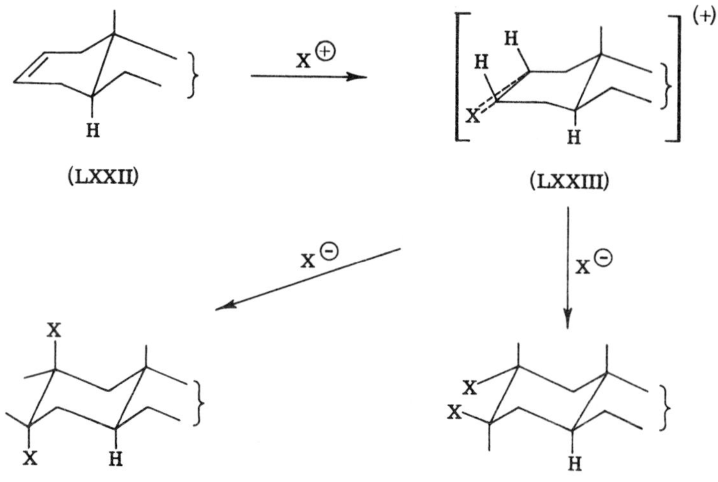

(LXXII) (LXXIII)

(LXXIV) (LXXV)

The addition of chlorine in carbon tetrachloride to (LXXII) yields 72 % of the *diaxial* 2β,3α compound (LXXIV, X=Cl) and 28 % of the *diequatorial* 2α, 3β-dichlorocholestane (LXXV, X=Cl).[43a] The addition of HOCl and HOBr also follows the same pattern, the *diaxial* 3α-halo-cholestan-2β-ol is formed predominantly.[43,104] Addition of bromine to cholest-3-ene (in carbon tetrachloride) give 97 % of the *diaxial*, 3α, 4β dibromide and only 3 % of the *diequatorial* 3β, 4α compound.[43a,105]

Also the addition of bromine or chlorine[106] to the 5,6 double bond of cholesterol leads mainly to the *diaxial* 5α, 6β-dihalide. In this case, the intermediate halonium ion results from attack on the less-hindered α side of the molecule (LXXVI).

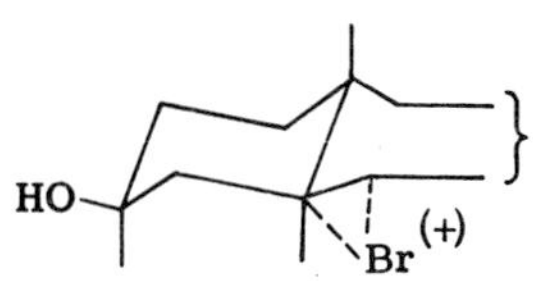

(LXXVI)

[104] N. L. Wendler and H. L. Slates, *Chem. & Ind.* (*London*) p. 167 (1955).

[105] For further examples see ref. 43.

[106] D. H. R. Barton, E. Miller, and H. T. Young, *J. Chem. Soc.* p. 2598 (1951); cf. L. F. Fieser, *Experientia* **6**, 312 (1950).

A further interesting example, which confirms the above transition state, has also been reported by Barton.[43,107] The addition of bromine to 3-methylcholest-2-ene (LXXVII) gave mainly the *diaxial* 2β, 3α-dibromo-3β-methylcholestane (LXXIX) and little of the 2α, 3β isomer in which the bromines are equatorial (LXXX).

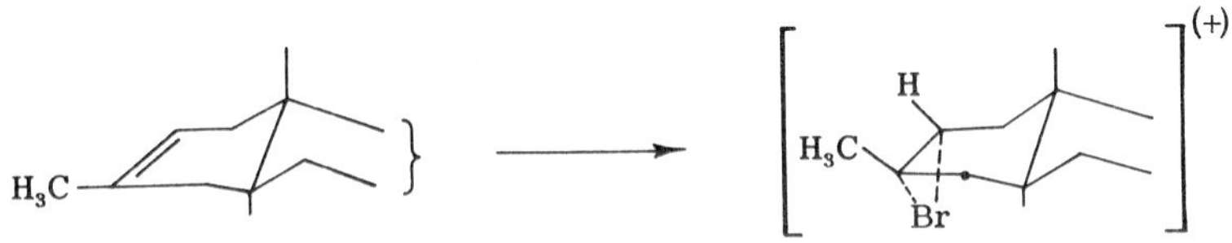

(LXXVII) (LXXVIII)

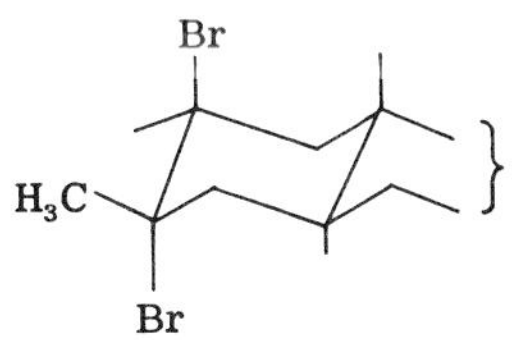

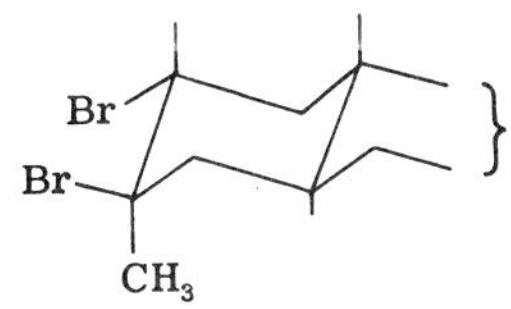

(LXXIX) (LXXX)

The following was assumed by Barton[43a]: Electrophilic addition to the carbon–carbon double bond follows Markownikoff's rule. The addition of a polar group proceeds in such a manner that the positive moiety adds to the carbon with the most hydrogens attached. The anion becomes attached to the carbon that has the fewest hydrogens. Because 3-methylcholest-2-ene (LXXVII) gave mainly the *diaxial* product (LXXIX) by way of the bromonium ion (LXXVIII), the rule of *diaxial* addition is here clearly preferred to Markownikoff's rule. The latter would have given more *diequatorial* 2α, 3β isomer (LXXX).

Only a few investigations are available concerning the conformations of products formed by the addition of hydrogen halides to steroids. Barton showed[107] that, in the additions of HCl or HBr to 3-methylcholest-2-en (LXXVII), 3α-methyl-3β-halocholestanes (LXXXIII, X=Cl or Br) were formed (*equatorial* halogens).

There are no indications that an intermediate corresponding to that of bromine addition is formed in proton additions. If the first step were the formation of the bridged carbonium ion (LXXXI), then *diaxial* opening would provide 2β-chloro-3β-methylcholestane (*axial* chlorine).

[107] D. H. R. Barton, A. da S. Campos-Neves, and R. C. Cookson, *J. Chem. Soc.* p. 3500 (1956).

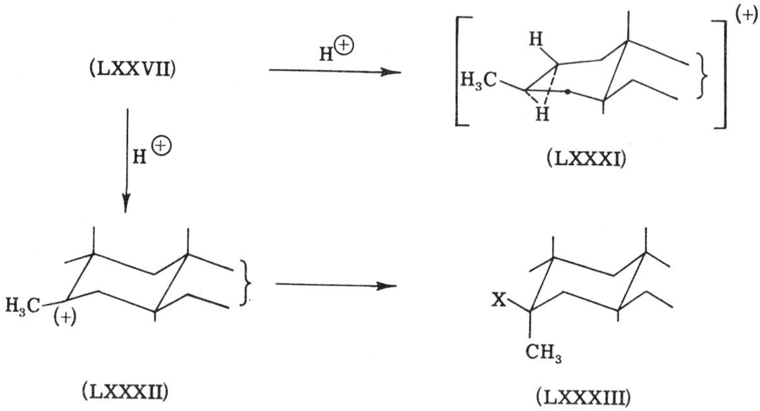

However, because the more stable product with *equatorial* chlorine is formed, the intermediate arrangement represented in (LXXXI) is not probable here. That the addition of hydrogen halide proceeds non-stereospecifically, differing from halogen additions, can be attributed to the fact that the classical carbonium ion (LXXXII) is formed before the addition of the halogen anion, leading to the stable isomer (LXXXIII). With the help of deuterium chloride, it was shown[56] that the addition to 3-methylcholest-2-ene takes place in such a way that the 2-α-deutero compound was formed. This indicates that the addition was *trans* and *diequatorial.*[108]

c. Epoxide Ring Openings

As has been known for quite some time, *trans*-glycols are always obtained from the acid-catalyzed hydration of epoxides of five and six-membered ring olefins. For example, cyclohexene oxide forms the *trans*-1,2-diol.[109] No *cis* isomer is formed. The opening of the oxide ring takes place, usually with configurational inversion (Walden inversion) at the point of fission.[110] There is no doubt that this is the normal path of an epoxide opening.

The ring opening of epoxides has been studied intensively in the steroid series. The rule of epoxide opening with configurational inver-

[108] For the stereochemistry of free radical additions to olefins see: B. A. Bohm and P. I. Abell, *Chem. Rev.* **62**, 599 (1962).

[109] J. Böeseken, *Rec. Trav. Chim.* **47**, 683 (1928); R. Criegee and H. Stanger, *Ber.* **69B**, 2753 (1936).

[110] For further examples see R. E. Parker and N. S. Isaacs, *Chem. Rev.* **59**, 737 (1959).

sion has been expanded in the following way.[69,111] Electrophilic, as well as nucleophilic opening of the epoxide ring leads preferably to a *diaxial* reaction product.

A few examples are given: The hydration of 2,3-epoxy-*trans*-decalin (LXXXIV) gave ~90% *diaxial trans* diol (LXXXV).[112]

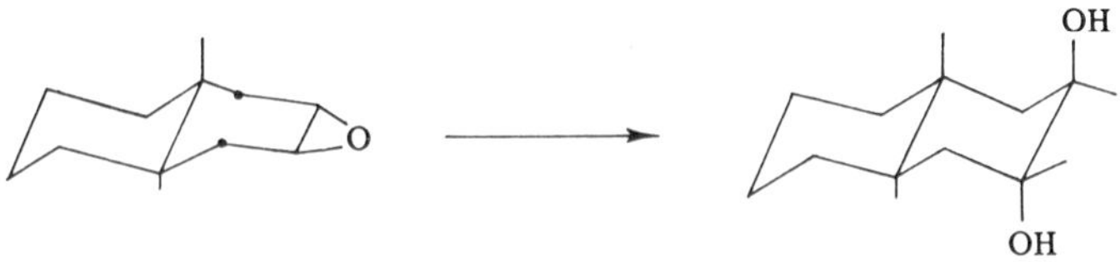

(LXXXIV) (LXXXV)

2α,3α-Epoxycholestane (LXXXVI) gave *diaxial* 2β-bromocholestan-3α-ol (LXXXVII) with hydrogen bromide. The *diequatorial* 3β,2α isomer, which also has a *trans* configuration, was not formed. The 2β,3β epoxide (LXXXVIII) gave the *diaxial* 3α-bromocholestan-2β-ol

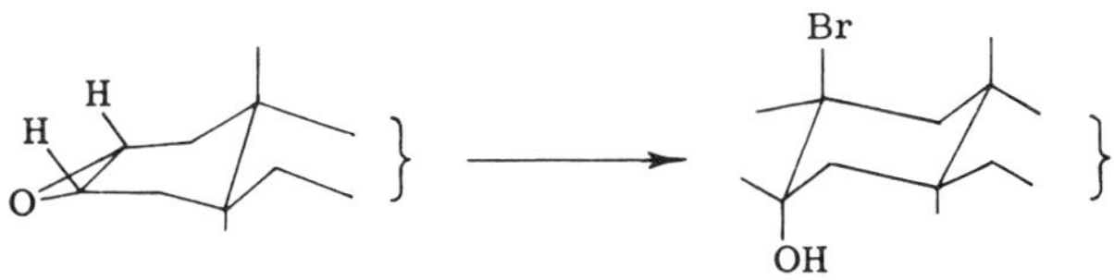

(LXXXVI) (LXXXVII)°

(LXXXVIII) (LXXXIX)

(LXXXIX).[43a] Some additional examples of the *diaxial* ring opening of steroid epoxides with various reagents have been compiled in Table V.

The predominant formation of *diaxial* reaction products from the opening of the epoxide ring cannot be readily understood. The opening of the ring following an S_N2-mechanism[110] could also proceed with

[111] A. Fürst and P. A. Plattner, *Abstr. Papers, 12th Intern. Congr. Pure Appl. Chem., New York*, 1951 p. 405; cf. also A. Fürst and P. A. Plattner, *Helv. Chim. Acta* **32**, 275 (1949); compare also: A. Hasner and C. Heathcock, *Tetrahedron Letters* p. 393 (1963).

[112] (a) Md. Erfan Ali and L. N. Owen, *J. Chem. Soc.* p. 2119 (1958); (b) H. B. Henbest, M. Smith, and A. Thomas, *ibid.* p. 3293 (1958).

TABLE VI

RING OPENING OF STEROID EPOXIDES

Configuration of the epoxides	Reagent	Configuration of the reaction product	References[a]
		Ring A/B trans	
3α, 4α	HX	3α–OH(a), 4β–X(a)	a
3β, 4β	HX	3α–X(a), 4β–OH(a)	a
4α, 5α	HX	4β–Br(a), 5α–OH(a)	b
5α, 6α	HX	5α–OH(a), 6β–X(a)	a, c
5β, 6β	HX	5α–X(a), 6β–OH(a)	a, c, d
8α, 9α	HX	8β–X(a), 9α–OH(a)	a
9β, 11β	HX	9α–X(a), 11β–OH(a)	e, f, g
16α, 17α	HX	16β–X(b), 17–OH(a′)[b]	h
1β, 2β	$Li/C_2H_5NH_2$	2β–OH(a)	i
2α, 3α	$Li/C_2H_5NH_2$	3α–OH(a)	j
5α, 6α	$Li/C_2H_5NH_2$	5α–OH(a)	i, j
5β, 6β	$Li/C_2H_5NH_2$	6β–OH(a)	i, j
6α, 7α	$Li/C_2H_5NH_2$	7α–OH(a)	j
7α, 8α	$Li/C_2H_5NH_2$	8α–OH	j
9α, 11α	$Li/C_2H_5NH_2$	9α–OH(a)	j
2α, 3α	$LiAlH_4$	3α–OH(a)	k, l
2β, 3β	$LiAlH_4$	2β–OH(a)	m
4α, 5α	$LiAlH_4$	5α–OH(a)	b, n
5β, 6β	$LiAlH_4$	6β–OH(a)[c]	o
6β, 7β	$LiAlH_4$	6β–OH(a)	d
2α, 3α	H_2SO_4	2β–OH(a), 3α–OH(a)	p
2β, 3β	H_2SO_4	2β–OH(a), 3α–OH(a)	p
2α, 3α	HOTs	2β–OTs(a), 3α–OH(a)	p
2β, 3β	HOTs	2β–OH(a), 3α–OTs(a)	p
		Ring A/B cis	
11α, 12α	HX	11β–X(a), 12α–OH(a)	a
11β, 12β	HX	11β–OH(a), 12α–X(a)	a

[a] Letters refer to parts of ref. 113, below.

[b] Cf. page 222.

[c] Hallsworth and Henbest, see 113j, found for the reduction of 5β,6β-expoxycholestane more (~60%) of the 5β-alcohol, in which the OH group is *equatorial* to ring B. For additional exceptions to the rule of the *diaxial* opening of epoxides see page 258.

[113] (a) According to D. H. R. Barton, ref. 69 and also the additional references there; (b) C. W. Shoppee, M. E. H. Howden, R. W. Killick, and G. H. R. Summers, *J. Chem. Soc.* p. 630 (1959); (c) H. B. Henbest and T. I. Wrigley, *J. Chem. Soc.* p. 4765 (1957); (d) C. W. Shoppee, R. H. Jenkins, and G. H. R. Summers, *J. Chem. Soc.* p. 1657 (1958); (e) J. Elks, G. H. Phillips, and W. F. Wall, *J. Chem. Soc.* p. 4001 (1958); (f) N. L. Wendler, R. P. Graber, C. S. Snoddy, Jr., and F. W. Bollinger, *J. Am. Chem. Soc.* **79**, 4476 (1957); (g) Cf. also R. F. Hirschmann, R. Miller, J. Wood, and R. E.

inversion of configuration by the formation of a *diequatorial* reaction product.

An explanation for the formation of the *diaxial* product can be found by examining the transition state of the ring opening: The attack of the nucleophile occurs *via* an S_N2 mechanism, on the side opposite to the leaving group, in this case oxygen. If we assume that at the transition state the chair form of the cyclohexane ring is regenerated, then the *diaxial* transition state must be more stable than the *diequatorial* one. Only in the former is a coplanar arrangement of the reaction centers possible.

For a certain few exceptions opening of the epoxide ring does not lead to a *diaxial* reaction product. However, other reasons are responsible for this. The 2β, 3β-epoxide of lanostane (XC) (and also of lanost-8-ene) would be expected to yield the 2β-OH-3α-substituted derivative from a *diaxial* ring opening (see Table VI). However, in both cases,[114] the *equatorial* 3β-alcohol was obtained by reduction with $LiAlH_4$. Hydrogen bromide also produced the *diequatorial* 2α-bromo-3β-hydroxy compound. The abnormal reaction course was explained as follows by Barton:[114]

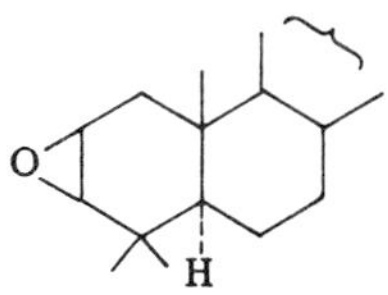

(XC)

The conformation of the 2β, 3β-epoxide in ring A does not correspond to a halfchair (XCI), but rather to a half-boat conformation (XCII) (see page 149). A *diaxial* opening of the epoxide (XCII) would yield the compound XCIII, in which the ring A exists in the boat conformation, which immediately is converted to the chair form giving the *diequatorial* 2α-bromo-3β-hydroxy compound.

Jones, *J. Am. Chem. Soc.* **78**, 4956 (1956); J. Fried and E. F. Sabo, *ibid.* **79**, 1130 (1957); (h) B. Ellis, D. Patel, and V. Petrow, *J. Chem. Soc.* p. 800 (1958); J. Romo, A. Zaffaroni, J. Hendrichs, G. Rosenkranz, C. Djerassi, and F. Sondheimer, *Chem. & Ind.* (*London*) p. 783 (1952); (i) A. S. Hallsworth and H. B. Henbest, *J. Chem. Soc.* p. 3571 (1960); (j) A. S. Hallsworth and H. B. Henbest, *J. Chem. Soc.* p. 4604 (1957); (k) R. C. Cookson and J. Hudec, *Proc. Chem. Soc.* p. 24 (1957);(l) cf. also A. Fürst and P. A. Plattner, *Helv. Chim. Acta* **32**, 275 (1949); (m) see ref. 111; (n) B. Camerino and D. Cattapan, *Farmaco* (*Pavia*) *Ed. Sci.* **13**, 39 (1958); *Chem. Abstr.* **52**, 13767 (1958); (o) P. A. Plattner, H. Heusser, and M. Feurer, *Helv. Chim. Acta* **32**, 587 (1949); (p) C. W. Shoppee, D. N. Jones, and G. H. R. Summers, *J. Chem. Soc.* p. 3100 (1957).

[114] D. H. R. Barton, D. A. Lewis, and J. F. McGhie, *J. Chem. Soc.* p. 2907 (1957).

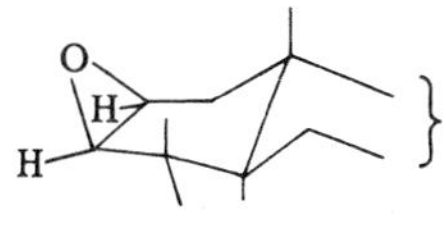

(XCI)

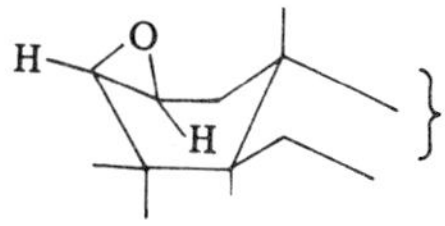

(XCII)

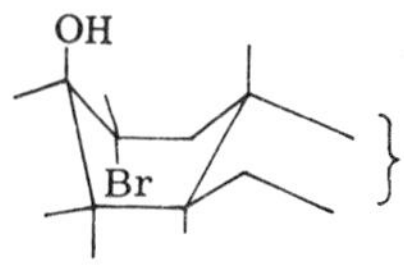

(XCIII)

A half-boat conformation can be accepted in this case. This is due to the interaction between the *axial* methyl groups at C-10 and C-4, as well as the β-epoxide ring. As can be seen from a model this interaction would be greater in the half-chair conformation (XCI) than in the half-boat form (XCII).

A ring opening with retention of configuration has been observed for the acid-catalyzed hydration of 2α,3α-epoxy-3β-phenyl cholestane (XCIV) with perchloric acid in aqueous acetone.[113k] The diol (XCV)

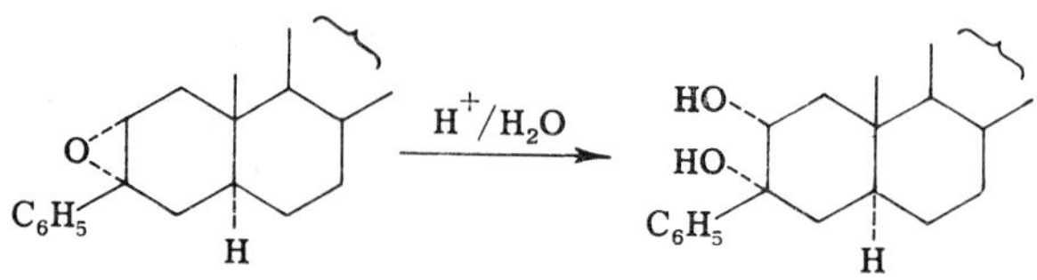

(XCIV) (XCV)

was formed (3β-phenylcholestane-2α, 3α-diol). However, reduction of the epoxide (XCIV) with $LiAlH_4$ proceeds normally with the formation of the *diaxial* product (3β-phenylcholestan-3α-ol) (see Table VI).

To explain this abnormal course, it was assumed[113k] that the protonated epoxide undergoes ring opening in the polar solvent in a unimolecular process with the formation of a carbonium ion. This is then stabilized by the solvent from the least-hindered α side. In this way, the stable *diequatorial* product is formed.[115,116]

The reverse of the *diaxial* opening of the epoxide ring with a hydrogen halide is found in the formation of epoxides from halohydrins with alkali. Barton[43a,56,70,114] and co-workers have shown that epoxides are formed more rapidly from *trans-diaxial* halohydrins than from *trans-diequatorial* ones.[117] From the examples[114] summarized by Barton it

---

[115] An S_Ni mechanism was also discussed.

[116] For further examples of abnormal epoxide openings see N. L. Wendler, D. Taub, S. Dohringer, and D. K. Fukushima, *J. Am. Chem. Soc.* **78**, 5027 (1956) and ref. 113j.

[117] A corresponding rule is valid for the *cis*-halohydrins (a, e): A ketone is formed readily with alkali from the *axial* halogen (This reaction involves elimination of *trans-diaxial* halogen and hydrogen), but with difficulty with alkali from the *equatorial* halogen (see ref. 114).

follows that *diaxial* halohydrins react $\sim 10^3$ times faster with alkali to form the epoxide than the *diequatorial* epimers.

d. Hydrolysis of Esters

The relationship between conformation and the hydrolysis of carboxylic esters, has been treated at various times (see page 120), and in detail in Chapter 4 (page 182f.) during the discussion of the decalols. It was shown that, as a result of differences in nonbonded interactions, variations in the rates of hydrolysis exist between *equatorial* and *axial* ester groups. The former are hydrolyzed more rapidly and the corresponding alcohols are esterified more rapidly than the latter. The difficulties inherent in attempts to find a correlation between the rates of hydrolysis and the conformation have also been covered.

This rule has been applied to steroids by Barton.[16b,68,69,70,117a] Here as well, *equatorial* ester groups are hydrolyzed (or esterified) more rapidly than the *axial* epimers. A comparison of the nonbonded interactions in the steroid nucleus (ring A/B *trans*) also revealed differences in the steric hindrance about *equatorial* hydroxyl groups. The *equatorial* positions at carbons 2 and 3 are the least hindered[69] in comparison to all other positions.

e. Solvolysis

Relative rates of solvolysis have been treated elsewhere (see page 115). It could be shown that *axial* groups suitable for the solvolysis reactions (such as halogens and sulfonates) react more rapidly than *equatorial* ones.[118] Various reasons have been presented for the different reaction rates of *axial* and *equatorial* substituents which, in essence, are determined by the mechanism of the solvolysis reaction.

If an S_N1 reaction is involved in the solvolysis, then the higher solvolysis rate of an *axial* group is due mainly to steric compression in this position. It must be kept in mind that pure S_N1 reactions are difficult to observe because elimination reactions (E 1) always take place concurrently. These have been considered in Section 5,a. The higher solvolysis rate in this latter case has been related to the fact that ionization of the *axial* group is facilitated by participation of the neighboring hydrogen, according to Winstein.[86] However, the view that a hydrogen participation is the only reason for the high solvolysis rate of

---

[117a] Cf. also N. B. Chapman, R. E. Parker, and P. J. A. Smith, *J. Chem. Soc.* p. 3634 (1960).

[118] For *cis*- and *trans*-4-chlorocyclohexyltosylates, in which the *trans*-tosylate reacts faster than the *cis*, see: D. S. Noyce, N. B. Bastian, and R. S. Monson, *Tetrahedron Letters* p. 863 (1962).

axial groups, is not accepted by all investigators (for example, Hückel *et al.*[64,75] and Goering and Reeves[83]).

If an S_N2 reaction is involved, then the *equatorial* groups should react more slowly than the *axial* isomers because attack of solvent at the carbon to which the substituent is attached is made more difficult by the *axial* hydrogens (XCVI). The attack from the rear is less hindered for an *axial* substituent (XCVII).[119]

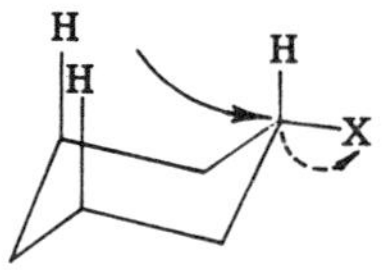

(XCVI)

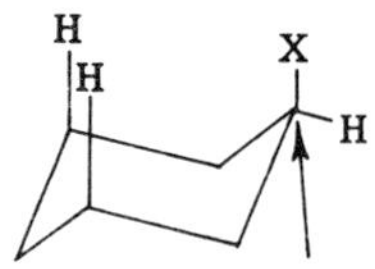

(XCVII)

A comprehensive investigation[64] has been made of the solvolysis of toluenesulfonates of substituted cyclohexanes. The solvolysis rates of decalyl toluenesulfonates have been treated in detail elsewhere (see p. 182f.). Few systematic investigations are available for steroids (see below). In the discussion of the following studies of the solvolysis of the toluenesulfonates, it cannot be positively decided which of the above-mentioned principles are responsible for the different rates of *equatorial* and *axial* substituents. It has been shown[120] that for solvolyses (ethanolyses were examined mostly[82]), no sharp distinction between an S_N1 and S_N2 mechanism is possible with regard to participation of the solvent.

The question of the magnitude of the rate differences between *axial* and *equatorial* substituents has been answered by Winstein and Holness,[86] using the example of the isomeric, conformationally homogeneous 4-*tert*-butyl- and 3-*tert*-butyl cyclohexyl toluenesulfonates (compare p. 114).

The *trans*-3 and *cis*-4 compounds with an *axial* tosyl group react at about the same rate in acetolysis. However, the rate is 3.3 times as fast as that of the *cis*-3 and *trans*-4 compounds with an *equatorial* tosyl group. The factor of 3.3 indicates the rate ratio for the solvolysis (acetolysis) in this case. A similar ratio appears for the ethanolysis of these toluenesulfonates: *Cis*-4-*tert*-butylcyclohexyl toluenesulfonate (*axial* tosyl group) reacts 3.9 times faster than the *trans* compound (*equatorial* tosyl group).[86]

---

[119] See E. L. Eliel, *in* "Steric Effects in Organic Chemistry," p. 73. Wiley, New York, 1956; cf. also: E. L. Eliel and R. S. Ro, *J. Am. Chem. Soc.* **79**, 5992 (1957); E. L. Eliel and R. P. Gerber, *Tetrahedron Letters* p. 473 (1961).

[120] W. Hückel and K. Tomopulos, *Ann. Chem.* **610**, 78 (1957). For a detailed discussion of the reaction mechanism see A. Streitwieser, Jr., *Chem. Rev.* **56**, 571 (1956); A. Streitwieser, Jr., "Solvolytic Displacement Reactions," Mc Graw-Hill, New York, 1962.

Trans-3-*tert*-butylcyclohexyl toluenesulfonate (*axial* tosyl group) reacts 3.7 times faster than the *cis* compound (*equatorial* tosyl group).[121]

The solvolysis (ethanolysis) rates of an additional series of 3-alkylated cyclohexyl toluenesulfonates was investigated. The values found are summarized in Table VII.

TABLE VII

SOLVOLYSIS OF 3-ALKYL CYCLOHEXYL TOLUENESULFONATES

Alkyl group	Rate constant k (sec^{-1}) $k_{50} \times 10^6$	Ratio
Methyl[a]	$\frac{trans\ (a)}{cis\ (e)} = \frac{2.25}{0.61}$	= 3.69
Isopropyl[a]	$\frac{trans\ (a)}{cis\ (e)} = \frac{2.83}{0.72}$	= 3.94
tert-Butyl[a]	$\frac{trans\ (a)}{cis\ (e)} = \frac{3.64}{0.98}$	= 3.73
4-*tert*-Butyl[b]	$\frac{cis\ (a)}{trans\ (e)} = \frac{4.18}{1.07}$	= 3.9

[a] See ref. 121.
[b] See ref. 86.

As is evident from Table VII, the ratio $k_{trans}: k_{cis}$ varies very little; it lies between 3.7 and 3.9 in all cases. Nevertheless, a definite trend can be recognized for the absolute rate constants for solvolysis. With increasing size of the alkyl groups at position 3, the rate constant increases, but by the same amount for the *cis* and *trans* isomers. Thus the ratio $k_{trans}: k_{cis}$ varies very little. This result also indicates, as mentioned previously in Chapter 3 (page 120), a specific influence on the solvolysis reaction by an alkyl group at position 3. The type of effect has not been determined as yet.

It is not possible to explain the observed rate ratios by a change in the ratio of *axial* to *equatorial* conformers due to increasing spatial requirements of the alkyl group in the series methyl, isopropyl, *tert*-butyl. If such were the case the *cis* and *trans* isomer from methyl to

[121] W. Hückel and K. Thiele, *Chem. Ber.* **94**, 2027 (1961).

tert-butyl should become conformationally more homogeneous. The rate constants should decrease for the *cis* isomer, increase for the *trans* isomer, and the ratio of $k_{trans} : k_{cis}$ should become larger.

The results from compounds alkylated at position 2 show that we cannot explain the solvolysis rates of toluenesulfonates on the basis of various conformations of the tosyl group alone. Other steric effects also have to be considered.

For this reason, the isomeric 2-*tert*-butylcyclohexyl toluenesulfonates will be considered.[122] The rate ratio between the *axial-cis* and the *equatorial-trans* compounds is about as large as that for acetolysis and ethanolysis of the isomeric 3- and 4-*tert*-butylcyclohexyl toluenesulfonates. However, the absolute reaction rates of the isomeric 2-compounds are markedly higher; the *axial cis*-2-tert butyl isomer is solvolyzed about 90 times faster than the *axial* 3- or 4-*tert*-butyl isomers. The *equatorial trans*-2-*tert*-butyl compound reacts ~145 times faster than the corresponding *equatorial* 3- or 4-tert-butyl compounds.

The reasons for the higher reaction rates of the isomeric 2-*tert*-butylcyclohexyl toluenesulfonates can be understand from a consideration of the following.[64,122]

On the basis of examination of a scale model, the *tert*-butyl group and the toluenesulfonate group are so close together in the 2-*tert*-butyl compound, that because of their bulk the rotation of both groups is strongly hindered, if not prevented altogether. The resulting steric strain (equivalent to H. C. Brown's B-strain[123]) leads to a steric acceleration,[123] which is also shown weakly by alkyl groups of medium size. However, the distance between the *tert*-butyl group and the toluenesulfonic ester group is about the same in the *cis* (a, e) and the *trans* (e, e) compounds. The *tert*-butyl group thus facilitates the ionization in both isomers in the same way, so that a higher reaction rate results in comparison to the 3- and 4-*tert*-butyl compounds. The unfavorable *axial* position of the toluenesulfonate group is an additional factor in the *cis* isomer. This second factor is the same as that involved in the difference in solvolysis rates between the *cis* and *trans* isomers of the 3- and 4-*tert*-butyl compounds. Thereby, the ratio between the solvolysis constants of *cis*-(a) and *trans*-(e)-2-*tert*-butylcyclohexyl toluenesulfonate is determined, and it is approximately the same as that between the isomeric 3- and 4-*tert*-butyl compounds.

[122] H. L. Goering and R. L. Reeves, ref. 83.

[123] H. C. Brown and R. S. Fletcher, *J. Am. Chem. Soc.* **71**, 1845 (1949); H. C. Brown, *Record Chem. Progr.* (*Kresge-Hooker Sci. Lib.*) **14**, 83 (1953).

If one now considers the cyclohexyl toluenesulfonates in which the alkyl group at position 2 is smaller, then it develops that the ratio $k_{cis} : k_{trans}$ is decidedly larger than that of the isomeric *tert*-butylcyclohexyl compounds. The following values for the ratio of $k_{cis} : k_{trans}$ were found for the ethanolysis of cyclohexyl toluenesulfonates alkylated at position 2 with various groups:

2-Methylcyclohexyl toluenesulfonate, ~90[64]; 2-isopropylcyclohexyltoluenesulfonate, ~75[124]; 2-cyclohexylcyclohexyl toluenesulfonate, ~80[124]; neomenthyl/menthyl toluenesulfonate, ~155[124]; neoisomenthyl/isomenthyl toluenesulfonate, ~180[124]; *trans*-2,5-dimethylcyclohexyl toluenesulfonate, ~140[64]; *cis*-2,5-dimethylcyclohexyl toluenesulfonate, ~23.[64]

The high solvolysis rates of the *cis*-toluenesulfonates can be based only in part on the fact that the sulfonate group is *axial*. As noted previously, this would give a ratio $k_{cis} : k_{trans}$ of about 4 at the most. A shift of the conformational equilibrium, which can be assumed for the alkyl groups with smaller spatial requirements, cannot be used for an explanation in this case. A specific neighboring effect of the alkyl group must be decisive in the high solvolysis rate of the *cis* isomer.

A plausible explanation for the large ratio of $k_{cis} : k_{trans}$, when the substituent is methyl, isopropyl, or cyclohexyl, is not easy to find. If one assumes that the high solvolysis rate of the *cis* isomer with a predominantly *axial* toluenesulfonate group is based on participation by the *β-axial* hydrogen, then it is difficult to understand why this is effective for methyl, isopropyl, and cyclohexyl substituents, but not for a *tert*-butyl[122] group.

An attempt at explanation was undertaken by Hückel.[64,125] The higher solvolysis rate of the *cis*-toluenesulfonate is attributed to the attainment of a "true *cis*" position during the conversion of one conformation of the ring into the other. During the process of flipping, the six-membered ring goes through the *twist* conformation (cf. page 46). In this conformation, the alkyl group and the tosyl group are nearly *eclipsed*, and ionization should be especially favorable due to the strong interaction. A typical case of steric acceleration exists, and the larger the spatial size of the substituents the greater is this action. The increase in solvolysis rate accompanying the increase in size of the substituent is thereby explained. A restriction of the mobility of the six-membered ring would then have to decrease the solvolysis rate. A con-

[124] Extrapolated from the values found by W. Hückel *et al.* in ref. 64.

[125] W. Hückel, *Bull. Soc. Chim. France* p. 1369 (1960).

firmation of this thesis can be found in the solvolysis rates of the toluenesulfonates of *trans-cis*-1-decalol and *cis-cis*-1-decalol (see page 184). As given in Chapter 4, the former reacted only about half as rapidly as the latter, even though the *axial* tosyl group in *trans-cis*-1-decalol (see XVII, page 188) suggests a high solvolysis rate.

However, the tosyl group in *cis-cis*-1-decalyl toluenesulfonate must first be converted into the unfavorable *axial* conformer. As can readily be seen from a model, the toluenesulfonate group and the methylene group of the other ring are momentarily in an *eclipsed* relationship in the transition state for this conversion. The higher solvolysis rate compared to that of *trans-cis*-1-decalyl toluenesulfonate would be explained by this.

Without going into this hypothesis any further, it may be stated that the *eclipsed* transition state indicated here is energetically very unfavorable. This transition state must form from an energetically favorable conformation and must nevertheless contribute to an increase in the reaction rate.

Only a few investigations concerning compounds capable of solvolysis are available in the steroid series.[126]

As for the rigid 1- and 2-decalyl toluenesulfonates (see page 182f.), the variation in solvolysis rates of toluenesulfonates in the cholestane series can be attributed to a variation in the size of the 1,3-*diaxial* interaction.[126a] Of the compounds studied, a few are illustrated.[126a]

The *equatorial* toluenesulfonate of cholestan-2α-ol (XCVIII) reacts 3.1 times as rapidly in acetolysis at 75° as the similarly *equatorial* 3β-toluenesulfonate (XCIX).

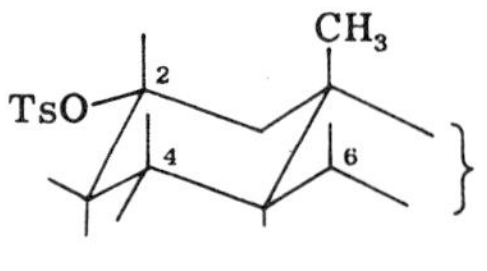

(XCVIII)

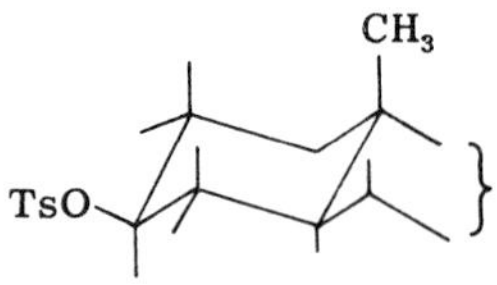

(XCIX)

The increased reaction rate of the 2α isomer over that of the 3β isomer cannot be explained by interactions with neighboring substituents. The case of attack by a solvent molecule should also be about the same in both cases. Therefore, another effect must be called upon to explain the different rates.[126a]

---

[126] (a) S. Nishida, *J. Am. Chem. Soc.* **82**, 4290 (1960); (b) Cf. also J. Mathieu, M. Legrand, and J. Valls, *Bull. Soc. Chim. France* p. 549 (1960); (c) C. W. Shoppee and G. A. R. Johnston, *J. Chem. Soc.* p. 3261 (1961); (d) J. F. Biellmann and G .Ourisson, *Bull. Soc. Chim. France* p. 331, 341 (1962).

The greater acetolysis rate of the 2α-tosylate (XCVIII) is attributed to the 1,3-interactions between the angular methyl group and the *axial* hydrogens at positions 2, 4, and 6. The possibility exists that the *axial* hydrogen can avoid steric interaction with the methyl group in this position by ionization of the 2α-tosyl group. Thereby the ionization is facilitated and the reaction rate increased. Such an effect is not present for the isomeric 3β compound (XCIX).

The toluenesulfonate of cholestan-2β-ol (C) reacts ~9.5 times more rapidly than the 2α isomer (XCVIII). This factor, greater than the usual ratio of 3 to 4 (see page 260), must be related to a steric acceleration caused by the angular methyl group.

Aside from the effect on the solvolysis rate of a 2α-sulfonate group, produced by an *axial* methyl group at position 10, a double bond in rings B or C exerts an effect of decreasing the rate of solvolysis of a 3β-sulfonate group. This was shown by a comparison of the rate constants of compounds with unsaturation in rings B or C, compared to the corresponding saturated compounds.[126b] Long range effects of this type (conformational transmission) have already been discussed (page 225).

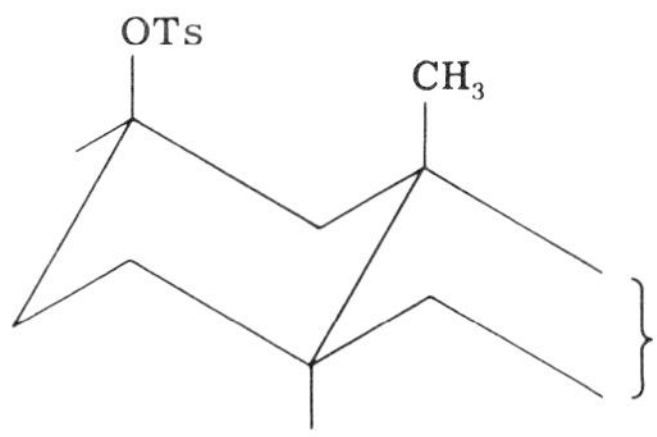

(C)

f. Reaction of Amines with Nitrous Acid

The reaction of cyclohexyl amines with nitrous acid, discussed in connection with the decalyl amines (Chapter 4, page 197), gave the following result, contrary to the original rule of Bose and Mills:

If the amino group is *axial*, the reaction with nitrous acid proceeds with the predominant formation of unsaturated hydrocarbon. In addition, some alcohol of the same configuration is formed without a Walden inversion. There was no deviation from the original rule observed for the *equatorial* amines. The alcohol of the same configuration is formed, in addition to a small amount of unsaturated hydrocarbon.

This deviation of the reaction of cyclohexylamines with nitrous acid from the original rule, had been observed earlier by Shoppee and co-workers in an investigation of the deamination of amino steroids.[127] Table VIII gives a summary of the amines studied by Shoppee and their reaction products.

TABLE VIII

REACTION OF AMINES WITH NITROUS ACID

Amine	Reaction products	
	Substitution	Elimination
Axial NH_2 group		
Cholestan-2β-yl	Cholestan-2β-ol (21%)	Cholest-1- and 2-ene (74.5%)
Cholestan-3α-yl	Cholestan-3α-ol (45%)	Cholest-2-ene (54%)
Coprostan-3β-yl	Coprostan-3β-ol (46%)	Coprost-3-ene (50%)
Cholestan-4β-yl	—	Cholest-4-ene (92%)
Cholestan-6β-yl	—	Cholest-5-ene (99%)
Cholestan-7α-yl	Cholestan-7α-ol (36.5%)	Cholest-7-ene (61%)
Equatorial NH_2 group		
Cholestan-2α-yl	Cholestan-2α-ol (96%)	—
Cholestan-3β-yl	Cholestan-3β-ol (99%)	—
Coprostan-3α-yl	Coprostan-3α-ol (92%)	—
Cholestan-4α-yl	Cholestan-4α-ol (82%)	—
Cholestan-6α-yl	Cholestan-6α-ol (97%)	—
Cholestan-7β-yl	Cholestan-7β-ol (95%)	—

The preferred formation of unsaturated hydrocarbon from an *axial* amino group has been covered previously (page 196). The predominant retention of configuration in the alcohol product from *equatorial* as well as *axial* amino groups after reaction with nitrous acid, cannot be easily explained. Contrary to the view of some investigators, that elimination of nitrogen from the intermediate diazonium ion (from the action of nitrous acid on the amine) follows a bimolecular process,[127,128,129] more recent investigations indicate a unimolecular process.[130,131] An

[127] (a) C. W. Shoppee, D. E. Evans, and G. H. R. Summers, *J. Chem. Soc.* p. 97 (1957); (b) C. W. Shoppee, R. J. W. Cremlyn, D. E. Evans, and G. H. R. Summers, *ibid.* p. 4364 (1957); (c) cf. also C. W. Shoppee and J. C. P. Sly, *ibid.* p. 345 (1959); compare also (d) G. Drehfahl and S. Huneck, *Chem. Ber.* **93**, 1961 (1960); **93**, 1967 (1960).

[128] J. A. Mills, *J. Chem. Soc.* p. 260 (1953).

[129] (a) A. Streitwieser, Jr., *J. Org. Chem.* **22**, 861 (1957); (b) cf. also A. Streitwieser, Jr. and C. E. Coverdale, *J. Am. Chem. Soc.* **81**, 4275 (1959).

[130] K. D. Thomas, Dissertation, Tübingen, 1961.

[131] Cf. also H. Felkin, *Bull. Soc. Chim. France* p. 20 (1959); J. H. Ridd, *Quart. Rev.* **15**, 418 (1961).

attempt to explain the path of the reaction with retention of configuration was made as follows[130]: *Equatorial* amines first form an "*equatorial*" carbonium ion from the decomposition of the diazonium ion. This carbonium ion slowly attempts to assume a flat structure. The further course of the reaction is determined by the rate of assuming a planar position, as well as by the rate of nitrogen departure. The removal of the nitrogen apparently proceeds much more rapidly than the transition to a planar position. Substitution can, therefore, take place mainly on the nearly tetrahedral carbon. Because the back side of this carbon is still shielded by the ring, substitution with Walden inversion is made difficult. As a result, the configuration is predominantly maintained.[132]

Intramolecular interactions (e.g., similar to hydrogen participation) can have an effect on the "*axial*" carbonium ion as soon as it is formed from *axial* amine. This can lead to elimination with hydrocarbon formation (or rearrangement), but also to an increase in the life-time of the carbonium ion. As a result, the departing nitrogen can withdraw far enough in this case for substitution to take place with retention of configuration. At the same time, an intermolecular influence (solvent) acts upon the back side that can lead to substitution with inversion. This will not be discussed in detail at this time.

Attention was called[130] (page 199) to the discrepancy in the results which were obtained from the deamination of *trans-cis*-1- and *trans-trans*-2-decalyl amine (both *axial* amino groups). No explanation can be found for this.

g. Halogenation of Ketones

Corey set up the following rule concerning the stereochemistry accompanying halogenation, particularly for the bromination of cyclohexanones and ketosteroids.[133,134,135,136] Kinetically controlled brominations of ketones always yield that product in which the bromine assumes the *axial* position. If no, or only minor steric interactions exist between the *axial* bromine and other substituents, such as an *axial* methyl group in the case of the steroids, then the kinetically controlled reaction product is also the thermodynamically more stable one. If the steric interaction is large, then the *axial* bromine is converted into the *equatorial* bromine under the influence of thermodynamic control (cf. Table II, page 230).

---

[132] Cf. also R. Huisgen and C. Rüchardt, *Ann. Chem.* **601**, 1 (1956).

[133] E. J. Corey, *J. Am. Chem. Soc.* **75**, 2301 (1953).

[134] E. J. Corey, *Experientia* **9**, 329 (1953).

[135] E. J. Corey, *J. Am. Chem. Soc.* **76**, 175 (1954).

[136] E. J. Corey and R. A. Sneen, *J. Am. Chem. Soc.* **78**, 6269 (1956); cf. also H. E. Zimmermann, *J. Org. Chem.* **20**, 549 (1955), and later investigations.

The preferred formation of *axial* bromoketones has been explained in various ways. During bromination, the bromine cation attacks the enolized ketone (CI) preferentially from the *axial* side because the arrangement of the *p* orbitals of the enol allows maximum overlap with the vacant orbital of the attacking species in the transition state.[135]

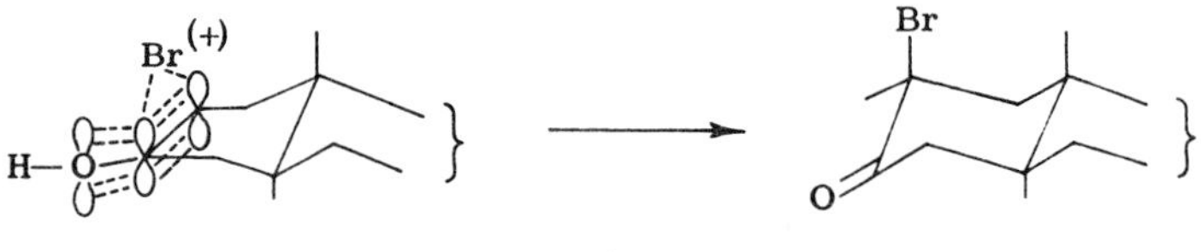

(CI)

Another mechanism for the bromination of cyclohexanone was proposed by analogy with the bromination of cyclohexene,[70] in which an intermediate bromonium ion is opened with the formation of a *diaxial* product (cf. page 251). The addition is said to take place on the double bond of the enol in the same way. The ketone then is formed by the elimination of hydrogen bromide. This proposal has been fairly well excluded by recent investigations of Djerassi and co-workers.[137]

Furthermore, Corey's rule was modified by Djerassi[137,138] as follows: The formation of an *axial* bromoketone, considered qualitatively under conditions of kinetic control, is possible if no steric hinderance blocks *axial* attack of the reagent. Viewed quantitatively, an *equatorial* bromoketone was isolated from some of the compounds studied by Djerassi, even though steric factors were absent. In the presence of steric hindrance, the product of kinetic control is not the *axial* but the *equatorial* isomer.

Of the examples studied by Djerassi[137,138] one may be singled out: The kinetically controlled bromination of the 2α-methyl-3-keto-5α-steroid (CII) or its enol acetate yields the 2α-bromo-2β-methyl-3-ketone (CIII), in which ring A has the boat conformation (see page 281). The product of thermodynamic control is the 2β-bromo-2α-methyl-3-ketone (CIV), with an *axial* bromine. According to Corey's rule one might expect that the product of kinetically controlled bromination of (CII) would be the 2β-bromo-2α-methyl-3-ketone (CIV).

At this point Djerassi assumed that a direct *equatorial* bromination of the enol of a 3-ketosteroid is possible. This is qualified by the fact that steric hindrance (by the angular methyl group) is present. The *equatorial* 2α-bromoketone (CV) is formed first.

---

[137] R. Villotti, H. J. Ringold, and C. Djerassi, *J. Am. Chem. Soc.* **82**, 5693 (1960).

[138] Cf. also C. Djerassi, N. Finch, R. C. Cookson, and C. W. Bird, *J. Am. Chem. Soc.* **82**, 5488 (1960); R. Mauli, H. J. Ringold, and C. Djerassi, *ibid.* **82**, 5494 (1960); compare also: E. W. Warnhoff, *J. Org. Chem.* **28**, 887 (1963) with further references.

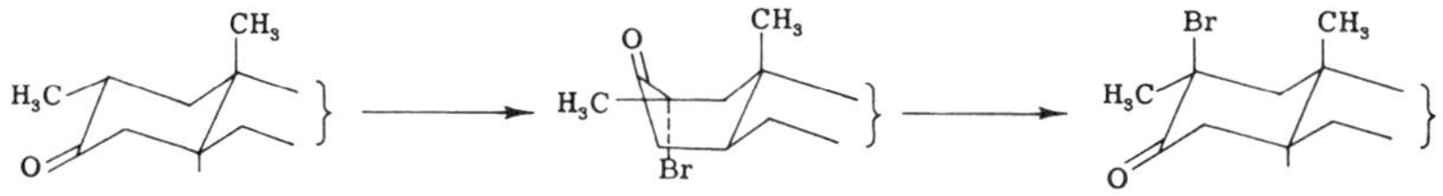

(CII) (CIII) (CIV)

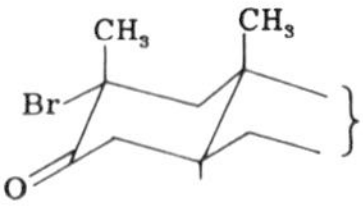

(CV)

This transition product (CV) flips over because of the steric interaction of the *axial* methyl groups at C-2 and C-10, as well as the unfavorable electrostatic interaction between the *equatorial* bromine and the carbonyl group. The boat conformation (CIII) is formed. That ring A actually exists in the boat conformation was shown from its optical rotatory dispersion curve (see page 282).

h. Reductions of Ketones[139]

A series of correlations between the stereochemical course of ketone reductions and the conformations of the alcohols obtained has been found by Barton.[69] Although some of these rules had to be supplemented in some respects by later investigations, they have been found to be basically correct.

The regularities found for the reduction of ketones will be discussed briefly. The reduction of ketones with sodium in alcohol yields predominantly the more stable *equatorial* alcohol. The quantitative composition corresponds to the equilibrium position between the isomers. The reaction products are thermodynamically controlled. Extensive experimental data are available concerning this reaction in steroids and also for simpler cyclohexanones.[66] No exceptions to this rule have been observed. However, in some cases, the ratio of isomers produced by the reduction of the ketone with sodium in alcohol does not seem to correspond exactly with the ratio of isomers at the point of equilibrium. The reduction of ketones with sodium in liquid ammonia corresponds to the reduction with sodium in alcohol for the few examples investigated.[66]

The catalytic hydrogenation of sterically hindered, and nonhindered keto groups in strongly acid solution (rapid hydrogenation) yields predominantly the less stable *axial* alcohol.[139a] This rule is valid in most of

---

[139] Cf. also the comprehensive article by A. V. Kamernitzky and A. A. Akhrem, *Tetrahedron* **18**, 705 (1962), including additional references there.

[139a] Cf. R. J. Wicker, *J. Chem. Soc.* p. 2165 (1956); p. 3299 (1957).

the cases investigated, but exceptions have been observed. For example in the cholestanone series, it could be shown[140] that ketones at C-1 and C-12, exclusively, and C-7, predominantly, yielded the *equatorial* alcohol by catalytic hydrogenation in acid medium.

Cis-1- and *cis*-2-decalone also yielded in greater amounts the corresponding *cis*-decalols with *equatorial* hydroxyl groups.[66,75] Nevertheless, in contrast to the above examples, this example cannot be considered an exception. As a result of the mobility of the *cis*-decalin system, the hydroxyl group preferentially assumes the *equatorial* position (see page 185).

Hydrogenation in neutral solution (slow hydrogenation) produced mainly the *axial* alcohol from sterically hindered ketones, and the *equatorial* alcohol from nonhindered ketones. Exceptions to this rule were also observed: The keto group in 2-methylcyclohexanone, which is not especially hindered, is reduced preponderantly to the *axial cis*-2-methylcyclohexanol under a variety of conditions.[66,141] 3-Methylcyclohexanone yields mainly *trans*-3-methylcyclohexanol (*axial* hydroxyl).[142,143]

The kinetically controlled Meerwein-Pondorf reduction with aluminum isopropylate produces a larger amount of *axial* alcohol than other methods, except for catalytic hydrogenations in acid solution.[144] The limitation that strongly hindered ketones cannot be reduced by this method, originally cited by Barton[70] does not hold.[66]

Reductions with complex hydrides, such as sodium borohydride or lithium aluminum hydride, in general produce mostly the *equatorial* alcohol from nonhindered ketones. Strongly hindered keto groups give mainly *axial* alcohols. Though these reactions have been studied thoroughly, no conformity exists in the view of the different investigators concerning the mechanism and the factors controlling the stereochemistry of the reduction products.[145]

Dauben[140,146] studied the factors which control the stereochemistry

---

[140] W. G. Dauben, E. J. Blanz, Jr., J. Jiu, and R. A. Micheli, *J. Am. Chem. Soc.* **78**, 3752 (1956).

[141] P. Anziani and R. Cornubert, *Bull. Soc. Chim. France* p. 359 (1945).

[142] W. Hückel and J. Kurz, *Chem. Ber.* **91**, 1290 (1958).

[143] Concerning events which take place on the catalyst surface leading to the formation of a preferred conformation, cf. W. Hückel *et al.* ref. 66 and J. H. Brewster, *J. Am. Chem. Soc.* **76**, 6361 (1954); also ref. 139a.

[144] Cf. also, L. M. Jackman, A. K. Macbeth, and J. A. Mills, *J. Chem. Soc.* p. 2641 (1949).

[145] For the stereochemistry of ketone reductions with trimethylamine borane see W. M. Jones, *J. Am. Chem. Soc.* **82**, 2528 (1960).

[146] W. G. Dauben, G. J. Fonken, and D. S. Noyce, *J. Am. Chem. Soc.* **78**, 2579 (1956); cf. also ref. 66.

of the products produced from the reduction of simple alkylcyclohexanones with complex hydrides. Essentially there are two factors involved. First is the ease with which the organometallic complex is formed from the carbonyl group and the complex hydride. This factor is known as "steric approach control." The second consists of the relative energetics of the formation of the products once the initial complex is formed. This factor is known as "product development control," and from this, the relative stabilities of the intermediate products is determined.

For an unhindered ketone, a mixture of isomeric alcohols is formed because the complex hydride anion can attack the keto group from both sides. This mixture approximates the equilibrium condition. If the keto group is hindered, then the attack of the complex hydride anion takes place from the less-hindered side and principally one isomer is formed.

If we compare the reductions carried out with lithium aluminum hydride and sodium borohydride, then the latter leads to more *axial* alcohol.[140,146,147] This is explained by Dauben[140,147a] by the fact that the BH_4^- ion in methanol solution is more bulky, due to solvation, than the AlH_4^- ion in ether solution. Therefore, the BH_4^- ion attacks the cyclohexane ring from the less-hindered *equatorial* side and an *axial* alcohol results (steric approach control). The addition of $AlCl_3$ during reduction with $LiAlH_4$ under kinetic control also promotes an increase in the ratio of *axial* alcohols, because the complex formed from $LiAlH_4$ and $AlCl_3$ possesses greater spatial requirements.[148]

This rule,[149] that sodium borohydride leads to more axial alcohol derived from the simple cyclohexanones, has proven useful in the steroid series.[140,145,150] Sodium borohydride yields more *axial* alcohol than lithium aluminum hydride from cholestan-2-one, cholestan-7-one and 11-ketosteroids.[150a]

Doubts exist as to the exact determination of the ratio of isomers in the reductions of simple cyclohexanones with lithium aluminum hydride and sodium borohydride. This led to a questioning of the rule of preferred formation of *axial* alcohol from sodium borohydride reductions.[151]

---

[147] Cf. also A. H. Beckett, N. J. Harper, A. D. J. Balon, and T. H. E. Watts, *Tetrahedron* **6**, 319 (1959); W. M. Jones and H. E. Wise, Jr., *J. Am. Chem. Soc.* **84**, 997 (1962).

[147a] Cf. also ref. 16a, p. 268.

[148] E. L. Eliel and M. N. Rerick, *J. Am. Chem. Soc.* **82**, 1367 (1960).

[149] Cf. also A. Hajös and O. Fuchs, *Acta Chim. Hung.* **21**, 137 (1959).

[150] N. L. Wendler, R. P. Graber, R. E. Jones, and M. Tishler, *J. Am. Chem. Soc.* **74**, 3630 (1952).

[150a] Concerning the hydride reductions of 17-keto-C/D-*cis*- and *trans*-steroids see L. J. Chinn, *J. Org. Chem.* **27**, 54 (1962) with further literature cited.

[151] K. D. Hardy and R. J. Wicker, *J. Am. Chem. Soc.* **80**, 640 (1958); cf. also W. G. Dauben and R. E. Bozak, *J. Org. Chem.* **24**, 1596 (1959).

It has also been shown that the reduction with lithium tri-*tert*-butoxy aluminum hydride [LiAlH(*tert*-BuO)$_3$], which is much larger than lithium aluminum hydride,[152] produced more *equatorial* alcohol than lithium aluminum hydride.[153] This occurred despite its greater size.

The original assumption that the greater stereospecificity of the BH_4^- ion could be attributed to a larger effective spatial requirement produced by solvation was critized by Wheeler and Huffman,[154] who claimed that neither the BH_4^- or the AlH_4^- ion is solvated. Then the latter would have a greater spatial requirement, as shown by the experimentally determined bond lengths.[154] Wheeler and Huffman[154] have proposed, therefore, another mechanism for the reduction with complex hydrides. Two reactions are possible. One consists of a direct attack of the complex hydride anion on the carbonyl group (CVI). The second involves formation of a complex between the metal hydride and the carbonyi group (CVII), which is subsequently reduced. In the first

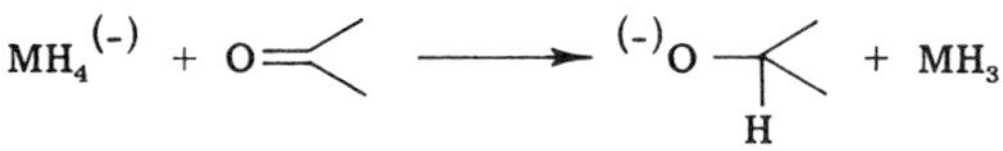

(CVI)

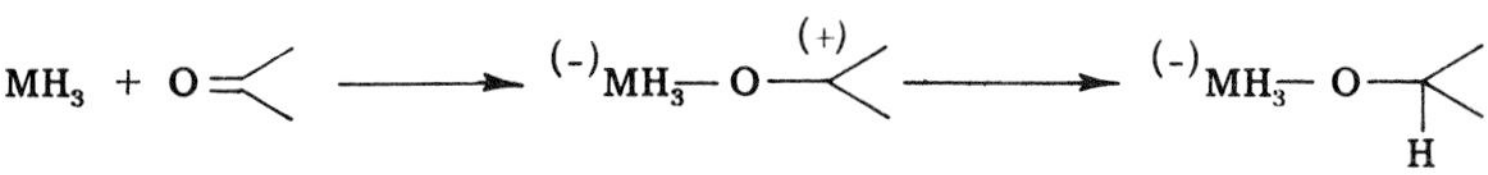

(CVII)

case, the unstable *(axial)* alcohol is formed, because the complex hydride anion approaches preferentially from the less-hindered *equatorial* side. In the second case, the complex produced preferentially assumes the *equatorial* position. Differences in the results from various reducing agents can be explained in this way without assuming solvation effects. An assumption for this type of mechanism is that lithium aluminum hydride

[152] H. C. Brown and R. F. McFarlin, *J. Am. Chem. Soc.* **78**, 252 (1956); *ibid.* **80**, 5372 (1958).

[153] O. H. Wheeler and J. L. Mateos, *Chem. & Ind. (London)* p. 395 (1957); *Can. J. Chem.* **36**, 1431 (1958).

[154] D. M. S. Wheeler and J. W. Huffman, *Experientia* **16**, 516 (1960) with further references; compare also M. G. Combe and H. B. Henbest, *Tetrahedron Letters* p. 404 (1961).

can produce LiH and AlH_3 by a homolytic scission[155] besides its normal dissociation into Li^+ and AlH_4^-.

The homolytic cleavage should be strongly hindered in the case of the complex borohydrides, which possess more ionic character.[155a] That is why the direct attack according to (CVI) is preferred. More of the unstable isomer is formed than from lithium aluminum hydride reductions. In the reduction of strongly hindered ketones, attack of the hydride anion is hindered from the *axial* side, as seen in (CVI). The *axial* alcohol is formed preferentially from $LiAlH_4$ as well as from $NaBH_4$. Here too more *axial* alcohol is formed from $NaBH_4$ for the reasons given above.

On the other hand Haubenstock and Eliel[156] showed that the stereochemistry of sodium borohydride reductions is strongly dependent on the solvent. A similar effect was found in lithium aluminum hydride reductions when the solvent was changed from diethyl ether to tetrahydrofuran. These findings, contrary to the view of Wheeler and Huffman,[154] show that both, the BH_4^- and AlH_4^--ions are solvated. Solvation of the borohydride and aluminum hydride ions would seem the most likely explanation for the variation in the stereochemistry of the reduction products, according to Eliel.[156,157]

Corresponding rules can also be set up for the reduction of oximes to amines.[69] The reduction with sodium and alcohol yields preferentially the more stable *equatorial* amine. The ratio of isomers during catalytic hydrogenation depends on the acidity of the solvent and the degree of steric hindrance[158] just as do the corresponding reductions of ketones. In comparison to the ketones, the experimental data are too sparse to allow definite statements.

Aside from the characteristic examples of reactions treated in this section, others can be found for which there are correlations between conformation and reactivity. For example: 1,2-*Diaxial* rearrangements of the Wagner-Meerwein type proceed readily if the leaving group and the

---

[155] (a) Cf. literature references in 154; (b) E. Wiberg and R. Bauer, *Z. Naturforsch.* **5b**, 397 (1950).

[156] H. Haubenstock and E. L. Eliel, *J. Am. Chem. Soc.* **84**, 2363 (1962); **84**, 2368 (1962).

[157] For further investigations in reduction of ketones with complex hydrides see: (a) H. Kwart and T. Takeshita, *Tetrahedron Letters* p. 404 (1962); (b) O. R. Vail and D. M. S. Wheeler, *J. Org. Chem.* **27**, 3803 (1962); (c) P. T. Lansbury and R. E. MacLeay, *ibid.* **28**, 1940 (1963); (d) C. D. Ritchie, *Tetrahedron Letters* p. 2145 (1963).

[158] Cf. references 59b in Chap. 3; 27i in Chap. 4; ref. 127d; A. Skita and W. Faust, *Ber.* **64**, 2878 (1931); G. Cauquil, R. Guizard, and R. Calas, *Bull. Soc. Chim. France* p. 252 (1942); P. Anziani and R. Cornubert, *ibid.* p. 857 (1948); M. M. Claudon, P. Anziani and R. Cornubert, *ibid.* p. 150 (1956); W. Hückel and K. D. Thomas, *Ann. Chem.* **645**, 177 (1961).

migrating group are *axial*.[69,70,159] Reactions which proceed with ring expansion and ring contraction are easily initiated when the participating centers assume a coplanar position.[160] The acid-catalyzed *Prins*-reaction between cyclohexenes and formaldehyde, leads to *trans*-2-hydroxymethylcyclohexanols by way of *diaxial* addition.[161] The oxidation of epimeric secondary alcohols proceeds in such a way that the alcohol with an *axial* hydroxyl is oxidized more easily than the epimer with an *equatorial* hydroxyl.[162] Also, variations in preferred bond migration during the rearrangement of 17-hydroxy-20-ketosteroids are related to the conformation of the resulting D-homosteroids.[163]

In conclusion one should remember that conformational analysis by chemical methods is limited for the many reasons stated in the introduction to Part II and Section 5. Nevertheless, interpretations of reaction mechanisms are not possible today without consideration of the stereochemical requirements.

[159] E. R. H. Jones, G. D. Meakins, and J. S. Stephenson, *J. Chem. Soc.* p. 2156 (1958).

[160] Ref. 70; J. F. Biellmann and G. Ourisson, ref. 126d and *Bull. Soc. Chim. France* p. 348 (1960); J. F. Biellmann, D. Kučan, M. Rajić, P. Witz, and G. Ourisson, *ibid.* p. 330 (1962); J. F. Biellmann, D. Kučan and G. Ourisson, *ibid.* p. 337 (1962).

[161] E. E. Smissman and R. A. Mode, *J. Am. Chem. Soc.* **79**, 3447 (1957); A. T. Blomquist and J. Wolinsky, *ibid.* **79**, 6025 (1957); E. E. Smissman and D. T. Witiak, *J. Org. Chem.* **25**, 471 (1960).

[162] J. Schreiber and A. Eschenmoser, *Helv. Chim. Acta* **38**, 1529 (1955); H. G. Kuivila and W. J. Becker, *J. Am. Chem. Soc.* **74**, 5329 (1952); S. Winstein and N. J. Holness, *ibid.* **77**, 5562 (1955); H. Kwart and P. S. Francis, *ibid.* **81**, 2116 (1959).

[163] I. Elphimoff-Felkin, *Bull. Soc. Chim. France* p. 1845 (1956); N. L. Wendler, D. Taub, and R. Firestone, *Experientia* **15**, 237 (1959).

CHAPTER 6

Compounds with Boat Conformations

In previous discussions we have limited ourselves essentially to the basic idea that the chair form is always the most stable conformation of a cyclohexane ring.

However, it was pointed out in the treatment of cyclohexane (Chapter 2, page 47) that recent investigations have indicated otherwise. In special cases, because of particular structural circumstances, the six-membered ring need not exist exclusively in the chair conformation. A boat conformation or, more probably, the twist, skew or stretched, conformation (see page 46) may be present. On occasion in the preceding chapters reference has been made to the occurrence of boat conformations. For example, *trans*-1,3-di-*tert*-butylcyclohexane (see pages 48 and 112) is forced into the boat conformation because of the bulkiness of the *tert*-butyl groups. One of the *tert*-butyl groups would have had to assume the *axial* position in the chair form because of the *trans*-1,3 configuration. Consideration of the case of the *trans-syn-trans*-perhydrophenanthrene (see page 213, XII) led to the conclusion that the middle ring must be in the boat conformation. Mention has also been made of the 2β,3β-epoxide of lanostane (see page 257) and of a 2α-bromo-2β-methyl steroids (see page 269), in which ring A exists in the boat conformation. 1,4-Cyclohexanedione was found to exist predominantly in the boat form (see page 157) and *cis*-2,5-di-*tert*-butyl-1,4-cyclohexanedione also exists in a boat form.[1]

In these examples and those to be mentioned in Section 1, the boat form is made possible due either to the interaction of the substituents or due to two trigonal carbon atoms, as in the case of 1,4-cyclohexanedione. However, the boat conformation can also be insured by a bridged ring formation. Compounds of the bicyclo[2,2,1]heptane (I) or bicylo-[2.2.2]octane (II) type seem, therefore, to be quite useful in studying the chemistry of the true boat conformations (see Section 2).

[1] R. D. Stolow and C. B. Boyce, *J. Am. Chem. Soc.* **83**, 3722 (1961); cf. also R. D. Stolow and M. M. Bonaventura, *Tetrahedron Letters* p. 95 (1964).

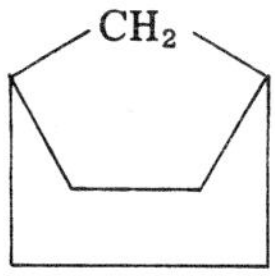

I

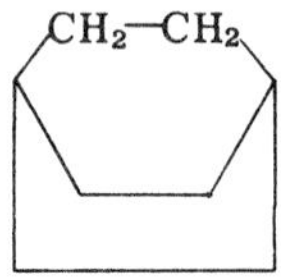

II

One can differentiate four types of substituents on the cyclohexane ring in the true boat form, which vary in their stereochemical behavior[2,3] in contrast to the two types on the chair form.

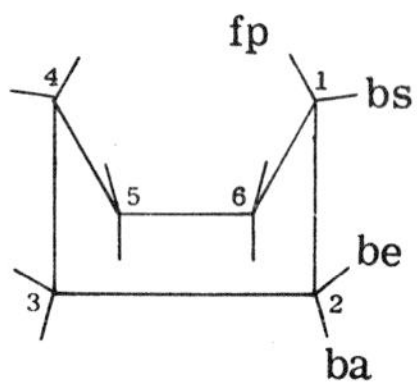

III

The substituents on carbons 1 and 4 are designated by "flagpole" (fp) and "bowsprit" (bs) (III). The substituents on carbons 2, 3, 5, and 6, in contrast to the chair form, assume a pair of *eclipsed* conformations, called boat-*equatorial* (be) and boat-*axial* (ba). They resemble the *equatorial* or *axial* substituents on a chair cyclohexane ring.

1. Compounds with Preferred Boat (Twist) Conformations[4]

a. Steroids and Related Compounds

In the following considerations, examples which are forced into a boat conformation because of geometric reasons, such as *trans-syn-trans*-perhydrophenanthrene, will be disregarded. Examples in which the molecule cannot exist in the *all-chair* conformation for this reason are also known among the steroids and triterpenes.[3b] Only those cases in which the boat conformation is assumed due to other causes will be treated.

Barton and co-workers were able to make the existence of a boat conformation seem probable for the first time from an example in the

[2] S. J. Angyal and J. A. Mills, *Rev. Pure Appl. Chem.* **2**, 185 (1952).

[3] Compare also ref. 6 in Chap. 2, p. 78.

[4] (a) Cf. J. Levisalles, *Bull. Soc. Chim. France* p. 551 (1960); (b) D. H. R. Barton and G. A. Morrison, *Progr. Chem. Org. Nat. Prod.* **19**, 212 (1961); (c) M. Balasubramanian, *Chem. Rev.* **52**, 591 (1962).

triterpenoid series.[5] Two products were obtained from the bromination of lanost-8-en-3-one (IV), the 2α-(V) and the 2β-bromolanost-8-en-3-one (VI).

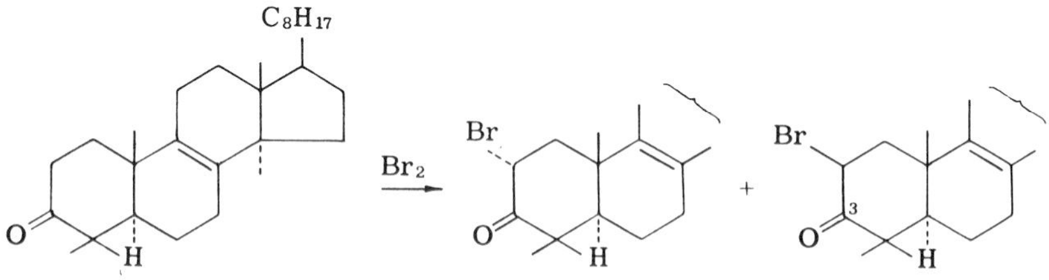

IV V VI

The configurations of the two bromoketones were unambiguously established by chemical methods. The 2α-bromoketone (V) should have an *equatorial* bromine; the 2β-bromoketone (VI) would have an *axial* bromine if ring A exists in the chair form (VII). The ultraviolet and infrared spectra of the two bromoketones (V) and (VI) showed, however[6] (see page 232), that the bromine exists in the *equatorial* position in both cases. An *equatorial* conformation is possible for the 2β-bromoketone only if ring A no longer has the chair form, but is in the boat or more precisely in a twist conformation (VIII).

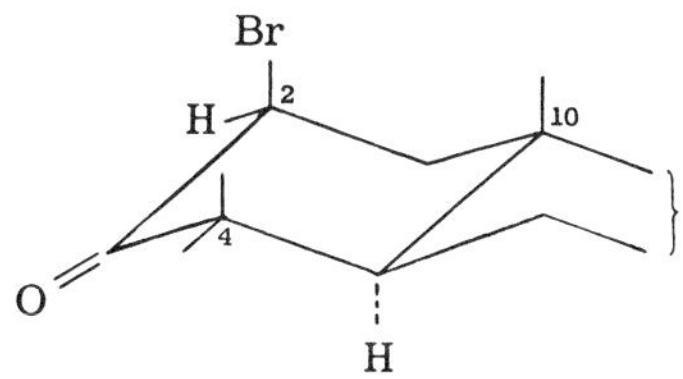

VII

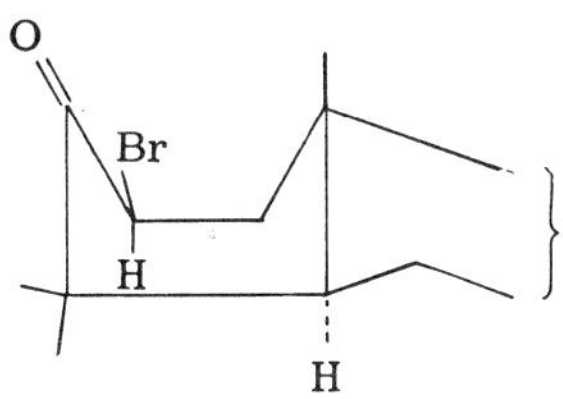

VIII

The spectrum of 2β-bromolanostan-3-one also indicated an *equatorial* position of the bromine. Ring A must assume the boat conformation in this case as well.

2β-Bromolanost-8-en-3β-ol (IX) was obtained by the reduction of the 2β-bromoketone (VI). The infrared spectrum indicated that the bromine of (IX) was *axial*. The rapid reaction rate for *diaxial* elimination with alkali also demonstrated this, and one can conclude from this that ring A of the bromohydrin (IX) exists in the chair conformation.

On the basis of these facts, the following explanation can be deduced

---

[5] D. H. R. Barton, D. A. Lewis, and J. F. McGhie, *J. Chem. Soc.* p. 2907 (1957).
[6] E. G. Cummins and J. E. Page, *J. Chem. Soc.* p. 3847 (1957).

for the origin of the boat conformation in bromoketone (VI)[5]: The bromoketone probably exists in the boat (or *twist*) conformation because strong 1,3-*diaxial* interactions appear between the *axial* bromine at C-2, the methyl group at C-4 and the methyl group at C-10 in the chair conformation (VII). These interactions can be avoided by changing to the boat conformation. In the boat conformation of cyclohexane, however, severe 1,4 interactions of the flagpole hydrogens appear. However, carbon 3 of bromoketone (VI) is trigonal, and this interaction is greatly reduced. This also explains why 2β-bromolanost-8-en-3β-ol (IX) exists in the chair conformation. A boat conformation of

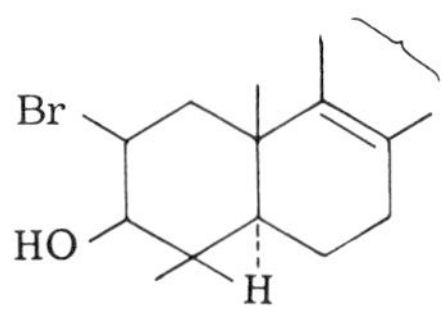

IX

ring A would lead to too great an interaction between the 3β-OH and the methyl group at C-10.

Aside from the case treated in some detail here, a series of additional examples of one ring in the boat conformation have been found recently among steroids and triterpenoids. Some of these are mentioned below. In agreement with the original view expressed by Barton[4b,5] all of these boat cyclohexane rings contain a trigonal carbon atom. As shown in Chapter 3 (page 150) the difference between the chair and boat forms of cyclohexanone is smaller (2.7 kcal./mole) than the difference between the chair and boat forms of cyclohexane (~6 kcal./mole).

A situation analogous to that in 2β-bromolanostan-3-one (page 277) is found with the epimeric 2-bromo-4,4-dimethylcholestane-3-ones (X, R=Br)[7] Spectroscopic data indicate that ring A exists in the chair conformation for the 2α-bromoketone (X, R=Br, *equatorial* bromine), but in the boat form for the 2β-bromoketone.

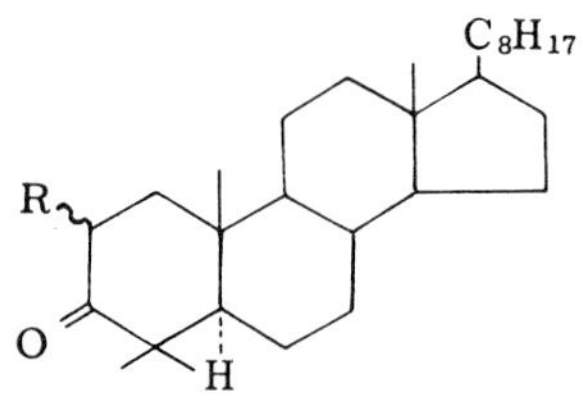

X

[7] G. R. Chaudhry, J. G. Halsall, and E. R. H. Jones, *J. Chem. Soc.* p. 2725 (1961).

2β-Methoxy-4,4-dimethylcholestan-3-one (X, R=OCH_3) also is an example with ring A in the boat form.[8] Another interesting case is 2-bromo-4,4-dimethylcholest-5-en-3-one (XI, R=Br), formed along with the 2,2-dibromo derivative in the bromination of 4,4-dimethylcholest-5-en-3-one (XI, R = H).

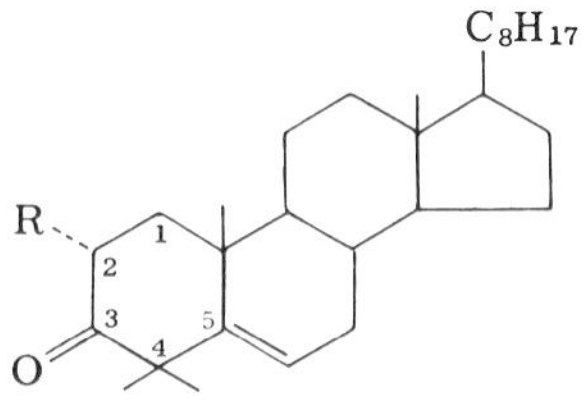

XI

With the aid of ultraviolet and infrared spectroscopy, as well as measurements of the optical rotatory dispersion, it seemed likely that the 2-bromo compound (XI, R=Br) has an α-oriented bromine and ring A assumes the boat conformation (see p. 232). The bromine would then exist in the *axial* position.[7,9] This assumption was supported by chemical methods.[10]

In addition, the NMR spectrum disclosed which boat conformation ring A assumes.[10] Ring A could take on two different boat conformations. First, there is the conformation shown as (XII), in which carbons 2 and 5 project out of the plane formed by the remaining carbon atoms. The other possible conformation is (XIII), in which carbons 3 and 10 project out of the plane. NMR measurements indicated that conformation (XIII) is probably the preferred one.

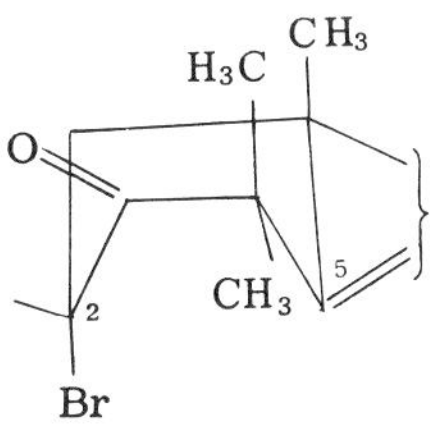

XII

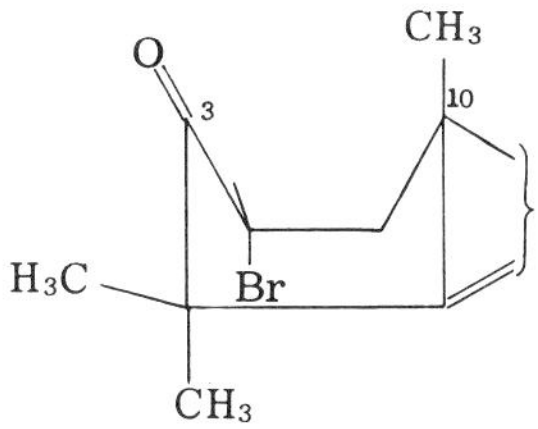

XIII

[8] H. P. Sigg and C. Tamm, *Helv. Chim. Acta* **43**, 1402 (1960).

[9] D. T. Cropp, B. B. Dewhurst, and J. S. E. Holker, *Chem. & Ind.* (*London*) p. 209 (1961).

[10] B. B. Dewhurst, J. S. E. Holker, A. Lablache-Combier, and J. Levisalles, *Chem. & Ind.* (*London*) p. 1667 (1961); see also R. J. Abraham and J. S. E. Holker, *J. Chem. Soc.* p. 806 (1963).

The existence of a boat conformation is not limited to ring A, as could be shown for a series of D-homosteroids. The existence of ring D in the boat form seemed probable based on chemical methods.[11]

Of the dibromides formed by the bromination of 5α-cholestan-4-one, 3α,5α-dibromocholestan-4-one (XIV) was investigated in more detail.[12] From spectroscopic data it was concluded that this compound exists as an equilibrium mixture of chair and boat conformations of ring A. If ring A exists in the chair form, then a strong *diaxial* interaction between the 3α- and 5α-bromines is present. In the boat form, aside from the interaction between the 3β-hydrogen and the C-10 methyl group, a dipolar repulsion exists between the 3α-bromine and the carbonyl groups at C-4.

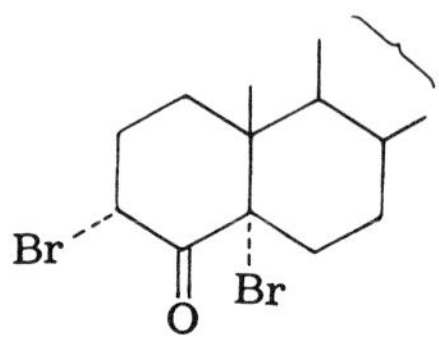

XIV

Optical rotatory dispersion led to the assumption of the boat conformation in a further series of steroidal ketones. This was demonstrated by application of the *axial* haloketone rule,[13] which is also valid when the

[11] N. L. Wendler, *Chem. & Ind. (London)* p. 1662 (1958); p. 20 (1959); N.L. Wendler, *Tetrahedron* **11**, 213 (1960).

[12] C. W. Shoppee, M. E. H. Howden, R. W. Killick, and G. H. R. Summers, *J. Chem. Soc.* p. 630 (1959).

[13] The *axial* haloketone rule [see C. Djerassi, "Optical Rotatory Dispersion, Application to Organic Chemistry." McGraw-Hill, New York, 1960; C. Djerassi and W. Klyne, *J. Am. Chem. Soc.* **79**, 1506 (1957); compare also C. S. Barnes and C. Djerassi, *ibid.* **84**, 1962 (1962)] represents a special case of the so-called "Octant Rule" [the Octant Rule is an empirical rule which permits the prediction of the sign of the Cotton effect of a certain compound on the basis of its structure, configuration, and conformation: Djerassi, *loc. cit.*; W. Klyne, *Bull. Soc. Chim. France* p. 1396 (1960); C. Djerassi, *Tetrahedron* **13**, 13 (1961); W. Moffitt, R. B. Woodward, A. Moscowitz, W. Klyne, and C. Djerassi, *J. Am. Chem. Soc.* **83**, 4013 (1961); C. Djerassi and W. Klyne, *Proc. Natl. Acad. Sci. U. S.* **48**, 1093 (1962); *J. Chem. Soc.* p. 4929 (1962); *ibid.* p. 2390 (1963).] and the following is a brief statement. As was shown before (page 232), the introduction of an *axial* or *equatorial* halogen in the position α to a keto group produces a shift in the ketone absorption. This shift is also observed for the peaks and troughs of the optical rotatory dispersion curve of α-halo cyclohexanones (and α-hydroxy and α-acetoxy cyclohexanones) because of the close relation between the optical rotatory dispersion curve and the ultraviolet wavelength at which the ketone absorbs (see ref. 258b in

cyclohexanone ring exists in the boat conformation.[14,15] As an example 2-bromo-2-methylcholestan-3-one (XVII), obtained by Djerassi and co-workers from the bromination of 2α-methylcholestan-3-one is cited (CII, page 269)[14]

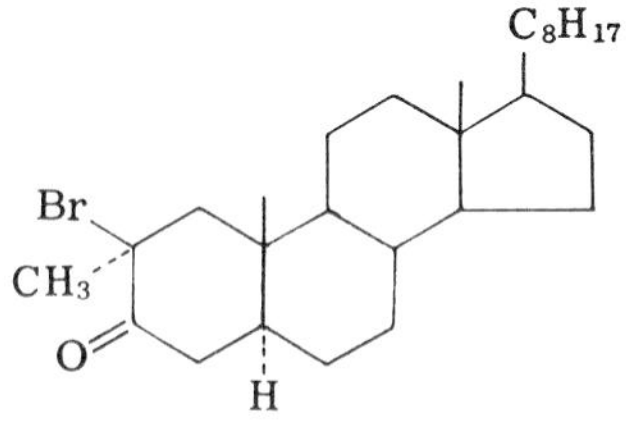

XVII

Infrared and ultraviolett spectral measurements (compare p. 232) indicated an *axial* bromine.[16] Therefore, the bromoketone was considered to be 2β-bromo-2α-methylcholestan-3-one (XVII) (*axial* bromine). In addition, the bromination was carried out under kinetically controlled conditions, which also suggests an *axial* position of the bromine (compare page 268). Examination of a model (XVIII) requires that this

Chapter 3). The introduction of an *equatorial* halogen α to a keto group does not change the sign of the Cotton effect of the halogen-free ketone. However, an *axial* halogen is capable of affecting the sign of the Cotton effect and the opposite sign may appear. The *axial* haloketone rule permits a prediction of direction of the sign. Looking in the direction of the C=O axis (arrow) of a cyclohexanone derivative with an *axial* α-chlorine, bromine, or iodine, then a negative Cotton effect results if the halogen is to the left side (XV). If the halogen is to the right, then positive Cotton effect (XVI) is observed. This rule does

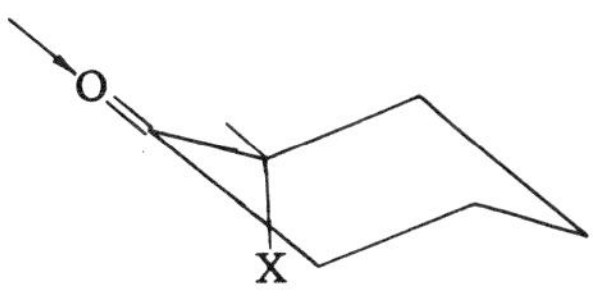

XV

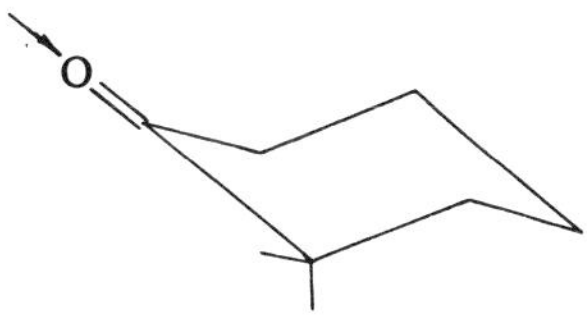

XVI

not hold for fluorine. Thus the rule permits one to determine if the halogen α to the keto group assumes the *equatorial* or the *axial* position. If the Cotton effect is reversed in comparison to the halogen-free ketone, then the halogen assumes the *axial* position. On the other hand, the halogen may also assume the *axial* position without a reversal of the Cotton effect. In this case, however, the *axial* position manifests itself through a greater amplitude of the rotatory dispersion curve and a bathochromic shift of the extrema (C. Djerassi and J. Staunton, *J. Am. Chem. Soc.* **83**, 736 (1961).

[14] C. Djerassi, N. Finch, and R. Mauli, *J. Am. Chem. Soc.* **81**, 4997 (1959).

[15] Cf. also C. Djerassi, O. Halpern, V. Halpern, and B. Riniker, *J. Am. Chem. Soc.* **80**, 4001 (1958).

[16] Y. Mazur and F. Sondheimer, *J. Am. Chem. Soc.* **80**, 5220 (1958).

compound have a positive Cotton effect according to the *axial* halogen ketone rule. However, a negative Cotton effect was found.[14]

As already noted in Chapter 5 (page 268), it is also possible for a kinetically controlled bromination to yield an *equatorial* brominated product. Thus, 2α-bromo-2β-methylcholestan-3-one (XIX), and not (XVII), was formed in the bromination.[14,17] The bromine would have to be *equatorial* in the chair form of this compound, which is contrary to the above-mentioned spectroscopic results.

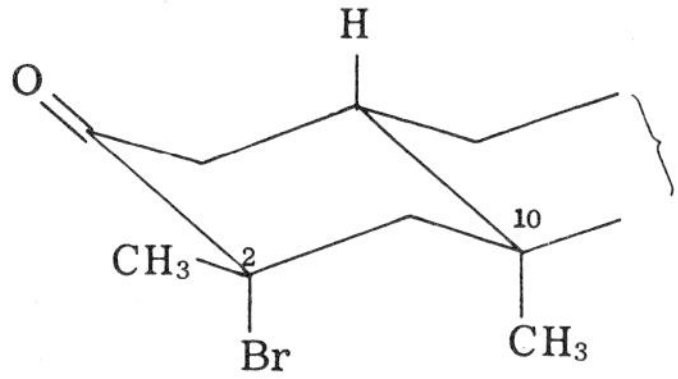

XVIII

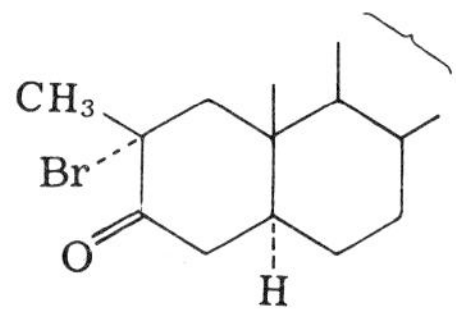

XIX

However, if ring A were to assume the boat conformation, the *axial* position of the bromine as well as the negative Cotton effect would be explained. The reason for ring A assuming the boat conformation is the strong *diaxial* methyl–methyl interaction which would arise in (XIX) if ring A were in the chair conformation. (The methyl group assumes the *axial* position in the chair conformation.) In addition an *equatorial* bromine adjacent to the keto group is unfavorable from an electrostatic viewpoint, and should be considered. These two unfavorable factors are absent in the boat conformation.

Djerassi and co-workers[18] have shown in the same way, that ring A exists in the boat conformation in 2α-bromo-2β-methylandrostan-17β-ol-3-one acetate (XX). A negative Cotton effect was measured, while the infrared and ultraviolet spectra indicated an *axial* bromine. Both results can be explained only by assuming a boat conformation for ring A.[19]

The boat conformation was also assigned to ring A in 2β-methyl-5α-cholest-6-en-3-one (XXI, R=H) and 2,2-dimethyl-5α-cholest-6-en-3-one (XXI, R=CH_3) on the basis of the optical rotatory dispersion measurements and chemical data.[20]

---

[17] C. Djerassi, N. Finch, R. C. Cookson, and C. W. Bird, *J. Am. Chem. Soc.* **82**, 5488 (1960).

[18] R. Mauli, H. J. Ringold, and C. Djerassi, *J. Am. Chem. Soc.* **82**, 5494 (1960).

[19] Compare also: C. Djerassi, E. Lund, and A. A. Akhrem, *J. Am. Chem. Soc.* **84**, 1249 (1962).

[20] F. Sondheimer, Y. Klibansky, Y. M. Y. Haddad, G. H. R. Summers, and W. Klyne, *Chem. & Ind. (London)* p. 902 (1960); *J. Chem. Soc.* p. 767 (1961); compare also J. S. E. Holker and W. B. Whalley, *Proc. Chem. Soc.* p. 464 (1961).

Both compounds, similar to the examples cited above, show a strongly negative Cotton effect curve. In comparison, the corresponding saturated 2β-methyl-3-ketone shows a normal positive Cotton effect, which agrees

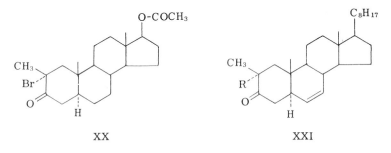

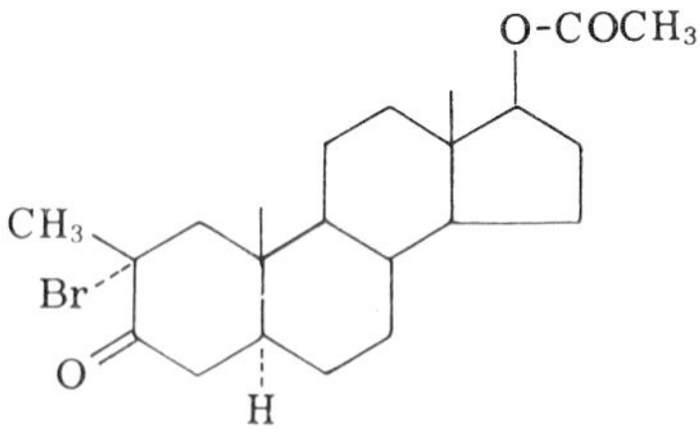

XX

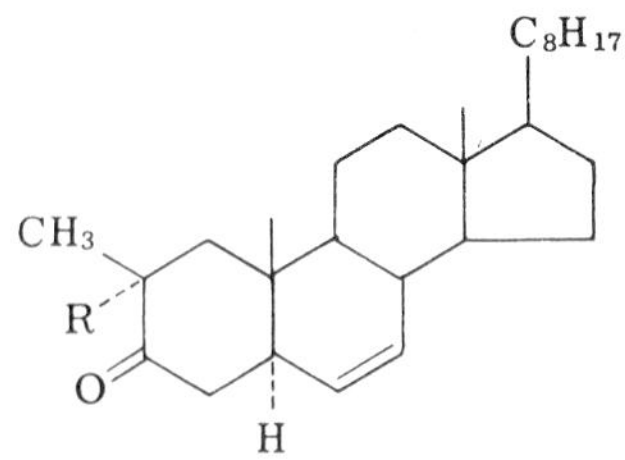

XXI

with a chair conformation of ring A in this compound. This observation is interesting because the preference of the boat conformation in the unsaturated compound (XXI, R=H) must be attributed to a change in stability as a result of the introduction a double bond at 6.

Ring A in all 8β-methyl-3-keto triterpenoids (XXII) is claimed to exist in the boat conformation, as deduced from rotatory dispersion measurements.[21]

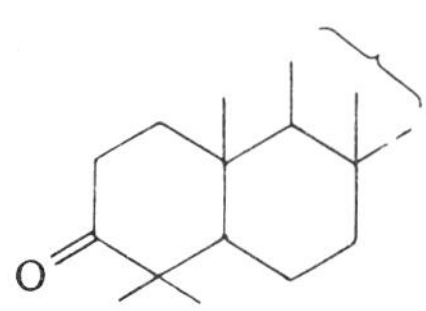

XXII

Because the three methyl groups at C-4, C-8, and C-10 are *axial* in an all chair conformation, the 1,3-*diaxial* interaction of the methyl groups would be decreased with ring A in the boat conformation.

At the end of this section, however, it should be pointed out that due to recent investigations with the help of dipole moment and NMR measurements as well as studies of rotatory dispersion, the proposal of boat conformations (or equilibria between chair and boat) in 3-ketosteroids containing the 2,2- and 4,4-gem dimethyl group as well as in certain 3-keto triterpenoids has been challenged by several workers.[21a] The conformation of ring A in this compounds is deduced as a flattened chair.

---

[21] R. Hanna, J. Levisalles, and G. Ourisson, *Bull. Soc. Chim. France* p. 1938 (1960).

[21a] N. L. Allinger and M. A. DaRooge, *Tetrahedron Letters* **19**, 676 (1961); J. M. Lehn, J. Levisalles and G. Ourisson, *ibid.* **19**, 682 (1961); N. L. Allinger and M. A. DaRooge,

b. Cyclohexane Derivatives of Simple Structure

The examples cited here from the steroid and triterpene series, in which boat (or *twist*) conformations were shown, can be expanded with additional examples of cyclohexane derivatives of simple structure. As mentioned previously (page 48) *trans*-1,3-di-*tert*-butylcyclohexane, and *cis*-2,5-di-*tert*-butyl-1,4-cyclohexanedione (page 157), exist in the *twist* conformation. 2,5-Di-*tert*-butyl-1,4-cyclohexanediol, in which the OH groups as well as the *tert*-butyl groups have a *cis* relationship to each other, has been studied with the help of infrared spectroscopy. The infrared spectrum showed an intramolecular hydrogen bond, which is only possible with the existence of a *twist* conformation.[21b]

Application of the octant rule[13] to the rotatory dispersion curve of *cis*-2-*tert*-butyl-5-methylcyclohexanone[22] led to the conclusion that the chair form (XXIII) was not the only conformation present. As a result of the interaction between the *axial* methyl group at C-5 and the *axial* hydrogen at C-3, in addition to the interaction of one of the methyl groups of the *tert*-butyl group and the *axial* hydrogen at C-3, the molecule flips into the *twist* form (XXIV).

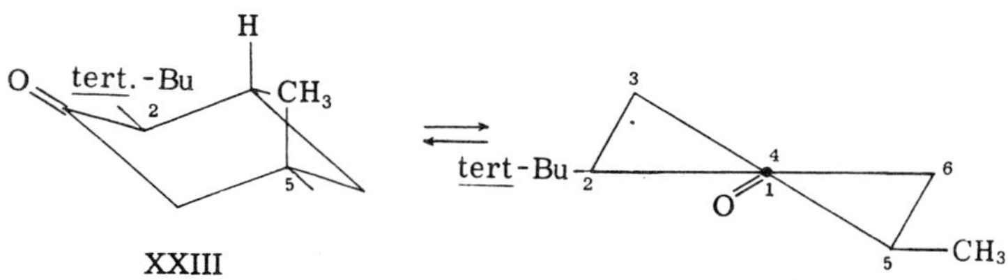

XXIII XXIV

From the optical rotatory dispersion results it seems probable that 2-methylcyclohexanone also contains a small amount in the *twist* form.[23] The same is true of 2,4-di-*tert*-butylcyclohexanone. The low

J. Am. Chem. Soc. **84**, 4561 (1962); A. K. Bose, M. S. Manhas, and E. R. Malinowski, *ibid.* **85**, 2795 (1963); J. M. Lehn, J. Levisalles, and G. Ourisson, *Bull. Soc. Chim. France* p. 1096 (1963); P. Witz, H. Herrmann, J. M. Lehn, and G. Ourisson, *ibid.* p. 1101 (1963); J. M. Lehn and G. Ourisson, *ibid.* p. 1113 (1963); A. D. Cross and I. T. Harrison, *J. Am. Chem. Soc.* **85**, 3223 (1963); compare also: M. Tichý, J. Šipoš, and J. Sicher, *Collection Czech. Chem. Commun.* **27**, 2907 (1962).

[21b] R. D. Stolow, *J. Am. Chem. Soc.* **83**, 2592 (1961); R. D. Stolow and M. M. Bonaventura, *ibid.* **85**, 3636 (1963); compare also R. D. Stolow and C. B. Boyce, *J. Org. Chem.* **26**, 4726 (1961); R. D. Stolow, *J. Am. Chem. Soc.* **84**, 686 (1962); M. Svoboda, M. Tichý, J. Fajkoš, and J. Sicher, *Tetrahedron Letters* p. 717 (1962).

[22] C. Djerassi, E. J. Warawa, J. M. Berdahl, and E. J. Eisenbraun, *J. Am. Chem. Soc.* **83**, 3334 (1961); C. Djerassi, P. A. Hart, and E. J. Warawa, *ibid.* **86**, 78 (1964).

[23] C. Beard, C. Djerassi, T. Elliott, and R. C. C. Tao, *J. Am. Chem. Soc.* **84**, 874 (1962); C. Beard and C. Djerassi, J. Sicher, F. Šipoš, and M. Tichý, *Tetrahedron* **19**, 919 (1963).

enthalpy change during the equilibration of the *cis* and *trans* isomers (2.39 kcal./mole) seems to indicate that the *trans* compound (in which one of the *tert*-butyl groups would have to assume the *axial* position in the chair form) exists in the *twist* form.[24] The appearance of *twist* forms must be considered also in the conformational equilibrium of α-hydroxy-cyclohexanone as determined by optical rotatory dispersion.[25]

Kinetic measurements have also been called upon to indicate the presence of a boat conformation. Oxazoline formation from *trans*-2-benzamidocyclohexylmethanesulfonate (XXV, R=H) and from the *cis* isomer (XXVI, R=H) was investigated.[26]

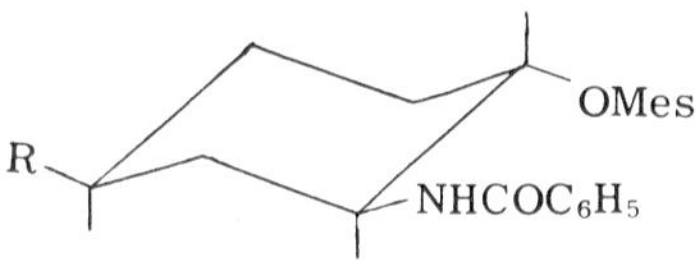

XXV

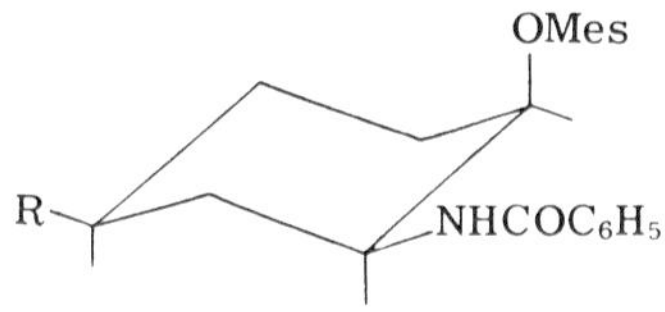

XXVI

Because the oxazoline is formed *via* a *diaxial* transition state (intramolecular S_N2 mechanism, see page 236), the *trans* compound (XXV, R=H) reacts much faster than the *cis* compound. The two substituents are easily brought into the *diaxial* position by interconversion of the chair forms. If a *tert*-butyl group is introduced into the *trans* compound (XXV, R=*tert*-butyl), then all substituents would have to become *axial* through conversion to the other chair. This is unlikely because of the extremely severe 1,3 interactions. Nevertheless, the *tert*-butyl derivative (XXV, R=*tert*-butyl) forms an oxazoline as rapidly as the compound (XXV, R=H) without the *tert*-butyl group. It is, therefore, necessary to assume that the *tert*-butyl compound (XXV, R=*tert*-butyl) reacts by way of the boat or *twist* form.

In this last example a question was raised which has been alluded to at various times. In the previous examples, the boat form was the conformation of the ground state, but in this last case, the boat form actually occurs only during the reaction path, i.e., in the transition state.[27] This possibility cannot be excluded, especially in compounds of suitable structure. We alluded to this possibility in the treatment of the relationship between conformation and reactivity. These concepts still possess a great deal of hypothetical character, but a series of reac-

[24] N. L. Allinger and H. M. Blatter, *J. Am. Chem. Soc.* **83**, 995 (1961).

[25] C. Djerassi, R. Records, and B. Bach, *Chem. & Ind. (London)* p. 258 (1961).

[26] J. Sicher, M. Tichý, F. Šipoš, and M. Ṕanková, *Proc. Chem. Soc.* p. 384 (1960); *Collection Czech. Chem. Commun.* **26**, 2418 (1961).

[27] Compare D. S. Noyce, B. R. Thomas, and B. N. Bastian, *J. Am. Chem. Soc.* **82**, 885 (1960).

tions are known which can only be explained by the assumption of a boat (or *twist*) conformation in the transition state.[28] (There is a detailed treatement of this problem in ref. 28.)

2. Compounds with Bridged Ring Structure

a. Bicyclo[2.2.1]heptane and Bicyclo[2.2.2]octane

In compounds of the bicycloheptane (I) or bicyclooctane type (II), the cyclohexane ring is fixed in the boat conformation because of the bridged ring structure.

There are various possibilities for the steric disposition of a substituent attached to bicyclo[2.2.1]heptane (norbornane) (I, XXVII, XXVIII) depending on the carbon atom to which the substituent is attached.

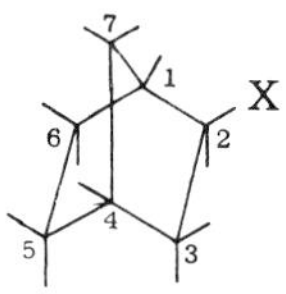

XXVII

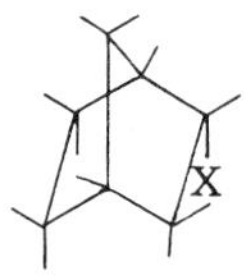

XXVIII

A substituent on carbons 2, 3, 5, and 6 can assume either the boat-*equatorial* position (XXVII) or the boat-*axial* position. In the former, the substituent is on the same side as the methylene (C-7) bridge. In the latter, the substituent is *trans* to the methylene bridge and is directed toward the inside of the semispherical molecule. The boat-*equatorial* position is also designated as *exo*-, and the boat-*axial* as the *endo* position. The position of a substituent on carbon atoms 1 and 4 is unequivocal. A geometrical isomerism is possible at carbon 7 only if there is a substituent on carbon atoms 2, 3, 5, or 6, or if there is a 2,3 or 5,6 double bond (bicyclo[2.2.1]heptene-2). The substituent at carbon 7 can then assume a *syn* (=*cis*) or *anti* (=*trans*) relationship to the second substituent or to the double bond.

The application of the rules of conformational analysis to such compounds is, nevertheless, subject to various restrictions. Experimental data for *simple* derivatives of bicycloheptane (I) or bicyclooctane (II), in comparison to cyclohexane compounds in the chair conformation, are relatively sparse.

In contrast to this, methyl-substituted stereoisomeric bicycloheptane

[28] J. Levisalles, *Bull. Soc. Chim. France* p. 551 (1960).

derivatives, such as 3-methylnorborneol, 3,3- and 5,5-dimethylnorborneols, and also compounds of the camphane (XXIX), fenchane (XXX), camphenilane (XXXI), and pinane (XXXII) types have been investigated much more thoroughly.

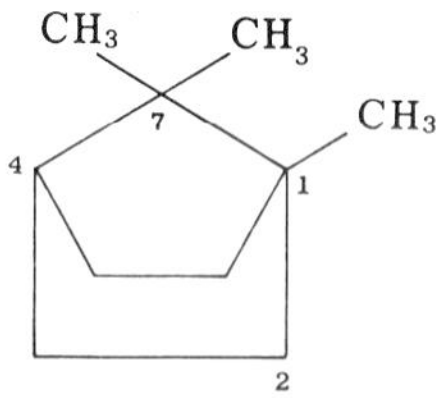

XXIX

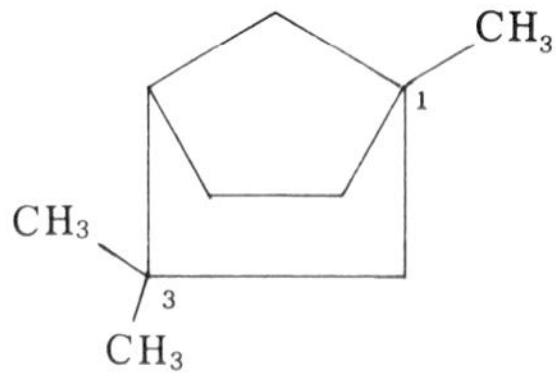

XXX

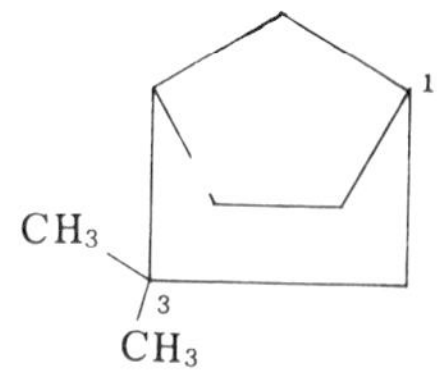

XXXI

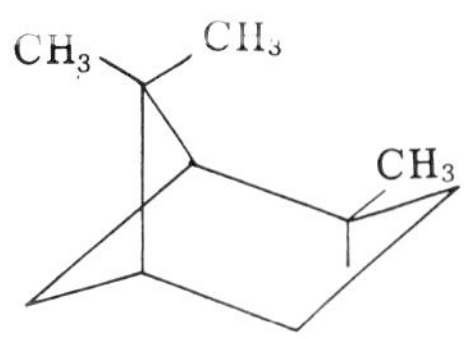

XXXII

As shall be described below with some examples, the reactivity of a substituent in these cases cannot be related with certainty to the boat-*equatorial* (be, see formula XXVII), *exo* position or boat-*axial* (ba), or *endo* position (XXVIII) on the cyclohexane ring. The reason lies in the fact that the methyl groups on carbon atom 7 of the bridge, the methyl groups at C-3 (XXX) and (XXXI), and the methyl group on the bridgehead at C-1 (XXIX) exert an influence on the reactivity of substituents in *exo* or *endo* positions at C-2. These effects of the methyl groups are due mainly to steric hindrance and make considerations of the reactivity of a substituent based only on its conformation very difficult. A conclusion as to whether the high or low reaction rate is due only to the boat-*axial* or boat-*equatorial* position of a substituent is not possible (compare also page 292).

The cyclohexane ring of the bicyclo[2.2.1]heptane system (norbornane) (I) is rigidly fixed in the boat form because of the methylene bridge.[29] Because the bridge consists of only one carbon atom, the

[29] (a) C. W. Shoppee, *Chem. & Ind. (London)* p. 86 (1952); (b) D. H. R. Barton, *J. Chem. Soc.* p. 1027 (1953); (c) For calculation of distances in bicyclo[2.2.1]heptane see C. F. Wilcox, Jr., *J. Am. Chem. Soc.* **82**, 414 (1960).

molecule is deformed. Carbon atom 1 and 4, to which the bridge is attached, are brought closer together than would be normal for a boat form without the methylene bridge. As a result, considerable Baeyer strain is introduced; it has been estimated at about 20 kcal./mole.[30] Other bicycloheptane systems, such as bicyclo[3.1.1]heptane (cf. pinane, XXXII) have been assigned a larger strain energy (~24 kcal./mole) because they contain a highly strained four-membered ring.[30] This also manifests itself in the well-known ionic rearrangement of bicyclo[3.1.1]-heptanes into the bicyclo[2.2.1]heptane series. This rearrangement proceeds with great facility, whereas reverse rearrangement, in which pinane derivatives are formed from camphane derivatives, have not been observed.

The introduction of a double bond into norbornane leads to norbornene (XXXIII). The strain of the molecule is thereby greatly increased.[30]

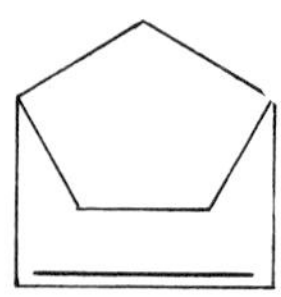

XXXIII

The great strain becomes evident in various special properties of the double bond. For example, the C=C stretching vibration in the infrared spectrum of norbornene lies at 1568 cm.$^{-1}$,[31] which corresponds approximately to the C=C stretching vibration in highly strained cyclobutene.[30,31] The olefinic C—H stretching vibration of norbornene has a decidedly higher value (3070 cm.$^{-1}$) than the normal one for *cis*-olefins (3010 cm.$^{-1}$).[32] R.B. Turner and co-workers were able to show that norbornene possesses an extraordinarily high heat of hydrogenation.[33] The value found for norbornene was 33.1 kcal./mole. A series of non-bonded interactions between the hydrogens vanishes with the introduction of a double bond in bicyclo[2.2.1]heptane. These are the interactions of the *eclipsed* methylene groups of carbon atoms 2 and 3 (XXXIV), and the 1,3-boat-*axial* interactions between the *endo* (ba)-hydrogens at C-2 and C-6, and at C-3 and C-5. A situation similar to the

[30] (a) P. v. R. Schleyer, *J. Am. Chem. Soc.* **80**, 1700 (1958); (b) Earlier estimates are lower, at about 6 kcal./mole; cf. R. P. Linstead, *Ann. Rep. Chem. Soc.* **32**, 315 (1935).

[31] R. C. Lord and R. W. Walker, *J. Am. Chem. Soc.* **76**, 2518 (1954).

[32] P. v. R. Schleyer and M. M. Donaldson, *J. Am. Chem. Soc.* **78**, 5702 (1956); (b) N. Sheppard and D. M. Simpson, *Quart. Rev.* (*London*) **6**, 1 (1952); (c) For the dipole moment of norbornene see N. L. Allinger and J. Allinger, *J. Org. Chem.* **24**, 1613 (1959).

[33] R. B. Turner, W. R. Meador, and R. E. Winkler, *J. Am. Chem. Soc.* **79**, 4116 (1957).

one comparing the heats of hydrogenation of cyclopentene and cyclohexene exists in this case (see Chapter 3, page 79). One might, therefore, expect that the heat of hydrogenation of norbornene would

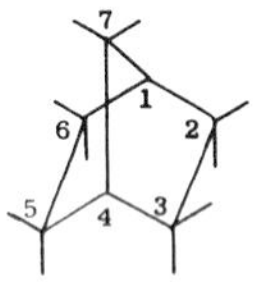

XXXIV

be lower than that of cyclohexene, as is true for cyclopentene.[30] The value for cyclohexene amounts to 27.1 kcal./mole.[33] Actually, the heat of hydrogenation of norbornene is 6 kcal./mole higher. The high heat of hydrogenation of norbornene is attributed to a greatly increased Baeyer strain (angle strain).[33] The spectroscopic data mentioned[31] and the facile isomerization of norbornene to nortricyclene[30] also indicate the high Baeyer strain.

As has been shown in various additional examples, the introduction of a double bond in the bicyclo[2.2.1]heptane system is difficult if the possibility exists for formation of an exocyclic double bond. For example, 2-methyl isofenchol (XXXV) yields mainly 1-methyl-β-fenchene (XXXVI) through dehydration, and only a very small amount of 1-methyl-α-fenchene (XXXVII) is formed.[34]

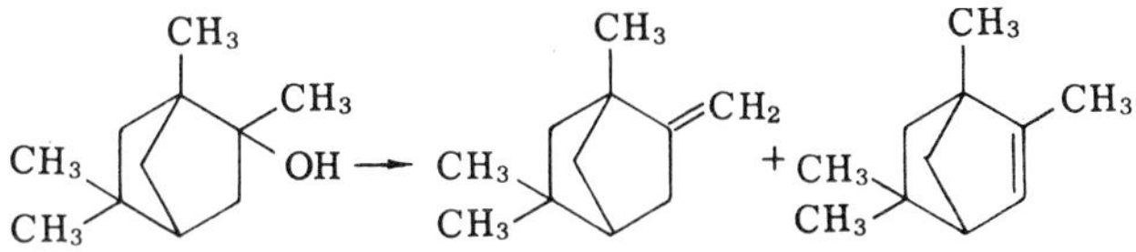

XXXV XXXVI XXXVII

The introduction of two double bonds into bicyclo[2,2,1]heptane leads to bicycloheptadiene-2,5 (norbornadiene, XXXVIII) which is much more strained then norbornene and shows a heat of hydrogenation of 68.1 kcal./mole.[33]

XXXVIII

[34] (a) G. Komppa and G. A. Nyman, *Ann. Chem.* **523**, 87 (1936); (b) Cf. also O. H. Wheeler, R. Cetina, and J. Z. Zabicky, *J. Org. Chem.* **22**, 1153 (1957) with further examples.

In contrast to bicyclo[2.2.1]heptane, bicyclo[2.2.2]octane has no Baeyer strain. Because the bridge is formed by two methylene groups, the valence angles were not deformed.[35] But it can be seen from a model that bicyclooctane (II or XXXIX) possesses a maximum of nonbonded interactions.

XXXIX

There are present six 1,2 and six 1,3 nonbonded hydrogen interactions. Because this arrangement, as mentioned above, represents the energy maximum for nonbonded interactions, the molecule will endeavor to avoid the strain. Turner, Meador, and Winkler[33] have claimed that this can be attained by rotation on the C-1–C-4 axis and is possible without distortion of the bond angles if the twist around the axis does not exceed 10°.[33] On the other hand, Schleyer and Nicholas[35b] showed that twisting increases rather than decreases the energy of the molecule. With infrared and Raman spectra the non twisted structure for bicyclo-[2.2.2]octane was confirmed.[35d]

The heats of hydrogenation for bicyclooctene (XL) and bicyclo-octadiene (XLI) amount to 28.2 and 56.2 kcal./mole, respectively.[33]

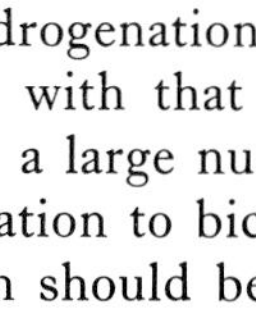

XL

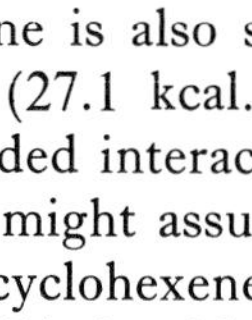

XLI

The heat of hydrogenation for bicyclooctene is also surprisingly high when compared with that of cyclohexene (27.1 kcal./mole). Because, as shown above, a large number of nonbonded interactions are formed in the hydrogenation to bicyclooctane, one might assume that the heat of hydrogenation should be below that of cyclohexene. An explanation may lie in the fact that bicyclooctene (XL) is able to form a twist

[35] (a) Cf. W. Hückel, *Ann. Chem.* **455**, 123 (1927); (b) P. v. R. Schleyer and R. D. Nicholas, *J. Am. Chem. Soc.* **83**, 2700 (1961); cf. also: P. v. R. Schleyer, K. R. Blanchard, and C. D. Woody, *ibid.* **85**, 1358 (1963).

[35d] J. J. Macfarlane and I. G. Ross, *J. Chem. Soc.* p. 4169 (1960).

conformation. The higher strain of the double bond developed thereby is responsible for the increased heat of hydrogenation.[33]

Whereas extensive material is available to show the relation between the conformation of cyclohexane compounds in the chair form and their physical properties (see Chapter 3, 4 and 5), experimental data is sparse in the series of compounds with bridged ring structure.

Hirsjärvi and Salo were able to show for various bicycloheptanols[36] that as in the case of cyclohexanols (see page 113), the OH stretching vibrations of *axial* OH groups occur at frequencies different than those of *equatorial* OH groups. *Endo*-1-norborneol (XXVIII, X=OH) with a boat-*axial* OH group, absorbs at 3623 cm.$^{-1}$. The isomeric *exo*-norborneol (XXVII, X=OH), with a boat-*equatorial* OH group, absorbs at 3616 cm.$^{-1}$.

In general, it was determined that in such alcohols the *exo* (boat-*equatorial*) compound absorbs at lower wave numbers than its *endo* (boat-*axial*) isomer by about 2–7 cm.$^{-1}$. The OH stretching vibration, therefore, can be used for a configurational determination. If vicinal methyl groups are present at position 3 (fenchols), then the rule is no longer valid.

In addition, it has been shown[37] that there is a difference in the shape of the OH absorption bands between *exo*-(boat-*equatorial*) and *endo*-(boat *axial*)bicycloheptanols. Alcohols of the *exo* configuration show a single band, while *endo* alcohols give rise to a doublet.

Some 3-halocamphors were studied by means of ultraviolet and infrared spectroscopy, measurement of dipole moments, and by NMR spectroscopy.[38] The infrared shifts (cf. page 233) are practically the same for *exo*-3-chlorocamphor (XLIIa, X=Cl, Y=H) and the *endo* isomer (XLIIa, Y=Cl, X=H) (also designated as α′ and α-chlorocamphor). No typical difference appears between the *exo* and *endo* positions employing this technique.[38b] It has been deduced from this that the chlorine atom assumes an intermediate position between boat-*axial* and boat-*equatorial* in the *exo* as well as the *endo* isomer (bisectional position, cf. page 75) (XLIIb).

No interaction occurs between the methyl group on C-7 and the *exo* chlorine. However, an interaction can be established between this

---

[36] (a) P. Hirsjärvi and K. Salo, *Suomen Kemistilehti* **32B**, 280 (1959); (b) cf. also J. Faivonen and P. J. Mälkönen, *Suomen Kemistilehti* **32B**, 277 (1959); (c) for infrared spectra of bicycloheptane-diols see P. Hirsjärvi, *Ann. Acad. Sci. Fennicae, Ser. A* **84**, 3 (1957).

[37] (a) R. Piccoloni and S. Winstein, *Tetrahedron Letters* **13**, 4 (1959); (b) P. Tuomikoski, E. Pulkkinen, P. Hirsjärvi, and N. J. Toivonen, *Suomen Kemistilehti* **23B**, 44 (1950); cf. also H. S. Aaron and C. P. Rader, *J. Am. Chem. Soc.* **85**, 3046 (1963).

[38] (a) R. C. Cookson, *J. Chem. Soc.* p. 282 (1954); (b) F. V. Brutcher, Jr., T. Roberts, S. J. Barr, and N. Pearson, *J. Am. Chem. Soc.* **81**, 4915 (1959); (c) W. D. Kumler, N. Pearson, and F. V. Brutcher, Jr., *ibid.* **83**, 2711 (1961).

methyl group and the bulkier bromine in α,α′-dibromocamphor (XLIIa, X=Y=Br).[38b,c] A bisectional position of the chlorine in the isomeric 3-chlorocamphors is also indicated by measurement of the dipole moments. The *exo* and *endo* isomers (XLIIa, X = Cl and Y=Cl) possess the same dipole moment.[38c]

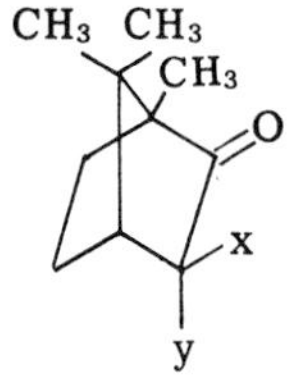

XLIIa

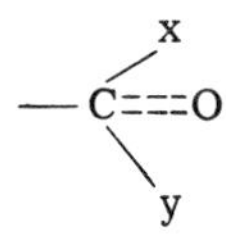

XLIIb

At the outset, the difficulties of relating reactivity and stability of methyl-substituted bicycloheptane derivatives with the boat-*equatorial* (*exo*) or the boat-*axial* (*endo*) position were mentioned. This is exhibited by many examples. In the non-methyl-substituted bicyclo[2.2.1]-heptanes (norbornanes) e.g., the boat-*equatorial exo*-alcohol (XXVII, X=OH) is more stable than the corresponding *endo* isomer (XXVIII, X=OH).[39] The stability conditions reverse themselves with the introduction of geminal dimethyl groups at position 7.[29a] As a result borneol (XLIII), with a boat-*axial* OH group is more stable than the isomeric boat-*equatorial* isoborneol (XLIV).[39c,40]

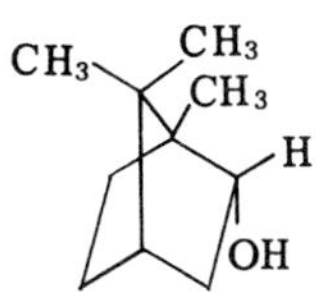

XLIII

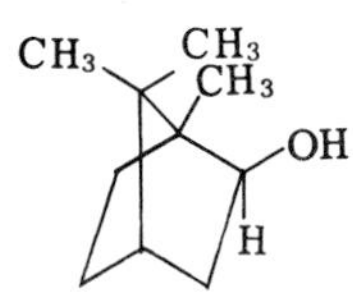

XLIV

The reversal of the stability is attributed mainly to the interaction of the *syn* methyl group at carbon atom 7 with the OH group in the *exo* position of isoborneol (XLIV).

The effect of the steric hinderance on the reactivity and the degree

---

[39] (a) K. Alder and G. Stein, *Ann. Chem.* **514**, 211 (1934); *ibid.* **525**, 183 (1936); (b) J. D. Roberts, C. C. Lee, and W. H. Saunders, Jr., *J. Am. Chem. Soc.* **76**, 4501 (1954); (c) T. Takeshita and M. Kitajima, *Nippon Kagaku Zasshi* **79**, 1472 (1958); *Chem. Abstr.* **54**, 5729 (1960); (d) C. F. Wilcox, Jr., M. Sexton, and M. F. Wilcox, *J. Org. Chem.* **28**, 1079 (1963).

[40] (a) G. Vavon, *Bull. Soc. Chim. France* [4] **49**, 937 (1931); (b) M. Lipp and E. Bund, *Ber.* **68**, 249 (1935); (c) Y. Asahina, M. Ishidate, and T. Sano, *Ber.* **69**, 343 (1936).

of association was investigated by Beckmann and co-workers for a large number of stereoisomeric bicyclo [2.2.1]-heptanes substituted at various positions by one or more methyl groups.[41] Saponification of the acid phthalates and measurements of the degree of association of the alcohols gave the following results.[41a]

The nonsubstituted stereoisomeric norborneols (*endo* and *exo*) differ very little in the saponification rates of their acid phthalates (acid phthalate of *endo*-norborneol 5.8; of *exo*-norborneol 5.6)[42] or in the degree of association. No specific difference between a boat-*axial* and a boat-*equatorial* position of the acid phthalate group can be determined.

However, if a methyl group is introduced at positions 1, 3, or 7, a large difference is found in the saponification rates of the acid phthalates and in the degree of association of an OH group in position 2, depending on whether the OH group is *endo* or *exo*.

If the methyl group is at position 1, (compare XXVII) then an *exo*-2-OH group (XXVII) is much more hindered than an *endo*-2-OH group. The saponification rate of the acid phthalate of the *exo* alcohol (0.59)[42] is much smaller than that of its *endo* epimer (XXVIII) (2.1).[42,43] If the methyl group is *exo* at position 3, (compare XXVII) then a *cis*-2-OH group is greatly hindered (0.62),[42] and is essentially not hindered (5.3)[42] when it is *trans* (*endo*). The reaction rate for the latter acid phthalate nearly corresponds to that of norborneol. A methyl group at position 7, which is *syn* to an *exo*-OH group at C-2, (compare XXVII) causes less steric hindrance than an *exo*-3-methyl group.

The influence of steric hinderance is also noticeable in other reactions. A series of methylated bicycloheptanones was prepared by Wheeler and co-workers[34b] and the dissociation constants of their cyanohydrins were determined. α-Fenchocamphorone (XLVI) is 13 times less reactive in cyanohydrin formation than norcamphor (XLV) as a result of its *gem*-dimethyl group on the methylene bridge. Camphor (XLVII) with an additional methyl group at the bridgehead (carbon 1) is even less reactive by a factor of 18.

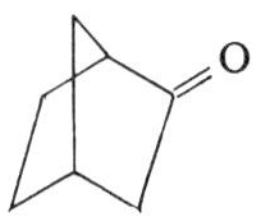

XLV

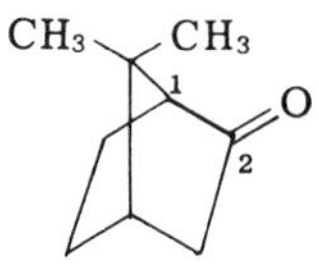

XLVI

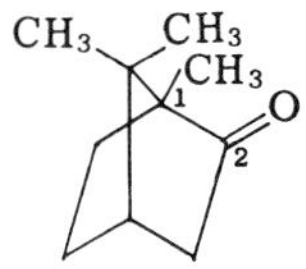

XLVII

[41] (a) Review: S. Beckmann, *Bull. Soc. Chim. France* p. 1319 (1960); (b) S. Beckmann, G. Eder, and H. Geiger, *Suomen Kemistilehti* **31B**, 56 (1958).

[42] $k \times 10^2$ (liters mole^{-1} min.$^{-1}$) at 40°C.

[43] Cf. also T. Kuusinen and M. Lampinen, *Suomen Kemistilehti* **32B**, 26 (1959).

If the *gem*-dimethyl group is introduced adjacent to the keto group, as in camphenilone (XLVIII) and fenchone (XLIX), then the effect of steric hinderance is still greater. Camphenilone (XLVIII) is 90 times and fenchone (XLIX) 130 times less reactive in cyanohydrin formation than norcamphor (XLV). There is practically no steric hinderance in

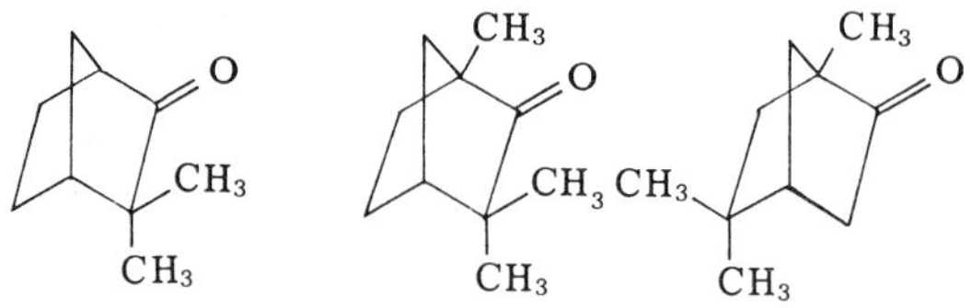

XLVIII XLIX L

isofenchone (L), in which the *gem* dimethyl groups are on the other side of the ring.

The reaction rate is characteristically different for substituents on the bicyclo[2.2.1]heptane nucleus which are capable of solvolysis reactions (halides, sulfonic acid esters). A substituent in the *exo* position (boat-*equatorial*) reacts decidedly faster than an *endo* substituent (boat-*axial*). The reaction rate of the following sulfonates and chlorides have been investigated[43a] *exo*- and *endo*-norbornyl[44,45] (XXVII; XXVIII) *endo*-bornyl and *exo*-isobornyl,[44,45,46] (compare XLIII and XLIV) and *endo*- and *exo*-camphenilyl (LI),[47] and *exo*-fenchyl (LII, X=OTs).[48]

LI LII

From the examples studied to date it follows that the ratio of the solvolysis rates $k_{exo} : k_{endo}$ is always much greater than the ratio $k_{cis} : k_{trans}$ of the corresponding esters of simple alkylated cyclohexanes

[43a] Further examples see: J. A. Berson, *in* " Molecular Rearrangements " (P. de Mayo, ed.), Vol 1, p. 111. Wiley (Interscience), New York, 1963.

[44] S. Winstein, B. K. Morse, E. Grunwald, H. W. Jones, J. Corse, D. Trifan, and H. Marshall, *J. Am. Chem. Soc.* **74**, 1127 (1952); S. Winstein and D. Trifan, *ibid.* **74**, 1174, 1154 (1952).

[45] W. Hückel and K. Tomopulos, *Ann. Chem.* **610**, 78 (1957), with further references.

[46] Only the isobornyl chloride has been investigated. It has not been possible to prepare an isobornyl sulfonate up to now.

[47] W. Hückel and D. S. Nag, *Ann. Chem.* **645**, 101 (1961).

[48] (a) W. Hückel and H. Rohrer, *Chem. Ber.* **93**, 1053 (1960); (b) Only the toluenesulfonate has been investigated.

(such as *cis*- and *trans*-2-methylcyclohexyl toluenesulfonate, etc.), and for fused bicyclic rings of the decalin and hydrindane types (see pages 173 and 182). The reaction rate of the *endo* isomers seem to fit into the cyclohexane series. Bornyl- and *endo*-norbornyl benzene sulfonates react about as rapidly in acetolysis as cyclohexyl benzene sulfonate.[45] Bornyl tosylate reacts about as rapidly as menthyl tosylate.[45] In the latter case, the *gem*-dimethyl group on the methylene bridge has a similar effect in reducing the reaction rate as does a *trans* neighboring group on cyclohexyl toluenesulfonate.[49] However, *endo*-camphenilyl (LI, X=OTs) and *endo*-fenchyl toluenesulfonates (LII, X=OTs) react much more slowly. Their rates are lower than those of the slowest reacting *trans*-toluenesulfonate of the cyclohexane series.

The high reaction rate of the *exo*-chlorides and sulfonates is, according to Winstein[44], a result of neighboring group participation by carbon during the rate-determining ionization step.

The stereochemistry of the *exo* compound is especially favorable for delocalization of the participating electron pair, as shown in (LIII). Carbon atoms 6, 1 and 2 and the leaving group X, lie nearly in one plane.

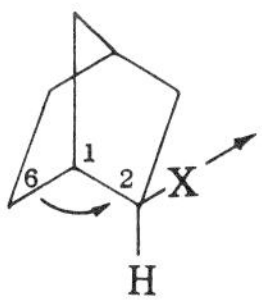

LIII

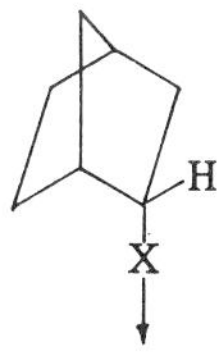

LIV

Rearrangements are well known for these compounds and have been thoroughly investigated. They proceed more easily in the *exo* series than for the isomeric *endo* compounds[50] for the above reason.

The ionization of the corresponding *endo* derivatives (LIV) cannot be assisted by carbon participation. The *p* orbital developed during ionization lies in about the same direction from which the anion leaves the molecule. The direction of the *p* orbital is very unfavorable for participation with the electrons of the C-6–C-1 bond.

The stereoselectivity of a series of reactions was studied with compounds of the bicyclo[2.2.1]heptane type.[51] Reductions of bicycloheptane ketones are especially pertinent. As was be shown previously (page 269), these reactions possess a certain significance for conformational analysis of ketones of the cyclohexane series. On the basis of new

---

[49] W. Hückel, *et al. Ann. Chem.* **624**, 142 (1959).

[50] Cf., for example, N. A. Abraham, *Ann. Chim.* (*Paris*) p. 961 (1960).

findings concerning the reduction reactions, the following statements have been made.[41a,47,51]

The reduction of ketones of the bicycloheptane series with sodium in alcohol produces a mixture of *exo*- and *endo*-alcohols. The amount of *endo*-alcohol is always larger. Norcamphor (XLV) yields 76% *endo*-norborneol (XXVIII, X=OH). Camphor (XLVII) yields 79% borneol (XLIII).

The influence of steric hinderance is again felt in the catalytic hydrogenation. About 95% *endo*-norborneol (XXVIII, X=OH) is formed from norcamphor (XLV). Camphenilone (XLVIII) and fenchone (XLIX), whose keto groups are hindered by neighboring *gem*-dimethyl groups, form ~25% *exo*-alcohol. Camphor (XLVII) gives 95% isoborneol (*exo*) (XLIV).

In the reduction of the ketones with lithium aluminum hydride, the *endo*-alcohol is always formed in greater amounts (~90%) if no methyl group is present at C-7 *syn* to the keto group. If such a group is present in a *syn* position on the bridge (such as in camphor, epicamphor, apocamphor, 7-*syn*-methylnorcamphor), then the *exo*-alcohol is formed predominantly (>90%).

From this it becomes evident that the AlH_4^- ion in lithium aluminum hydride reductions (compare page 270) attacks from the side of the methylene bridge (C-7). This leads to *exo*-addition of the hydride hydrogen and the formation of an *endo*-OH group. If the attack of the AlH_4^- ion is hindered on this side by a methyl group at position 7, then approach to the carbonyl group is from the *endo* side. This leads to the formation of an *exo*-alcohol.[41a,51a]

b. Bicyclo[3.1.1]heptane

Compounds of the bicyclo[3.1.1]heptane series are more difficult to treat from the standpoint of conformational analysis. Their most important representative is pinane (LV). Two stereoisomeric forms of pinane are possible. The methyl group at carbon atom 1 can be *cis* (LVa) or *trans* (LVb) to the bridge. The two configurational isomers are distinguished as *cis*- and *trans*-pinane.

With the introduction of a methylene bridge from C-4 to C-6 of

---

[51] P. Hirsjärvi, *Ann. Acad. Sci. Fernicae* **Ser. A 81**, 7 (1957); S. Beckmann and R. Mezger, *Chem. Ber.* **89**, 2738 (1956); W. Hückel and O. Fechtig, *Ann. Chem.* **652**, 81 (1962); compare also J. A. Berson, J. Singh Walia, A. Remanick, S. Suzuki, P. Reynolds-Warnhoff, and D. Willner, *J. Am. Chem. Soc.* **83**, 3986 (1961); G. Ourisson and A. Rassat, *Tetrahedron Letters* **21**, 16 (1960).

[51a] For the conformation of the bicylo[3.3.1]nonane system see: W. A. C. Brown, G. Eglington, J. Martin, W. Parker, and G. A. Sim, *Proc. Chem. Soc.* p. 57 (1964); M. Dobler and J. D. Dunitz, *Helv. Chim. Acta*, **47**, 695 (1964).

the cyclohexane ring, a chair and a boat form are formally present in the molecule at the same time. By the introduction of this bridge a four-membered ring is formed, and a significant deformation of the molecule is introduced.[52] Nevertheless, a substituent at C-1, C-2, or

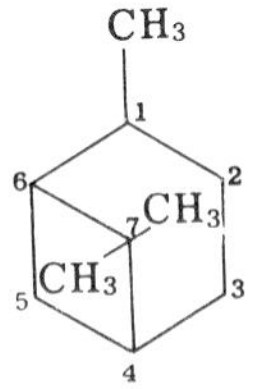

LV

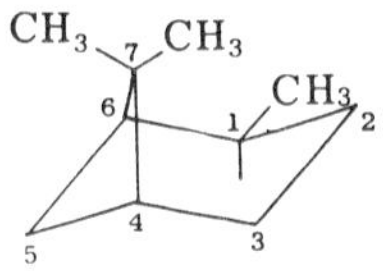

LVa

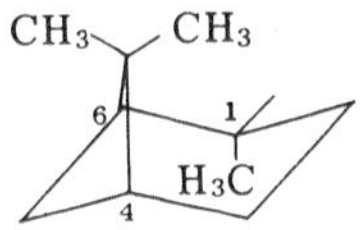

LVb

C-3, unlike those of bicyclo[2.2.1]heptane, is not fixed in its conformation. The introduction of the 4,6-methylene bridge interferes only partly with the mobility of the six-membered ring, so that a substituent on the carbon atoms named can assume either the *axial* or *equatorial* position without changing its relative configuration. It is understandable that a conformational analysis for such substituents is difficult because the interaction of the *gem*-dimethyl group at C-7 with the substituents must be considered.

Conformational analysis of the pinane system will be briefly considered using the isomeric pinocampheols as an example.[53] The four possible stereoisomeric forms of pinocampheol are known from the investigations of H. Schmidt.[54] By analogy to the menthol series (cf. page 131), they were designated as pinocampheol (LVI), neopinocampheol (LVII), isopinocampheol (LVIII) and neoisopinocampheol (LIX). Their configurations were assigned as shown in formulae (LVI–LIX).[54] Pinocampheol and neopinocampheol are derived from *trans*-pinane, isopinocampheol and neoisopinocampheol from *cis*-pinane.

The original configuration for isopinocampheol (LVIII) given by Schmidt[54] has been challenged by Bose[53] and Hückel.[47] Both assigned the extreme cis configuration for isopinocampheol. But recent investigations of G. Zweifel and H. C. Brown[54a] confirmed the original assignment of Schmidt.

---

[52] O. H. Weeler, *Chem. & Ind. (London)* p. 1020 (1954).

[53] (a) A. K. Bose, *J. Org. Chem.* **20**, 1003 (1955); (b) Cf. also Y. R. Naves, *Bull. Soc. Chim. France* p. 1020 (1956); (c) For the conformations of the corresponding ketones, pinocamphone and isopinocamphone, see M. V. Bhatt, *Chem. & Ind. (London)* p. 1452 (1959).

[54] H. Schmidt, *Ber.* **77**, 544 (1944); **80**, 520 (1947).

[54a] G. Zweifel and H. C. Brown, *J. Am. Chem. Soc.* **86**, 393 (1964).

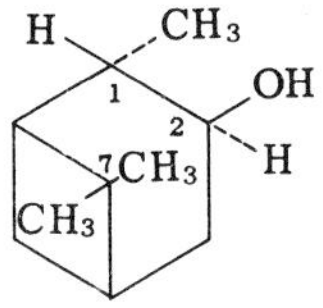

LVI

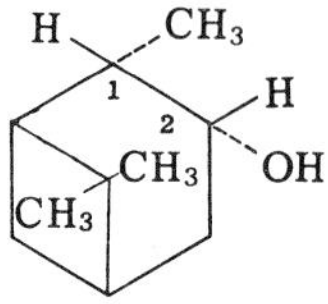

LVII

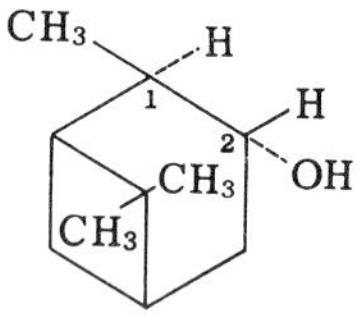

LVIII

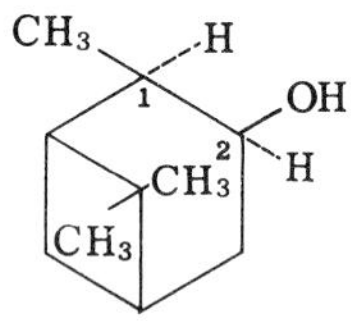

LIX

As can be seen from a model, the methyl group at C-1 and the OH group at C-2 can be brought into an *axial* or an *equatorial* position in all isomeric pinocampheols. This results from the partially retained mobility of the ring. The assignment of the conformations of the isomeric pinocampheols was provided by Bose[53a] on the basis of chemical data. He used a combination of the stability of the alcohols, their rates of acetylation,[54] as well as the solvolytic behavior of their toluenesulfonates.[54] Because pinocampheol (LVI) was esterified and also the esters were saponified rapidly, Bose designated an *equatorial* hydroxyl group for it (LX). An *axial* hydroxyl group was assumed for the slower reacting

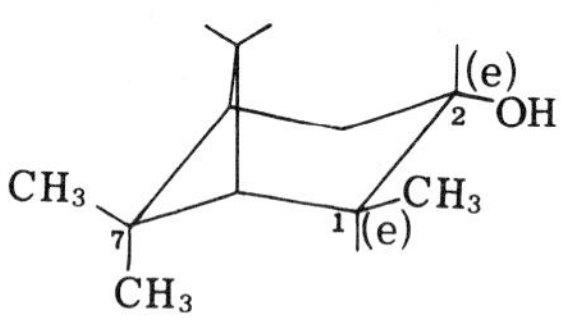

LX

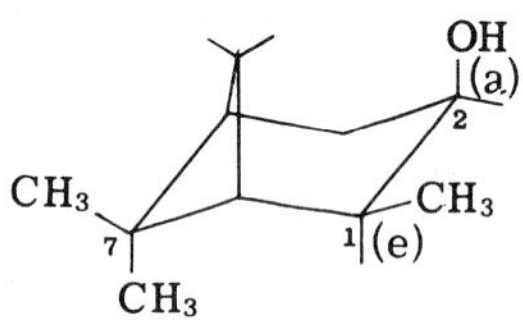

LXI

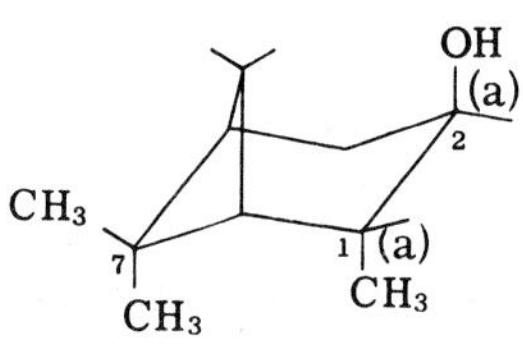

LXII

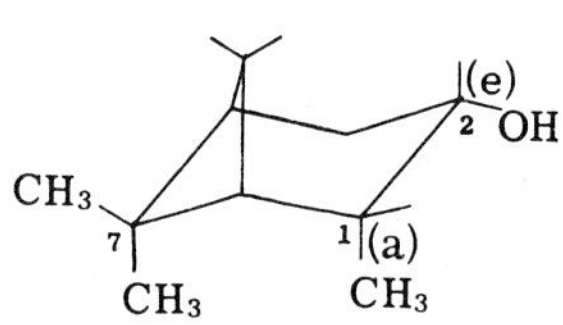

LXIII

neopinocampheol (LVII); see (LXI). Isopinocampheol (LVIII) and neoisopinocampheol (LIX) probably do not have the extreme conformations shown in (LXII) and (LXIII). Due to the spatial requirements of the methyl groups the cyclohexane ring is deformed, producing a planar conformation.[54a]

CHAPTER 7

Heterocyclic Compounds

The use of models for the study of heterocyclic compounds is helpful, but certain limiting conditions must be remembered. For one thing, one must consider the difference in distance between a hetero atom and carbon in comparison to the distance between two carbon atoms. If we limit ourselves first of all to the hetero atoms nitrogen and oxygen, no great differences are involved, as shown in Table I.

Therefore, the question of whether or not the normal valence angles of nitrogen and oxygen vary greatly from the tetrahedral angles of carbon is of greater importance. We can conclude from the limited experimental data, which include the size of these angles as well as the ease of formation and the stability of the preferred five- and six-membered rings, that the valence angles of heterocyclic compounds do not differ very much from tetrahedral angles (see Table I). The NH-group and oxygen are essentially sterically equivalent to the methylene group. The small differences in interatomic distances are hardly noticeable (see Table I).[1]

TABLE I

ATOM DISTANCES

Bond length (Å)			Bond angle (in degrees)	
C–C	1.52	(in cyclohexane)	C–C–C	111.5[a]
C–N	1.47	(in piperazine)	C–N–C	112.6[a]
C–O	1.42	(in 1,4-dioxane)	C–O–C	112.4[a]
C–S	1.82		C–S–C	~100[b]

[a] See ref. 2a below; [b] see ref. 2b below.

[1] Compare O. Hassel and B. Ottar, *Acta Chem. Scand.* **1**, 929 (1947).

[2] (a) M. Davis and O. Hassel, *Acta Chem. Scand.* **17**, 1181 (1963); (b) A. Maccoll, *Progr. Stereochem.* **1**, 361 (1954).

Therefore, O, NH, and CH_2 can replace each other in cyclic compounds without any essential change in the Baeyer strain.[3]

However, a difference can arise more easily in the case of sulfur because of the greater C—S bond distance (Table I) and the smaller C—S—C angle.

Care must be used in the usual rules of conformational analysis to heterocyclic compounds because knowledge of quantitative strain relations in heterocyclic systems is incomplete. It should be especially noted that the introduction of a hetero atom in place of a CH_2 group decreases the interaction with other groups. This can lead to conformations that would be unfavored in corresponding carbocyclic compounds.

1. Five-Membered Rings

Of the heterocyclic five-membered ring compounds, we shall first consider pyrrolidine (I), which has been investigated quite thoroughly.

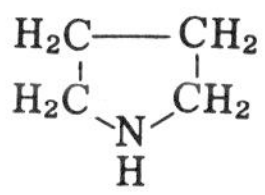

I

From calorimetric and spectroscopic data it was determined that pyrrolidine is not planar, but possesses a puckered conformation similar to cyclopentane (see Chapter 3, page 73).[4] As in the case of cyclopentane, a pseudorotation arises. The best agreement between calculated and determined thermodynamic data is obtained if the assumption is made that the pseudorotation is hindered by a potential barrier amounting to about 0.3 kcal./mole.[4] A hindered pseudorotation has also been postulated for tetrahydrothiophene.[5]

In agreement with older investigations[6,7] LeFèvre and LeFèvre by measuring the molecular polarizability were able to show[8] that tetrahydrofuran (II) also possesses a conformation which roughly corresponds

[3] Cf. H. C. Brown, J. H. Brewster, and H. Schechter, *J. Am. Chem. Soc.* **76**, 467 (1954).

[4] J. P. McCullough, D. R. Douslin, W. N. Hubbard, S. S. Todd, J. F. Messerly, I. A. Hossenlopp, F. R. Frow, J. P. Dawson, and G. Waddington, *J. Am. Chem. Soc.* **81**, 5884 (1959).

[5] W. N. Hubbard, H. L. Finke, D. W. Scott, J. P. McCullough, C. Katz, M. E. Gross, J. F. Messerly, R. E. Pennington, and G. Waddington, *J. Am. Chem. Soc.* **74**, 6025 (1952).

[6] Compare S. A. Barker and R. Stephens, *J. Chem. Soc.* p. 4550 (1954).

[7] H. Tschamler and H. Voetter, *Monatsh. Chem.* **83**, 302, 835, 1228 (1952).

[8] C. G. LeFèvre and R. J. W. LeFèvre, *J. Chem. Soc.* p. 3549 (1956).

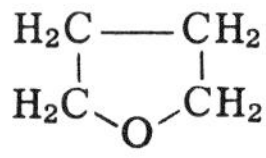

II

to the half-chair form of cyclopentane (see page 75). The dipole moment also indicates a puckered conformation for tetrahydrofuran.[9]

A similar conclusion has been reached for 1,3-dioxolane (III), in

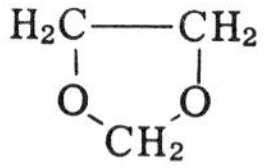

III

which the dipole moment is consistent with a nonplanar conformation.[9,10] Analogies have been found between substituted 1,3-dioxolanes and correspondingly substituted cyclopentanes in respect to their physical properties[11] (boiling point, stability).

2. Six-Membered Rings

By analogy with cyclohexane, tetrahydropyran should exist in the chair conformation (IV)[12,13] (see also the pyranose forms, p. 324f). Oxygen is able to replace a methylene group without causing a noticeable distortion of the ring. This becomes even more obvious if we assume that the *p* orbitals roughly retain their tetrahedral distribution (IV).[12]

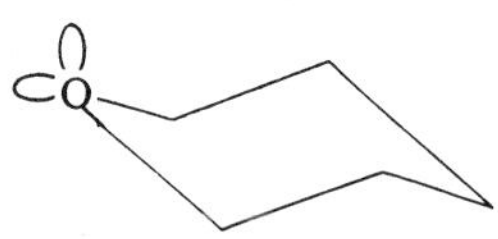

IV

---

[9] B. A. Arbousow, *Bull. Soc. Chim. France* p. 1311 (1960) with further literature cited there; cf. also F. Oehme and M. Feinauer, *Chem. Ztg.* **86**, 71 (1961); D. Gagnaire and P. Vottero, *Bull. Soc. Chim. France* p. 2779 (1963).

[10] Cf. also S. A. Barker, E. J. Bourne, R. M. Pinkard, and D. H. Whiffen, *J. Chem. Soc.* p. 802 (1959); B. A. Arbousow and L. K. Yuldaschewa, *Izv. Akad. Nauk SSSR* pp. 1728, 1734 (1962).

[11] S. A. Barker, E. J. Bourne, R. M. Pinkard, M. Stacey, and D. H. Whiffen, *J. Chem. Soc.* p. 3232 (1958).

[12] D. H. R. Barton and R. C. Cookson, *Quart. Rev. (London)* **10**, 44 (1956).

[13] J. A. Mills, *Advan. Carbohydrate Chem.* **10**, 1 (1955).

The infrared spectrum shows that ~65% of cyclohexanol (see page 99) exists in the conformation with an *equatorial* OH group. The comparable tetrahydropyran-3-ol, however, possesses about equal amounts of molecules with *equatorial* (V) and *axial* (VI) OH groups. The less preferred conformation with the *axial* OH group, can be stabilized by an intramolecular hydrogen bond between the OH group and the ring oxygen,[14] while no such stabilization is possible in cyclohexanol.

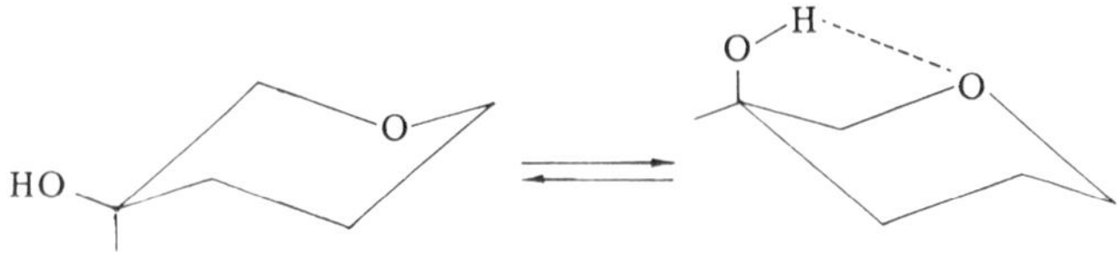

V VI

The introduction of a second oxygen into the ring, as in 1,3-dioxane-5-ol (VII), further increases the proportion of *axial* OH groups present, because of the stronger hydrogen bond.[14] However, the strength of

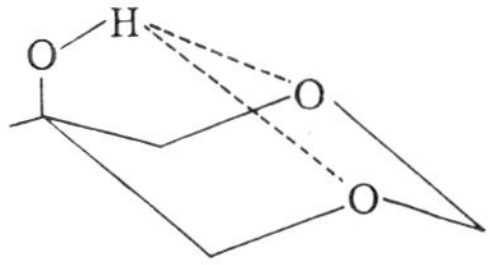

VII

these hydrogen bonds is not sufficient (see page 309) for complete fixation of conformations (VI) and (VII).[15]

The conformation of piperidine[16,16a] was investigated by measuring the molecular polarizability. Two chair conformations are possible, one with the hydrogen of the NH group in an *equatorial* position (VIII), the other with this hydrogen in the *axial* position (IX). In addition, four boat forms can be visualized, in which the hydrogen of the NH group

---

[14] S. A. Barker, J. S. Brimacombe, A. B. Foster, D. H. Whiffen, and G. Zweifel, *Tetrahedron* **7**, 10 (1959).

[15] Concerning the conformations of dihydropyran derivatives in flavanones, dihydroflavanols, and catechols see E. A. H. Roberts, *Chem. & Ind.* (*London*) p. 737 (1956); E. M. Philbin and T. S. Wheeler, *Proc. Chem. Soc.* p. 167 (1958); J. W. Clark-Lewis and L. M. Jackman, *ibid.* p. 165 (1961).

[16] M. Aroney and R. J. W. LeFèvre, *Proc. Chem. Soc.* p. 82 (1958); *J. Chem. Soc.* p. 3002 (1958).

[16a] Cf. the review of N. S. Prostakov and N. N. Mikheeva, *Russian Chem. Rev.* **31**, 556 (1962) [*Uspekhi Khimi* **31**, 1191 (1962].

can assume either the flagpole position (cf. page 276) (X), the bowsprit position (XI), the boat-*equatorial* position (XII), or the boat-*axial* position (XIII).

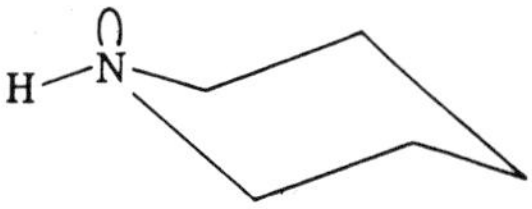

VIII

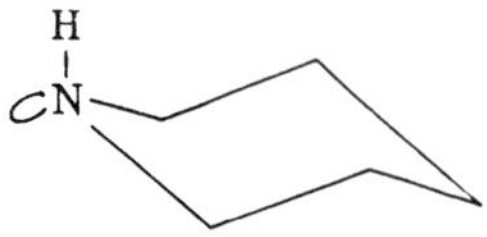

IX

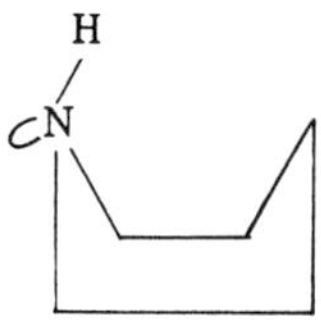

X

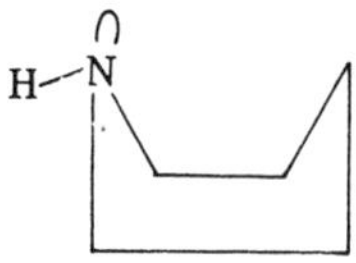

XI

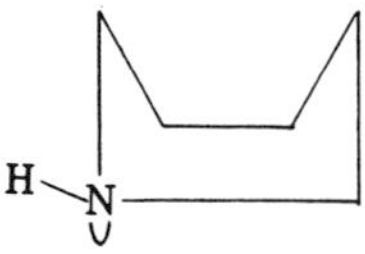

XII

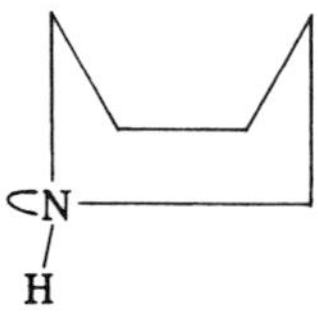

XIII

By analogy with cyclohexane the boat conformations (X) to (XIII) appear improbable. Measurement of the molecular polarizability indicated that piperidine exists predominantly (80% or more) in conformation (IX), with the hydrogen of the NH groups in the *axial* position.[10]

By application of the same method LeFèvre determined the preferred conformation of *N*-methylpiperidine. Although the methyl group has greater spatial requirements than hydrogen, the measurements revealed no special preference for that chair conformation in which the methyl group of the nitrogen assumes the *equatorial* position. This would have been expected. According to these measurements *N*-methylpiperidine exists in both chair conformations, with *axial* and *equatorial* methyl groups in a ratio of 1 : 1.[16]

The conformation of morpholine, for which six forms can also be constructed from models, has been determined by molecular polarizability.[16] The chair conformation with the hydrogen on the NH group in the *axial* position predominates (XIV).

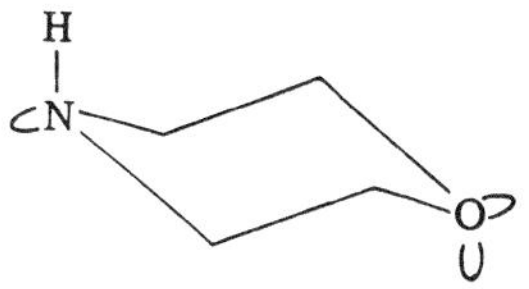

XIV

These results confirm a prediction originally made by Barton[17] concerning the preferred conformation of piperidine. The polarizability measurements indicate that the spatial requirements of a lone electron pair, such as those present on the nitrogen of piperidine, are greater than the spatial requirements of a covalently bound hydrogen. From the conformational equilibrium found for *N*-methylpiperidine (ratio between *axial* and *equatorial* methyl group on the nitrogen about 1 : 1), it follows that the spatial requirements of a lone electron pair are about the same as for a methyl group. As Aroney and LeFèvre were able to show with *N*-phenylpiperidine,[18] bulkier substituents on the nitrogen exist mainly in the *equatorial* position.

With this assumption in mind, that the lone electron pair of the NH group has a larger spatial requirement than the covalently bound hydrogen, piperazine was investigated with the help of molecular polarizability measurements.[18] It was found to exist in a 1 : 2 mixture of conformations (XV) and (XVI). The hydrogens of the NH groups in these conformations always exist in the *axial* or boat-*axial* position.

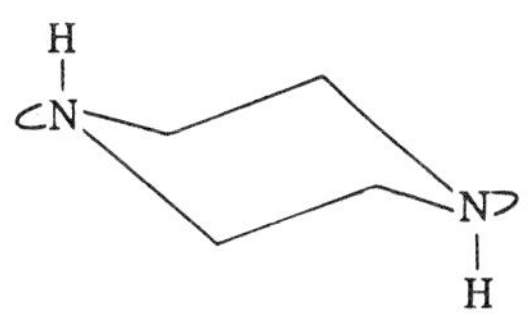

XV

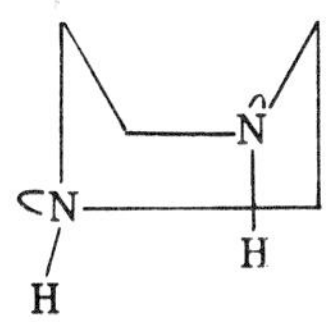

XVI

The conformations of 1,4-dimethylpiperazine are more difficult to derive.[18] From the results for *N*-methylpiperidine, it was assumed that dimethylpiperazine is a mixture of various conformers. The methyl groups are located in the *equatorial* (XVII) as well as the *axial* (XVIII) position of the chair conformation. A considerable portion of the molecules, however, must also be in a boat conformation, probably the

[17] D. H. R. Barton and R. C. Cookson, *Quart. Rev.* (*London*) **10**, 73 (1956); cf. also G. Fodor, "Stereochemistry of Alkaloids," in press.

[18] M. Aroney and R. J. W. LeFèvre, *J. Chem. Soc.* p. 2161 (1960).

predominant conformation (XIX).[18,19] More recent investigations, on the other hand, show that (XVII) is the predominant conformation.[2a]

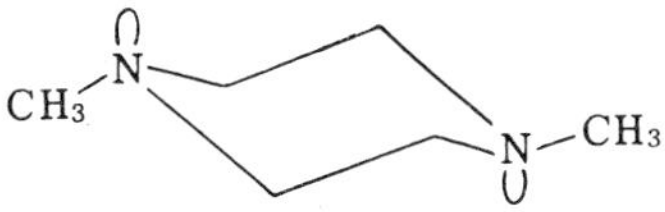

XVII

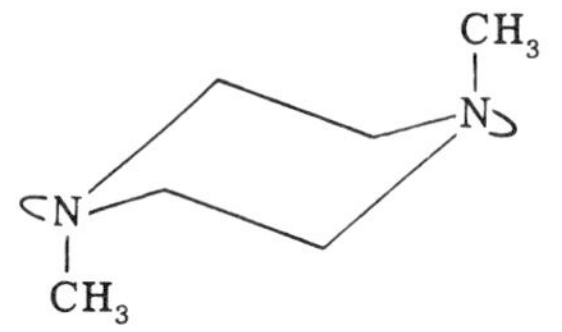

XVIII

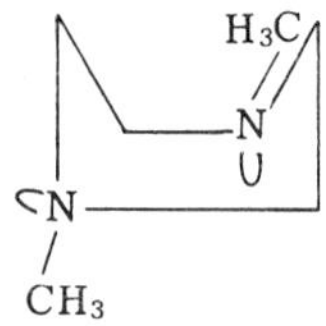

XIX

It has been demonstrated by X-ray analysis of a solid dimethylpiperazine-palladiumchloride complex that dimethylpiperazine can be stabilized in the boat conformation.[20]

On the basis of the results obtained from dimethylpiperazine, it is probable that other piperazine derivatives exist not only in the chair conformation, for instance 1,4-dichloropiperazine, but also that a considerable amount is involved in a conformational equilibrium as the boat form.[18,19,21] The results obtained for 1,4-diphenyl- and 1,4-di-*p*-tolylpiperazine also indicate this. They probably have a large proportion of boat conformation with boat-*equatorial* position of the phenyl groups.[22]

Although piperidine exists exclusively in the chair conformation, as shown above, it is possible, as it was for cyclohexane (compare page 284), to force the piperidine ring into the boat conformation with suitable *axial*-oriented substituents. 1,2,2,6,6-Pentamethyl-4-phenyl-4-piperidinol (XX)[23] is a piperidine derivative in which the assumption of a boat conformation is sterically possible, due to *axial* interactions of the substituents. This compound is unique in its inertness toward reactions

[19] Cf. also M. V. George and G. F. Wright, *Canad. J. Chem.* **36**, 189 (1958).

[20] O. Hassel and B. F. Pedersen, *Proc. Chem. Soc.* p. 394 (1959).

[21] Cf. also P. Andersen and O. Hassel, *Acta. Chem. Scand.* **3**, 1180 (1949).

[22] For the conformation of some hexahydropyrimidines see H. Piotrowska and T. Urbánski, *J. Chem. Soc.* p. 1942 (1962).

[23] R. E. Lyle, *J. Org. Chem.* **22**, 1280 (1957).

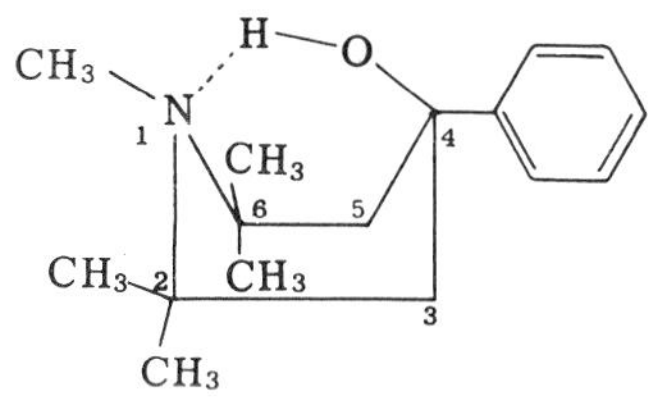

XX

at the hydroxyl group. The infrared spectrum shows a hydrogen bond between the nitrogen and the hydroxyl group. This bond is not present in the absence of the geminal methyl groups on carbon atoms 2 and 6. Thus it is apparent that (XX) exists mainly in the boat conformation, which seems to have been especially stabilized by the hydrogen bond. In other piperidinols, as well, intramolecular hydrogen bonds have been observed between the nitrogen and the hydrogen of the OH group. Examples are 3-piperidinol (XXI, R=H) and 1-methyl-3-piperidinol (XXI, $R{=}CH_3$).[24]

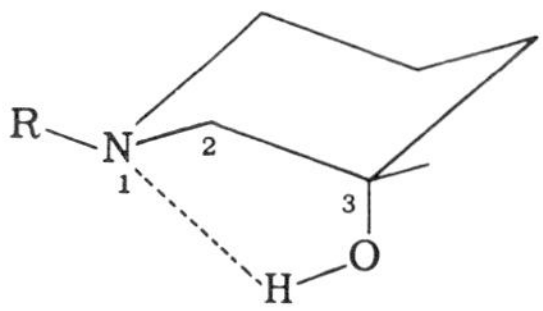

XXI

Investigation of the infrared spectra of both revealed a hydrogen bond in addition to a free OH group in the region of the OH stretching vibration. To make this possible, piperidinol has to assume the less favorable conformation with an *axial* OH group. In addition to the conformation with an *axial* OH group (XXI), the conformation with an *equatorial* OH group is also present. Apparently the hydrogen bond is not strong enough to form (XXI) exclusively. If, in addition, a phenyl group is introduced into position 3, this group assumes the *equatorial* position, exclusively. Then all the molecules are capable of a hydrogen bond formation because of the *axial* OH group. On the other hand, 1-methyl-4-piperidinol shows no sign of a hydrogen bond formation.[23,25]

Some investigations based on physical measurements are available concerning the conformations of the isomeric dioxanes. The activation energy required to transform one chair conformation of 3,3,6,6-tetra-

[24] G. Hite, E. E. Smissman, and R. West, *J. Am. Chem. Soc.* **82**, 1207 (1960).

[25] Cf. also M. Balasubramanian and N. Padma, *Tetrahedron Letters* **14**, 23 (1960); *ibid.* p. 49 (1963); *Tetrahedron* **19**, 2135 (1963).

methyl-1,2-dioxane (XXII) into the other was determined to be 18.5 kcal./mole by NMR spectroscopy. The corresponding dithio-cyclohexane (XXIII) requires an activation energy of 16.1 kcal./mole.[26]

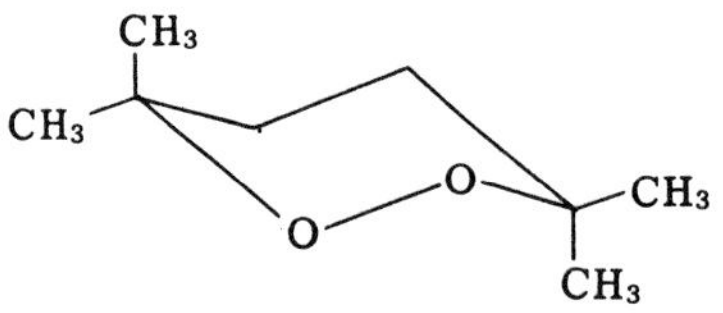

XXII

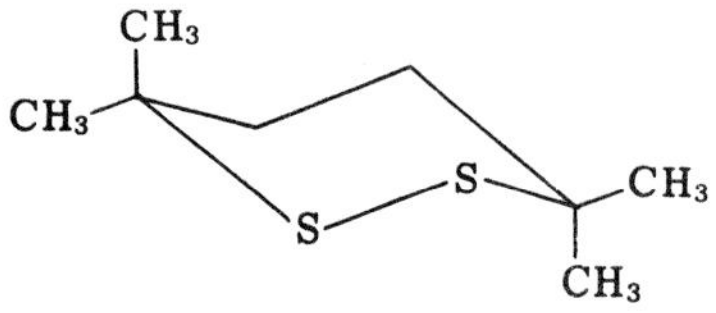

XXIII

The difference in the activation energy of conversion is attributed to different transition forms of 1,2-dioxane and 1,2-dithiocyclohexane.[26] In contrast to reactive dithiocyclopentane,[27] which possesses a high ring strain because two longer C—S bonds (see Table I, page 300) and an S—S bond (~1.81 Å) are forced into a five-membered ring,[28] 1,2-dithiocyclohexane has relatively high stability. This is another indication that it exists in the nonstrained chair conformation.[28] This is true of 1,2-dioxane, which is also relatively stable.[29,30,30a]

Otto[31] has assumed that the chair form is the most probable conformation of 1,3-dioxane. A study of a model shows that the boat form, as in the case of cyclohexane, is not favored, due to the interaction of the hydrogens. The dipole moment determined experimentally, is in best agreement with the calculated value for the chair form[9,32]. The molecular polarizability of dimethylenepentaerythrol (XXV),[33] verifies that 1,3-

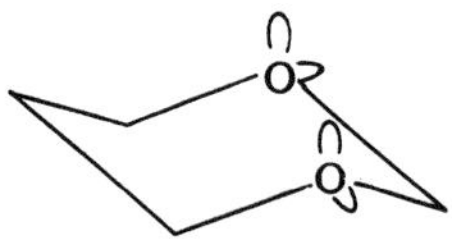

XXIV

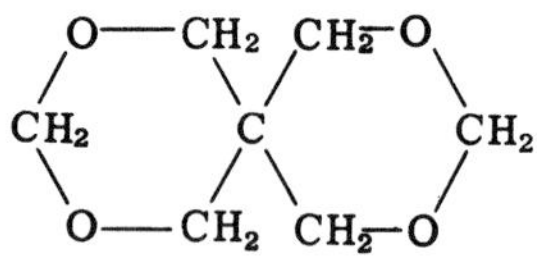

XXV

[26] G. Claeson, G. Androes, and M. Calvin, *J. Am. Chem. Soc.* **83**, 4357 (1961).
[27] J. A. Barltrop, P. M. Hayes, and M. Calvin, *J. Am. Chem. Soc.*, **76**, 4348 (1954).
[28] A. Schöberl and H. Gräfje, *Ann. Chem.* **614**, 66 (1958).
[29] R. Criegee and G. Müller, *Chem. Ber.* **89**, 238 (1956).
[30] Concerning the conformation of further cyclic disulfides, see ref. 28.
[30a] Further investigations on 1,2-dioxanes and 1,2-dithiocyclohexanes see: A. Lüttringhaus, S. Kabuss, W. Maier, and H. Friebolin, *Z. Naturforsch.* **16b**, 761 (1961); H. Friebolin and W. Maier, *ibid.* **16a**, 640 (1961).
[31] M. M. Otto, *J. Am. Chem. Soc.* **59**, 1590 (1937).
[32] Cf. also R. Walker and D. W. Davidson, *Can. J. Chem.* **37**, 492 (1959).
[33] C. G. LeFèvre, R. J. W. LeFèvre, and M. R. Smith, *J. Chem. Soc.* p. 16 (1958).

dioxane possesses the chair conformation (XXIV). The measured molar Kerr constants indicate a double chair conformation of the spiro compound (XXV). The activation energy required to transform one chair conformation of 1,3-dioxane into the other was determined to be 9.7 kcal./mole by NMR spectroscopy.[33a] This indicates that the enery barrier is only slightly lower than the barrier for cyclohexane (p. 44).

The conformations of a series of substituted, 1,3-dioxanes has been investigated with the aid of infrared spectroscopy. Of these, 1,3-dioxan-5-ol has already been mentioned (page 303). In this case, the influence of the hydrogen bond becomes significant and conformations which would otherwise be unfavorable become more predominant. Infrared spectra revealed *trans*-5-hydroxy-2-phenyl-1,3-dioxane (*trans*-1,3-O-benzylidene glyceritol) may exist in two conformations (XXVI and XXVII). This conclusion was reached not only because the free OH absorption (~3630 cm.$^{-1}$) was found in carbon tetrachloride solution, but also because the OH absorption was found shifted to ~3580 cm.$^{-1}$ by intramolecular hydrogen bond formation. This is possible only for conformation (XXVI).

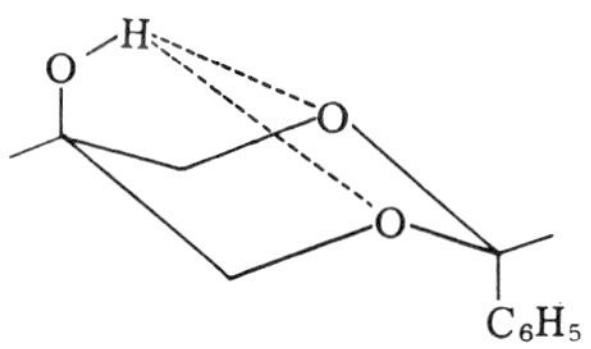

XXVI

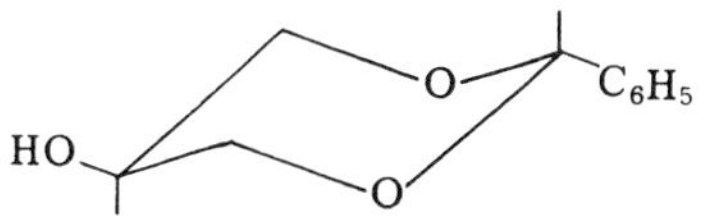

XXVII

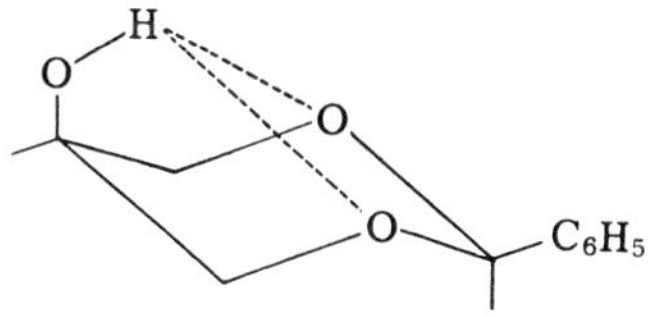

XXVIII

Conformation (XXVI) is apparently stabilized by the hydrogen bond, even though the phenyl group is brought into the unfavorable *axial* position.[34] Similar influences have been observed in other 1,3-dioxa-

[33a] H. Friebolin, S. Kabuss, W. Maier, and A. Lüttringhaus, *Tetrahedron Letters* p. 683 (1962).

[34] (a) J. S. Brimacombe, A. B. Foster, and M. Stacey, *Chem. & Ind.* (*London*) p. 1228 (1958); (b) N. Bagett, J. S. Brimacombe, A. B. Foster, M. Stacey, and D. H. Whiffen, *J. Chem. Soc.* p. 2574 (1960); compare also: N. Baggett, M. A. Bukhari, A. B. Foster, J. Lehmann, and J. M. Webber, *ibid.* p. 4157 (1963).

nes.[34b,35] The intramolecular hydrogen bond also exerts itself through an increase in the rate of esterification (cf. page 182, ref. 33) of *cis*-5-hydroxy-2-phenyl-1,3-dioxane, which exists exclusively in the conformation with *axial* hydroxyl group (XXVIII) and, as a result, has a strong hydrogen bond. In comparison to 4-phenylcyclohexanol the esterification rate is increased.[34b]

The conformation of 1,4-dioxan has been investigated intensively by various laboratories. Aside from the nonpolar chair conformation (XXIX), an unsymmetrical boat form (XXX) and a symmetrical boat form (XXXI) are possible.

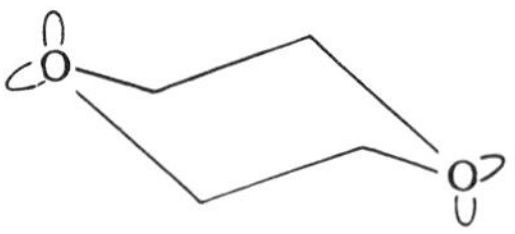

XXIX

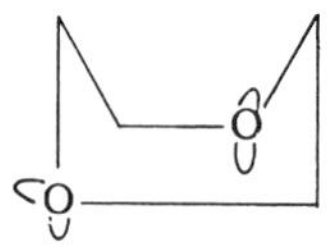

XXX

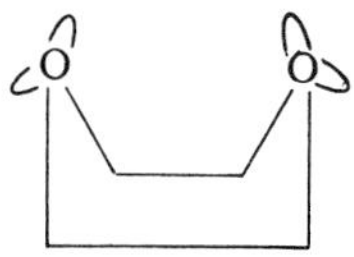

XXXI

Measurements of the dipole moment by various authors[36] showed that no permanent dipole moment can be measured under 70°. The dioxane, therefore, exists in the chair conformation (XXIX). On the basis of infrared and Raman spectra, a boat form in dioxane could also be excluded with certainty.[37] 1,4-Dioxane was also investigated by electron diffraction and the chair conformation (XXIX) confirmed.[2a,38,39] According to LeFèvre[40] dioxane in the gaseous state above 200°, can exist partly in a boat conformation.

[35] (a) B. Dobinson and A. B. Foster, *J. Chem. Soc.* p. 2338 (1961); (b) For the conformation of some 1,3-oxazines see D. Gürne and T. Urbánski, *J. Chem. Soc.* p. 1912 (1959).

[36] (a) See for example J. Böeseken, F. P. A. Tellegen, and P. C. Henriquez, *Rec. Trav. Chim.* **54**, 733 (1935); (b) W. C. Vaughan, *Phil. Mag.* **27**, 669 (1939); (c) M. Yasumiu and M. Shirai, *Bull. Chem. Soc. Japan* **28**, 193 (1955).

[37] (a) See for example D. A. Ramsay, *Proc. Roy. Soc.* **A190**, 562 (1947); (b) S. C. Burket and R. M. Badger, *J. Am. Chem. Soc.* **72**, 4397 (1950); (c) F. E. Malherbe and H. J. Bernstein, *ibid.* **74**, 4408 (1952).

[38] O. Hassel and H. Viervoll, *Acta Chem. Scand.* **1**, 149 (1947).

[39] (a) L. E. Sutton and L. O. Brockway, *J. Am. Chem. Soc.* **57**, 473 (1935); (b) M. Kimura and K. Aoki, *J. Chem. Soc. Japan* **72**, 169 (1951).

[40] (a) R. S. Armstrong, R. J. W. LeFèvre, and J. Yates, *Australian J. Chem.* **11**, 147 (1958); (b) cf. also J. H. Gibbs, *Discussions Faraday Soc.* **10**, 122 (1951).

1,4-Dithiocyclohexane does not exist exclusively in the chair conformation, according to electron diffraction data. A twist form in equilibrium has been discussed.[38]

A series of 1,4-dioxane derivatives has been investigated with respect to their stability. The same regularities were noted as in the case of cyclohexane. Because of an *axial* carboxyl group, *cis*-1,4-dioxane-2,5-dicarboxylic acid (XXXII) is less stable than the corresponding *trans*-dicarboxylic acid with both carboxyl groups in *equatorial* position (XXXIII). (XXXII) can be rearranged to (XXXIII).

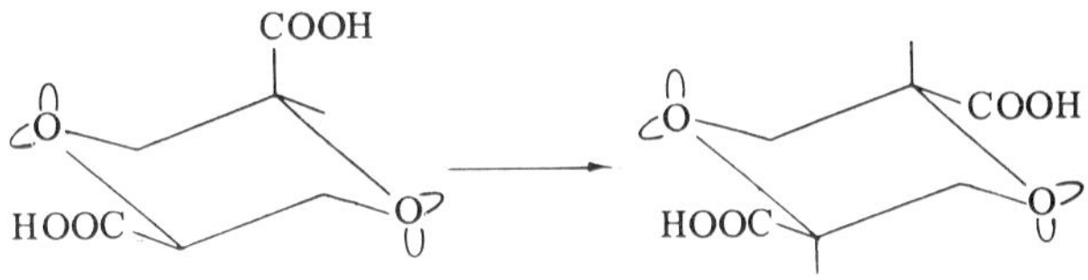

XXXII XXXIII

A similar situation applies to the isomeric 1,4-dioxane-2,3-dicarboxylic acids and the isomeric 1,4-dioxane-2,6-dicarboxylic acids. In the rearrangement of the *cis*-2,3-dicarboxylic acid (one *equatorial* and one *axial* COOH group), the *trans*-2,3-dicarboxylic acid (both COOH groups *equatorial*) predominates at equilibrium. For the 2,6-dicarboxylic acid, the *cis* form predominates at equilibrium and thereby proves to be the more stable one.[41,42]

3. Bicyclic Compounds

Although no detailed investigations concerning the conformations of bicyclic hetero compounds are available, on the basis of the results obtained for monocyclic hetero compounds, no essential differences from the carbocyclic compounds should be expected (see page 172f.). This is illustrated by the following examples:

The *cis*-[43] and *trans* form[44] of perhydroindole are known. The confor-

---

[41] R. K. Summerbell and J. R. Stephens, *J. Am. Chem. Soc.* **76**, 731, 6401 (1954); R. K. Summerbell and G. J. Lestina, *ibid.* **79**, 3878 (1957).

[42] For further investigations on substituted 1,4-dioxanes, see for example: R. K. Summerbell and H. E. Lunk, *J. Am. Chem. Soc.* **79**, 4802 (1957); C. Altona, C. Romers, and E. Havinga, *Tetrahedron Letters* **10**, 16 (1959); E. Caspi, Th. A. Wittstruck, and D. M. Piatak, *J. Org. Chem.* **27**, 3183 (1962); C. Altona and C. Romers, *Rec. Trav. Chim.* **82**, 1080 (1963); C. Altona, C. Knobler, and C. Romers, *ibid.* **82**, 1089 (1963).

mations (XXXIV) and (XXXV) respectively would be obtained for these forms, by analogy with the carbocyclic compounds.[45,46]

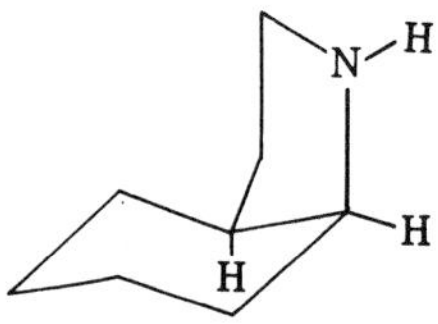

XXXIV

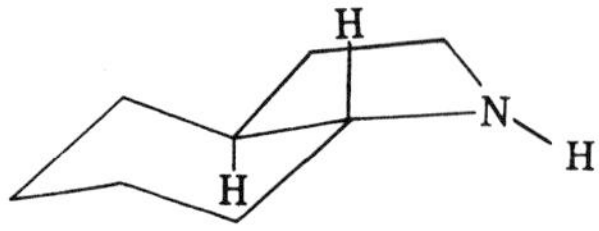

XXXV

Similar to decalin, decahydroquinoline also exists in a *cis*- and a *trans* configuration[47] so that the conformations (XXXVI) and (XXXVII) correspond to *cis*- and *trans*-decalin.

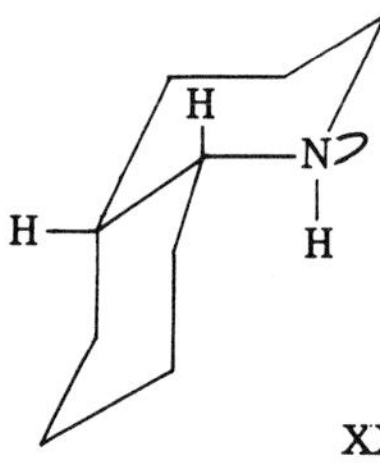

XXXVI

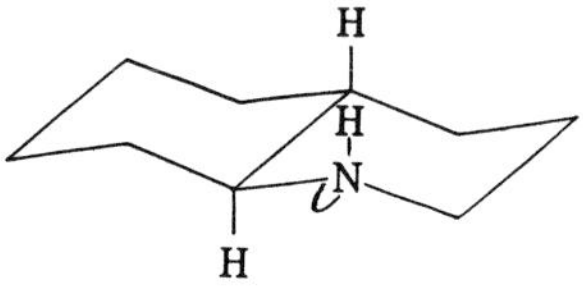

XXXVII

The same applies to *cis*-[48] and *trans*-decahydroisoquinoline.[45,49]

If the nitrogen is located at a ring juncture, as in quinolizidine (octahydropyridocolin, norlupinane) (XXXVIII), then two stereoisomeric quinolizidines (XXXIX and XL) are theoretically possible; they correspond to *cis*- and *trans*-decalin.[50]

According to the experience gained so far, it is assumed[51] that the *trans*-

---

[43] R. Willstätter and D. Jaquet, *Ber.* **51**, 767 (1918); R. Willstätter, F. Seitz, and J. v. Braun, *ibid.* **58**, 385 (1925).

[44] F. E. King, D. M. Bovey, K. G. Mason, and R. L. St. D. Whitehead, *J. Chem. Soc.* p. 250 (1953).

[45] F. E. King and H. Booth, *J. Chem. Soc.* p. 3798 (1954).

[46] For cyclopentanotetrahydropyran cf. M. Procházka and J. V. Cerny, *Tetrahedron* **16**, 25 (1961).

[47] (a) W. Hückel and F. Stepf, *Ann. Chem.* **453**, 163 (1927); (b) A. Albert, D. J. Brown, and H. Duewell, *J. Chem. Soc.* p. 1284 (1948).

[48] A. Skita, *Ber.* **57**, 1977 (1924).

[49] B. Witkop, *J. Am. Chem. Soc.* **70**, 2617 (1948).

[50] (a) R. C. Cookson, *Chem. & Ind.* (*London*) p. 337 (1953); (b) cf. also N. J. Leonard and B. L. Ryder, *J. Org. Chem.* **18**, 598 (1953).

[51] (a) F. Galinovsky and H. Nesvadba, *Monatsh. Chem.* **85**, 1300 (1954); (b) N. J. Leonard and W. K. Musker, *J. Am. Chem. Soc.* **82**, 5148 (1960).

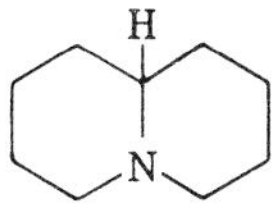

XXXVIII

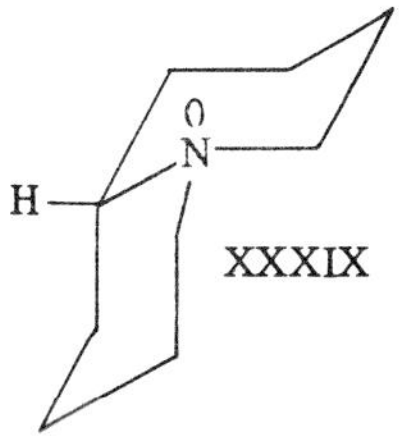

XXXIX

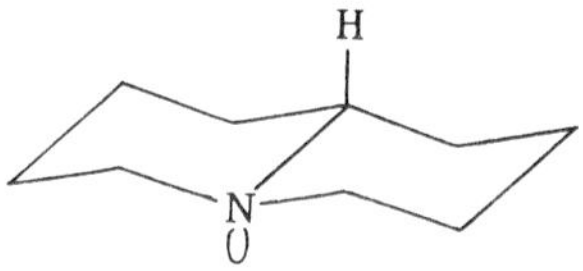

XL

fused two-chair conformation (XL) is the more stable one. However, it could be shown by infrared[52] and NMR spectroscopy, that the *trans*-two-chair conformation is not always preferred.[53] In the case of certain monosubstituted quinolizidines, such as *trans*-4-methylquinolizidine (the prefix *trans* here applies to the configuration of the hydrogen at C-10 with respect to the methyl group), the *cis*-fused two-chair conformation (XXXIX) is preferred (cf. also page 323). Considering the investigations of Aroney and LeFèvre[16] (described on page 305), in which the spatial requirements of the lone electron pair of nitrogen was found to exceed those of the hydrogen atom and approach those of the methyl group, quinolizidine has been compared to the corresponding 9-methyldecalin.[54] An appraisal of the relative stabilities of the *cis* and the *trans* form of quinolizidine, using this assumption, leads to an even greater predicted stability of *cis*-fused quinolizidine (XXXIX).

The presence of two nitrogens in one ring, such as in decahydroquinoxaline (XLI), does not change the possibility of forming configurational isomers. Both the *cis* (*meso*) and *trans*-(*d,l*)-decahydroquinoxaline have been prepared.[55] The double chair conformation is the most probable for *trans*-decahydroquinoxaline.

---

[52] Cf. in this connection especially F. Bohlmann, *Chem. Ber.* **91**, 2157 (1958).

[53] T. M. Moynehan and K. Schofield; R. A. Y. Jones, and A. R. Katritzky, *Proc. Chem. Soc.* p. 218 (1961); *J. Chem. Soc.* p. 2637 (1962); compare also K. Schofield and R. J. Wells, *Chem. & Ind.* (*London*) p. 572 (1963); S. F. Mason, K. Schofield, and R. J. Wells, *Proc. Chem. Soc.* p. 337 (1963).

[54] S. V. Kessar, *Experientia* **18**, 56 (1962).

[55] H. S. Broadbent, E. L. Allred, L. Pendleton, and C. W. Whittle, *J. Am. Chem. Soc.* **82**, 189 (1960); E Brill and H. P. Schultz, *J. Org. Chem.* **28**, 1135 (1963).

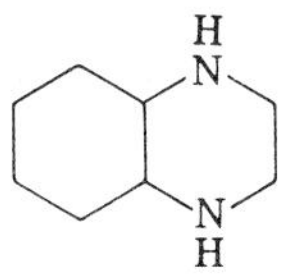

XLI

We may also mention 2,3-dithiodecalin (XLII), which could be prepared in the *cis* and *trans* forms.[56] The *cis* compound displays a somewhat higher strain but both isomers most probably exist in the two-chair form.

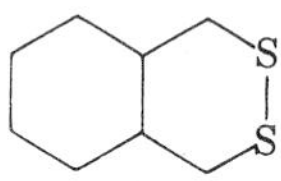

XLII

4. Some Examples from Alkaloid Chemistry

a. Tropane Alkaloids

Among the alkaloids which have received the most thorough investigations with respect to their stereochemistry, belong the tropane alkaloids, especially as a result of Fodor's elegant investigations.[57]

The tropane alkaloids are derivatives of nortropane (XLIII, R=H), which represents an 8-azabicyclo[3.2.1]octane. The piperidine ring contained in the tropane nucleus (XLIII, R=CH_3) can theoretically, exist in the chair conformation (XLIV) as well as in the boat conformation (XLV). We can think of (XLIV) as a piperidine distorted by an

[56] A. Lüttringhaus and A. Brechlin, *Chem. Ber.* **92**, 2271 (1959).

[57] See the following reviews which cite further references: (a) G. Fodor, *Acta Chim. Acad. Sci. Hung.* **5**, 380 (1955); (b) G. Fodor, *Experientia* **11**, 129 (1955); (c) G. Fodor, *Bull. Soc. Chim. France* p. 1032 (1956); (d) G. Fodor, *in* "Anniversary Publication for Arthur Stoll," p. 682. Birkhäuser, Basel, 1957; (e) G. Fodor *in* "The Alkaloids" (R. H. F. Manske, ed.), Vol. VI, p. 145. Academic Press, New York, 1960; (f) G. Fodor, *Chem. & Ind.* (*London*) p. 1500 (1961); (g) G. Fodor, "Stereochemistry of Alkaloids," in press; (h) cf. also A. Stoll and E. Jucker, *Angew. Chem.* **66**, 376 (1954); (i) A. Heusner, *Arzneimittel-Forsch.* **6**, 105 (1956); (k) A. Sekera, *Ann. Chim.* (*Paris*) **7**, 537 (1962).

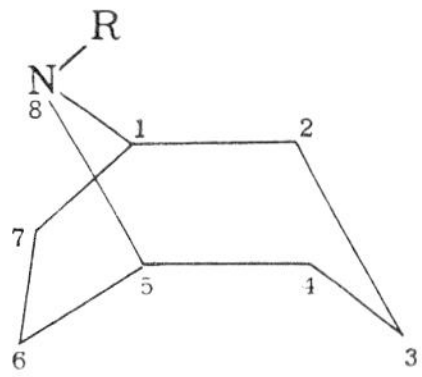

XLIII

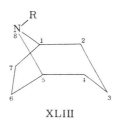

ethylene bridge or as an *N*-methyl-bridged cycloheptane[57b] which exists in the boat conformation. Whichever of the two conformations is preferred, this can be illustrated using the isomeric pair, tropine

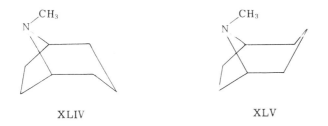

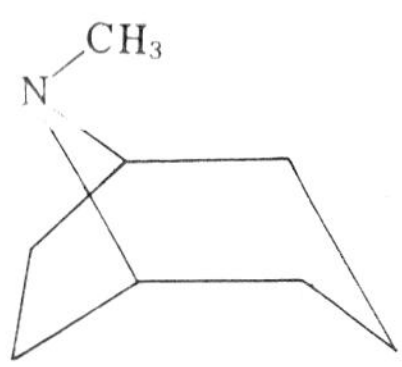

XLIV

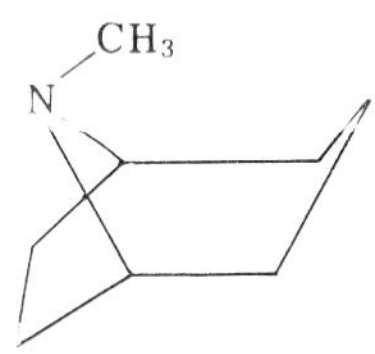

XLV

and pseudotropine (XLVI),[58] as examples. They only differ by the position of the OH group relative to the bridge.

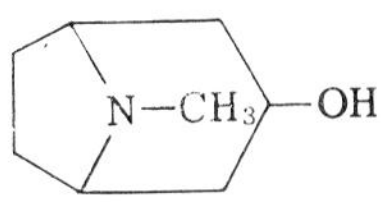

XLVI

On the basis of the hydrolysis rate measurements of the stereoisomeric benzoyl- and *p*-nitrobenzoyl tropines and their corresponding methiodides, the piperidine ring was assigned the boat conformation.[59] The reason that the tropine ester (XLVII) has a slower saponification rate than the corresponding pseudotropine derivative (XLVIII) was explained by the fact that the ester group in (XLVII) is hindered by the $N-CH_3$ group. In pseudotropine (XLVIII) the ester group is not subject to such steric hindrance.[59]

---

[58] According to Fodor, the substituents in the *cis* position to the bridge are designated as β, the *trans* ones as α. G. Fodor and O. Kovács, *J. Chem. Soc.* p. 724 (1953).

[59] F. L. J. Sixma, S. M. Siegmann, and H. C. Beyerman, *Koninkl. Ned. Akad. Wetenschap., Proc. Ser. B* **54**, 452 (1951).

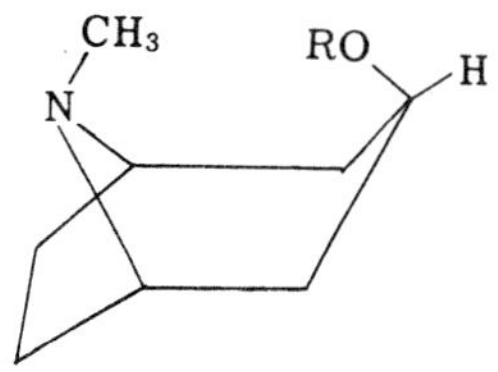

XLVII

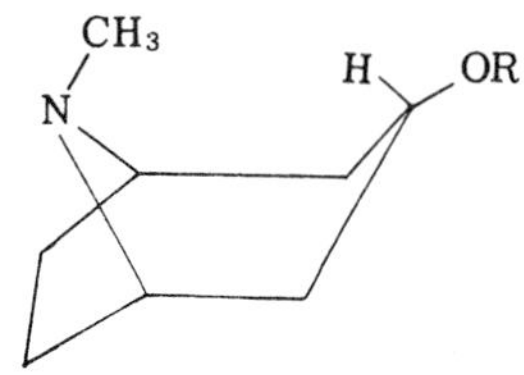

XLVIII

With the provision of the boat conformation for the piperidine ring, the *cis* configuration was assigned to tropine and the *trans* configuration to pseudotropine. The *trans* configuration on the basis of a boat conformation of the piperidine ring was also assigned to pseudotropine by another method.[60]

The questions of the configurations of tropine and pseudotropine was finally settled by the investigations of Fodor.[57,61] Fodor was able to show by means of the so-called acyl migration method, in which an acyl group migrates from nitrogen to oxygen, that tropine must have the *trans* configuration and pseudotropine the *cis* configuration, in contrast to that reported above. Whereas *N*-acetylpseudonortropine (XLIX) readily undergoes a reversible N → O acyl migration, this is not the case for N-acetylnortropine (L). The necessary spatial proximity

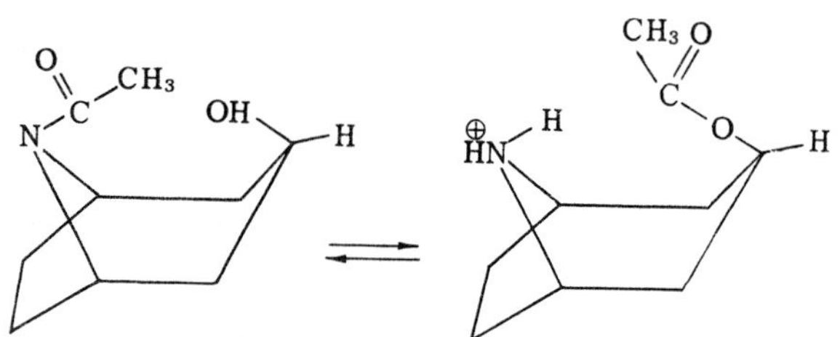

XLIX

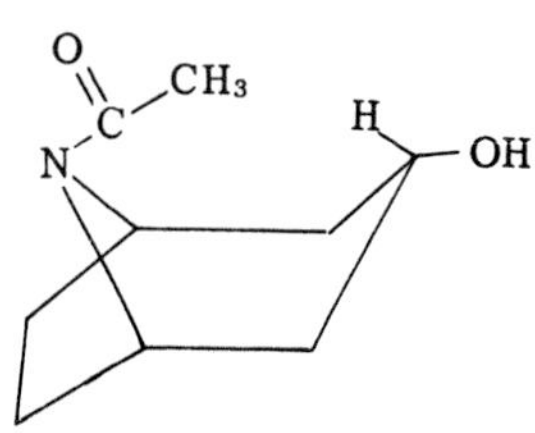

L

[60] N. L. Paddock, *Chem. & Ind.* (*London*) p. 63 (1953).

[61] (a) G. Fodor and K. Nádor, *Nature* **169**, 462 (1952) and ref. 58; (b) cf. also A. Nickon and L. F. Fieser, *J. Am. Chem. Soc.* **74**, 5566 (1952).

is lacking for nortropine because of the *trans* position of the bridge and the OH group.

As an intermediate of the acyl migration, a ring compound is assumed. This is confirmed by the isolation of a tricyclic tetrahydro-*m*-oxazine from norpseudotropinyl carbamate and *p*-nitrobenzaldehyde.[62]

The *cis* position of the OH group in pseudonortropine can also be applied to all *N*-methylated and O-acylated derivatives, such as tropacocaine and tigloidine. The *trans* position of the OH group applies to the nortropine esters, atropine and hyocyamine, etc. The measurements of the dipole moments,[63] the study of the infrared spectra,[63b] and the X-ray analysis[64] of tropine hydrobromide are in agreement with the *trans* configuration of tropine and the *cis* configuration of pseudotropine. The pK values of tropine and pseudotropine, however, cannot be used for a determination of configuration.[65]

The correct assignment of configuration by Fodor makes the existence of the piperidine ring in the boat conformation improbable.[57a] Tropin and pseudotropin must exist in conformations (LI, R=H) and (LII, R=H).[50,57i,66] On the basis of the chair conformation for the piperi-

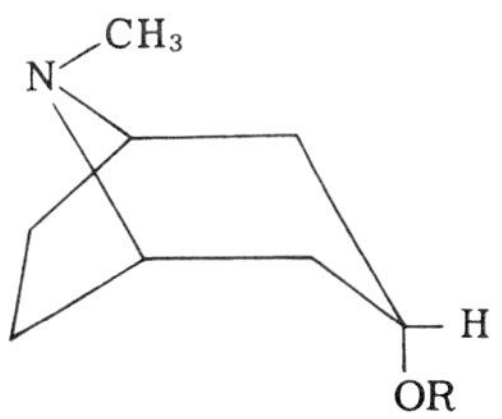

LI

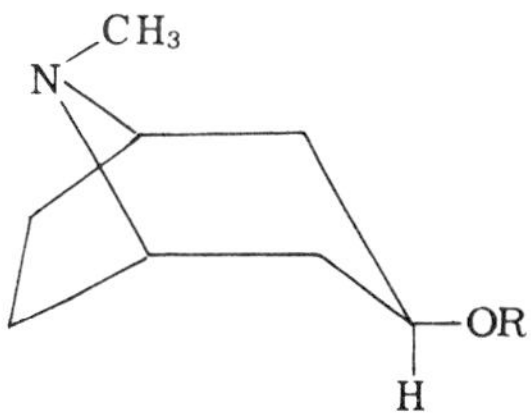

LII

dine ring, the above-mentioned experimental results agree (see page 315); the lower hydrolysis rate of the tropine ester (LI) results from the more hindered *axial* position of the ester group, whereas the *equatorial* ester group of the pseudotropine ester is more easily hydrolyzed (see page 315). With a boat conformation, the hydrolysis rates should be reversed.[67]

[62] E. Hardegger and H. Ott, *Helv. Chim. Acta* **36**, 1186 (1953); **37**, 685 (1954).

[63] (a) G. R. Clemo and K. H. Jack, *Chem. & Ind. (London)* p. 195 (1953); (b) B. L. Zenitz, C. M. Martini, M. Priznar, and F. C. Nachod, *J. Am. Chem. Soc.* **74**, 5564 (1952).

[64] J. W. Visser, J. Manassen, and J. L. deVries, *Acta Cryst.* **7**, 288 (1954).

[65] (a) P. F. Smith and W. H. Hartung, *J. Am. Chem. Soc.* **75**, 3859 (1953); (b) F. A. Geissmann, B. D. Wilson, and R. B. Medz, *ibid.* **76**, 4182 (1954).

[66] (a) M. B. Sparke, *Chem. & Ind. (London)* p. 749 (1953); (b) A. K. Bose and D. K. R. Chaudhuri, *Nature* **171**, 652 (1953).

[67] (a) Cf. Also O. Hromatka, C. Csoklich, and I. Hofbauer, *Monatsh. Chem.* **83**, 1321 (1952); (b) H. C. Beyerman, C. M. Siegmann, F. L. J. Sixma, and J. H. Wisse, *Rec. Trav. Chim.* **75**, 1445 (1956).

Various other reactions also fit well into this scheme. The reduction of tropinone with sodium and alcohol in various solvents yields mainly pseudotropine.[68] Tropine is rearranged into pseudotropine during its equilibration with sodium amylate or aluminum isopropylate.[68] The latter is thermodynamically more stable and has the OH group in an *equatorial* position. The reduction of tropinone with lithium aluminum hydride also yields more pseudotropine,[68b] whereas the reduction with aluminum isopropylate, as in the case of cyclohexanones (see page 270), yields more *axial* tropine.[68b] The infrared spectra of tropine and pseudotropine also indicate the conformations (LI) and (LII). The C—(OH) stretching vibration, as in the case of *axial* cyclohexanols, lies lower (1040 cm.$^{-1}$) for tropine than for pseudotropine with an *equatorial* OH group (1057 cm.$^{-1}$).[63a,69]

In consideration of these results we will have to assume that the more stable form of the tropane alkaloids is that with the piperidine ring in the chair conformation. The boat form of the piperidine ring cannot be excluded because it has to be present for the acyl migration (XLIX and L). A shift of the conformational equilibrium toward the boat form of the piperidine ring takes place only at the higher temperatures necessary for initiation of acyl migration.[57a,70]

If a phenyl group is introduced into position 3 of tropane, as in 3α-phenyl-3β-tropanyl phenyl ketone (LIII), then, on the basis of spectroscopic findings, the piperidine ring appears to exist in the boat conformation.[71] Conversion into the boat conformation facilitates the nitrogen-carbonyl interaction and avoids the steric repulsion between the *axial* phenyl group and the ethylene group present in the chair form of the piperidine ring.

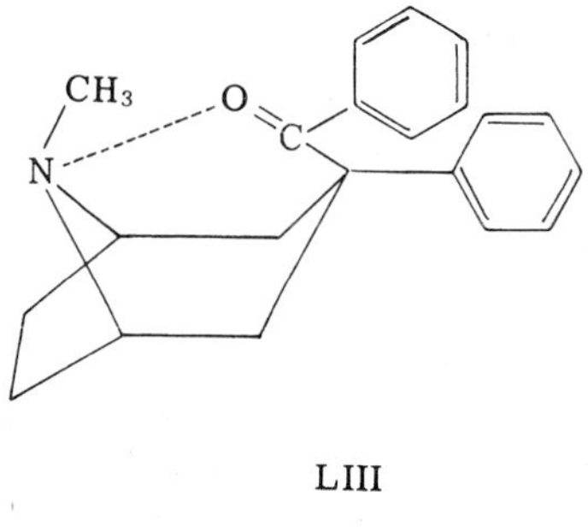

LIII

[68] (a) R. Willstätter, *Ber.* **29**, 393, 936 (1896); (b) A. H. Beckett, N. J. Harper, A. D. J. Balon, and T. H. E. Watts, *Tetrahedron* **6**, 319 (1959).

[69] S. Archer and T. R. Lewis, *Chem. & Ind. (London)* p. 853 (1954).

[70] Cf. also G. Drehfahl and K. Braun, *Chem. Ber.* **93**, 514 (1960).

[71] M. R. Bell and S. Archer, *J. Am. Chem. Soc.* **82**, 151 (1960).

In pseudopelletierine (LIV), which also belongs to the piperidine alkaloids, the piperidone ring assumes the chair conformation, as determined on the basis of dipole measurements.[72] The reduction of pseudo-

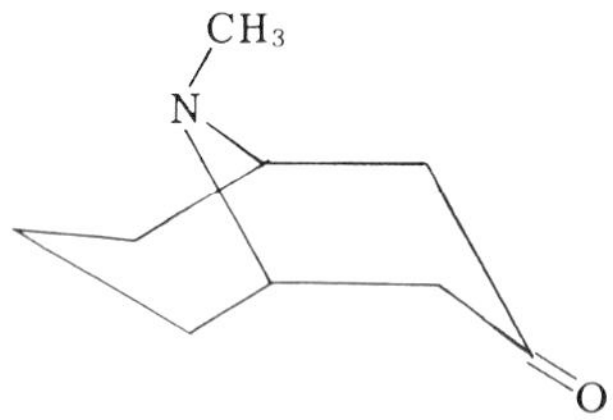

LIV

pelletierine (LIV) under various conditions leads to *N*-methyl-granatoline (LV) or to *N*-methyl-pseudogranatoline (LVI). Most likely the piperidine ring also exists in the chair form in these compounds, although a hydrogen bond absorption band in the infrared spectrum of *N*-methyl-pseudogranatoline gives indication of a possible boat form.[73]

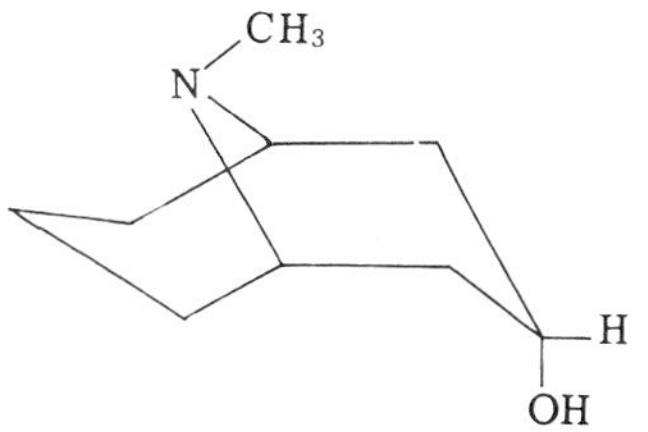

LV

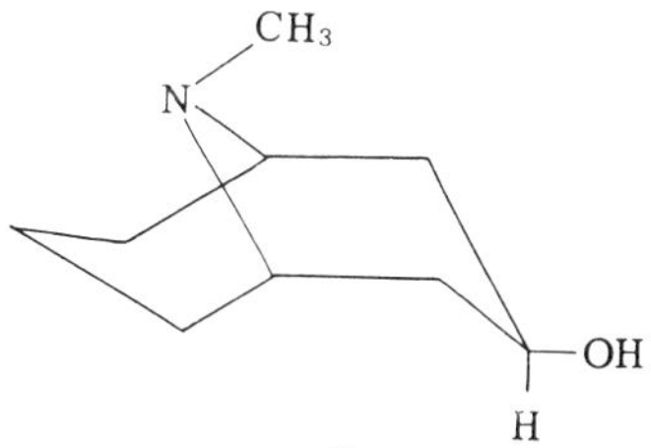

LVI

Four stereoisomeric forms are possible for ecgonine (LVII) (2-carboxy-3-tropanol), each of which can exist in two mirror image isomers. Of these, ecgonine and pseudoecgonine, and a third isomer are known. The configurations of ecgonine and pseudoecgonine were established by the investigations of Fodor.[73a] Ecgonine is a 2β-carboxy-3β-tropanol, with carboxyl and hydroxyl groups in *cis* position (LVIII). Pseudoecgonine is a 2α-carboxy-3β-tropanol with *trans* carboxyl and hydroxyl groups (LIX). Both belong in the same series as pseudotropine. It may be assumed that the piperidine ring assumes the chair conformation in both ecgonine (LVIII) and pseudoecgonine (LIX). Pseudoecgonine is the more stable isomer, because ecgonine can be rearranged into it.[74]

[72] N. J. Leonard, D. F. Morrow, and M. T. Rogers, *J. Am. Chem. Soc.* **79**, 5476 (1957); cf. also J. M. Eckert and R. J. W. LeFèvre, *J. Chem. Soc.* p. 358 (1964).

[73] K. Alder and H. A. Dortmann, *Chem. Ber.* **86**, 1544 (1953).

[73a] G. Fodor, *Nature* **170**, 278 (1952); G. Fodor and O. Kovács, *J. Chem. Soc.* p. 724 (1953).

[74] St. P. Findlay, *J. Am. Chem. Soc.* **75**, 4624 (1953), **76**, 2855 (1954).

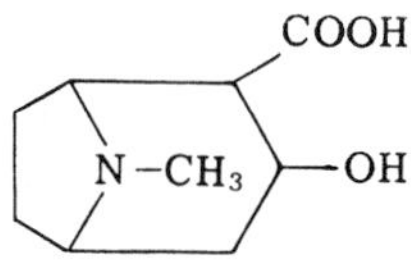

LVII

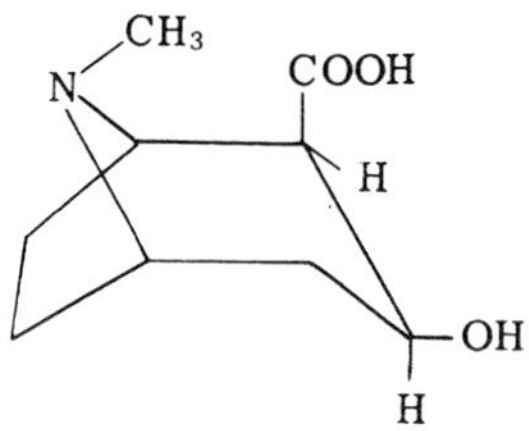

LVIII

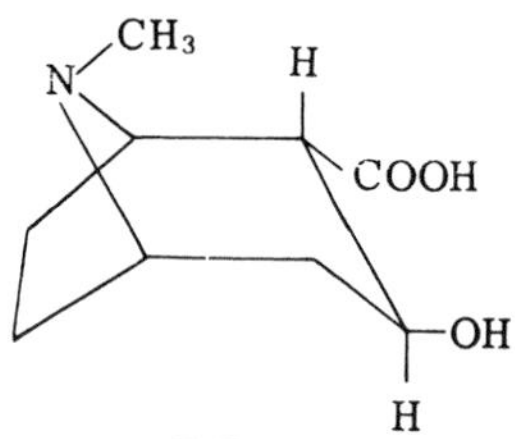

LIX

The third isomer mentioned above, which probably has the configuration shown in (LX), is especially easily dehydrated.[73a] The ease of dehydration indicates a *diaxial* coplanar arrangement of H and OH, which is only possible if the piperidine ring assumes the chair conformation. Of the three isomers, pseudoecgonin (LIX) is the most difficult to dehydrate since a coplanar arrangement of H and OH is possible neither in the chair nor the boat form. A coplanar arrangement of H and OH in the transition state for dehydration of ecgonine is only possible with the boat form of the piperidine ring.

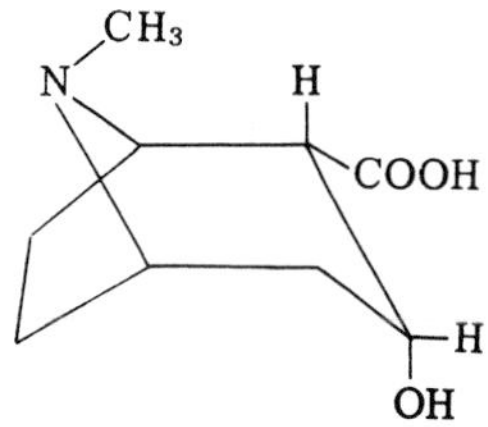

LX

The spontaneous rearrangement of scopine (LXI) into scopoline (LXII), for which the chair form of the piperidine ring is necessary,[73a] may also be mentioned.

In conclusion we shall also briefly consider the position of the methyl group on the nitrogen. Theoretically the methyl group in tropane can assume the position in which it is situated above the pyrrolidine ring (LXIII) or next to the piperidine ring (LXIV).

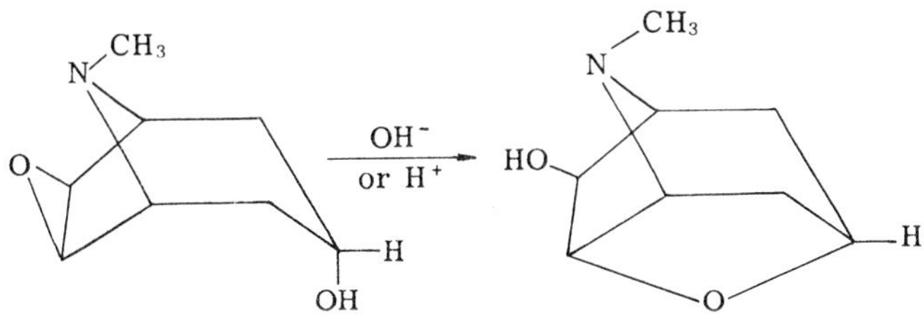

LXI LXII

In this way a pair of geometric isomers would result, but as in the case of asymmetric substituted *tert*-amines, they cannot be resolved. Fodor[57a,57b,75,75a] has attempted to answer this question on the basis of chemical evidence. It could be shown that in the formation of a quaternary salt from a tropane derivative, the substituent added to the nitrogen last was situated over the pyrrolidine ring. From the steric course of the quaternization, it may be concluded that the methyl group in tropane

LXIII LXIV

exists preferably in the position shown in (LXIV). The fixation of the methyl group in this position is mainly attributed by Fodor[57a,76] to the Pitzer strain of the deformed pyrrolidine five-membered ring. However, these results are not in agreement with the NMR spectra of pseudotropine deuterochloride, which indicates the presence of both *N*-epimeric deuterochlorides, i.e., both stereoisomers with different positions of the methyl group.[77] This would signify that not only (LXIII), but also (LXIV), is present at equilibrium. Because the NMR spectra were not taken on the free base, but on the salts in acid solution, and the activation energy of the conversion of the methyl group on the nitrogen from one side to the other is unknown, the question of the relative position of the methyl group still cannot be answered with certainty.

Measurements of the molecular polarizability, leads to the conclusion, that conformation (LXIII) is the preferred one.[77a]

[75] Cf. also K. Zeile and W. Schulz, *Chem. Ber.* **88**, 1078 (1955).

[75a] Cf. also C. H. Mac Gillavry and G. Fodor, *J. Chem. Soc.* p. 597 (1964).

[76] G. Fodor, J. Tóth, and T. Vincze, *J. Chem. Soc.* p. 3504 (1955).

[77] G. L. Closs, *J. Am. Chem. Soc.* **81**, 5456 (1959).

[77a] J. M. Eckert and R. J. W. LeFèvre, *J. Chem. Soc.* p. 3991 (1962).

b. Morphine

The conformations of other alkaloids have also been investigated more thoroughly in recent years. Of the many investigations available in the literature, only a few will be considered.[78]

The piperidine ring of the morphine structure exists in the chair form. This has been demonstrated by X-ray analysis of the morphine hydroiodide dihydrate[79] and the codeine hydrobromide dihydrate[80]

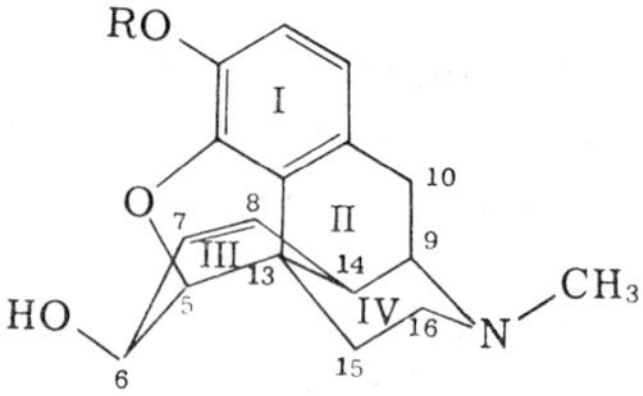

LXV

(LXV, R=H). The methyl group on the nitrogen is oriented toward the C-9–C-10 bond. In ring III, the atoms C-6, C-7, C-8 and C-14 lie approximately in a plane. The half-boat conformation is assumed by this ring because of the 7,8 double bond. The OH group on C-6 is in a *cis* position to the oxide ring. Rings II and III have a *cis* juncture. The atoms of the morphine skeleton are, therefore, distributed approximately in two planes perpendicular to each other. One of these planes is formed by the octahydroisoquinoline system (rings III and IV). The other plane has a tetralin system (rings I and II) as well as a dihydrofuran ring.

In reference to the position of the hydroxyl group in ring III, various epimeric morphine derivatives have been investigated. The saponification rate of dihydrocodeine and dihydroisocodeine acetates[81] (LXV, R=CH_3; with double bond in ring III hydrogenated), as well as the oxidation rate of the epimeric alcohols[82] demonstrated that the OH group on C-6 assumes an *equatorial* position in dihydrocodeine.[83]

---

[78] (a) Cf. R. H. F. Manske and H. L. Holmes, eds., "The Alkaloids," Vols. I-VII Academic Press, New York, 1950-1960; (b) cf. also H. G. Boit, "Ergebnisse der Alkaloidchemie bis 1960," p. 288ff. Akademie-Verlag, Berlin, 1961.

[79] M. Mackay and D. C. Hodgkin, *J. Chem. Soc.* p. 3261 (1955).

[80] J. M. Lindsey and W. H. Barnes, *Acta Cryst.* **8**, 227 (1955).

[81] D. Elad and D. Ginsburg, *J. Am. Chem. Soc.* **78**, 3691 (1956).

[82] H. Rapoport, R. Naumann, E. R. Bissel, and R. M. Bonner, *J. Org. Chem.* **15**, 1103 (1950).

[83] (a) Cf. also D. Ginsburg, *Bull. Narcotics U.N. Dep. Soc. Affairs* **9**, 18 (1957); (b) Cf. also G. Stork, *in* "The Alkaloids" (R. H. F. Manske and H. L. Holmes, eds.), Vol. II, pp. 171 ff.; Vol. VI, p. 233.

c. Quinolizidine Alkaloids

The ring arrangement of quinolizidine (norlupinane), (mentioned on page 312) could well exist in the energetically preferred *trans*-two-chair conformation. This ring structure exists in the so-called quinolizidine alkaloids, either alone in lupinine and epilupinine (LXVI), or condensed with a piperidine or a second quinolizidine system. The sparteines (LXVII) and their derivatives, an example of the latter case, have been well investigated, especially by F. Galinovsky and co-workers and F. Bohlmann and co-workers, with respect to their conformations.

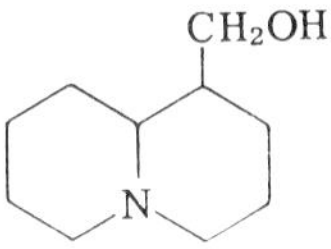

LXVI

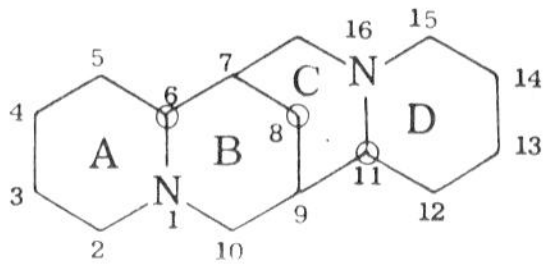

LXVII

For the two lupinines, the hydroxymethylene group is assumed to be *equatorial* in epilupinine, the more stable isomer, and *axial* in lupinine.[50a,51a,84]

Three stereoisomeric, racemic forms of sparteine (LXVII) are possible; they differ in the *cis* or *trans* position of the H atoms at C-6 and C-11 in relation to C-8. Both H atoms are *cis* to C-8 in α-isosparteine, both *trans* to C-8 in β-isosparteine. In sparteine, the H atom at C-6 is *cis*, at C-11 *trans* to C-8. The conformations of the three sparteines have been extensively studied.[85] (*Cis-cis*)-α-isosparteine possesses the conformation shown in (LXVIII), as determined by X-ray analysis.[86] All rings, accordingly, assume the chair conformation, i.e. two *trans*-norlupinane rings are connected in the 1,3 position.

(*Cis-trans*)-sparteine can only be constructed from a *trans*-norlupinane (rings A and B) and a *cis*-norlupinane (rings C and D) with the retention of the chair forms for the individual rings. This would lead to conformation (LXIX) for sparteine for which experimental evidence is also available.[85a]

---

[84] (a) A. F. Thomas, H. J. Vipond, and L. Marion, *Canad. J. Chem.* **33**, 1290 (1955); (b) J. Ratsky, A. Reiser, and F. Sorm, *Chem. Listy* **48**, 1794 (1954); (c) cf. also W. D. Crow, *Australian J. Chem.* **11**, 366 (1958).

[85] (a) F. Galinovsky, P. Knoth, and W. Fischer, *Monatsh. Chem.* **86**, 1014 (1955); (b) N. J. Leonard, P. D. Thomas, and V. W. Gash, *J. Am. Chem. Soc.* **77**, 1552 (1955); (c) F. Bohlmann, W. Weise, H. Sander, H. G. Hanke, and E. Winterfeldt, *Chem. Ber.* **90**, 653 (1957) and later investigations concerning lupine alkaloids.

[86] M. Przybylska and W. H. Barnes, *Acta Cryst.* **6**, 377 (1953).

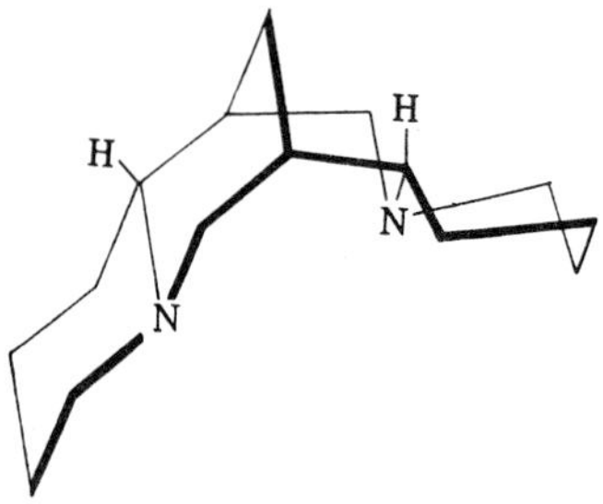

LXVIII

The relatively great stability of the *cis*-norlupinane ring in sparteine (LXIX) can be attributed to the rigidity of the entire ring system. Sparteine is readily rearranged into α-isosparteine.[85a] A structure of two

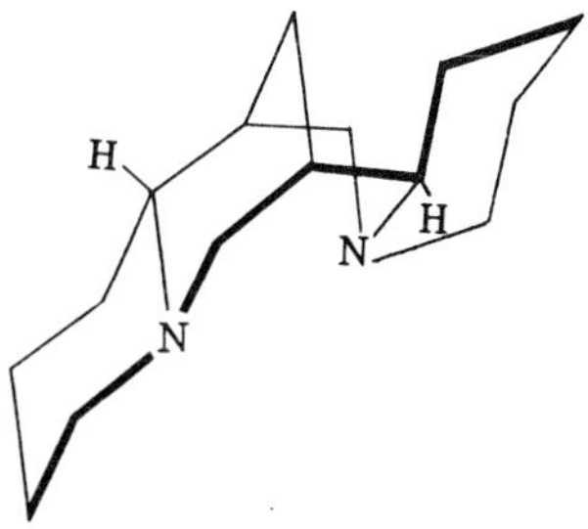

LXIX

cis-norlupinanes in the chair form is assumed for β-isosparteine, because both H atoms at C-6 and C-11 are in the *trans* position to the methylene bridge.

5. Carbohydrates[87,88]

The sugars assume a special place among heterocyclic compounds because their conformations, having drawn the early[87] interest of chemists, have been thoroughly investigated. The conformations of many sugars can be regarded as settled, due especially to the extensive investigations by Reeves.

---

[87] Compare W. N. Haworth, "The Constitution of Sugars," p. 90. Arnold, London, 1929.

[88] (a) Cf. the review by R. J. Ferrier and W. G. Overend, *Quart. Rev. (London)* **13**, 265 (1959); (b) G. Michel, *Bull. Soc. Chim. France* p. 2173 (1960); B. N. Stepanenko, *Russian Chem. Rev.* **31**, 681 (1962); *Usp. Khim.* **31**, 1437 (1962).

It is not our intention to treat the conformations of individual compounds, but we shall consider essential general cases.

In the year 1929 Haworth[87] first discussed various conformations of pyranoses. A decision concerning a preferred conformation of the possible chair and boat forms was not yet possible because no criterion was available for the stability of the conformers.

Not until 1947 did Hassel and Ottar[88d] demonstrate that of the various conformations which the six-membered pyranose ring can assume, the chair form is the most stable one (cf. also page 44), just as for cyclohexane. In a series of basic investigations, Reeves[89,90,91,92,93] was able to show that the pyranose ring does assume the chair conformation when geometrically possible. However, if the nonbonded interactions between the OH groups in the chair conformation become too large, due to a special carbohydrate configuration, then the boat or twist conformations are possible.[89,90]

To differentiate among the various theoretically possible conformations of pyranose sugars, various conventions have been proposed. Reeves distinguished[89,91,92,93,94] eight different conformations for the pyranose ring (Fig. 1) and designated them with the symbols C (chair) and B (boat) (Fig. 1a)

One chair conformation is designated C1 and the one obtained by conversion of this, by 1C. The substituents which assume the *equatorial* position in C1, assume the *axial* position in 1C. The mirror image of C1, however, is also designated as 1C. The three boat conformations are B1, B2, and B3; their inverse conformations are 1B, 2B, and 3B. The numbering of the carbon atoms (1, 2, 3, 4, 5) follows the convention established for the nomenclature of the carbohydrates.[95] In addition to

[88d] O. Hassel and B. Ottar, *Acta Chem. Scand.* **1**, 929 (1947).

[89] R. E. Reeves and F. A. Blouin, *J. Am. Chem. Soc.* **79**, 2261 (1957).

[90] (a) R. E. Reeves, *Ann. Rev. Biochem.* **27**, 15 (1958); (b) R. E. Reeves, *Advan. Carbohydrate Chem.* **6**, 107 (1951).

[91] R. E. Reeves, *J. Am. Chem. Soc.* **71**, 2116 (1949).

[92] R. E. Reeves, *J. Am. Chem. Soc.* **73**, 957 (1951).

[93] (a) R. E. Reeves, *J. Am. Chem. Soc.* **71**, 215 (1949); (b) *ibid.* **71**, 1737 (1949); (c) *ibid.* **72**, 1499 (1950).

[94] Cf. also H. S. Isbell, F. A. Smith, E. C. Creitz, H. L. Frush, J. D. Moyer, and J. E. Stewart, *J. Res. Natl. Bur. Stand.* **59**, 41 (1957).

[95] During the formation of the ring form from an open-chain sugar a new asymmetric center is produced at carbon atom 1. The resulting isomers (also called anomers) are designated with α and β. The cyclic structures may be five-membered rings (furanoses) or six-membered rings (pyranoses), whose configurations are represented by the well-known perspective formulae, introduced by Drew and Haworth [H. D. K. Drew and W. N. Haworth, *J. Chem. Soc.* p. 2303 (1926).] α-Forms are those in which the configuration on the anomeric carbon is the same as that on a certain reference carbon (see below). If the two carbons possess opposite configuration, then the isomer is designated as β.

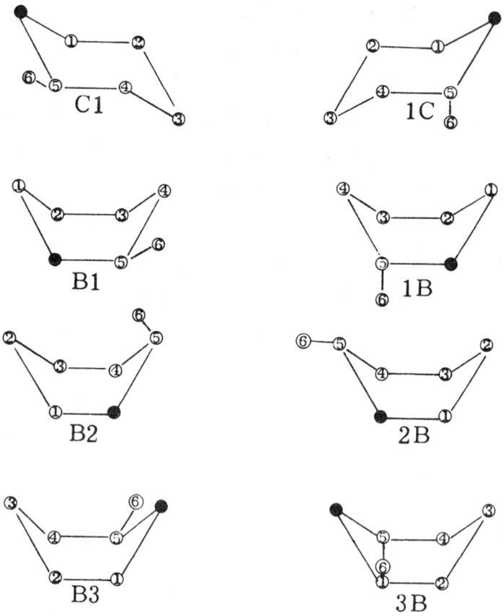

FIG. 1a.

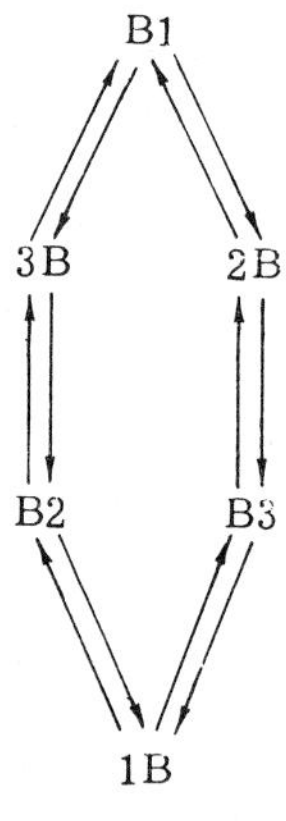

FIG. 1b.

the boat forms show in Fig. 1, an infinite number of twist conformations which are free of angle strain (see Chapter 2, page 46) are possible.[89,90] These are passed through as the boat conformations are interchanging, one into the other. According to Reeves[89,90] the conversion of the

For assignment into the D or L series, the configuration at that asymmetric carbon which is farthest removed from the anomeric carbon is related to D- or L-glyceraldehyde. The α-L isomer is the mirror image of the α-D isomer, just as the β-L isomer is the mirror image of the β-D isomer.

individual boat forms into each other follows a specific order, as shown in Fig. 1b.

As mentioned above, according to Reeves' nomenclature the symbols, such as C1 and 1C, are not only used for different conformations, in this case the chair conformations of any pyranose sugar, but are also employed to designate enantiomorphic forms in which the same arrangement of substituents exists in relation to their *axial* and *equatorial* position.

Another nomenclature system for sugar conformers has been developed independently by Guthrie[96] and by Isbell and Tipson.[97] Only the symbols C (chair), B_1, B_2, and B_3 (boat), and $S_{1,3}$; $S_{1,5}$; $S_{2,0}$, $S_{2,4}$, $S_{3,5}$ and $S_{4,0}$ (skew) were used. The symbols for the chair and boat forms are used as in Reeves' nomenclature; C combines the symbols C1 and 1C. B_1 combines B1 and 1B, B_2 combines B2 and 2B, etc. The newly introduced symbol S for the skew (or twist) form is supplied with indices which indicate the atoms which are located exoplanar, i.e. they protrude out of the plane of the other four atoms.

The anomeric carbon atom 1, present in all pyranose sugars, is used as the reference carbon atom. The anomeric group[98] located on this carbon atom is used to differentiate between the two conformations of a ring type. If the reference group of carbon 1 of an α anomer is *axial* (D or L series), then the symbol A is added to the symbol of the ring type. If this group is *equatorial*, then symbol E is added to the symbol of the ring type. α-D-Glucopyranose-CA indicates that the compound exists in the chair conformation, which possesses an *axial* anomeric OH group. Formulas (LXX) and (LXXI) show the CA and CE conformations of the D- and L-series. The corresponding boat conformations may be represented in a similar manner.[97a]

As has been mentioned above, the chair conformations are preferred throughout for the pyranose forms. According to Reeves,[90b,93c] which of the two chair conformations a pyranose of a certain configuration preferably assumes depends on three factors. The following tend to promote the instability: (a) all *axial* groups with the exception of hydrogen; (b) an *axial* OH group on C-2, where the C—O valence on C-2 bisects the two C—O valences on C-1 (Δ_2-factor) (this instability factor can probably be attributed to an unfavorable dipole interaction[88a]; (c) an *axial* hydroxymethyl group on C-5 and an *axial* OH group, which are on the same side of the ring (Hassel-Ottar effect).

---

[96] R. D. Guthrie, *Chem. & Ind.* p. 1593 (1958).

[97] (a) H. S. Isbell and R. S. Tipson, *Science* **130**, 793 (1959); *J. Res. Natl. Bur. Stand.* **A64**, 171 (1960); (b) cf. also ref. 94 and H. S. Isbell, *J. Res. Natl. Bur. Stand.* **57**, 171 (1956).

[98] The anomeric reference group on carbon 1 may be any atom except hydrogen or any group except an alkyl or polyhydroxyl alkyl group

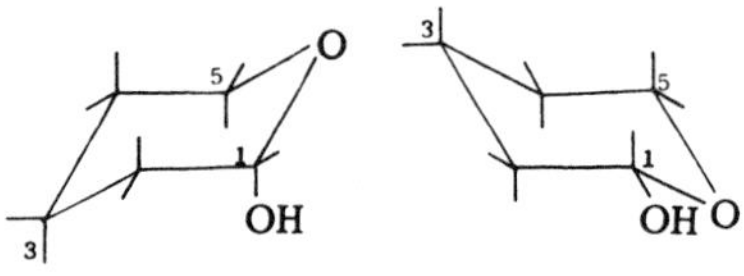

D-Series
CA

L-Series
CE

LXX

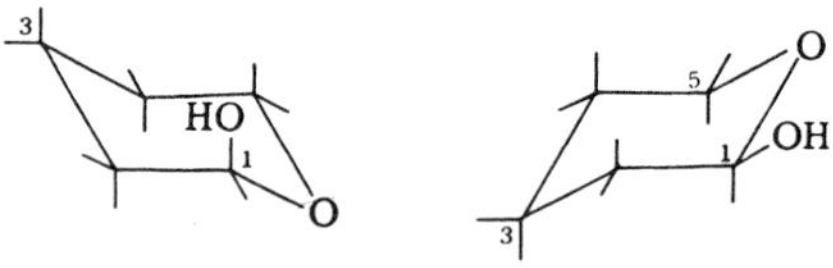

CA

CE

LXXI

Reeves[90b] investigated the complex formation with ammoniacal copper solution[99] in order to confirm the predicted conformations. As mentioned previously (Chapter 3, page 125), formation of a complex with ammoniacal copper solution requires a certain geometric arrangement. If sugar molecules with a 1,2-diol system are used, complex formation is possible only if the dihedral angle of the OH groups amounts to 0° or 60°. If the dihedral angle is larger than 60° no complex formation takes place.

Reeves was also able to show that complex formation with ammoniacal copper solution takes place only on the OH groups at 2,3- or 3,4-positions of substituted D-glucopyranosides. Other OH group pairs do not react. On the basis of the conformations shown in fig. 1, one can state the dihedral angle of the 2,3- and 3,4-hydroxyls for every conformation of the D-glucopyranose ring.[90b] In conformation C1 the dihedral angle amounts to 60° for the 2,3-hydroxyl groups, and 60° for the 3,4-hydroxyl groups. In conformation 1C, the dihedral angle for the 2,3-hydroxyl groups and for the 3,4-hydroxyl groups is 180°. For all the other possible conformations, with the exception of 3B, the dihedral angle is larger than 60° and does not take part in a complex formation in accordance with the above statement. These conformations are excluded. In conformation 3B the two angles also amount to 60°, so only the C1 and 3B conformations permit the formation of complexes. Conformation 3B, however, will be excluded for energetic reasons as well as by the fact that the formation of 4,6-benzylidene and 4,6-ethylidene derivatives is not possible in conformation 3B. The D-glucopyrano-

[99] Cf. also R. Bentley, *J. Am. Chem. Soc.* **82**, 2811 (1960).

sides therefore exist in the C1 form, in agreement with the predictions made by Reeves.

In this way and on the basis of the above-mentioned conformational analytical considerations Reeves derived the C1 conformation of the methylpyranosides of D-galactose, D-glucose, D-gulose, D-mannose, D-ribose, and D-xylose (LXXII–LXXVII).

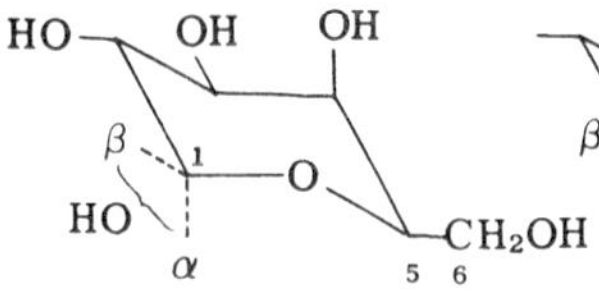

D-Galactose

LXXII

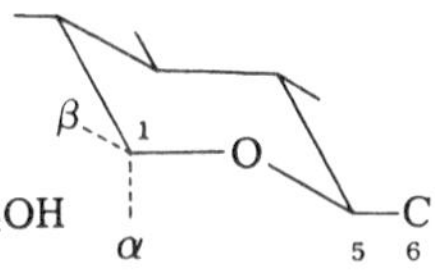

D-Glucose

LXXIII

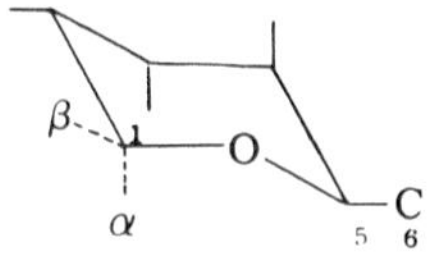

D-Gulose

LXXIV

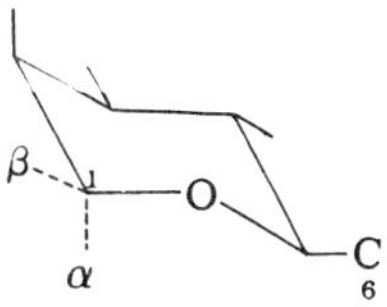

D-Mannose

LXXV

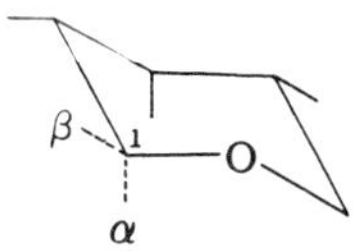

D-Ribose

LXXVI

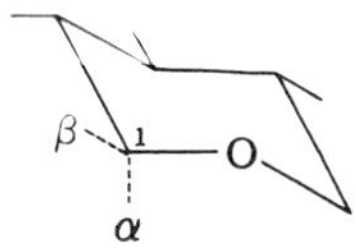

D-Xylose

LXXVII

The methylpyranosides of D-arabinose (LXXVIII) and L-rhamnose (LXXIX) assume the 1C conformation.

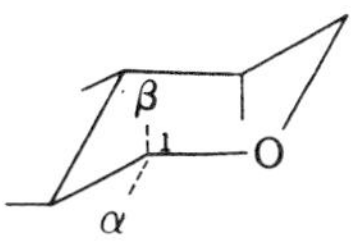

D-Arabinose

LXXVIII

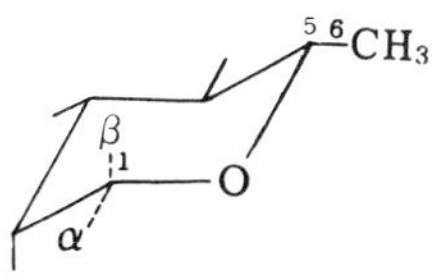

D-Rhamnose

LXXIX

D-lyxose, D-altrose, and D-idose exist in a conformational equilibrium, as is shown in (LXXX) for altrose.

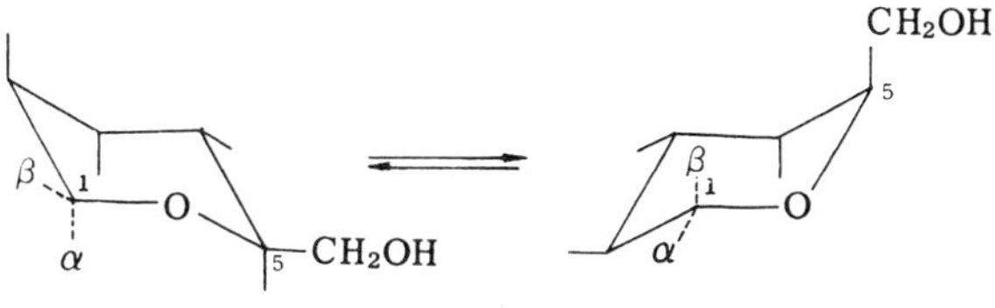

D-Altrose

LXXX

From this it follows that in each case the conformation is assumed in which the greatest number of substituents are equatorial. The most stable compound of all according to this view is β-D-glucose, in which all hydroxyl groups and the hydroxymethyl group exist in the *equatorial* position. In the case of altrose or idose either a number of hydroxyl groups must be *axial* in the C1 conformation or the spatially large CH_2OH group on C-5 is *axial*, which is the reason for the conformational equilibrium.

Nevertheless, 1-substituted derivatives of α-D-glucose are generally more stable than the β-form. The equilibrium mixture of the methyl-D-glucopyranoside contains more of the α than the β form.[100] From this it appears that an *axial* alkoxyl group on C-1 of the pyranose ring is more stable than an *equatorial* one. This is explained by Edward[101] as follows: The dipole–dipole interaction between the alkoxyl group and the ring oxygen is especially large in the β position, so that the α position is preferred (LXXXI).

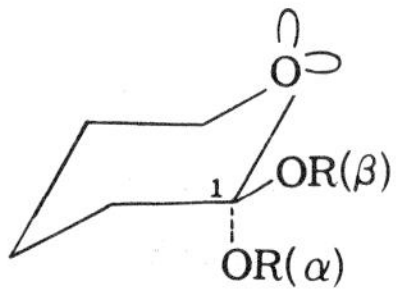

LXXXI

Other methods already mentioned frequently in other connections for the determination of the conformation have been used recently for the pyranoses. Infrared[102] and nuclear magnetic resonance[103] spectroscopic techniques have confirmed the conformations postulated by Reeves. A check on the above conformations by Reeves has also been possible by X-ray analysis of a series of sugars in the crystalline state.[104]

[100] G. N. Bollenback, "Methyl Glucosides," p. 12. Academic Press, New York, 1958.

[101] J. T. Edward, *Chem. & Ind. (London)* p. 1102 (1955).

[102] (a) W. Brock-Neely, *Advan. Carbohydrate Chem.* **12**, 13 (1957); (b) Cf. also S. A. Barker, R. H. Moore, M. Stacey, and D. H. Wiffen, *Nature* **186**, 307 (1960).

[103] R. U. Lemieux, R. K. Kullnig, H. J. Bernstein, and W. G. Schneider, *J. Am. Chem. Soc.* **79**, 1005 (1957); *ibid.* **80**, 6098 (1958); R. J. Abraham, K. A. McLauchlan, L. D. Hall, and L. Hough, *Chem. & Ind. (London)* p. 213 (1962); L. D. Hall, L. Hough, K. A. McLauchlan, and K. Pachler, *ibid.* p. 1465 (1962); R. J. Abraham, L. D. Hall, L. Hough, and K. A. McLauchlan, *J. Chem. Soc.* p. 3699 (1962).

[104] (a) See for example T. R. R. McDonald and C. A. Beevers, *Acta Cryst.* **5**, 654 (1952); S. Furberg and A. Hordvik, *Acta Chem. Scand.* **11**, 1594 (1957); (b) Concerning the conformation of amino sugars cf. for example G. Fodor and L. Ötvös, *Chem. Ber.* **89**, 701 (1956).

CHAPTER 8

Acyclic Diastereomers

Admittedly the conformations of cyclic compounds have been stressed up to now, but in this final chapter the acyclic diastereomeric compounds will be considered briefly.

To begin with, certain questions of nomenclature that were not mentioned in Chapter 2 will be treated.

If two asymmetric carbon atoms are present in a molecule, as for example in tartaric acid, two diastereomeric compounds are possible. If the substituents on the two asymmetric C atoms are the same (I; A, B,C), then one of the diastereomers cannot be resolved into its optical antipodes. This is so because the two halves of the molecule behave as mirror images, and an intramolecular compensation takes place.[1] Such compounds are designated as *meso-* (I) (see page 7). The second diastereomer can be resolved into its optical antipodes and is called the *d,l* or racemic isomer (II).[2] As an example, structure (III) would be the *meso* diastereomer of 2,3-dibromobutane and (IV) the *d,l* or racemic isomer.

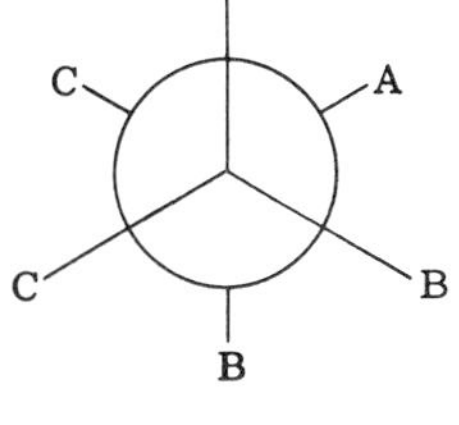

meso

(I)

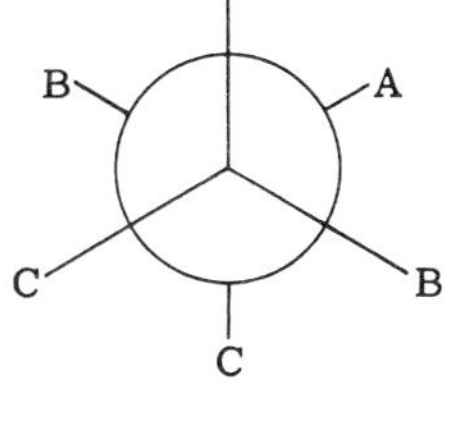

d,l or *racem*

(II)

The *meso* and *d,l* isomers, therefore, are not conformers. They cannot be converted into each other by rotation around the C—C bond axis. They are configurational isomers which can only be obtained from each

[1] For a more detailed discussion, cf. C. R. Noller, *Science* **102**, 508 (1945).

[2] Here as well as in the following, only one of the enantiomers will be shown.

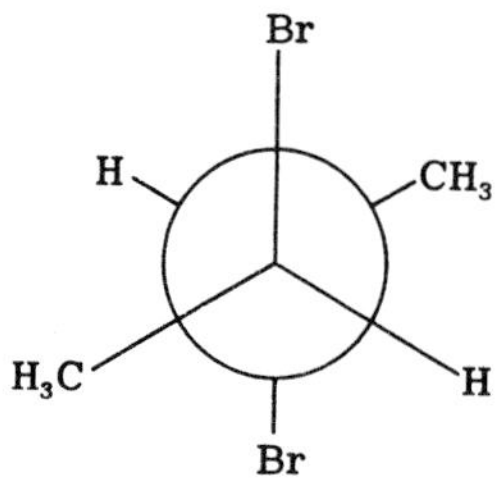

meso

(III)

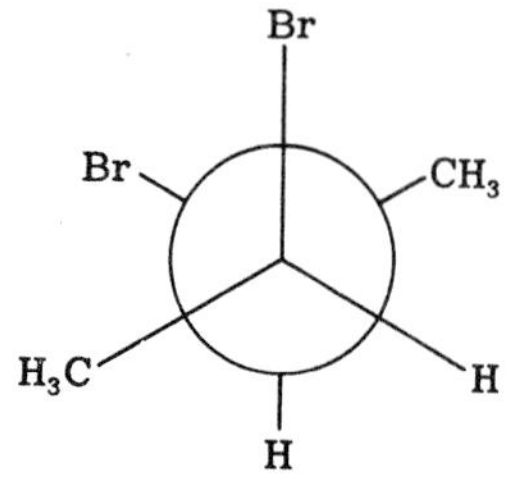

d,l or *racem*

(IV)

other by the exchange of two substituents. They correspond to the *cis* and *trans* isomers of cyclic compounds. However, there are various conformations possible for each configuration, as has been shown in Chapter 2.

If the substituents on the two asymmetric C atoms are different, then both diastereomers can be resolved into optical antipodes. The two diasteromers are designated as the *erythro* and *threo* forms. These are also configurational isomers, whereby assertion concerning their absolute configurations is made. Only the relative configurations of two reference atoms or groups on the asymmetric C atom is given.

When it comes to nomenclature, certain difficulties not encountered with the *meso* and *d,l* isomers arise. First, the following definition of the *erythro* and *threo* isomers will suffice.[3] It is assumed that at least two substituents on each of the adjacent asymmetric C atoms are similar or identical, which is the case for many compounds. The *erythro* isomer corresponds to that configuration in which at least two similar or identical substituents are superimposed on each other in the projection of one of the three *eclipsed* conformations (V and VI). If this is not the case, then a *threo* isomer (such as VII and VIII) is present.

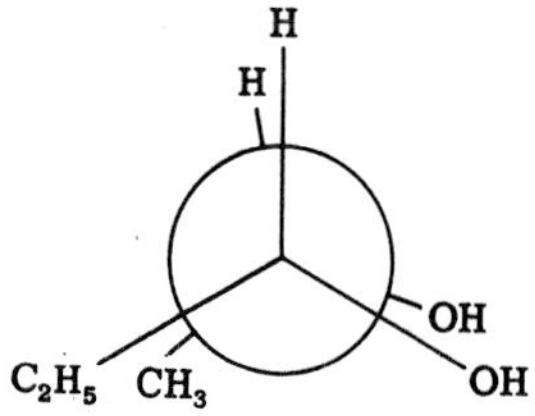

erythro

(V)

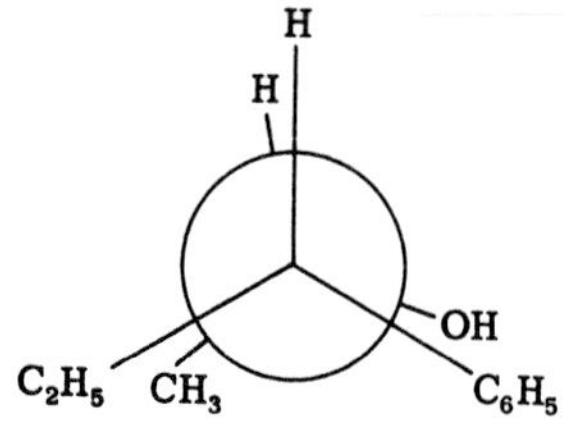

erythro

(VI)

[3] Cf. D. J. Cram and F. A. Abd. Elhafez, *J. Am. Chem. Soc.* **74**, 5828 (1952).

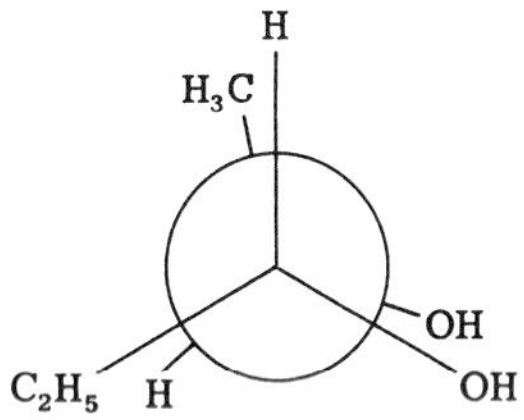

threo

(VII)

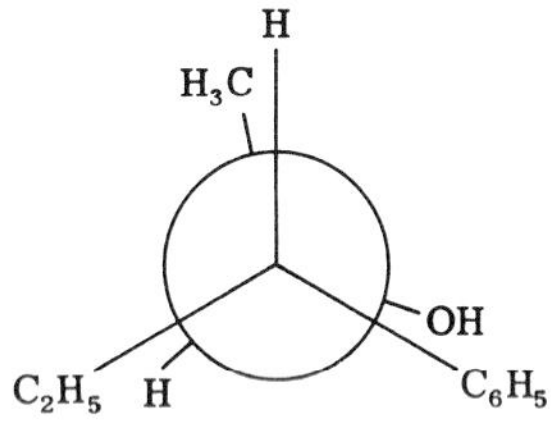

threo

(VIII)

On the other hand, examples can be found which do not fit into this definition.[4] For example, the *erythro* or *threo* configuration of alcohol (IX) cannot be definitely defined.

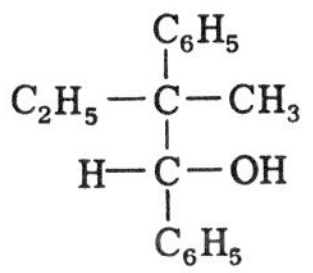

(IX)

The difficulties in nomenclature can be circumvented if description of the configuration of asymmetric compounds is based on the convention proposed by Cahn *et al.*[5] (see page 68, ref. 144). The *erythro* isomer is the one in which the configuration of the two asymmetric C atoms is the opposite. The *threo* isomer is the one in which the configuration of the two asymmetric C atoms is the same.

1. The Relative Stability of Diastereomers[3,6,7]

Differences in the stability of diastereomeric compounds have already been discussed. With 1,2-disubstituted cyclohexanes for example the *trans* compound proved to be more stable than the *cis* compound (see page 117). This was attributed to different interactions between the substituents or with other atoms in the *cis* or *trans* configurations.

---

[4] Cf. H. Felkin, *Bull. Soc. Chim. France* p. 1050 (1956).

[5] R. S. Cahn C. K. Ingold, and V. Prelog, *Experientia* **12**, 81 (1956).

[6] Cf. D. H. R. Barton and R. C. Cookson, *Quart. Rev.* (*London*) **10**, 44 (1956).

[7] Cf. L. D. Bergelson, *Usp. Khim.* **27**, 817 (1958).

The same considerations can now be applied in a qualitative way to diastereomeric acyclic compounds. A primary assumption is that acyclic diastereomers may adopt a large number of conformations as a result of the rotation around the C—C axis. As with cyclic compounds, the various possible conformations are not equally probable, but because of nonbonded interactions, certain preferred conformations will predominate. These conformations are determined by the effective size of the various substituents on the two asymmetric C atoms.[6,8]

Consider first the *erythro* (X) and the *threo* configurations (XII) of a certain compound and designate the substituents on the asymmetric C atoms as L (large), M (medium), and S (small), thereby expressing the effective bulk. The *erythro* isomer, shown in the eclipsed conformation (X), will now assume the most stable conformation (XI). In the latter the two largest substituents (L) will be most distant from each other and will assume a *trans* (or *anti*) position relative to each other (XI).

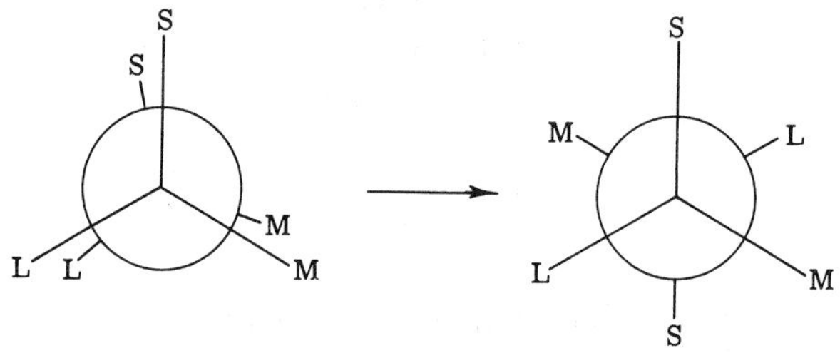

(X) (XI)

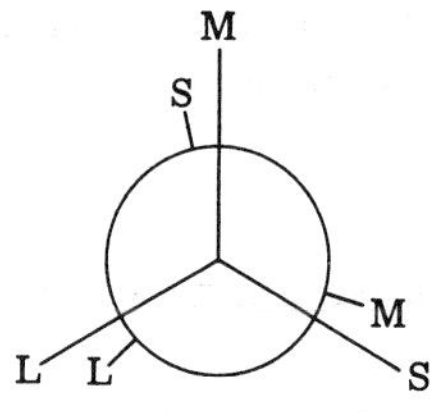

(XII)

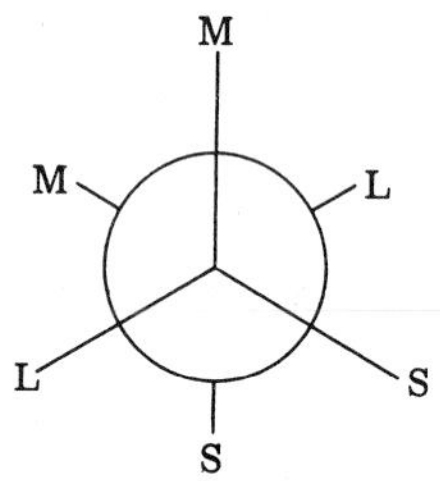

(XIII)

The same is true of the *threo* isomer (XII), which will also assume the conformation in which the L groups are *trans* (XIII). It is quite obvious that the interactions between nonbonded groups are different in the

[8] (a) D. J. Cram and F. D. Greene, *J. Am. Chem. Soc.* **75**, 6005 (1953); (b) D. J. Cram, H. L. Nyquist, and F. A. Abd Elhafez, *J. Am. Chem. Soc.* **79**, 2876 (1957); (c) J. L. Mateos and D. J. Cram, *J. Am. Chem. Soc.* **81**, 2756 (1959).

most stable conformations of the two diastereomers. In the *erythro* isomer (XI), all substituents of the same size have a *trans* relationship, from which the following *gauche* interactions result (see page 30): 2 LS + 2 LM + 2 MS. On the other hand the following interactions result in the *threo* isomer (XIII): 2 LS + 2 LM + MM + SS. Since the MM interaction is decidedly greater than the MS interaction, the *erythro* isomer is more stable than the *threo* isomer, and the *meso*-isomer correspondingly is more stable than the *d,l* isomer. The assumption for these qualitative arguments is that the differences in energy between the diastereomers is determined mainly by steric factors (by the non-bonded interactions). This rule is not always applicable for polar compounds. We have already seen in Chapter 2 that the latter can lead to the stabilization of other, seemingly less stable, conformations through inter- or intramolecular interactions (i.e. hydrogen bonds) (see also page 337).

The greater stability of the *erythro* and *meso* isomers over the *threo* and *d,l* isomers has been confirmed experimentally with a number of compounds.[6] This data can be obtained as we have already learned in cyclic compounds, through a direct equilibration of the diastereomers. For example, in equilibration of the disubstituted *d,l*-succinic acids (XIV, R=R'=alkyl, phenyl, halogen, OH), the more stable *meso* acids (XV, R=R') result.[9]

Correspondingly, the *threo* acids (XIV, R≠R'=alkyl, aryl) yield the *erythro* acids (XV, R≠R').[10]

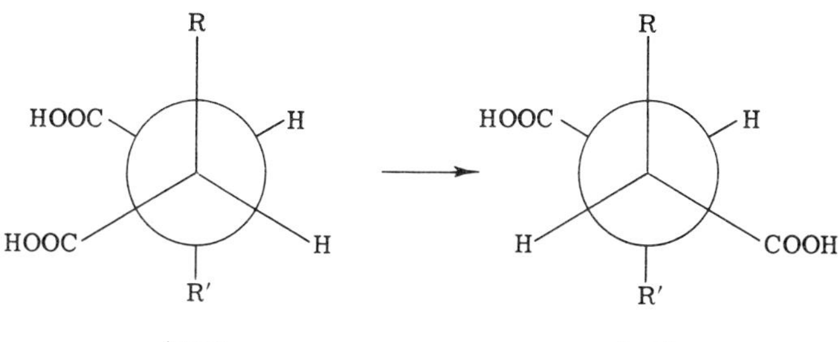

(XIV) (XV)

An additional method for experimentally testing the greater relative stability of *meso* and *erythro* isomers starts with an analogous compound which possesses only one asymmetric C atom, such as a ketone or an

[9] K. L. Wolf "Leipziger Vorträge 1931" p. 17. Hirzel Leipzig 1931; see also Th. Wagner Jauregg *in* "Stereochemie" (K. Freudenberg, ed.). Deuticke Leipzig 1933.

[10] (a) Cf. also R. P. Linstead and M. Whalley, *J. Chem. Soc.* p. 3722 (1954); (b) Cf. also F. A. Abd Elhafez and D. J. Cram, *J. Am. Chem. Soc.* **75**, 339 (1953); for further examples see ref. 6.

oxime. A second asymmetric center can be introduced by a reduction of the ketone or oxime. Those reducing agents are chosen for which the composition of the reaction products would correspond to that of the equilibrium mixture. Sodium in alcohol (see page 269) is suitable for the reduction of cyclic ketones.

Cram and Abd Elhafez[3] have shown, from examples in the literature, that the *meso* and *erythro* isomers (alcohol or amine) (XVII, R=R'= aryl, X=OH or NH_2) are formed in larger amounts from the reduction of ketones or oximes (XVI, R=R'=aryl, X=OH or NH_2) with sodium and alcohol or with sodium amalgam.

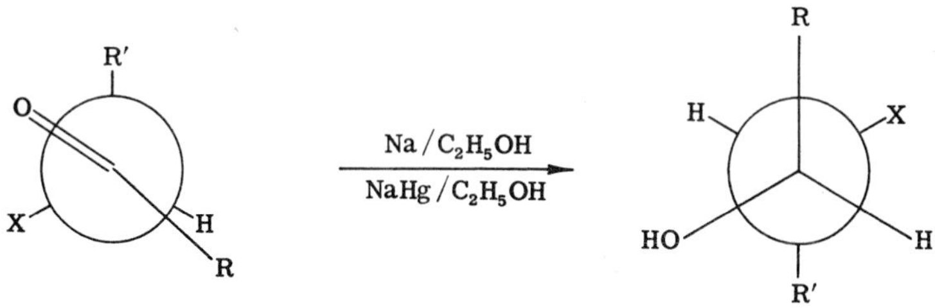

(XVI) (XVII)

2. Conformations and Physical Properties

Physical methods have repeatedly been cited as being well suited for conformational analysis, and have been quite useful in solving problems of conformation among cyclic compounds. The preferred conformations of acyclic compounds have been determined almost exclusively through physical methods (cf. Chapter 2).

In comparison with cyclic compounds, less experimental data involving physical methods of analysis have been forthcoming concerning the conformations of acyclic diastereomers. Some of the available examples will be considered.

Mateos and Cram[8c] studied the infrared spectra of the diastereomeric pairs of the general formula (XVIII). Here a, b, c, and d represent hydrogen, alkyl, or phenyl groups.

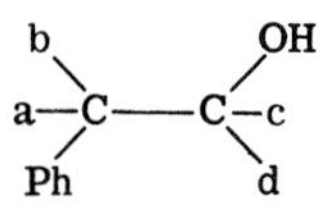

(XVIII)

In the OH stretching region (3400–3600 cm.$^{-1}$), the diastereomers show differences in the position and the intensity of the bands for the free and hydrogen-bonded OH groups. Using the relationship of Kuhn (see page 124), the length of the intermolecular O···H hydrogen bond was determined for the various dimeric alcohols. The diastereomeric pairs [XVIII, a = alkyl (methyl, ethyl, isopropyl), b = H, c = H, d = alkyl (methyl, ethyl, isopropyl)] showed a shorter relative length of the intermolecular hydrogen bond for the *erythro* isomers than for the *threo* isomers. A shorter length for the hydrogen bond of the *erythro* isomer indicates that the OH group is less hindered by the other groups of the molecules than in the case of the *threo* isomer. This was confirmed by the lower intensity of the band which corresponds to the nonassociated OH group in the *erythro* isomer. In each of the diastereomers one conformation predominates, due to the above-mentioned reasons (see page 335). Here, as in other cases discussed below, the preferred conformation is not only stabilized by the nonbonded interactions of neighboring groups, but also by the hydrogen bonds. In compounds of structure (XVIII, a = alkyl=R, b=H, c=H, d=alkyl=R′), the preferred conformation was obtained as follows:

The effective size of the substituents on one asymmetric C atom is $C_6H_5 > R > H$, on the other it is $R' > OH > H$. For the *threo* isomer, (XIX) is then the most stable conformation because the two largest groups on the adjacent asymmetric C atoms (C_6H_5 and R′) are *trans* and are located between the smallest groups of the other asymmetric C atom. In addition, the four largest groups in conformation (XIX) are separated in two pairs by a hydrogen.

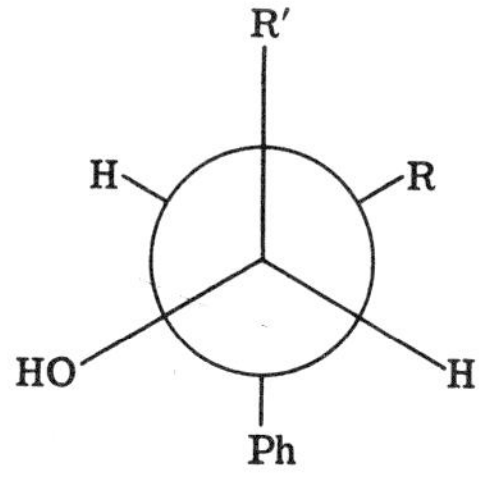

(XIX)

The determination of the most stable conformation for the *erythro* isomer of structure XVIII (a=alkyl=R, b=H, c=H, d=alkyl=R′) is not possible simply on the basis of steric factors. Of the three conformations (XXa), (XXb), and (XXc), (XXc) is the least stable for steric reasons. Which of the two conformations (XXa) and (XXb) is the more

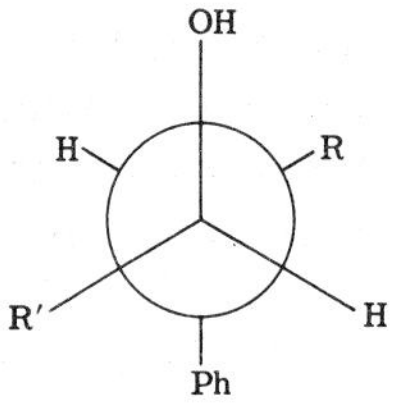

(XXa)

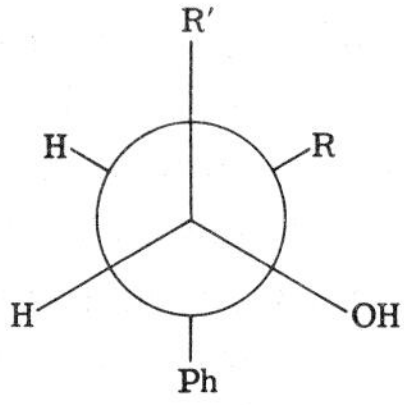

(XXb)

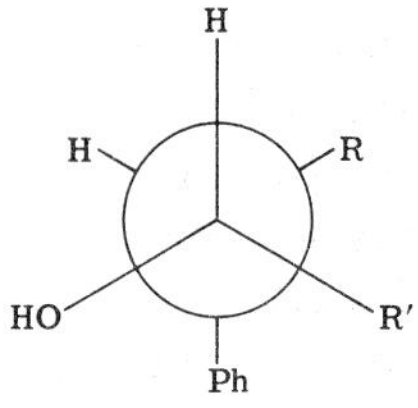

(XXc)

stable, however, cannot be determined with certainty. Because the *erythro* isomers are able to form stronger hydrogen bonds than the *threo* isomers, the OH groups of the *erythro* isomers must be less hindered. In conformation (XXa), the OH group of the *erythro* isomer is located between H and R, in (XXb) between phenyl and R. In conformation (XIX) of the *threo* isomer, the OH group is located between H and phenyl. From this it appears that the OH group in (XXa) is less hindered than in (XIX) and the OH group in (XIX) less than in (XXb). The results of infrared spectroscopy lead to the conclusion that (XXa) is the most stable of the three conformations of the *erythro* compound.

Kuhn[11] also investigated a series of *meso* and *d,l* 1,2-disubstituted diethylene glycols RCHOH–CHOHR, where R=methyl, *n*-pentyl, isopropyl, and *tert*-butyl, with the help of infrared spectroscopy. The bands corresponding to the free and intramolecular hydrogen-bonded OH groups were measured. Two bands were found for all *d,l* compounds, of which one corresponds to the free OH group and the other to the intramolecular bonded one. These results indicate that the *d,l* compounds exist predominantly in conformation (XXI) as a result of the hydrogen bonding.

For the *meso* isomers, conformations (XXIIa) and (XXIIb) must be distinguished. If conformation (XXIIa) is stabilized by intramolecular hydrogen bonding, then the two alkyl groups (R) have an unfavorable *gauche* position. If R is an alkyl group of normal size (methyl or *n*-amyl), strong hydrogen bonding is noted in the infrared spectrum. These compounds do prefer conformation (XXIIa). The stabilization produced by the hydrogen bonding is stronger than the repulsive interactions of the alkyl groups. However, if R is an isopropyl group, then the band corresponding to the bonded OH appears with diminished intensity; conformation (XXIIa) is present in only minor amounts. The bonded

[11] L. P. Kuhn, *J. Am. Chem. Soc.* **74**, 2492 (1952); **76**, 4323 (1954); **80**, 5950 (1958). For a review see M. Tichý, *Chem. Listy*, **54**, 506 (1960); cf. also P. von R. Schleyer, *J. Am. Chem. Soc.* **83**, 1368 (1961); L. P. Kuhn, P. von R. Schleyer, W. F. Baitinger, Jr., and L. Eberson, *ibid.* **86**, 650 (1964).

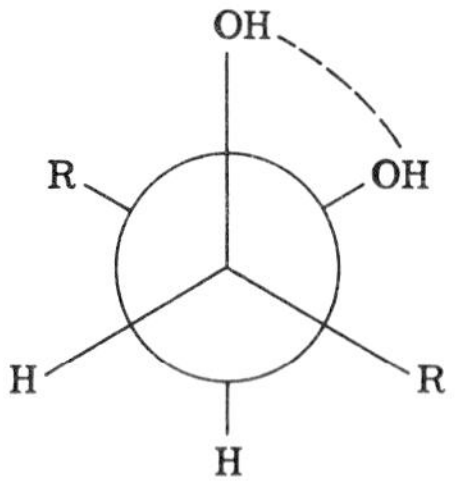

(XXI)

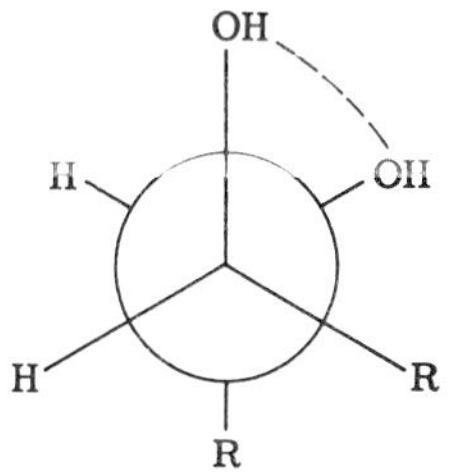

(XXIIa)

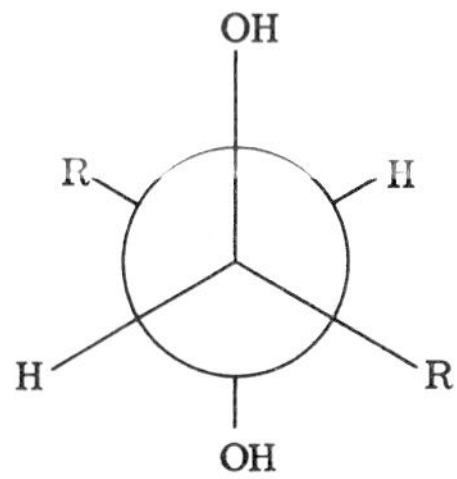

(XXIIb)

OH absorption bond vanishes entirely when R=*tert*-butyl, indicating that conformation (XXIIb) with a *trans* relationship of the bulky tert-butyl groups is assumed exclusively.

Chiurdoglu and co-workers came to the same conclusion[12] in studies of a series of diastereomeric diols, such as butane-2,3-diol, pentane-2,3-diol, and hexane-3,4-diol. For the *threo* compounds, the band which corresponds to the intramolecularly hydrogen bonded OH was always stronger than the band of the free OH group. The opposite was true for the *erythro* compound. Therefore, the conformation with a *trans* position of the OH groups appear more preferable for the *erythro* isomers than in the *threo* isomers.

In this connection, the diastereomeric ephedrines will be considered in more detail (see also page 342). The configurations of both isomers are established. Ephedrin (XXIII) possesses the *erythro* configuration, pseudoephedrin (XXIV) the *threo* configuration.

The conformations of both diasteromers have been investigated by various workers.[13] It is generally agreed that that conformation in which the

[12] G. Chiurdoglu, R. DeGroote, W. Masschelein, and H. vanRisseghem, *Bull. Soc. Chim. Belges* **70**, 342 (1961).

[13] (a) Cf. H. Emde, *Helv. Chim. Acta* **12**, 365 (1929); (b) W. N. Nagai and S. Kanao,

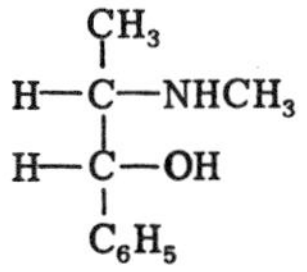

(XXIII)

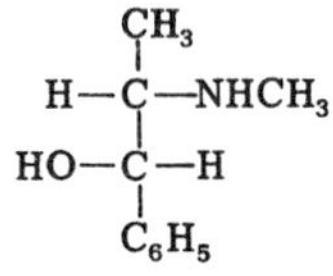

(XXIV)

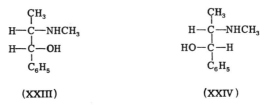

OH and $NHCH_3$ groups assume the *trans* position (XXV) is preferred for ephedrin (XXIII). For pseudoephedrine, the *gauche* conformation of the OH and $NHCH_3$ groups is preferred (XXVI).

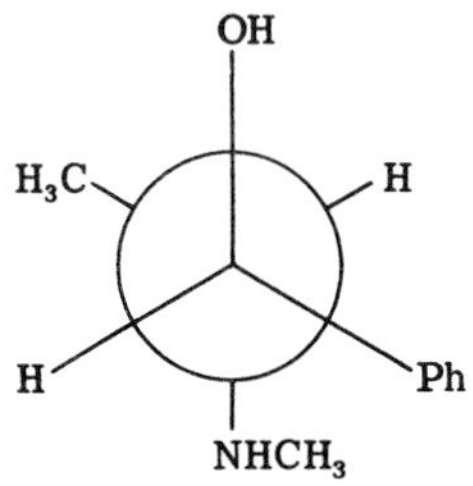

(-)-Ephedrine

(XXV)

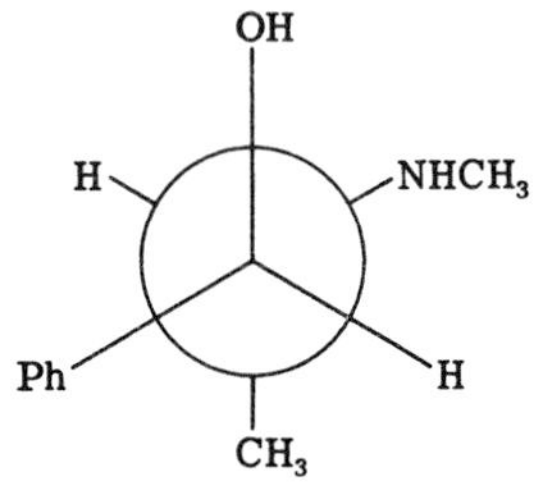

(+)-Pseudoephedrine

(XXVI)

Based on the repulsive interactions between nonbonded groups on the two asymmetric carbon atoms, the preferred conformations have been calculated approximately for ephedrin and pseudoephedrin.[14] According to these calculations, the OH and $NHCH_3$ groups in ephedrin are *trans* (XXV). In pseudoephedrin they are *gauche* (XXVI), and the methyl and phenyl groups have a *gauche* relationship (XXVI). However, in the crystal lattice of ephedrin and pseudoephedrin hydrochlorides, the OH and $NHCH_3$ groups assume the *gauche* position in both cases. This may be attributed to the chloride ion in the lattice.[15]

A comparison of the infrared spectra of ephedrin and pseudoephedrin in nonpolar solvents shows a strong intramolecular hydrogen bonding for pseudoephedrin, which indicates a preference for conformation (XXVI).[16] For the diastereomeric *N*-methylephedrins, the *trans* relation-

---

Ann. Chem. **470**, 157 (1929); (c) L. H. Welsh, *J. Am. Chem. Soc.* **71**, 3500 (1949); (d) W. J. Close, *J. Org. Chem.* **15**, 1131 (1950); (e) G. Fodor, J. Kiss, and J. Sallay, *J. Chem. Soc.* p. 1858 (1951); (f) G. Fodor and K. Koczka, *ibid.* p. 850 (1952); (g) Cf. also G. Fodor, V. Bruckner, J. Kiss, and G. Ohegyi, *J. Org. Chem.* **14**, 337 (1949).

[14] D. H. Everett and J. B. Hyne, *J. Chem. Soc.* p. 1636 (1958).

[15] D. C. Phillips, *Acta Cryst.* **7**, 159 (1954).

[16] T. Kanzawa, *Bull. Chem. Soc. Japan* **29**, 398, 479 (1956).

ship of the OH and $N(CH_3)_2$ groups again proved to be more stable in the *erythro* compound.[17]

The conformations illustrated in (XXV) and (XXVI) have been questioned recently, a point that shall be considered again (page 342).[18]

An additional physical method for the solution of conformational problems, is nuclear magnetic resonance spectroscopy, although to date only a few acyclic diastereomers have been examined by this technique.[19] Brownstein[20] for example was able to show that the rotation about the C-3–C-4 bond is strongly hindered in the diastereomeric 2,5-dimethyl-2,5-dimethoxy-3,4-diphenylhexanes. The *meso* and *d,l* isomers exist preferentially in one conformation, The two conformations of the *meso* isomer differ by about 1.7 kcal./mole, and that conformation with the phenyl groups *trans* is the more stable one. For the *d,l* isomers, conformation (XXVIIa) is more stable than (XXVIIb) by about 1.9 kcal./mole. The third possible conformation of the *d,l* isomer is not present in detectable amounts.

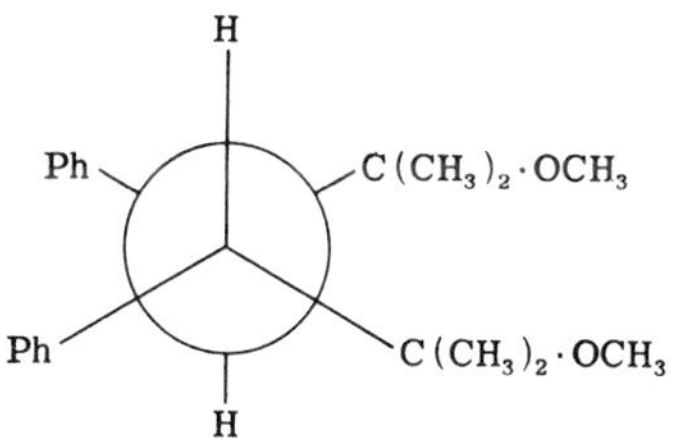

(XXVIIa)

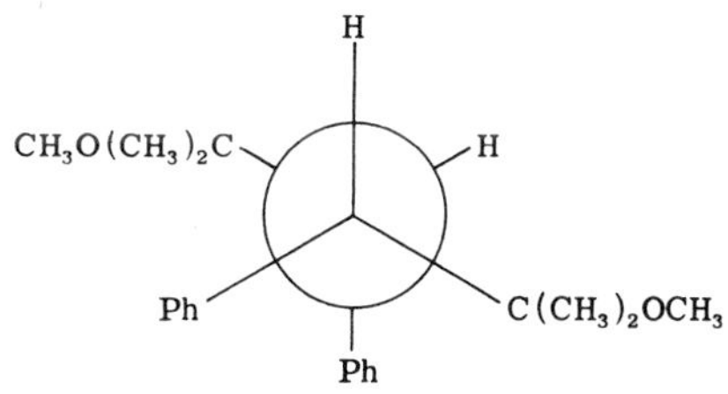

(XXVIIb)

Various investigations in the field of NMR spectroscopy have shown that the chemical shift of the hydroxyl hydrogens depends upon the degree in which the hydrogens are involved in hydrogen bonding.[21] The

---

[17] T. Kanzawa, *Bull. Chem. Soc. Japan* **29**, 604 (1956).

[18] For additional studies of the conformations of acyclic diastereomeric 1,2-amino alcohols *via* infrared spectroscopy see: (a) J. Sicher *Collection Czech. Chem. Commun.* **24**, 2727 (1959); (b) G. Drehfahl and G. Heublein *Chem. Ber.* **94** 922 (1961). For the conformations of diastereoisomeric α-amino-β-hydroxy acid esters see: (c) G. Drehfahl and H. Zimmerman *Chem. Ber.* **94** 2011 (1961).

[19] (a) Cf. J. A. Pople, *Mol. Phys.* **1**, 3 (1958); (b) cf. J. D. Roberts, "Nuclear Magnetic Resonance." McGraw-Hill, New York, 1959; (c) J. A. Pople, W. G. Schneider, and H. J. Bernstein, "High-resolution Nuclear Magnetic Resonance," p. 377. McGraw-Hill, New York, 1959; (d) A. A. Bothner-By and C. Naar-Colin, *J. Am. Chem. Soc.* **84**, 743 (1962); (e) F. A. L. Anet, *ibid.* **84**, 747 (1962).

[20] S. Brownstein *Can. J. Chem.* **39**, 1677 (1961).

[21] Cf. for example (a) E. D. Becker, U. Liddel, and J. N. Shoolery, *J. Mol. Spect.* **2**, 1 (1958); (b) M. Saunders and J. B. Hyne, *J. Chem. Phys.* **29**, 1319 (1958).

relative strength of the hydrogen bonds in diastereomeric alcohols can be determined in this way, and conclusions as to the preferred conformations are possible. The alcohols (VIII) mentioned previously (page 336) were investigated in this manner[8c], and the results of infrared spectroscopy were confirmed. The *erythro* isomers (XXa) showed stronger intermolecular hydrogen bonding than the *threo* isomers.

With the diastereomeric ephedrins, a stronger hydrogen bond was determined by the same method[22] for pseudoephedrin, in agreement with the results described previously (page 340).

Despite the large amount of data available in the literature (see page 339), the problem of the preferred conformation of the diastereomeric ephedrins has recently been investigated again by Hyne.[23] Results were obtained which differ considerably from the previous results. Intra- and intermolecular hydrogen bonding, anisotropic properties of the phenyl group, and the spin-spin coupling constants were investigated with the aid of NMR. Because the spin–spin coupling constants between the neighboring C—H bonds depend on the dihedral angle[19b,c,23a] a determination of the preferred conformation was made possible.[23]

The results are reasonably compatible with a conformation corresponding to (XXVI) in the case of pseudoephedrin, but conformation (XXV) is not supported from the results obtained for ephedrin. Because the experimental results for both ephedrins cannot be brought into agreement with those expected for purely *staggered* conformations, Hyne assumed[23] that these cannot represent the most stable conformations. "Pure" *staggered* conformations only represent the energy minimum in the case of simple, symmetric systems such as ethane (see page 27). For asymmetric substituted compounds, where the interactions between the groups are not alike, the "pure" *staggered* conformation does not represent the energy minimum. Therefore, the so-called "*off-staggered*" conformations are supposed to represent the preferred conformations, and these are shown in (XVIII) for ephedrin and (XXIX) for pseudoephedrin.

In each of the two conformations, the bulky phenyl and methyl groups repel each other. This leads to an enlargement of the dihedral angle between the OH and $NHCH_3$ groups in the *erythro* isomer (ephedrin, XXVIII), and to a decrease of the dihedral angle betweeen these groups in the *threo* isomer (pseudoephedrin, XXIX). The tendency to form an

---

[22] J. B. Hyne, *Can. J. Chem.* **38**, 125 (1960).

[23] J. B. Hyne, *Can. J. Chem.* **39**, 2536 (1961); cf. also J. B. Hyne, *J. Am. Chem. Soc.* **81**, 6058 (1958).

[23a] M. Karplus, *J. Chem. Phys.* **30**, 11 (1959); cf. also C. N. Banwell, A. D. Cohen, N. Sheppard, and J. J. Turner, *Proc. Chem. Soc.* p. 266 (1959); N. Sheppard and J. J. Turner, *Proc. Roy. Soc.* **A252**, 506 (1959).

intramolecular hydrogen bond is thus greater for pseudoephedrin,[18b] in agreement with the results discussed above (page 340).

Aside from infrared and NMR spectroscopy, attempts have been made to use the ultraviolet absorption spectra[8c] and measurement of dipole moments[24] to determine the preferred conformations of acyclic diastereomers.

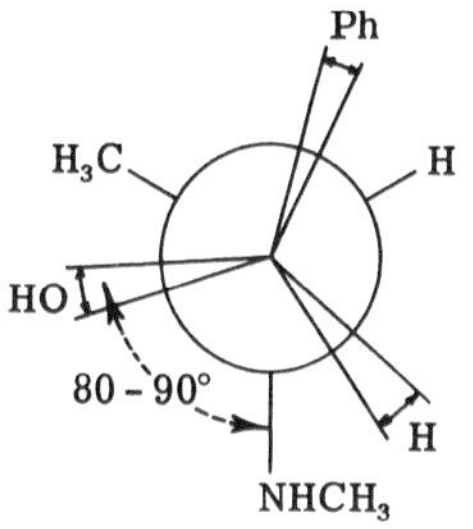

(−)-Ephedrine
(XXVIII)

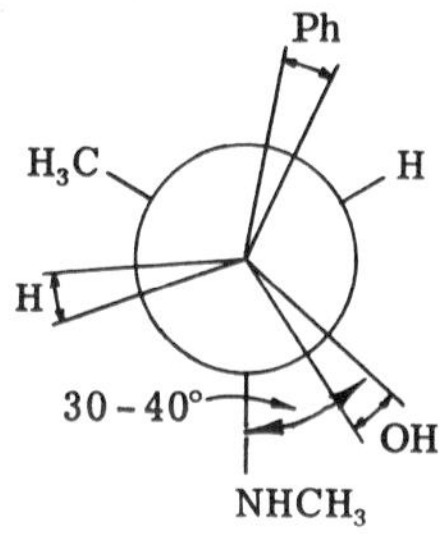

(+)-Pseudoephedrine
(XXIX)

Few investigations are available concerning other physical properties of acyclic diastereoisomers. Mateos and Cram[8c] were able to show that, for the alcohols investigated by them (see page 336), the refractive indices of the *erythro* isomers were always higher than those of the *threo* compounds. No regularity could be found, however, in the density (cf. in this connection the Auwers-Skita rule, page 142).

3. Conformation and Reactivity of Acyclic Diastereomers: Some Examples

In this section a few characteristic examples of studies of the relation of conformation and reactivity of acyclic diastereomers will be discussed.[24a] In principle, the same is true here as was described in Chapter 5 (Section II,5). The reactivity of a compound is not necessarily dependent on the conformation in the ground state, the thermodynamically most stable conformation. The mechanism of the reaction, which must be known, is most important. The conformations which can arise in the transition state during the reaction which in the case of diastereomers can differ by a relatively large amount of energy must also be known.

[24] For example, see H. Sawatzky, G. K. White, and G. F. Wright, *Can. J. Chem.* **37**, 1132 (1959) and ref. 18b.

[24a] Cf. also: E. L. Eliel, "Stereochemistry of Carbon Compounds," p. 139ff. McGraw-Hill, New York, 1962.

An *erythro* isomer, for example, can react more much rapidly than the less stable *threo* isomer. The relative rates of the reactions depend on the steric interactions in the transition states of both compounds.

a. Cram's Rule of Asymmetric Induction

It has been shown in some fundamental studies by Cram[3,8a,25] and by Prelog (see page 346), of the so-called "asymmetric induction" that the conformation of an acyclic compound is of decisive significance in determining the configurations of the products in a reaction.

The phenomenon of asymmetric induction, by which one diastereomer predominates in the reaction products of compounds with one asymmetric carbon atom, has been known for a long time, especially from investigations by McKenzie.[26] The presence of an asymmetric carbon atom in a compound influences the ratio of isomers (*threo* and *erythro*) during the formation of a second asymmetric carbon atom.

Compounds which have a carbonyl group adjacent to the asymmetric carbon atom (XXX) were considered by Cram[3,8a,25]. By determining the configurations of the reaction products which result from the addition of a reagent R′Z to the carbonyl group, the following rule was derived[3]:

"In reactions of the following type (XXX → XXXI), that diastereomer (XXXI) will predominate which would be formed by the approach of the entering group (R′) from the least-hindered side of the double bond (of the C=O group) when the rotational conformation of the C—C bond is such that the double bond is flanked by the two least bulky groups attached to the adjacent asymmetric center."

The effective size of the substituents in (XXX) and (XXXI) is $L > M > S$ (see page 335). However, it must be assumed that the effective size of the substituents in relation to each other is definitely known.

Cram[3,8a,25] derived this rule from the reaction of ketones and aldehydes of type (XXX), which he had studied, and from examples available in the literature. The ketones and aldehydes (XXX) were reacted with the following reagents: lithium aluminum hydride, sodium borohydride, sodium in alcohol, sodium amalgam and water, Grignard reagents, and

[25] (a) D. J. Cram and J. D. Knight, *J. Am. Chem. Soc.* **74**, 5835, 5839 (1952); (b) D. J. Cram, F. A. Abd Elhafez, and H. Weingartner, *ibid.* **75**, 2293 (1953); (c) D. J. Cram, F. A. Abd Elhafez, and H. LeRoy Nyquist, *ibid.* **76**, 22 (1954); (d) D. J. Cram and J. E. McCarty, *ibid.* **76**, 5740 (1954); (e) D. J. Cram and K. R. Kopecky, *ibid.* **81**, 2748 (1959); (f) D. J. Cram and D. R. Wilson, *ibid.*, **85**, 1245 (1963); (g) Cf. also D. Y. Curtin, E. E. Harris, and E. K. Meislich, *ibid.* **74**, 2901 (1952).

[26] (a) A. McKenzie, *J. Chem. Soc.* **85**, 1249 (1904) and later investigations; (b) cf. also E. E. Turner and M. M. Harris, *Quart. Rev. London* **1**, 299 (1947).

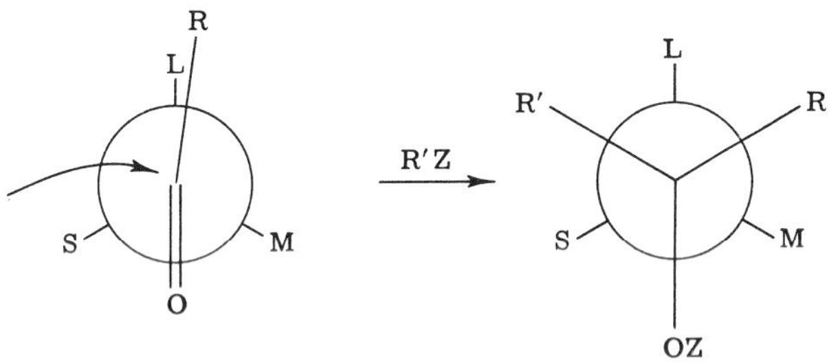

(XXX) (XXXI)

methyl lithium. The oximes corresponding to the ketones were treated with sodium amalgam. In all cases studied the predominating isomer was the one predicted by the rule, provided that the reaction was kinetically controlled.

The explanation for the preferred formation of one diastereomer, according to Cram, lies in the following[3,25e]:

Consider first the reduction of a ketone (type XXX) to the corresponding alcohol with a Grignard reagent or lithium aluminum hydride. Without going into detail concerning the mechanism (see page 272), the first step probably is the formation of a complex of the carbonyl group with the metallic atom of the attacking reagent. In this way the oxygen of the carbonyl group becomes the largest group in the molecule. This large group will then assume a position between the two smallest groups (M and S in XXXII) of the neighboring asymmetric C atom (XXXII). The attack of the nucleophile R′ then proceeds from the least-hindered side of the molecule, and the diastereomer (XXXIII) predominates.[25e]

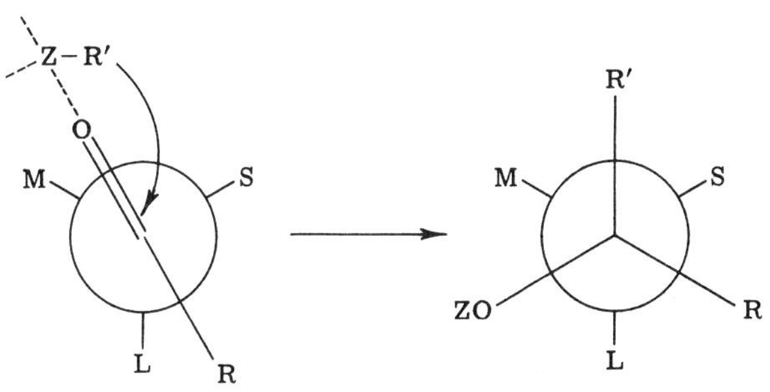

(XXXII) (XXXIII)

A second model (XXXIV) was discussed by Cram[3,25e,25f,27] for compounds which have an OH or NH_2 group on the asymmetric C atom. A complex formation between these groups and the organometallic reagent is possible.

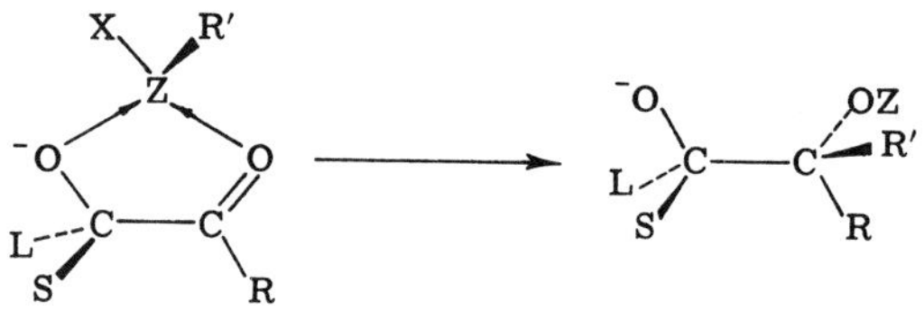

(XXXIV) (XXXV)

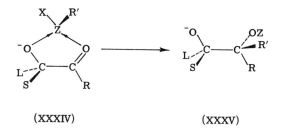

The conformation is fixed by the intermediate, relatively rigid, five-membered ring. Attack of R′ on the carbonyl group from the least-hindered side (from above the plane of the paper) leads to the preferred formation of diastereomer (XXXV).

Cram's rule is not applicable to the thermodynamically controlled reductions of ketones of type (XXX) with aluminum isopropylate, as noted by various investigators.[4,8a,28] Temperature,[28c] solvent,[28d] and catalyst[28e] also have an effect on the course of asymetric syntheses.

When the asymmetric C atom is not adjacent to the keto group, the stereochemistry of addition reactions is also controlled. Prelog and co-workers[29] have demonstrated this based on McKenzies' early investigation (reviewed in ref. 26b). An example will be given and the possibility of asymmetric syntheses will be illustrated for such compounds:

The phenylglyoxylic acid ester of an alcohol having an asymmetric C atom (XXXVI) is treated with a Grignard reagent. Of the two resulting

[27] Cf. also B. M. Benjamin, H. J. Schaeffer, and C. J. Collins, *J. Am. Chem. Soc.* **79**, 6160 (1957); J. H. Stocker, D. Sidisunthorn, B. M. Benjamin, and C. J. Collins, *ibid.* **82**, 3943 (1960).

[28] (a) J. Sicher, M. Svoboda, M. Hrda, J. Rudinger, and F. Sorm, *Collection Czech. Chem. Commun.* **18**, 487 (1953); (b) Cf. especially: Y. Gault and H. Felkin, *Colloq. Intern. Chim. Org. Probl. Stereochem., Montpellier*, 1959 in *Bull. Soc. Chim. France* p. 1342 (1960); (c) cf. H. Pracejus, *Ann. Chem.* **634**, 9 (1960); (d) Y. Inouye, S. Inamasu, M. Ohno, T. Sugita, and H. M. Walborsky, *J. Am. Chem. Soc.* **83**, 2962 (1961); (e) H. M. Walborsky, L. Barash, and T. C. Davis, *J. Org. Chem.* **26**, 4778 (1961).

[29] See for example (a) V. Prelog, *Helv. Chim. Acta* **36**, 308 (1953); (b) V. Prelog and H. L. Meier, *ibid.* **36**, 320 (1953); (c) W. G. Dauben, D. F. Dickel, O. Jeger, and V. Prelog, *ibid.* **36**, 325 (1953); (d) V. Prelog, M. Wilhelm, and D. B. Bright, *ibid.* **37**, 221 (1954); (e) V. Prelog, O. Ceder, and M. Wilhelm, *ibid.* **38**, 303 (1955); (f) V. Prelog, E. Philbin, E. Watanabe, and M. Wilhelm, *ibid.* **39**, 1086 (1956); (g) V. Prelog and H. Scherrer, *ibid.* **42**, 2227 (1959).

diastereomeric α-hydroxy esters, one (XXXVII) predominates, and an optically active α-hydroxy acid (XXXVIII) can be obtained after hydrolysis.

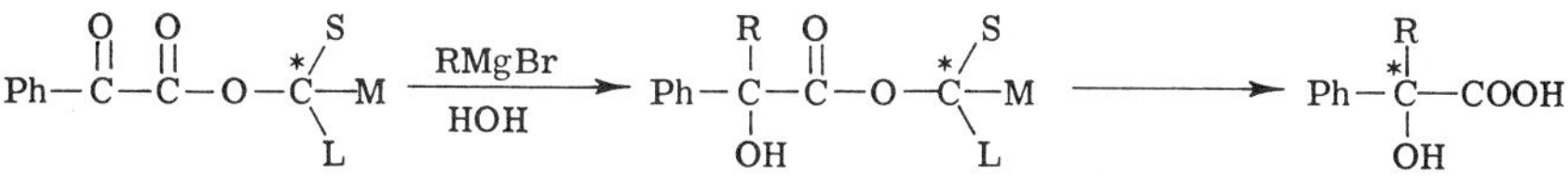

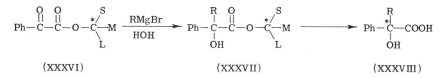

(XXXVI) (XXXVII) (XXXVIII)

Again designating the size of the substituents by S, M and L in (XXXVI) and (XXXVII), the explanation for the preponderance of one diastereomer from the addition to the keto group is similar to Cram's rule.

The preferred conformation of the phenylglyoxylic acid ester is the one shown in (XXXIX), where the two keto groups assume a *trans* position. The attack of the alkyl group of the Grignard reagent again takes place from the least-hindered side, i.e., from the side having the medium-sized group, M. As a result, the diastereomer (XL) predominates. From this the optically active α-hydroxy acid may be obtained by hydrolysis. The configuration of the asymmetric C atom in the α-hydroxy acid (XXXVIII) can be related in this way to the configuration of the original asymmetric alcohol. Prelog and co-workers[29d] were

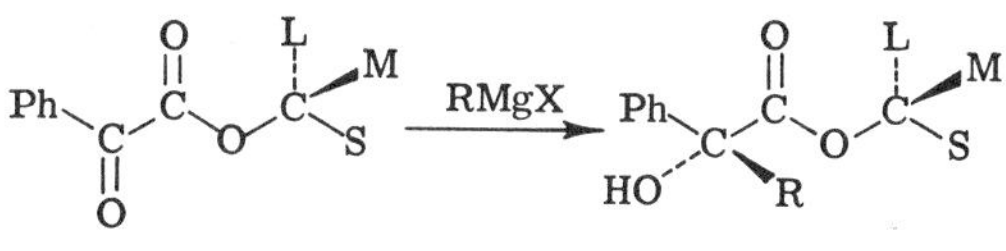

(XXXIX) (XL)

also able to show that the reduction of phenylglyoxylic acid esters with lithium aluminum hydride proceeded in an asymmetric sense and optically active phenylethane-1,2-diols resulted.

In order for asymmetric synthesis to take place with a high degree of optical purity, i.e., with one diastereomer predominating, the substituents S, M, and L must differ as much as possible in their effective bulk[29e,f] (e.g., S=H, M=CH_3, L=bornyl).

b. Elimination Reactions

Bimolecular elimination reactions are highly stereospecific and have already been described in detail for cyclic compounds (page 236). A few characteristic examples of acyclic diastereomers should, therefore, suffice.[30]

[30] Cf. also the detailed description by D. J. Cram *in* "Steric Effects in Organic Chemistry" (M. S. Newman, ed.), p. 304 ff. Wiley, New York, 1956.

Clear-cut differences in the reaction rate have been found in the debromination of acyclic diastereomeric *vicinal* dibromides, with iodide ions (see page 81).[31] *Meso*-2,3-dibromobutane and *meso*-stilbene dibromide react more rapidly than the *d,l* isomers. In addition, olefin formation is quite stereospecific: The *meso* compounds (XLI) produce the *trans* olefins (XLII), while the corresponding *d,l*-dibromides (XLIII) yield *cis* olefins (XLIV) from the elimination. The stereochemical requirement of an E2 reaction is, as frequently shown (see page 81), that it takes place through a transition state with the four atoms involved in a coplanar orientation. The bromine atoms must be *trans* to each other. For the *meso* isomers this would be conformation (XLI), for the *d,l* isomers conformation (XLIII).

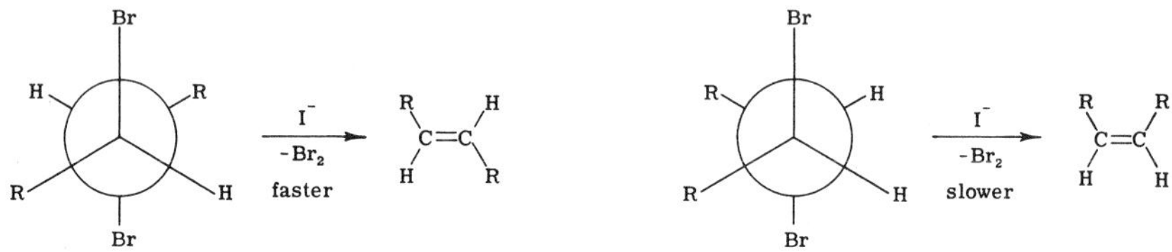

meso (XLI) *trans* (XLII) *d,l* (XLIII) *cis* (XLIV)

Conformation (XLI) for the *meso* isomer is preferred to (XLIII) for the *d,l* isomer, because the two substituents R are not *trans* in the latter, as they are in (XLI). These groups are in the *gauche* position to each other. The transition state necessary for formation of the *cis* olefin (XLIV) from the *d,l*-dibromide (XLIII) leads to an *eclipsed* interaction of the bulky R groups (the so-called "eclipsing effect"[30]). On the other hand, the transition state from (XLI), leading to the *trans*-olefin (XLII), has *R* and *H eclipsed*.[32] The transition state from (XLIII) is, therefore, energetically less favored that the one from (XLI), relative to the corresponding ground states. (The ground states also do not have the same energy, but the difference is small and may be neglected in this qualitative consideration.) The greater reaction rate of the *meso* isomer (XLI) is explained on this basis.

The dehydrohalogenations of diastereomers proceed in an analogous manner, and a coplanar *trans* position of the hydrogen and halogen leaving

[31] (a) W. G. Young, D. Pressman, and C. D. Coryell, *J. Am. Chem. Soc.* **61**, 1640 (1939). (b) S. Winstein, D. Pressman, and W. G. Young, *ibid.* **61**, 1645 (1939).

[32] The term "eclipsing effect" not only refers to differences in nonbonded interactions of the substituents, but also to differences in steric inhibition of resonance and differences in the degree of solvation. (a) D. J. Cram, F. D. Greene, and C. H. DePuy, *J. Am. Chem. Soc.* **78**, 790 (1956). The expression, "*cis* effect" was introduced by Curtin. (b) Cf. D. Y. Curtin and D. B. Kellom, *J. Am. Chem. Soc.* **75**, 6011 (1953).

groups is necessary. An example is the diastereomeric stilbene dibromides; the *meso* isomer (XLV) yields the *cis*-olefin (XLVI) and the *d,l* isomer (XLVII) leads to the *trans*-olefin (XLVIII)[33]

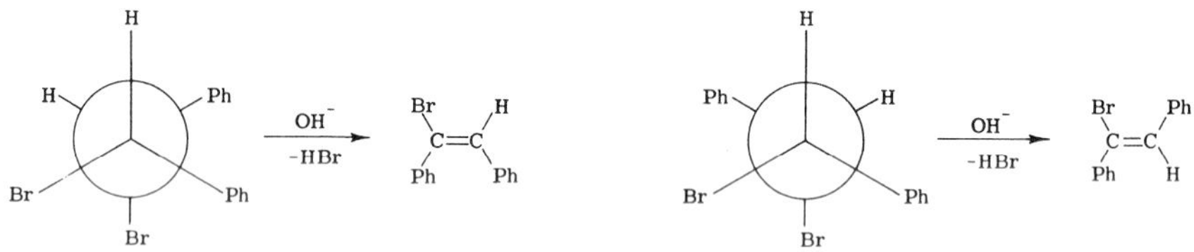

meso (XLV) *cis* (XLVI) *d,l* (XLVII) *trans* (XLVIII)

The 1,2-diphenylpropyl system [XLIX, X=Cl, Br, and $N^{+}(CH_3)_3$] was investigated by Cram and co-workers.[32a] E2 elimination reactions (such as with sodium ethylate), always proceeded more rapidly with the *threo* isomers than with the *erythro* compounds. The rate ratio depends on the substituent *X*. For example, this ratio was 57 for the diastereomeric 1,2-diphenyl-1-propyl trimethylammonium iodides [XLIX, X= $N^{+}(CH_3)_3$]. As before, the *threo* isomers yield the *trans*-olefin, the *erythro*

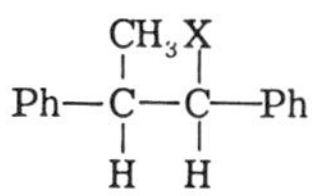

(XLIX)

isomers the *cis* olefin. The slower reaction rate of the *erythro* isomers (L) compared to the *threo* isomers (LI) can again be related essentially to eclipsing effects in the olefin-forming transition state.

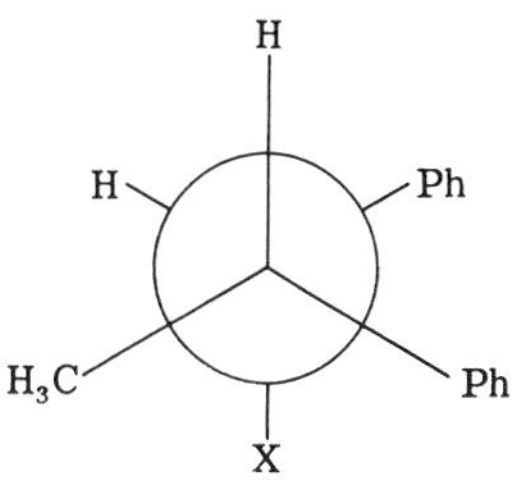

erythro
(L)

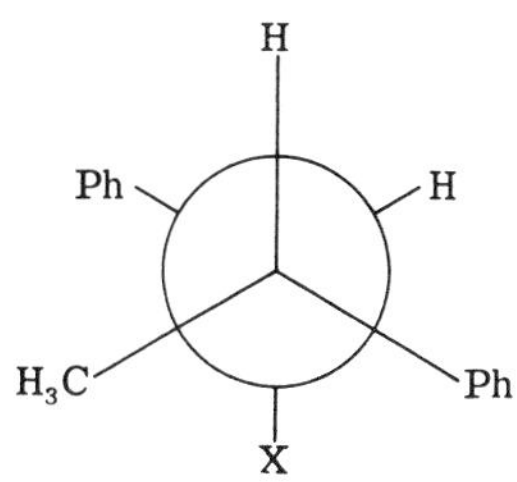

threo
(LI)

[33] (a) P. Pfeiffer, *Z. Physik. Chem.* **48**, 40 (1904); (b) Cf. also S. J. Cristol and R. S. Bly, Jr., *J. Am. Chem. Soc.* **82**, 142 (1960), with further references.

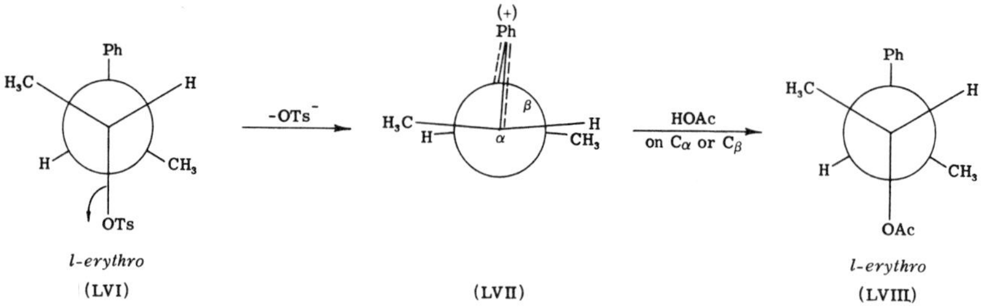

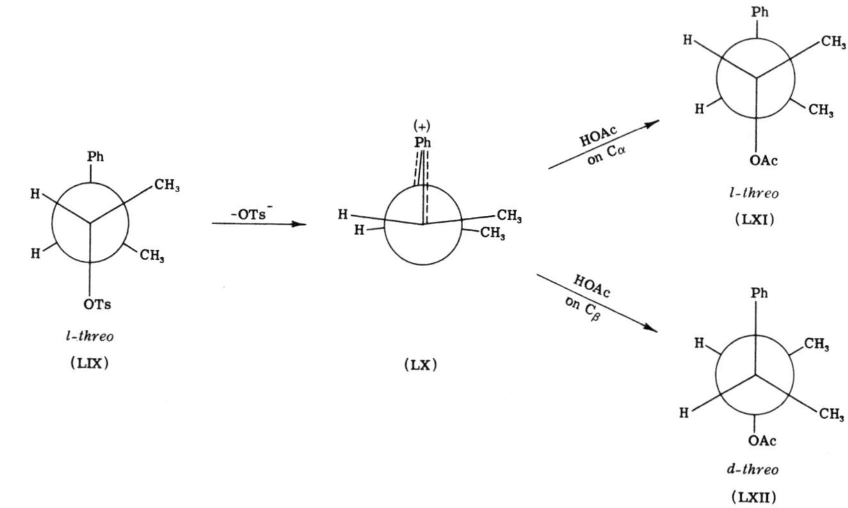

group (in this case phenyl) during the formation of the phenonium ion (LVII) or (LX). The transition state (LXIV) necessary for this is most favored when the departing and migrating groups form an angle of 180°. In the phenonium ion (LVII) or (LX), the substituents then assume an *eclipsed* arrangement toward each other.[30]

However, Cram was able to demonstrate[8b,37] that for acyclic diastereomers which undergo a Wagner-Meerwein rearrangement with a migrating phenyl, the importance of eclipsing effects depends mainly on the relative position of the transition state. This can lie at any point between the starting material, in which the substituents assume the *staggered* position (LXIII), and the phenonium ion intermediate, where the substituents are *eclipsed* (LXV). For

[37] (a) Cf. also D. J. Cram and J. Allinger, *J. Am. Chem. Soc.* **79**, 2858 (1957); (b) D. J. Cram and J. E. McCarty, *ibid.* **79**, 2866 (1957).

example, the relative rates of acetolysis of the diasteromeric 3-phenyl-2-butyl tosylates and the 4-phenyl-3-hexyl tosylates do not vary for each diastereomeric pair.[8b] Because, as mentioned above, the phenonium ion requires an *eclipsed* position of the substituents (LXV), the transition state which leads to the removal of the tosyl group seems to correspond more closely to the conformation of the starting material (LXIV). The *eclipsed* position of the substituents is probably assumed only after the molecule has passed through this transition state.

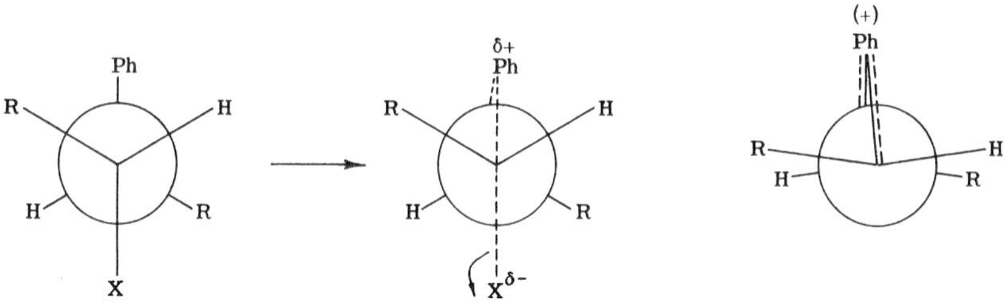

Transition state | Phenonium ion

(LXIII) (LXIV) (LXV)

In contrast, eclipsing effects play a greater role in other compounds. From the acetolysis of optically active 1,1-diphenyl-2-propyl-*p*-bromobenzenesulfonate (LXVI) there is formed, in addition to partially racemic 1,1-diphenyl-2-propyl acetate, nearly optically pure *threo*-1,2-diphenyl-1-propyl acetate (LXVIII) and a small amount of optically pure *erythro*-1,2-diphenyl-1-propyl acetate.[38]

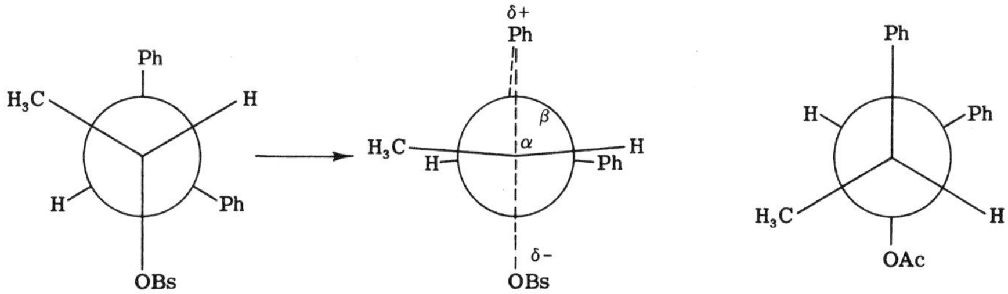

(LXVI) (LXVII) (LXVIII)

The high stereospecificity accompanying this rearrangement was rationalized by Cram[30,32a] as follows: Essentially only one of the two phenyl groups migrates because the *threo* isomer is formed preponderantly. It is the one which has the bulky, nonmigrating methyl and phenyl

[38] D. J. Cram and F. A. Abd Elhafez, *J. Am. Chem. Soc.* **76**, 28 (1954).

groups *trans* in the transition state. During the development of the transition state (LXVII), the migrating phenyl group is located approximately halfway between C_α and C_β, and the substituents assume the *eclipsed* position in relation to each other. When the phenonium ion is produced, the phenyl group is more strongly bound to C_α than to C_β. The formation of the bridged phenonium ion proceeds through the point of maximum eclipsing in contrast to the cases described previously.

Eclipsing effects also play a decisive role in the pinacolic rearrangement, which was studied by Curtin and co-workers.[39,40]

[39] (a) Cf. P. I. Pollak and D. Y. Curtin, *J. Am. Chem. Soc.* **72**, 961 (1950); (b) D. Y. Curtin and P. I. Pollak, *ibid.* **73**, 992 (1951); (c) D. Y. Curtin, E. E. Harris, and E. K. Meislich, *ibid.* **74**, 2901 (1952); (d) D. Y. Curtin and E. K. Meislich, *ibid.* **74**, 5518, 5905 (1952); (e) D. Y. Curtin and M. C. Crew, *ibid.* **77**, 354 (1955).

[40] Cf. also the extensive investigations of C. J. Collins and co-workers concerning the deaminations of aliphatic amines and amino alcohols. C. J. Collins, M. M. Staum and B. M. Benjamin, *J. Org. Chem.* **27**, 3525 (1962), with further references there.

Author Index

Numbers in parentheses are footnote numbers and are inserted to enable the reader to locate a reference when the authors' names do not appear in the text.

Subject Index